T0183118

Graduate Texts in Mathematics

Graduate Texts in Mathematics bridge the gap between passive study and creative understanding, offering graduate-level introductions to advanced topics in mathematics. The volumes are carefully written as teaching aids and highlight characteristic features of the theory. Although these books are frequently used as textbooks in graduate courses, they are also suitable for individual study.

Wolfgang Arendt • Karsten Urban

Partial Differential Equations

An Introduction to Analytical and Numerical Methods

Translated from the German by James B. Kennedy

 Springer

Wolfgang Arendt
Institute of Applied Analysis
Ulm University
Ulm, Baden-Württemberg, Germany

Karsten Urban
Institute of Numerical Mathematics
Ulm University
Ulm, Baden-Württemberg, Germany

Translated by
James B. Kennedy
Department of Mathematics
University of Lisbon
Lisbon, Portugal

English translation of the original German edition published by Springer Spektrum, Berlin, Germany, 2018

ISSN 0072-5285   ISSN 2197-5612   (electronic)
Graduate Texts in Mathematics
ISBN 978-3-031-13381-7   ISBN 978-3-031-13379-4   (eBook)
https://doi.org/10.1007/978-3-031-13379-4

Mathematics Subject Classification: 35-00, 35-04, 35Axx, 35Jxx, 35Kxx, 35Lxx, 65Mxx, 65M60, 65Nxx, 65N30

This Springer imprint is published by the registered company Springer Nature Switzerland AG
The registered company address is: Gewerbestrasse 11, 6330 Cham, Switzerland

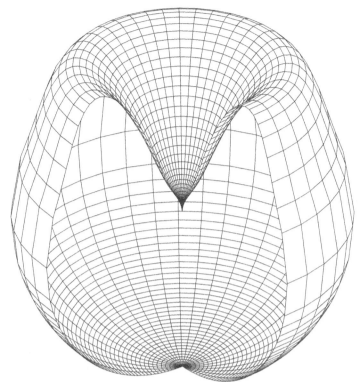

Lebesgue's cusp, see Section 6.9: The Dirichlet Problem.

For Frauke and Almut

Foreword by the Translator

When the authors approached me to ask whether I would be interested in translating their book on partial differential equations into English, as it turns out one of my first thoughts was exactly what the authors had written in the preface of the book: given the large number of excellent textbooks on PDEs available out there, why the need for another?

Of course, it must also be said that given the enormous breadth of the subject there are about as many different approaches to PDEs as there are books on them, and no two books are quite the same. This particular book also seemed to be enjoying some popularity in Germany, which may or may not be related to the authors' original stated motivation of combining material from different undergraduate PDE courses within the scope of the Bologna reforms to higher education as applied to their home country.

In fact, while this motivation may stem from what to an outsider looks a rather arcane educational reform somewhere in Europe, the resulting book is, if not unique, then certainly very rare. Especially in the Anglosphere, there is still an unhealthy tendency in many places to divide the world of PDEs rather sharply into pure topics and applied ones. Triangulations of domains for finite elements are all too often kept at rather more than arm's length from anyone dealing with Banach (or even Hilbert!) spaces, and vice versa.

But of course this is nonsense. To do numerics well, to understand what one is doing, one needs to understand the theory. Yet in a subject whose raison d'être is to be found in its broad applicability to problems coming from everything from physics, via chemistry and biology to economics and finance, a theory with no grounding in this reality with all its accompanying intuition will be not just less useful, but also less complete.

While there are many good books which intertwine PDE theory with applications to, or motivations coming from, one or more of the sciences, I would wager that it is extremely rare to intertwine all three: the modeling, a modern (weak) solution theory of PDEs, and both theoretical and practical aspects of numerical analysis of PDEs. But as the title indicates, that is exactly what this book offers.

I will not attempt an in-depth summary of the contents here, which the authors provided in a much better form than I could in the preface. A couple of words on the prerequisites and particularities as seen by an anglophone outsider might nevertheless be helpful. A strong grounding in elementary analysis, preferably as far as the Lebesgue integral, is essential, as is the sort of linear algebra one needs to be able to do functional analysis; as for the rest, in all essential points the book is generally self-contained. As is usual for the subject, a level of mathematical maturity corresponding to about a third or fourth-year level is implicitly assumed.

The first three chapters, on elementary modeling and methods, should be accessible to a diligent third-year student. The theory in earnest starts in the fourth chapter on the essentials of operators in Hilbert space, followed by a treatment of Sobolev spaces. A welcome point of distinction is the rather thorough treatment of Sobolev spaces in one dimension before moving on to the theory of weak solutions, in particular for elliptic equations, in higher dimensions. The book is rounded out with an extensive chapter on numerical methods and then a short chapter on computer-aided calculation (concretely using Maple®). When treating the numerical methods, as before the philosophy is to start with very simple cases, finite difference and finite element methods in one dimension, and gradually build up to more general situations.

The equations most comprehensively treated, and which form something of a leitmotif throughout the different stages of the book, are Poisson's equation (the Laplacian), the heat equation, and the wave equation on Euclidean domains: they appear in the modeling phase, as inspiration for the development of the Hilbert space theory, the Sobolev space theory, the weak solution theory, and then the numerics. Another particularity, however, is worthy of note: the treatment of the Black–Scholes equation from mathematical finance.

In translating the book I have tried to maintain the spirit and flavor of the original as much as possible. Most of the terminology and notation used here are more or less standard and should be easily recognizable to practitioners in the area, even if some choices are not canonical (for example *inhomogeneous* rather than *nonhomogeneous*, *coercivity* rather than *coerciveness*, the Lax–Milgram *theorem* rather than *lemma*). Care should be taken with exactly what is meant by Gauss's theorem, but this is perhaps always the case.

I would like to extend my thanks to the authors for their enthusiastic and far-reaching support throughout the process, as well as the publisher for taking the initiative to propose an English version in the first place.

Lisbon, Portugal   James Kennedy
December 2020

Preface

Numerous processes in nature, medicine, economics, and technology can be described by partial differential equations (PDEs). This alone is enough to explain the enormous interest in PDEs, as witnessed by the huge number of publications in the area. It is thus no surprise that there are so many textbooks on the market. So why write another one?

Since more than a decade, we observe a trend towards introductory lectures which combine multiple topics. Analysis and numerics of PDEs are well suited to such a union. Yet there are virtually no introductory textbooks for such courses available on the market. This combination is also well justified in practical terms. It is a fact of life that PDEs cannot be solved exactly on general domains (as is necessary for applications). Exactly what is meant by this will be explained in due course in the book. In such cases, one is reliant on computer-based approximation procedures, that is, on numerical methods. In recent years, it has become increasingly apparent that good numerical methods rest upon modern insights from the analysis of PDEs. As such, the unification of analytic and numerical methods is of prime importance both within mathematics and in applications.

We have tried to write an introductory textbook which brings together and intertwines the analytic and numerical aspects of PDE theory. The choice of contents is marked by this objective. At the same time, we attempt to build a bridge to the realm of applications. The book begins with a chapter on modeling, that is, the translation of a given problem from applications into the language of mathematics, here into a PDE. We describe the categorization of PDEs and discuss elementary solution methods. It turns out that both modeling and numerical processes lead to the use of Hilbert spaces. Our introduction shows that the associated methods are also the "correct" ones in a mathematical sense. We describe Hilbert space methods as simply as possible. Statements about the maximal regularity of solutions are particularly important for numerical methods. In addition to classical numerical methods (finite differences and finite elements), we give a few pointers on how to solve at least some PDEs using computer-based formula manipulation systems such as Maple®.

Finally, we have included a topic in the book which is of particular interest for students of financial mathematics. The pricing of risky products on financial and insurance markets requires a deeper mathematical analysis. These days, PDE-based models are increasingly finding application, especially in the finance and insurance industries. In this book, we treat the Black–Scholes equation as a key example of such a PDE.

A couple of words about the structure of the book might assist in its use. It consists of three parts:

A Elementary Methods and Modeling (Chapters 1–3)
B Hilbert Space Methods (Chapters 4–8)
C Numerical and Computer-Based Methods (Chapters 9 and 10)

Part A can be read completely independently of the others; it conveys the essence of PDEs in concrete situations. Only basic knowledge of calculus is needed here. The results on Fourier series used in this part are presented with detailed proofs.

Part B is of gradually increasing difficulty in terms of contents and exposition. It contains an introduction to Hilbert and Sobolev spaces, the latter of which are first considered in one dimension. A particularity is the systematic use of Sobolev spaces when treating harmonic functions and the Dirichlet problem. As a most efficient tool, the spectral theorem for self-adjoint compact operators is used to treat the prototypes of elliptic, parabolic, and hyperbolic equations in an elegant way. In the more advanced sections of Part B, properties at the boundary of a domain and integration by parts are used. Again, the needed material is presented with detailed proofs.

In Part C, we describe the finite difference method and give an introduction to the finite element method. By restricting to linear elements on triangles in two space dimensions, we have chosen a very simple situation in which the essential ideas should nevertheless become transparent. We treat elliptic, parabolic and hyperbolic problems. All proofs are presented in a detailed and self-contained manner, which builds upon the material presented in Part A and B. Numerical experiments are presented.

Each chapter ends with a set of exercises, which in many cases give information going beyond what was handled in the chapter. Solutions can be found on the book's homepage. Sections marked with an asterisk ∗ provide useful additional information.

The book is organized so as to be gradually increasing in difficulty, especially in terms of the theory, and we have tried to keep it as elementary as possible. Thus it should be accessible to students even at an early stage of their studies.

Ulm, Germany   Wolfgang Arendt
Ulm, Germany   Karsten Urban
May 2022

The address of the book's webpage containing the codes used in Chapters 9 and 10 as well as solutions to the exercises (in German only) reads:
sn.pub/HVTfL5

Acknowledgments

This book never could have been written without the help of numerous colleagues and friends. Our thanks go to the publisher, Springer Verlag (at that time Spektrum Akademischer Verlag), and in particular Dr. Andreas Rüdinger and Bianca Alton, for the opportunity to write the original German version of this book, and for their support in realizing it.

We also wish to thank many colleagues for a huge number of valuable tips and suggestions, in particular Tom ter Elst, Wilhelm Forst, Stefan Funken, Rüdiger Kiesel, Werner Kratz, Stig Larsson, and Delio Mugnolo. We also thank Thomas Richard (Scientific Computers) as well as our colleagues at Ulm, Markus Biegert, Iris Häcker, Daniel Hauer, Sebastian Kestler, Michael Lehn, Robin Nittka, Mario Rometsch, Manfred Sauter, Kristina Steih, Timo Tonn, and Faraz Toor, and our students, for many remarks, additions, and careful proofreading of the manuscript.

Very special thanks go to Daniel Daners (Sydney) who produced the wonderful image of Lebesgue's cusp by using Maple® (Figure 6.4 and reference [10]).

We are thankful to a number of colleagues for suggestions for improvement to the previous German editions: Ferdinand Beckhoff and Silke Glas (University of Twente), Stefan Hain (Ulm University), Julian Henning (Ulm University), Norbert Köckler (University of Paderborn), Markus Kunze (University of Konstanz), Tobias Nau (Ulm University), Patrick Winkert (Technical University of Berlin), and Marco Zank (University of Vienna).

We are immensely grateful to Petra Hildebrand for her excellent work in preparing the manuscript in LATEX and for producing numerous figures and graphics.

When finishing the second German edition, we did not expect to have the chance to work on another edition so soon. This was possible mainly due to two people, namely Dr. Thomas Hempfling from Springer, who made the English version possible from the editorial side, and Dr. James B. Kennedy (University of Lisbon), who undertook the enormous task of translating the German version into English. But James did much more than taking care of language and style with the highest competence. He also gave us many suggestions concerning the mathematical content which significantly improved the book. We are extremely grateful for the great work that James did!

About the Authors

Wolfgang Arendt studied mathematics and physics in Berlin, Nice, and Tübingen. After obtaining his doctorate and *habilitation* in Tübingen, he spent eight years as a professor at the University of Besançon. From 1995 to 2018, he was in charge of the Institute of Applied Analysis of Ulm University, where he is now an emeritus professor. His areas of expertise are centered around functional analysis and the theory of partial differential equations. Wolfgang Arendt has had research stays in Berkeley, Nancy, Oxford, Zurich, Sydney, Auckland, New Delhi, Stanford University, and the University Gustave Eiffel in Paris. He is editor-in-chief of the *Journal of Evolution Equations*.

Karsten Urban studied mathematics and computer science in Bonn and Aachen, obtaining his doctorate and *habilitation* at RWTH Aachen. He took up a professorship at Ulm University in 2002, where he has led the Institute of Numerical Mathematics since 2003. His research is focused on numerical methods for partial differential equations, with a particular emphasis on concrete applications in science and technology. In addition to guest professorships at Pavia, Utrecht, and M.I.T. he has had research stays in Turin, Cambridge, Marseilles, Göteborg, and Cornell University. In 2005, he was awarded the teaching prize of the German state of Baden-Württemberg. He is editor-in-chief of the journal *Advances in Computational Mathematics*.

About the Translator

James Kennedy studied economics and mathematics in Sydney, completing his doctorate in mathematics there in 2010. In addition to postdoctoral positions in Lisbon and Stuttgart, where he obtained his *habilitation*, he spent two years as a Humboldt postdoctoral fellow in Ulm. Since 2017 he has been in Lisbon, first as a research fellow and, since 2020, as a professor. His interests are focused around the theory of partial differential equations, in particular its intersection with mathematical physics, operator theory, and spectral theory.

Contents

List of Figures

Chapter 1
Modeling, or where do differential equations come from

Partial differential equations describe numerous phenomena in nature, technology, medicine, economics, ... In this first chapter we shall describe the derivation of the partial differential equations associated with several prominent examples, using the laws of nature and mathematical facts. One calls such a derivation (mathematical) *modeling*. The examples should also illustrate the variety of such partial differential equations as they arise in diverse *real problems*. A first, somewhat rough classification will be given at the end of the chapter.

Chapter overview

© The Author(s), under exclusive license to Springer Nature Switzerland AG 2023
W. Arendt, K. Urban, *Partial Differential Equations*, Graduate Texts in
Mathematics 294, https://doi.org/10.1007/978-3-031-13379-4_1

1.1 Mathematical modeling

We begin with a few basic remarks about modeling with partial differential equations. One should however exercise caution with the use of the term "modeling", since modeling is undertaken in many scientific disciplines, sometimes with different meanings. Even within mathematics the term is sometimes used in a different way from here, such as in stochastics or financial mathematics.

1.1.1 Modeling with partial differential equations

Modeling with partial differential equations typically proceeds in three steps:
1. Specification of the process or phenomenon to be modeled
2. Application of (natural) laws
3. Formulation of the mathematical problem

In the first step, the *specification of the process to be modeled*, one first has to clarify precisely which real-world process should be modeled. As an example, in this section we will consider gas flow in the exhaust system of a vehicle. Let us assume that one is interested in the volume and the spatial distribution of CO_2 for a given engine output. Obviously here a number of processes play a role, such as

- chemical reactions between various compounds in the air and the exhaust;
- combustion processes in the engine;
- flow of the exhaust gases through the exhaust pipe.

It should quickly become clear that the modeling of the whole process is very complex. Thus one often makes simplifying assumptions. For example, one could start by ignoring the reaction processes and the combustion and concentrate purely on the flow problem.

The second step is the *application of (natural) laws*: we need to establish the relationships between the quantities (e.g., physical ones) which are necessary to describe the process. This is usually done via the application of principles which are known from the relevant discipline, for example natural laws (e.g. *force is mass times acceleration*, Newton's second law).

However, often the application of such laws alone is not enough to yield a partial differential equation; generally, some mathematical theory is still required. This could be in the form of integral transforms, passage to the limit or basic theorems from analysis. Only after this step does one (sometimes with considerable creativity) reach a *formulation of the mathematical problem*, whose solutions describe the process being modeled. Here, once again, one often needs to make simplifying assumptions, for example that functions arising in the problem are differentiable sufficiently often.

In the example of the exhaust gas flow mentioned above, by invoking the principles of conservation of mass and conservation of momentum as well as Newton's third law (the action-reaction law), one obtains the *Navier–Stokes equations*, the fundamental equations of fluid and gas dynamics, cf. Section 1.7.4.

Our derivations are not only designed to illustrate the utility of the theory of partial differential equations in understanding natural processes; we are also interested in the other direction. Understanding the physical situation described by a certain equation can help enormously with understanding its mathematical properties and with developing a mathematical intuition for the problem.

One of the equations which we will consider does not describe a natural process, but rather a problem from finance. The solutions of the *Black–Scholes equation* specify the fair price of an option which gives the right to buy a given stock for a determined price at a future point in time, cf. Section 1.5. Our derivation here will use the same approach used by Black, Scholes and Merton, namely via stochastic differential equations. It will not be possible to give a complete derivation of all the necessary mathematics here; however, the arguments given here should help to make the model more transparent.

1.1.2 Modeling is only the first step

Of course, one is not finished after deriving a partial differential equation to describe a process. The equation might have infinitely many solutions, or it may have none at all. In order to answer these and other natural questions, one needs to perform a mathematical analysis of the equation. In addition to the question of the well-posedness of the equation (in the sense of Hadamard, that is, existence, uniqueness and continuous dependence on the data) one generally wishes to describe the qualitative behavior of the solution(s), or to reduce the problem or the equation to one which is already understood. This mathematical analysis of partial differential equations is a central topic of this book.

If one wishes to understand the process more fully or, say, optimize certain parameters in the problem, then one needs, in addition to the mathematical analysis of the problem, some concrete representation of the solution. In some cases this can be given explicitly via a formula (one then speaks of an *analytic* solution), but often the equation cannot be solved explicitly. One can then use approximation methods (usually on the computer), that is, *numerical methods*. In this book we will also deal with this aspect of partial differential equations.

Let us return to the example of exhaust gas flow one last time. The Navier–Stokes equations cannot be solved analytically. It turns out that if one uses suitable numerical iteration methods, one very often ends up having to solve linear partial differential equations of second order, in particular the *Poisson equation*. Linear partial differential equations of second order will be treated intensively in this book. They also represent a springboard to more complex equations, such as those which we will at least briefly introduce in Section 1.7*.

1.2 Transport processes

Consider a narrow pipe P with constant cross-section of area $A \in \mathbb{R}^+$ (in m^2), orientated so as to be parallel to the x-axis. In what follows we will consider a section of this pipe with x-values of the form $[a, b]$ with $a, b \in \mathbb{R}$, $a < b$, cf. Figure 1.1.

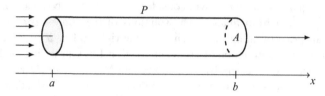

Fig. 1.1 Horizontal pipe with constant cross-section of area $A \in \mathbb{R}^+$.

Suppose that the pipe is filled with flowing water. We assume that the pipe is very thin, that is, the cross-sectional area A is taken as being "small". This means that we only need to consider the flow in the horizontal direction; all other directions may be ignored.

1.2.1 Conservation laws

We wish to describe the flow of water through the pipe P mathematically. To this end we denote by $u = u(t, x)$ the *density* (measured in kg/m^3) of the water at position $x \in (a, b)$ and at time t. The total mass of water in the interval $[x, x + \Delta x] \subset [a, b]$ (with Δx sufficiently small) at time t is thus given by

$$\int_x^{x+\Delta x} u(t, y)\, A\, dy. \tag{1.1}$$

We now want to describe the flow of water in the time period from t to $t + \Delta t$, $\Delta t > 0$, $t \in \mathbb{R}$. The difference in the mass of water in the section of pipe $[x, x + \Delta x]$ at the two points in time is clearly

$$\int_x^{x+\Delta x} (u(t + \Delta t, y) - u(t, y))\, A\, dy. \tag{1.2}$$

How can the mass of water to change between times t and $t + \Delta t$? There are two possible causes, namely
- the incoming and outgoing flow of water;
- any sources or sinks.

In physics, the term *flux* is used to describe the amount (and direction) of any quantity such as mass, energy, number of particles etc., which passes through a given surface per unit of time. In our case, we denote the flux of water by $\psi(t, x)$, that is, this is how much water is flowing through the cross-section of pipe, per second and square meter, at time t and at the point x. Hence

$$\int_t^{t+\Delta t} A \psi(\tau, x) \, d\tau$$

is the total mass of water which flows through the pipe in the time interval $[t, t + \Delta t]$. We can describe any sources and sinks via a function $f = f(t, x)$ which specifies how much water is added to (or removed from) the pipe, per second and square meter, in a section of length 1 at time t and at the point x. Thus

$$\int_t^{t+\Delta t} \int_x^{x+\Delta x} f(\tau, y) A \, dy \, d\tau \qquad (1.3)$$

is the total mass of water which is added (or removed) in the section of pipe $[x, x+\Delta x]$ in the time interval $[t, t + \Delta t]$. If $f > 0$, then we speak of a *source*; if $f < 0$, then f represents a *sink*.

Now one of the fundamental principles of physics asserts that in a closed system mass can neither be created nor destroyed; this is the principle of *conservation of mass*. We can formulate this mathematically in the form of a *conservation law*. In words, this principle states that

change in total mass of water = inflow − outflow + sources.

For all terms in this equation we have derived corresponding mathematical expressions, in (1.2) and (1.3). If we bring everything together, then we obtain the following equation in the section of pipe $[x, x + \Delta x]$:

$$\int_x^{x+\Delta x} A(u(t + \Delta t, y) - u(t, y)) dy =$$
$$= \int_t^{t+\Delta t} (A \psi(\tau, x) - A \psi(\tau, x + \Delta x)) \, d\tau + \int_t^{t+\Delta t} \int_x^{x+\Delta x} f(\tau, y) A \, dy \, d\tau \qquad (1.4)$$

On the left-hand side we have the difference in the mass of water in the section of pipe $[x, x + \Delta x]$ between times $t + \Delta t$ and t; on the right-hand side we have the total inflow (or outflow) through the cross-section x, minus the inflow (or outflow) through the cross-section $x + \Delta x$, during the time interval $[t, t + \Delta t]$, to which is added the mass of water coming from sources (or removed by sinks) in the time interval $[t, t + \Delta]$. Hence (1.4) is, as intimated, a *conservation law*.

1.2.2 From a conservation law to a differential equation

Now u and ψ can vary strongly in time and space. The equation (1.4) becomes more precise as the time and space intervals become smaller. Let us suppose that the functions u and ψ are continuously differentiable. We start by dividing (1.4) by $(A\Delta t)$ and passing to the limit as $\Delta t \to 0$. We may then exchange the order of integration and differentiation to obtain

$$\int_x^{x+\Delta x} \frac{\partial}{\partial t} u(t, y)\, dy = \psi(t, x) - \psi(t, x + \Delta x) + \int_x^{x+\Delta x} f(t, y)\, dy.$$

If we next divide by Δx and pass to the limit as $\Delta x \to 0$, then our equation becomes

$$\frac{\partial}{\partial t} u(t, x) = -\frac{\partial}{\partial x} \psi(t, x) + f(t, x), \tag{1.5}$$

provided f is continuous in t and x. We often write (1.5) in the following abbreviated form:

$$u_t + \psi_x = f. \tag{1.6}$$

It is clear from the derivation that this is a pure *conservation equation*. It should also be clear that we can argue completely analogously when u does not represent water density but rather the density of some other quantity such as energy, charge, bacteria, particles, automobiles, molecules, etc. The pipe could then be, for example, a conductor, an artery, a road or a nerve tract. For this reason, in general $u = u(t, x)$ is often referred to as a *state variable*, as it describes the state of the system.

In many models the flux $\psi(t, x)$ depends in a particular way on the density $u(t, x)$, that is, $\psi(t, x) = \phi(t, x, u(t, x))$ for some function $\phi : [0, \infty) \times \mathbb{R} \times \mathbb{R} \to \mathbb{R}$. In this case, (1.6) becomes a partial differential equation in the unknown function u. In the following three sections we will meet examples of such a dependency. A further, particularly important, example is the case of diffusion, where ψ depends on the derivative in the space variable(s), u_x; see Section 1.3. But there can very easily be other kinds of dependence, for example, on temperature, velocity, acceleration, concentration etc.

1.2.3 The linear transport equation

Let us return to the example of the water pipe. We will suppose that $f = 0$, that is, that there are no sources and no sinks. We will also assume that the water is moving with constant velocity $c \in \mathbb{R}$. The flux thus has the form $\psi(t, x) = c \cdot u(t, x)$, and the flux function in this case takes the form

$$\phi = \phi(t, x, u) := c \cdot u(t, x), \quad c \in \mathbb{R}. \tag{1.7}$$

This function is linear in u, that is, the flux is proportional to the density u. Hence (1.6) reduces to

$$u_t + c\, u_x = 0. \tag{1.8}$$

This equation is known as a *linear transport equation* (also *convection equation* or *advection equation*). Clearly, (1.8) is a *homogeneous* differential equation.

1.2.4 The convection-reaction equation

Let us assume that the state u decays in time at a constant rate $\lambda < 0$, perhaps via a radioactive decay process. This corresponds to the ordinary differential equation $u_t = \lambda u$, whose solution is $u(t, x) = u(0, x)\, e^{\lambda t}$. This can be expressed as a source term via

$$f(t, x, u(t, x)) := \lambda \cdot u(t, x) \tag{1.9}$$

In this case, that is, taking (1.9) as the flux function, (1.6) becomes

$$u_t + c u_x - \lambda u = 0. \tag{1.10}$$

This equation (which is still linear and homogeneous in u) is sometimes called a *convection-reaction equation*. All terms of zeroth order in the unknown state u are referred to as *reaction terms*.

A notable feature of the equations (1.8) and (1.10) is that both are linear in u. This is of course the case in only very few realistic problems; often, linear equations represent a highly simplified model of reality.

1.2.5* Burgers' equation

Here we wish to introduce a first nonlinear model. Let $u(t, x)$ be the density of traffic, that is, the number of vehicles at time t and location x on a single-lane road. In order to be able to model traffic jams, we may take as a simple first *ansatz* for the flux function

$$\psi = \psi(u) = \alpha \cdot u \cdot (\beta - u),$$

where $\alpha, \beta > 0$ are certain constants. The following considerations show that this is indeed a first, simple model for traffic congestion. If the density u of traffic is not too high, that is, $0 \leq u \ll \beta$, then ψ is approximately proportional to u, $\psi \approx \alpha\beta u$, so that the flow is approximately the same as in the linear transport equation (1.8) with transport velocity $c = \alpha\beta$. This matches our intuition: if there are few vehicles underway, then the flow of traffic is smooth and uniform. However, if the density of traffic is higher, say, $u \approx \beta$, then $\psi \approx 0$; the increasing density of vehicles causes the flow of traffic to break down. In the absence of source terms (which means that there are no bifurcations or intersections along the road) the equation for the traffic density is

$$u_t + \alpha(u \cdot (\beta - u))_x = 0. \tag{1.11}$$

In this form the equation is somewhat unwieldy. If, however, we introduce the change of variables

$$v(t, x) := \beta - 2u\left(\frac{t}{\alpha}, x\right), \tag{1.12}$$

then (we omit the arguments to improve readability)

$$v_t + vv_x = -\frac{2}{\alpha}u_t + (\beta - 2u)(-2u_x) = \left(-\frac{2}{\alpha}\right)(u_t + \alpha\beta u_x - 2\alpha uu_x)$$

$$= \left(-\frac{2}{\alpha}\right)(u_t + \alpha(u(\beta - u))_x).$$

By (1.11), this means that

$$v_t + vv_x = 0. \tag{1.13}$$

This equation is called *Burgers' equation*, named after the Dutch physicist Johannes Martinus Burgers (1895–1981). It was originally introduced in 1915 by the English mathematician Harry Bateman (1882–1946). Now we have $vv_x = \frac{1}{2}\frac{\partial}{\partial x}v^2 = \frac{1}{2}(v^2)_x$, which means that the flux is given by $\psi(t, x, v) := \frac{1}{2}v(t, x)^2$. The equation (1.13) is also known as a *nonlinear transport equation*.

Additional viscosity. Of course it is possible to refine this model. One such possibility is to make the assumption that drivers will not just reduce their speed when the traffic density is already very high (that is, when $u \approx \beta$), but already whenever the traffic density increases, that is, when $u_x > 0$. We may thus add the term $-\bar{\varepsilon}u_x$ to ϕ in the above model, where $\bar{\varepsilon} > 0$, so that now $\phi(t, x, u, u_x) = \alpha u(t, x)(\beta - u(t, x)) - \bar{\varepsilon}u_x(t, x)$. As above we transform u into the function v given by (1.12) and obtain

$$v_t + vv_x = \varepsilon v_{xx} \tag{1.14}$$

with the *viscosity coefficient* $\varepsilon := \frac{\bar{\varepsilon}}{\alpha} > 0$. We call (1.14) the *viscous Burgers' equation*.

The difference between (1.14) (also known as the *inviscid* Burgers' equation) and the previous equations is that in (1.14) a term of second order, namely v_{xx}, appears, which means we are dealing with a second-order partial differential equation.

1.3 Diffusion

Instead of the pipe of Figure 1.1 we will now consider a solid rod R whose cross-section again has very small area $A \in \mathbb{R}^+$; see Figure 1.2.

We are interested in the temperature $\theta = \theta(t, x)$ at time t and at the point $x \in [a, b]$. We will assume that the rod is homogeneous, that is, that its *density* $\rho \in \mathbb{R}^+$ is constant. As usual, density means mass per unit of volume; here, density is measured in kg/m^3.

To derive the heat equation we need the notion of the *specific heat capacity* c of a body. This is defined as the energy (measured here in joules) that must be added to one unit of mass (here 1 kg) in order to increase its temperature by one unit of

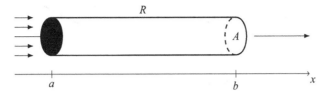

Fig. 1.2 Solid horizontal thin rod with constant cross-section of area A.

temperature (here 1 Kelvin); it is measured in J/(kg K). The temperature is thus proportional to the amount of heat per unit of mass, and so it corresponds to the density from Section 1.2.1.

The flux ψ describes how much heat is transferred ("flows") through the cross-section of the rod per second and square meter; as before, we have the conservation law $\theta_t(t, x) + \psi_x(t, x) = 0$. The law of heat conduction (Fourier's law) states that at every point x the rate of heat transfer ψ at x is proportional to the rate of decay of the temperature, $-\theta_x$, that is, $\psi(t, x) = -k(x)\theta_x(t, x)$. If the factor of proportionality $k(x)$, also known as the *thermal conductivity*, is independent of x, then we obtain the heat equation

$$\theta_t - k\,\theta_{xx} = 0. \tag{1.15}$$

This equation represents the simplest possible model of heat diffusion. A first refinement can be made by dropping the assumption that k is constant on the whole rod R. If k depends on x, then (1.15) becomes

$$\theta_t - (k(x)\theta_x)_x = \theta_t - k'(x)\theta_x - k(x)\theta_{xx} = 0. \tag{1.16}$$

Thus both first and second order derivatives in the space variable(s) appear, although the equation is still linear in θ. If we assume that the thermal conductivity is additionally dependent on the temperature, $k = k(x, \theta)$, then (1.16) becomes a nonlinear equation of the form $\theta_t - (k(\theta)\theta_x)_x = 0$.

1.4 The wave equation

In order to derive our third (and final) type of spatially one-dimensional equation we consider a perfectly elastic string with constant density ρ_0, which is fixed at both ends, cf. Figure 1.3. We denote by S the constant tension of the string.

We bring the string from its resting position (for example, by plucking it) and wish to determine the vertical displacement $u = u(t, x)$ in dependence on time and horizontal position. In doing so, we assume that the displacement is "small", so that horizontal movement can be neglected. We also assume that the string can return

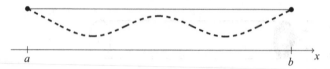

Fig. 1.3 An elastic string fixed at both ends.

to its original position, that is, that no plastic changes to its shape occur. We now consider a small part $[x, x + \Delta x]$ of the string, as depicted in Figure 1.4.

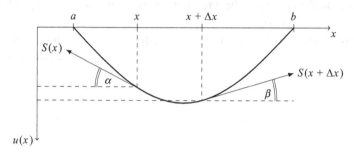

Fig. 1.4 Tension in a small section $[x, x + \Delta x]$ of the string.

All we need for our derivation is *Newton's second law*

$$\text{force} = \text{mass} \times \text{acceleration},$$

named after Sir Isaac Newton (1642–1727). The acceleration in the vertical direction is equal to the second derivative of the vertical displacement of u with respect to the time variable, that is,

$$\text{acceleration} = \frac{\partial^2}{\partial t^2} u(t, x) = u_{tt}(t, x).$$

Thus the force acting on the section of the string with length Δx is given by $(\rho_0 \Delta x) u_{tt}(t, x)$. In order to obtain a differential equation for the displacement $u(t, x)$, we first need to derive a relationship between the force and the tension S. Figure 1.4 shows the tangential components $S(x)$ and $S(x + \Delta x)$ of S at the points $x, x + \Delta x \in [a, b]$. From these, we can also obtain the horizontal components of S, which due to the assumption that S is constant have the form

$$S(x + \Delta x) \cos \beta = S(x) \cos \alpha = S. \tag{1.17}$$

The vertical components can equally be ascertained using Figure 1.4, and the difference of the two agrees with the force which we determined using Newton's second law:

$$S(x + \Delta x) \sin \beta - S(x) \sin \alpha = (\rho_0 \, \Delta x) \, u_{tt}(t, x). \tag{1.18}$$

We wish to derive a relationship with $u(t, x)$. Once again using Figure 1.4, we see that $\tan \alpha = u_x(t, x)$ and $\tan \beta = u_x(t, x + \Delta x)$. If we divide (1.18) by S and invoke (1.17), then we are led to

$$\frac{\rho_0 \, \Delta x}{S} u_{tt}(t, x) = \tan \beta - \tan \alpha = u_x(t, x + \Delta x) - u_x(t, x).$$

Division by $\frac{\rho_0 \Delta x}{S}$ and passing to the limit as $\Delta x \to 0$ finally yields

$$u_{tt} - c^2 \, u_{xx} = 0, \tag{1.19}$$

the *wave equation* with *wave speed*

$$c^2 = \frac{S}{\rho_0}, \qquad c > 0.$$

This equation was originally discovered by Jean-Baptiste le Rond d'Alembert in 1746.

1.5 The Black–Scholes equation

The majority of the partial differential equations considered thus far have their origin in the natural sciences. We now wish to briefly describe a famous equation from another discipline, namely economics, or more precisely financial mathematics. The underlying model was first introduced in 1973 by the American economist Fischer Sheffey Black and the Canadian economist Myron Samuel Scholes. The American mathematician and economist Robert Carhart Merton was also involved in their work but was not a co-author of the seminal paper from 1973, and hence his name is often, wrongly, omitted in this context. In 1997, Merton and Scholes were awarded the Nobel Memorial Prize in Economic Sciences "for a new method to determine the value of derivatives", to quote the official citation. Fischer Black had died in 1995, two years before the prize was awarded; however, he was recognized posthumously during the award ceremony.

An *option* depends first of all on an *underlying* such as for example an exchange rate, a share or block of shares, or the price of an asset which is traded on an exchange. In any case, this underlying has a market value, which we will denote by $S(t)$. This function is an example of a *stochastic process* (see Appendix A.4).

An option on this underlying is a financial product (a contract between a provider such as a bank, and a client) which guarantees the owner of the option the right (but not the obligation) to buy or sell the underlying at a given time T (the *maturity*) for a previously agreed price K (the *strike*). If the right is to buy, then we speak of a *call option*; if the right is to sell then it is a *put option*. Here we will only consider the case where the underlying can *only* be bought or sold at the maturity date; such an option is called a *European option*, although the adjective has no geographical significance whatsoever. These days there is a large number of complex financial products, whose study would take us far too far afield; for details we refer for example to [15].

We wish to determine the "fair" price $V(0, y)$ of an option in dependence on the current share (or asset) price y, where "fair" means that at this price there are no riskless profit opportunities: no-one can make gains from trade without incurring the risk of a loss; in this case we speak of an "arbitrage-free" valuation. In doing so it will be opportune to consider, for each point in time $t \in [0, T]$, which price $V(t, y)$ at that point in time $t \in [0, T]$ and share price y would in fact be fair.

We shall consider the case of a European call on an underlying, say a share, that is, the right to buy the share for an agreed price K at time T. One is hence speculating on rising share prices. If the share price at time T is lower than K, then the option is worthless, since one could buy the same share on the free market for a more favorable price. If, however, the price $S(T)$ of the underlying is above K at time T, then the underlying can be resold immediately on the stock exchange for the current price $S(T) > K$. Hence at the time of maturity the option has the value $V(T, S(T))$, where

$$V(T, y) = (y - K)^+ := \begin{cases} y - K & \text{if } y > K, \\ 0 & \text{otherwise.} \end{cases} \tag{1.20}$$

We also call this function the *payoff* (or *payoff function*). Clearly, (1.20) is a *terminal condition*, as opposed to an *initial condition* of the type we saw for instance in the case of the transport equation. We also consider that the Black–Scholes equation may be interpreted as running "backwards in time".

The modeling requires some background in financial mathematics and is described in more detail in Appendix A.4. Finally, one arrives at the *Black–Scholes equation*

$$V_t + \frac{1}{2}\sigma^2 y^2 V_{yy} + ry V_y - rV = 0, \qquad (t, y) \in (0, T) \times \mathbb{R}, \tag{1.21}$$

for the unknown function $V : (0, T) \times \mathbb{R}_+ \to \mathbb{R}_+$, where $\sigma > 0$ is a constant related to the volatility of the financial product and $r > 0$ is a fixed interest rate, see Section A.4.3. We will see in Section 3.5 that the equation (1.21) admits a unique polynomially bounded solution V which takes on the final value prescribed in (1.20). The initial value $V_0 := V(0, S_0)$ of this solution V at the point $y = S_0$ with the known share price S_0 at initial time $t = 0$ is the fair price of the option. In Section 10.1.5, we will give a Maple®-sheet for this problem.

1.6 Let's get higher dimensional

In both the above examples of the pipe and the string we neglected the vertical component and thus obtained a partial differential equation in one space dimension. The Black–Scholes equation, too, is a time-dependent *one-dimensional* equation since $y \in \mathbb{R}$ is a one-dimensional variable. Of course, in many real situations this approach represents an oversimplification. We thus now wish to consider the case of more than one space variable.

1.6.1 Transport processes

As above, we begin by describing transport processes. Here, however, we replace the pipe P from Figure 1.1 by a general domain $\Omega \subset \mathbb{R}^d$, $d = 2, 3$, in which the transportation should take place. We again let $u = u(t, x) : [t_1, t_2] \times \Omega \rightarrow \mathbb{R}$ denote the density, which is allowed to depend on both time and space. We shall consider a conservation law on a *control volume* $V \subset \Omega$, which may be for example a rectangle or box parallel to the coordinate axes, or a ball. Then just as in (1.1) the mass in V is given by

$$\int_V u(t, x)\, dx. \tag{1.22}$$

In the absence of sources and sinks, the principle of conservation of mass states that any change in mass can only be caused by in and outflows. In terms of V this means in and outflows through its boundary ∂V. We express this change as a flux function $\varphi = \varphi(t, x)$, $\varphi = (\varphi_1, \dots, \varphi_d)$, which specifies what quantity of the substance in question (such as water) exits per second and square meter through a small part of the surface. More precisely, $\varphi : \mathbb{R}_+ \times \Omega \rightarrow \mathbb{R}^d$ is a (continuously differentiable) function for which

$$\int_t^{t+\Delta t} \int_{\partial V} \varphi(s, z)\, v(z)\, d\sigma(z)\, ds$$

gives the mass of the substance which exits through the surface ∂V of a small control volume V during the time interval $[t, t + \Delta]$. Here V should be a small set with C^1-boundary such that $\overline{V} \subset \Omega$. We denote by $v(z)$ the outer unit normal vector to V at the point $z \in \partial V$, and by $d\sigma$ the surface measure on ∂V (see Chapter 7 for the precise definitions). Our conservation law (in the absence of sources and sinks) then reads

$$\int_V \big(u(t + \Delta t, x) - u(t, x)\big) dx = -\int_t^{t+\Delta t} \int_{\partial V} \varphi(s, z)\, v(z)\, d\sigma(z)\, ds.$$

Here the left-hand side gives the difference between the quantity of the substance in V at time $t + \Delta$ and at time t; the right-hand side expresses the quantity of the substance which flows through the surface during this time. If we divide this equation by Δt and let Δt tend to 0, then we obtain (assuming sufficient differentiability of the functions involved)

$$\int_V u_t(t, x) \, dx + \int_{\partial V} \varphi(t, z) \, v(z) \, d\sigma(z) = 0.$$

By the divergence theorem (see Corollary 7.6),

$$\int_{\partial V} \varphi(t, z) \, v(z) \, d\sigma(z) = \int_V \operatorname{div} \varphi(t, x) \, dx,$$

where $\operatorname{div} \varphi := \sum_{i=1}^d \frac{\partial}{\partial i} \varphi_i(x)$ refers to the space variables. We thus obtain

$$\int_V u_t(t, x) \, dx + \int_V \operatorname{div} \varphi(t, x) \, dx = 0$$

for every control volume V. Now any continuous function $f : \Omega \to \mathbb{R}$ satisfies

$$f(x) = \lim_{\varepsilon \downarrow 0} \frac{1}{|B(x, \varepsilon)|} \int_{B(x,\varepsilon)} f(y) \, dy,$$

where $x \in \Omega$, $B(x, \varepsilon) := \{y \in \mathbb{R}^d : |x - y| < \varepsilon\}$, and $|B(x, \varepsilon)|$ is the volume of $B(x, \varepsilon)$. We finally arrive at the infinitesimal version

$$u_t(t, x) + \operatorname{div} \varphi(t, x) = 0 \tag{1.23}$$

of the conservation law. This passage to the limit is called *localization*. We now obtain a partial differential equation for u, provided φ can be expressed as a function of u. If for example a liquid is flowing in the direction $\mathbf{b} = (b_1, \ldots, b_d)^T$ (where $|\mathbf{b}| = (b_1^2 + \cdots + b_d^2)^{1/2} = 1$) with constant speed $c > 0$, then the flux has the form $\varphi(t, x) = c \, \mathbf{b} \cdot u(t, x)$. In this case, the conservation law reads

$$u_t(t, x) + c \, \mathbf{b} \cdot \nabla u(t, x) = 0,$$

where $\nabla u(t, x) = \left(\frac{\partial u}{\partial x_1}, \ldots, \frac{\partial u}{\partial x_d} \right)^T$ is the *gradient* in the space variables. This is the transport equation, which describes a pure transport process in a fixed direction and with constant speed.

1.6.2 Diffusion processes

In the case of diffusion processes, the flux φ in the conservation law (1.23) depends on the rate of change of the particle density. In the simplest case this function is of the form $\varphi(t, x) = -c(x) \nabla u(t, x)$. This means that the particles move away from locations of higher density towards sets where the density is lower. If we substitute this expression into (1.23), then we obtain $u_t(t, x) = \text{div}(c(x) \nabla u(t, x))$. If the factor of proportionality c is independent of the space variable x, $c \equiv c(x)$, then this becomes the *heat equation*

$$u_t(t, x) = c \Delta u(t, x).$$

This equation describes chemical diffusion processes (for example ink in water) as well as the diffusion of heat. But in population models as well (such as for bacteria) it is possible to observe such a diffusion: each individual observes the population density in its vicinity and has the tendency to migrate in the direction of the greatest decrease in this density (given by the gradient of u).

1.6.3 The wave equation

In Section 1.4 we met the wave equation in one space dimension as a model for the displacement of a string. If instead of a string we consider a taut elastic membrane, then we can derive the corresponding higher-dimensional wave equation with similar arguments. The components of the tension in (1.17) and (1.18) now need to be considered in every direction, and so u_{xx} must be replaced by Δu. Hence the general form of the wave equation reads

$$u_{tt} - c^2 \Delta u = f. \tag{1.24}$$

1.6.4 Laplace's equation

We now turn to the derivation of Laplace's equation, an equation that will accompany us for a large part of the book.

We consider an elastic membrane (without bending strength), such as the skin of a drum. Suppose that the membrane has the form of a two-dimensional domain $\Omega \subset \mathbb{R}^2$, and that the membrane is fixed, and held taut, along its boundary. Now suppose that a vertical force $f : \Omega \to \mathbb{R}$, measured in N/m^2, acts on the membrane; we are interested in the vertical displacement $u : \Omega \to \mathbb{R}$, measured in m. Since the membrane is fixed at the boundary, we obtain the *boundary condition*

$$u_{|\partial\Omega} = 0, \tag{1.25}$$

which means that $u(x) = 0$ for all $x \in \partial\Omega$ and we can thus implicity assume that $u \in C(\overline{\Omega})$, where $\overline{\Omega} = \Omega \cup \partial\Omega$ (Figure 1.5).

Now we will apply Newton's first law, the *principle of inertia*, which states that a body remains at rest as long as the sum of all forces acting on it is zero. In other

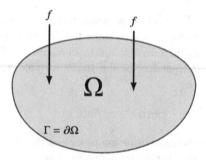

Fig. 1.5 Displacement of a membrane Ω subject to a vertical force.

words, a body behaves in such a way as to minimize its internal energy; for this reason a fixed membrane does not have any "bumps".

The *total energy* J is the sum of the *strain energy* J_1 and the *potential energy* J_2, $J = J_1 + J_2$. We will derive expressions for these two components separately. By *Hooke's law*, to deform an elastic body a force F is necessary which is proportional to the deformation s, that is, $F = \alpha s$, where α is often called *elasticity* (in the case of a spring, this is the spring constant). The energy stored in a body results from work performed on the body. Work, in turn, is the product of force and displacement,

$$\text{work} = \text{force} \times \text{displacement}. \tag{1.26}$$

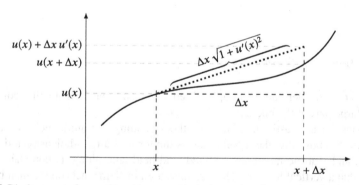

Fig. 1.6 Displacement of a membrane (cross-section): the section of curve given by the graph of u in $[x, x + \Delta x]$ is approximated by the dotted line segment of length $\Delta x \sqrt{1 + u'(x)^2}$.

Thus the strain energy is proportional to the deformation of the surface area, with factor of proportionality α. The surface area of the membrane at rest is $\int_\Omega 1\, dx$, that is, the Lebesgue measure of the domain Ω. We start by considering the one-dimensional case (which may also be thought of as a cross-section of the higher-dimensional case) and an interval $[x, x + \Delta x]$, cf. Figure 1.6. Here the length of the string at rest is Δx, or

$\int_x^{x+\Delta x} \sqrt{1 + u'(s)^2} \, ds$ after displacement. This curve integral may be approximated by the line through $(x, u(x))$ with slope $u'(t)$, that is,

$$\int_x^{x+\Delta x} \sqrt{1 + u'(s)^2} \, ds \approx \Delta x \sqrt{1 + u'(x)^2},$$

cf. Figure 1.6. For functions of several variables this approximation is given analogously by $dx\sqrt{1 + |\nabla u(x)|^2}$ (with the usual Euclidean norm $|x|^2 := x_1^2 + \cdots + x_d^2$, $x = (x_1, \ldots, x_d)^T \in \mathbb{R}^d$). We now integrate over Ω in order to obtain the change in the surface area and hence the strain energy:

$$J_1 = J_1(u) \approx \alpha \int_\Omega \left(\sqrt{1 + |\nabla u(x)|^2} - 1 \right) dx. \tag{1.27}$$

Since J_1 clearly depends on the function u (the displacement), we sometimes refer to J_1 as an *energy functional*. We wish to simplify this expression further. For x close to zero we can use the linear Taylor approximation $\sqrt{1 + x} = 1 + \frac{1}{2}x + O(x^2)$. When the distortion $|\nabla u(x)|$ is small we thus have

$$\sqrt{1 + |\nabla u(x)|^2} - 1 \approx \frac{1}{2}|\nabla u|^2$$

for the gradient ∇u. We thus obtain the following expression for the *strain energy* (at least approximately, here within the scope of the theory of *linear* elasticity)

$$J_1(u) \approx \frac{\alpha}{2} \int_\Omega |\nabla u|^2 \, dx. \tag{1.28}$$

The *potential energy* is generated by the external force F and determined by (1.26) as the product of force and displacement as

$$J_2(u) = - \int_\Omega f(x) \, u(x) \, dx. \tag{1.29}$$

Our overall goal is thus to minimize the energy functional

$$J(u) = \frac{\alpha}{2} \int_\Omega |\nabla u|^2 \, dx - \int_\Omega f(x) \, u(x) \, dx. \tag{1.30}$$

As usual, to determine a minimum (or maximum) we consider the zeros of the first derivative. In the case of a functional, this means that the first variation should vanish, that is,

$$\frac{d}{d\varepsilon} J(u + \varepsilon v)|_{\varepsilon=0} = 0 \tag{1.31}$$

for every possible displacement v for which $v_{|\partial\Omega} = 0$. So we wish to determine the *first variation*. For the second term in (1.30) we have

$$\frac{d}{d\varepsilon} \int_\Omega f(x)\,(u(x) + \varepsilon v(x))\,dx = \int_\Omega f(x)\,v(x)\,dx$$

independently of ε. For the first term we obtain

$$\frac{d}{d\varepsilon} \int_\Omega |\nabla(u + \varepsilon v)|^2\,dx = 2 \int_\Omega \nabla(u + \varepsilon v) \cdot \nabla v\,dx \xrightarrow{\varepsilon \to 0} 2 \int_\Omega \nabla u \cdot \nabla v\,dx,$$

whence

$$\frac{d}{d\varepsilon} J(u + \varepsilon v)|_{\varepsilon=0} = \alpha \int_\Omega \nabla u \cdot \nabla v\,dx - \int_\Omega f(x)\,v(x)\,dx = 0. \qquad (1.32)$$

We then modify the first term by integrating by parts and exploiting the fact that $u_{|\partial\Omega} = v_{|\partial\Omega} = 0$:

$$\int_\Omega \nabla u \cdot \nabla v\,dx = \sum_{i=1}^d \int_\Omega \frac{\partial}{\partial x_i} u \frac{\partial}{\partial x_i} v\,dx$$

$$= \sum_{i=1}^d \int_{\partial\Omega} v \frac{\partial}{\partial x_i} u\, v_i\,ds - \sum_{i=1}^d \int_\Omega \frac{\partial^2}{\partial x_i^2} u\, v\,dx = \int_\Omega (-\Delta u)\, v\,dx,$$

where $v = (v_1, \ldots, v_d)$ is the outer unit normal to Ω. Hence the condition that

$$-\alpha \int_\Omega (\Delta u)\, v\,dx = \int_\Omega f v\,dx \qquad (1.33)$$

for all (sufficiently smooth) functions $v : \Omega \to \mathbb{R}$ such that $v_{|\partial\Omega} = 0$, is necessary for u to minimize the energy functional J. This implies that

$$-\alpha \Delta u = f \quad \text{in } \Omega \qquad (1.34)$$

with what are known as *Dirichlet boundary conditions*

$$u_{|\partial\Omega} = 0 \quad \text{on } \partial\Omega. \qquad (1.35)$$

We have shown that the *boundary value problem* (BVP) (1.34), (1.35) is the Euler–Lagrange equation of the minimization problem

$$u = \arg\min\{J(v) : v \text{ displacement with } v_{|\partial\Omega} = 0\}.$$

We call the equation (1.34) *Poisson's equation*; if $f = 0$, then

$$\Delta u = 0 \quad \text{in } \Omega \qquad (1.36)$$

is *Laplace's equation*; we call C^2-solutions of (1.36) *harmonic functions*. We refer to Theorem 4.24 for an abstract treatment of Poisson-type problems, to Section 6.5

for a systematic treatment of Poisson's equation, as well as to Sections 9.1 and 9.2 for its numerical approximation.

1.7* But there's more

The equations which we have introduced to date will be studied over the course of this book. Of course, these equations are only a small sample of the immeasurable number of partial differential equations which arise in real problems. This section is devoted to collecting a few additional prominent examples; however, we will neither go into any detail as regards the modeling leading to them, nor will we give a mathematical analysis of these equations at any later point.

1.7.1 The KdV equation

This equation in one spatial variable was first suggested by Joseph Boussinesq in 1877. It was rediscovered in 1895 by Diederik Korteweg and Gustav de Vries as a means to describe and analyze waves on shallow water in narrow channels. It reads

$$u_t - 6u\,u_x + u_{xxx} = 0 \tag{1.37}$$

and, as one sees, is a nonlinear third-order equation. The equation originally derived by Korteweg and de Vries had a somewhat different form; however, their version can be transformed into the now far more common form (1.37). The equation is often referred to under the abbreviation *KdV equation*. As noted above, we will forgo a description of its derivation.

The KdV equation delivers a mathematical explanation of an experimental observation. We have all seen two effects when watching waves: the wave can spread out (like ripples in a pond; in this case we speak of *dispersion*), and it can break. Dispersion of waves is a linear effect, while breaking can only be explained as a nonlinear phenomenon. The two effects would seem to be mutually exclusive; it was thus all the more surprising when in 1834 the young British engineer John Scott Russell observed that the effects can balance each other and lead to waves which propagate without changing their shape. Such waves are called *solitons* and can be described mathematically by the profile

$$u(t, x) = A\left(\operatorname{sech}\left(\frac{x - vt}{L}\right)\right)^2,$$

where L is the breadth of the wave, v its velocity and A its amplitude.[1] These solitons are in fact solutions of the KdV equation, meaning that it does indeed represent a mathematical justification of the phenomenon observed by Russell. Such standing waves can occur, for example, in tsunamis.

A further area of application of the KdV equation surrounds what is often called the *Fermi–Pasta–Tsingou–Ulam experiment*, named after the American mathematician Mary Tsingou (born 1928), the Italian nuclear physicist Enrico Fermi (1901–1954), the American physicist and computer scientist John R. Pasta (1918–1984) and the Polish mathematician Stanisław Marcin Ulam (1909–1984). Up until 1955 people were convinced that the energy of a system of coupled oscillators subject to a small nonlinear perturbation would distribute uniformly among all of the natural frequencies of the system. Hence the results of the computer experiments of Fermi, Pasta, Tsingou and Ulam in 1955 were highly surprising. The three showed that the energy distribution displays quasi-periodic

---

[1] Here $\operatorname{sech}(x) = \cosh^{-1}(x)$ denotes the hyperbolic secant function.

behavior, that is, the energy distribution almost always returns back to the initial distribution. It took until 1965 for the groundwork of the explanation of this phenomenon to be laid, when Martin David Kruskal and Norman J. Zabusky managed to show that the Fermi–Pasta–Tsingou–Ulam experiment can be described by the KdV equation.

The mathematical solution of the KdV equation is due to Clifford Gardner, John M. Greene, Martin D. Kruskal and Robert Miura in 1967, who used inverse scattering theory from quantum mechanics and thus connected two previously completely unrelated areas of mathematics. This also led to the recognition of the connection between the KdV equation and the Schrödinger equation. Peter David Lax would eventually develop a unified mathematical approach to the two, which could also be applied to other soliton equations.

1.7.2 Geometric differential equations

Differential equations whose origin is in a geometric variational problem are often called geometric partial differential equations. Such equations are generally nonlinear, and the mathematical analysis of such equations is a subject of ongoing research. Here we wish to give a brief introduction to two examples.

The Monge–Ampère equation

Gaspard Monge (1746–1818) is, among other things, credited with the invention of descriptive geometry. He studied the problem of soil (or more generally mass) transportation, as arose for example in public works involving excavation and/or filling of earth. It was this context that Monge introduced the first form of the partial differential equation in 1784 that would later bear the name Monge–Ampère equation. The French physicist André-Marie Ampère (after whom the SI base unit for electric current is named) considered this nonlinear partial differential equation in 1820 when studying the geometry of surfaces.

The general form of the equation reads

$$\det(H(u)) = f, \tag{1.38}$$

where

$$H(u) = (u_{x_i, x_j})_{i,j=1,\ldots,d} = \begin{pmatrix} u_{x_1, x_1} & \cdots & u_{x_1, x_d} \\ \vdots & & \vdots \\ u_{x_d, x_1} & \cdots & u_{x_d, x_d} \end{pmatrix} = D^2(u)$$

is the *Hessian matrix* (or *Hessian*, for short) of the unknown function $u : \Omega \to \mathbb{R}$ (which may for example be a parametrization of a surface). The right-hand side

$$f = f(x, u, u_{x_1}, \ldots, u_{x_d}) : \Omega \times \mathbb{R} \times \mathbb{R}^d \to \mathbb{R}$$

is a given function (for example, the curvature of a surface). In the two-dimensional case ($d = 2$), with the notation $(x_1, x_2) =: (x, y)$ the equation can be simplified to $u_{xx} u_{yy} - u_{xy}^2 = f$. We say that this equation is *fully nonlinear*, since it is nonlinear (here quadratically) in all terms of the highest order derivatives (here the second derivatives).

If we interpret f as a given curvature, then the solutions of the Monge–Ampère equation describe surfaces with this curvature (this problem is also known as the Minkowski problem). The problem was solved in 1953 by Louis Nirenberg. A further, unexpected, application of the complex

Monge–Ampère equation was found in 1978 in the area of *string theory*, in the context of what are known as Calabi–Yau manifolds.

The minimal surface equation

A surface $M \subset \mathbb{R}^3$ is said to be a *minimal surface* with boundary $\partial M = \Gamma$ if M has minimal surface area among all surfaces with the same, given, boundary Γ. One might imagine a soap film which does not enclose any air (that is, without bubbles). If u again denotes the displacement, this time of a surface from a flat horizontal position, then a minimal surface is by definition a minimizer of the surface area functional

$$A(x) = \int_\Omega \sqrt{g(u(x))}\,dx, \qquad g(u) := \det H(x), \qquad \partial\Omega = \Gamma, \quad u_{|\Gamma} = 0,$$

where $H(x)$ is the Hessian, as above. The Euler–Lagrange equation associated with this functional is called the *minimal surface equation* and reads

$$(1 + u_y^2)u_{xx} - 2u_x u_y u_{xy} + (1 + u_x^2)u_{yy} = 0.$$

1.7.3 The plate equation

When we derived Laplace's equation we were modeling an elastic membrane. Since the membrane may be regarded as being "thin", we could ignore any and all bending resistance in the material. If instead of a membrane we consider a clamped plate (with a certain thickness), then we can no longer justify this simplification. Let us again assume that there is a vertical force acting on the plate which may be described by a function $f : \Omega \to \mathbb{R}$, and that the geometry of the plate may be described by the domain Ω.

A derivation analogous to the derivation of Laplace's equation leads to the partial differential equation

$$\Delta^2 u := \Delta\Delta u = f, \tag{1.39}$$

which we call the *plate equation*; the expression $\Delta^2 u$ is referred to as the *Bilaplacian* of the function u. Observe that this is a fourth-order problem.

The problem of the fixed membrane leads, via the fact that the membrane is fixed at the boundary, to a boundary value problem. This is naturally the case here as well. But the non-negligible thickness of the plate leads to an additional boundary condition; a clamped plate may be specified by

$$u_{|\partial\Omega} = 0, \qquad \frac{\partial}{\partial\nu}u = 0 \quad \text{on } \partial\Omega. \tag{1.40}$$

Thus in addition to the Dirichlet boundary condition we see the appearance of *Neumann boundary conditions* which feature the *outer normal derivative* (see (7.4)). However, in the context of fourth-order problems these two conditions together are known as Dirichlet conditions. This model is also sometimes called a *Kirchhoff plate*.

1.7.4 The Navier–Stokes equations

The Navier–Stokes equations are the foundation of fluid and gas dynamics; they describe the flow of Newtonian fluids such as water, air, and many oils and gases.

We consider a domain $\Omega \subset \mathbb{R}^d$ which is filled with a fluid. This fluid may be described by its *density* $\rho = \rho(t, x)$, its *velocity vector* $\mathbf{u} = \mathbf{u}(t, x) = (u_1, \ldots, u_d)^T$ (where u_i gives the velocity in the ith coordinate direction) and the *energy* $e = e(t, x)$, where "energy" refers to the total energy (internal and kinetic). Now it is possible to combine ρ, $\mathbf{u}$ and e into a $(d + 2)$-dimensional state vector; this leads to the recognition that we are dealing with a *system* of equations (as opposed to a *scalar* equation, as we have always considered until now). A difficulty that often arises when treating systems is that the individual components may be coupled in a very complicated manner.

Instead of ρ, $\mathbf{u}$ and e, it is more convenient to work with the vector

$$\mathbf{U} = \mathbf{U}(t, x) := \begin{pmatrix} \rho \\ \rho \mathbf{u} \\ \rho e \end{pmatrix} \in \mathbb{R}^{d+2}, \tag{1.41}$$

where $\rho \mathbf{u} = \rho(t, x) \mathbf{u}(t, x)$ is the density of the mass flow, that is, the impulse per unit of volume, and $\rho e = \rho(t, x) e(t, x)$ gives the *total energy* per unit of volume. We call the vector $\mathbf{U}$ the *state vector*. If we denote by

$$\delta_{i,j} := \begin{cases} 1 & \text{if } i = j, \\ 0 & \text{otherwise,} \end{cases} \tag{1.42}$$

the *Kronecker delta* for $i, j \in \mathbb{N}$, then $e_i := (\delta_{1,i}, \ldots, \delta_{d,i})^T = (0, \ldots, 0, 1, 0, \ldots, 0)^T$ is the ith canonical unit vector. With this notation we can define

$$\mathbf{F}_i = \mathbf{F}_i(\mathbf{U}) := \begin{pmatrix} \rho u_i \\ (\rho u_i)\mathbf{u} + p e_i \\ u_i (\rho e + p) \end{pmatrix},$$

where $p = p(t, x)$ is the pressure of the fluid. The vectors $\mathbf{F}_i$ model the *convective terms*; the *diffusive terms* are represented by the vector

$$\mathbf{G}_i = \mathbf{G}_i(\mathbf{U}) := \begin{pmatrix} 0 \\ -\boldsymbol{\tau}_i \\ -\sum_{j=1}^d u_j \tau_{i,j} + q_i \end{pmatrix},$$

where $\boldsymbol{\tau} = \boldsymbol{\tau}(\mathbf{u}) = (\tau_{j,i})_{i,j=1,\ldots,d}$, and $\tau_{i,j} := \mu\left(\frac{\partial u_j}{\partial x_i} + \frac{\partial u_i}{\partial x_j}\right) - \delta_{j,i}\frac{2}{3}\mu \operatorname{div} \mathbf{u}$ is the *viscous stress tensor*, with $\boldsymbol{\tau}_i$ the ith column vector. The number $\mu \in \mathbb{R}^+$ denotes the *dynamic viscosity*, and $\mathbf{q} = (q_1, \ldots, q_d)^T$, $q_i = -\lambda \frac{\partial T}{\partial x_i}$ the *heat flow* with *temperature* T and *heat conductivity* λ. Putting this all together, the (compressible) *Navier–Stokes equations* read

$$\mathbf{U}_t + \sum_{i=1}^d \frac{\partial}{\partial x_i} \mathbf{F}_i(\mathbf{U}) + \sum_{i=1}^d \frac{\partial}{\partial x_i} \mathbf{G}_i(\mathbf{U}) = 0. \tag{1.43}$$

Due to the dependence of the convective flows $\mathbf{F}_i$ on $\mathbf{U}$, this is a system of nonlinear partial differential equations.

Incompressible Navier–Stokes equations

If the density ρ is constant in space and time, $\rho = \rho(t, x) \equiv$ const, then we call the fluid *incompressible*. Strictly speaking there are no incompressible fluids in real life, but in the case of water or air (for example) with low velocity, one may assume constant density as a reasonable approximation. In this case the first component in (1.43), namely $\rho_t + \operatorname{div}(\rho \mathbf{u}) = 0$, known as the *continuity equation*, can be simplified to

$$\operatorname{div} \mathbf{u} = 0. \tag{1.44}$$

One can then check that the remaining equations reduce to

$$\rho \mathbf{u}_t + \rho (\mathbf{u} \cdot \nabla)\mathbf{u} - \eta \Delta \mathbf{u} + \nabla p = \mathbf{f}. \tag{1.45}$$

Here the Laplacian is to be understood componentwise, $\Delta \mathbf{u} = (\Delta u_1, \ldots, \Delta u_d)^T$, and the abbreviation of the convective terms is to be read as

$$(\mathbf{u} \cdot \nabla)\mathbf{u} = \left(\sum_{j=1}^{d} u_j \frac{\partial}{\partial x_j} u_i \right)_{i=1,\ldots,d}.$$

Thus we obtain d equations in (1.45) and one in (1.44) for the $d + 1$ unknowns $\mathbf{u}$ and p. Both equations together, that is, the pair (1.44, 1.45)

$$\begin{aligned} \rho \mathbf{u}_t - \eta \Delta \mathbf{u} + \rho (\mathbf{u} \cdot \nabla)\mathbf{u} + \nabla p &= \mathbf{f}, \\ \operatorname{div} \mathbf{u} &= 0, \end{aligned} \tag{1.46}$$

are known as the *Navier–Stokes equations* for incompressible fluids. In the form (1.46) the equations are also called *non-steady state*; if the velocity is constant in time, then the derivative in the time variable vanishes and we obtain the *steady-state* (or *stationary*) Navier–Stokes equations

$$\begin{aligned} -\nu \Delta \mathbf{u} + (\mathbf{u} \cdot \nabla)\mathbf{u} + \nabla p &= \mathbf{f}, \\ \operatorname{div} \mathbf{u} &= 0 \end{aligned} \tag{1.47}$$

in the unknowns $\mathbf{u} = \mathbf{u}(x)$, $p = p(x)$ and with right-hand side $\mathbf{f} = \mathbf{f}(x)$. The quantity $\nu = \eta \rho^{-1}$ is called the *kinematic viscosity* and is often identified with the inverse of the *Reynolds number*, $\nu = \mathrm{Re}^{-1}$. The Reynolds number captures the relationship between the forces of inertia and viscosity.

Although the Navier–Stokes equation "only" has a quadratic nonlinearity in the term $(\mathbf{u} \cdot \nabla)\mathbf{u}$ and the coupling via the incompressibility condition $(\mathbf{u} \cdot \nabla)\mathbf{u}$ appears to be weak, the mathematical theory of this equation is exceptionally difficult. The Navier–Stokes equations feature among the *Millennium problems*; for the solution of any of these problems the Clay Mathematics Institute will award a prize of one million US dollars [20]. For the incompressible Navier–Stokes equations, even the proof of the existence of a local solution for arbitrarily small times is an open problem.

The Stokes problem

A simplification in the equations results if the nonlinear convection term is suppressed; this is reasonable in the case of extremely viscous fluids. In the resulting *Stokes problem* the diffusivity of the impulse, that is, the kinematic viscosity, is many orders of magnitude larger than the thermal diffusivity; consequently, the convective term (inertia) may be neglected. An area of application is the study of currents on the surface of planets, in geodynamics. The equations read $-\nu \Delta \mathbf{u} + \nabla p = \mathbf{f}$ and $\operatorname{div} \mathbf{u} = 0$.

1.7.5 Maxwell's equations

Maxwell's equations are a system of four equations at the heart of classical electromagnetism and the theory of electric circuits. These four equations describe the generation of electric and magnetic fields by charges and currents, as well as the interaction of these fields. These electric and magnetic fields are regarded as non-stationary, that is, time dependent, and the interactions over time are part of the model.

The essential scientific contribution of Maxwell was to create a unified theory which united the following laws:
- Ampère's law (electrodynamic law),
- Faraday's law (magnetodynamic law),
- Gauss's law (electrostatic law), and
- the magnetostatic law.

In order to obtain the mathematical consistency of the continuity equation in electrodynamics $\rho_t + \nabla \cdot j = 0$ with the *current density* $j = j(t, x) = (j_1, j_2, j_3)^T$, Maxwell inserted an additional term, which he called *displacement current*, to Ampère's law. There is a clear analogy to continuity equations from fluid mechanics, where instead of j the expression $\rho \mathbf{u}$ is involved. In electrodynamics, vector fields are usually typeset in bold; in fluid mechanics on the other hand they are marked with an arrow.

We now wish to give the equations. We denote by $E = E(t, x) = (E_1, E_2, E_3)^T$ the *electric field strength* and by $B = B(t, x) = (B_1, B_2, B_3)^T$ the *magnetic field strength*, as well as by ρ the *charge density*, which is a given constant. Then the four equations read

$$B_t + \operatorname{rot} E = 0 \qquad \text{(magnetodynamic law, Faraday)} \qquad (1.48a)$$

$$\operatorname{div} E = 4\pi \rho \qquad \text{(electrostatic law, Gauss)} \qquad (1.48b)$$

$$E_t - \operatorname{rot} B = -4\pi j \qquad \text{(electrodynamic law, Ampère)} \qquad (1.48c)$$

$$\operatorname{div} B = 0 \qquad \text{(magnetostatic law)} \qquad (1.48d)$$

These four equations together are *Maxwell's equations*. Here we have used the notation $\operatorname{rot} E :=$ $\nabla \times E$, that is,

$$\operatorname{rot} E = \begin{pmatrix} \frac{\partial}{\partial x_2} E_3 - \frac{\partial}{\partial x_3} E_2 \\ \frac{\partial}{\partial x_3} E_1 - \frac{\partial}{\partial x_1} E_3 \\ \frac{\partial}{\partial x_1} E_2 - \frac{\partial}{\partial x_2} E_1 \end{pmatrix}.$$

Clearly, (1.48) constitutes a coupled system of non-stationary linear partial differential equations. A particularity of these equations is the appearance of the two differential operators div and rot together.

1.7.6 The Schrödinger equation

The Schrödinger equation is the foundational equation of non-relativistic quantum mechanics. It was first formulated as a wave equation by Erwin Schrödinger (1887–1961), in 1926. Schrödinger's work would later earn him the Nobel prize in physics, in 1933. Solutions of the Schrödinger equation, which are also known as *wave functions*, describe the evolution of the state of a quantum system in space and time.

The Schrödinger equation is a postulate (similar to Newton's axioms in classical physics) and as such cannot strictly speaking be mathematically derived from other laws. The equation was postulated taking into account certain basic principles of physics; Schrödinger drew upon knowledge of quantum-mechanical phenomena which were already known in his time. There are also numerous parallels between his theory and the theory of optics.

Unlike all the equations we have previously considered, the Schrödinger equation is complex-valued, a necessary consequence of the quantum-mechanical modeling. The equation for a single particle (such as an elementary particle or an atom) subject to a potential $V = V(t, x)$, whose state is described by the wave function ψ, reads

$$i\hbar \psi_t = -\frac{\hbar^2}{2m}\Delta\psi + V(t, x)\psi. \tag{1.49}$$

Here $i = \sqrt{-1}$ is the imaginary unit, $\hbar = h/2\pi$ is the reduced Planck constant (where Planck's constant, or the quantum of electromagnetic action, is given by $h = 6.626 \cdot 10^{-34}$ Js) and m is the mass of the particle. The unknown is the complex-valued wave function $\psi = \psi(t, x) : \Omega \times [0, T] \to \mathbb{C}$. The right-hand side of (1.49) can also be written as

$$\left(-\frac{\hbar^2}{2m}\Delta + V(t, x)\right)\psi =: \hat{H}\psi$$

with *Hamiltonian* $\hat{H}$.

1.8 Classification of partial differential equations

As previously announced, we wish to attempt to *classify* partial differential equations according to certain properties. The following properties allow us to define a first set of categories:
1.) Dimension (of the space variables)
2.) Order of the equation (with respect to both space and time)
3.) Algebraic type of the equation
We will meet a further classification somewhat later (in Chapter 2), which is of more practical importance for the mathematical study of such equations. There we will also give an overview of the partial differential equations introduced above and order these according to the criteria listed here (see Table 2.1 on page 47).

1.) Dimension
This category is clear, but often not particularly significant. It is, however, possible to find examples of partial differential equations whose behavior is strongly dependent on the space dimension.

2.) Order of the equation
The order of a partial differential equation is the highest order derivative appearing in the equation. The viscous Burgers' equation, for example, is of second order (due to the term εu_{xx}).

3.) Algebraic type of the equation
Here we mean, firstly, the division between *linear* and *nonlinear* equations. "Linear" means that the equation(s) is (are) linear in the unknown function(s). This is also

a purely algebraic property of the equation(s). Nonlinear equations can be divided into further subcategories (as done in [28]):

- Semilinear equations:
  These are linear in the term in which the highest-order derivative appears; more precisely, if the order is k, say, then they must have the form

$$\sum_{|\alpha|=k} a_\alpha(x) D^\alpha u + a_0(D^{k-1}u, \ldots, Du, u, x) = 0,$$

  where $\alpha = (\alpha_1, \ldots, \alpha_d)^T \in \mathbb{N}^d$ is a multi-index with $|\alpha| := \alpha_1 + \cdots + \alpha_d$, $x \in \mathbb{R}^d$, and $D^\alpha u := \frac{\partial^{|\alpha|}u}{\partial x_1^{\alpha_1} \cdots \partial x_d^{\alpha_d}}$ denotes the corresponding derivative. All the equations introduced above are semilinear with the exception of the minimal surface equation and the Monge–Ampère equation.

- Quasilinear equations:
  These are partial differential equations whose variable coefficients may depend on the solution and its derivatives, but only up to a degree strictly less than the order of the equation; the equation must still be linear in the derivatives of highest order. If the order of the equation is k, then these have the form

$$\sum_{|\alpha|=k} a_\alpha(D^{k-1}u, \ldots, Du, u, x) D^\alpha u + a_0(D^{k-1}u, \ldots, Du, u, x) = 0.$$

  All of the equations considered thus far except the Monge–Ampère equation are quasilinear. The minimal surface equation, in particular, is quasilinear but not semilinear.

- Fully nonlinear equations:
  Here all terms involving the highest-order derivatives are nonlinear. The Monge–Ampère equation is an example of such a fully nonlinear (here quadratic) equation.

The real background (that is, the natural or technical process which should be modeled by a differential equation) is of course essential for determining the properties of the differential equation modeling it. We will return to this point later, cf. Table 3.1 on page 110.

1.9* Comments

Jean Le Rond d'Alembert (1717–1783) is considered one of the pioneers of the modeling of physical phenomena via partial differential equations. His first contribution *Réflexions sur la cause générale des vents* was awarded the prize of the Prussian academy in 1747. The physical assumptions he made in this work were however relatively unrealistic and led to a certain amount of dispute. In the same year d'Alembert derived the wave equation from the Newtonian laws of motion, as a description of a vibrating string. We have attempted to reproduce his elegant and physically accurate derivation above. We refer to [38, end of Ch. VI] for a description of the rather fascinating history of this problem, which was also decisive for our modern understanding of functions. Indeed, every continuous initial position of the string is physically meaningful, and we will see later that in this case there is always a unique solution (see Section 3.1.2). In this example there are physical reasons for requiring the introduction of *weak solutions*; see Exercise 5.13. D'Alembert would also become well known for a completely different activity: in 1751, together with Denis Diderot he published the first volume of what was probably the most famous early encyclopedia. This would end up totaling 35 volumes, the last of which was published in 1780.

The study of heat diffusion may be traced back to Fourier, who incidentally was also responsible for the term greenhouse effect (l'effet de serre), which he invented in his groundbreaking work *Théorie analytique de chaleur* (The Analytic Theory of Heat) in 1822. Fourier participated in Napoleon's Egyptian campaign and even became the secretary of the Institut d'Egypte there. After his return to France, he was appointed Prefect of the Departement Isère in 1802. His tenure is considered to have been a success; there he was responsible, for example, for (successfully) draining swamps. Not long after his return to Paris in 1815, Fourier became secretary of the Académie des sciences, in 1817.

Louis Bachelier (1870–1946) is regarded as the founder of financial mathematics. He obtained his doctorate in 1900 under Henri Poincaré with a dissertation on the topic *Théorie de la Spécu-lation*. He was well ahead of his time, already working with Brownian motion five years before Albert Einstein and a full 23 years before Norbert Wiener would provide a rigorous mathematical construction. He also gave a formula for the price of options 73 years before the publication of the famous Black–Scholes formula. His book *Le jeu, la chance et le hasard*, published in 1914, was very successful, but his work remained essentially unknown for a long period of time. After having held academic positions in Paris, Dijon and Rennes, Bachelier was a professor in Besançon (Franche-Comté) from 1927 until his retirement.

1.10 Exercises

Exercise 1.1 Derive the equations of motion for the undamped double oscillator (spring/mass system) depicted in Figure 1.7. In doing so, you should work out the balance of forces on each of the oscillators.
Suggestion: Use Hooke's law: if a spring is stretched by a force, then its change in length is proportional to the strength of the force, $F = k\,s$ (where s is the change in length and k is a constant depending on the spring).

Exercise 1.2 Suppose that the displacement u of a vibrating string is restricted by a certain obstacle in such a way that $u \geq g$ pointwise. Find the corresponding differential *inequality* that the displacement must satisfy.
Suggestion: Consider the vertical displacement $S(x)$ analogous to the derivation of the wave equation and formulate the restriction by the obstacle pointwise.

Exercise 1.3 Let a given fluid have density $\rho(t, x)$ and velocity vector $\mathbf{u}(t, x)$. Derive the continuity equation $\rho_t + \mathrm{div}(\rho\mathbf{u}) = 0$ from physical principles.
Suggestion: Use the principle of conservation of mass and the divergence theorem.

Exercise 1.4 For the equation $u_t = u_{xx} + u$ determine all solutions of the form $u(t, x) = \varphi(x - ct)$ (the *traveling waves*).

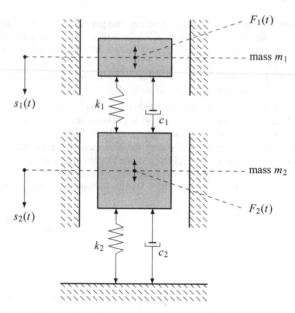

Fig. 1.7 Double oscillator with masses m_1, m_2, damping c_1, c_2, spring constants k_1, k_2 and external forces $F_1(t)$, $F_2(t)$. To be determined are the vertical displacements $s_1(t)$, $s_2(t)$.

Exercise 1.5 Derive the equation $u_{tt} + a^2 u_{xxxx} = 0$, $a \in \mathbb{R}$ for the displacement $u(t, x)$ of a rod of length ℓ resting at both ends.
Suggestion: Use the relation $Q = \frac{\partial M}{\partial x}$ between the bending moment M and the shear force Q. The relation between M and the desired vertical displacement u reads $M = -EIu_{xx}$, where E is the modulus of elasticity and I is the moment of inertia. These two quantities enter into the equation via the constant a.

Chapter 2
Classification and characteristics

In the preceding chapter we encountered a whole series of partial differential equations arising from natural processes as well as economic models. Given the huge variety of such equations, it should not be surprising that there is no unified mathematical theory and no unified method for solving partial differential equations, and indeed such a method cannot exist. It is however natural to ask whether they can be divided into categories for which a more or less complete theory can be developed.

Here we will introduce the principal equation types *elliptic*, *parabolic* and *hyperbolic*. In Chapter 6 we will then develop a method with which we can investigate elliptic equations systematically. For parabolic and hyperbolic equations we will use the same method in Chapter 8 (separation of variables and spectral decomposition); however, it will turn out that these equations have very different properties from each other. Numerical schemes for all three types will be introduced in Chapter 9.

For first-order equations there is in fact a general solution method, the *method of characteristics*. We will start this chapter with an introduction to this method, which is based on the idea of reducing the problem to an ordinary differential equation.

Chapter overview

© The Author(s), under exclusive license to Springer Nature Switzerland AG 2023
W. Arendt, K. Urban, *Partial Differential Equations*, Graduate Texts in
Mathematics 294, https://doi.org/10.1007/978-3-031-13379-4_2

Notation

We will use the following notation. If Ω is a set in $\mathbb{R}^d$, then we denote by $C(\Omega)$ the space of continuous real-valued functions on Ω. If Ω is open, then $C^1(\Omega)$ is the space of continuous functions $u : \Omega \to \mathbb{R}$ whose partial derivatives exist and are all continuous. The space of functions whose partial derivatives admit continuous extensions in $C(\overline{\Omega})$ will be denoted by $C^1(\overline{\Omega})$. If $u \in C^1(\Omega)$, then we write $\frac{\partial u}{\partial x_j}$ for its partial derivatives and also for their continuous extensions in the case that $u \in C^1(\overline{\Omega})$. Spaces of functions which admit derivatives of higher order can be defined similarly: if $\Omega \subset \mathbb{R}^d$ is open, then we set

$$C^2(\Omega) := \left\{ u \in C^1(\Omega) : \frac{\partial u}{\partial x_j} \in C^1(\Omega), \ j = 1, \ldots, d \right\},$$

that is, $u \in C^2(\Omega)$ if all its partial derivatives of second order $\frac{\partial^2}{\partial x_i \partial x_j}$ are in $C(\Omega)$. We may also define inductively

$$C^{k+1}(\Omega) := \left\{ u \in C^1(\Omega) : \frac{\partial u}{\partial x_j} \in C^k(\Omega), \ j = 1, \ldots, d \right\}, \quad k \geq 2.$$

Finally, $C^\infty(\Omega) := \bigcap_{k=1}^\infty C^k(\Omega)$ will be the space of infinitely differentiable real-valued functions on Ω. We likewise define $C^k(\overline{\Omega})$ inductively by

$$C^{k+1}(\overline{\Omega}) := \left\{ u \in C^1(\overline{\Omega}) : \frac{\partial u}{\partial x_j} \in C^k(\overline{\Omega}), \ j = 1, \ldots, d \right\}, \quad k \geq 0,$$

and $C^\infty(\overline{\Omega}) := \bigcap_{k \in \mathbb{N}} C^k(\overline{\Omega})$. We will sometimes have occasion to consider mixed spaces of the form $C^k(\Omega) \cap C(\overline{\Omega})$; this is the space of all continuous functions on $\overline{\Omega}$ whose restriction to Ω belongs to $C^k(\Omega)$. If $u \in C(\Omega)$, then the set

$$\operatorname{supp} u := \overline{\{x \in \Omega : u(x) \neq 0\}} \tag{2.1}$$

is called the *support* of u. The set $\Omega \setminus \operatorname{supp} u$ is the largest open subset of Ω on which u vanishes identically. We denote by $C_c(\Omega)$ the space of those functions $u \in C(\Omega)$ whose support is a compact subset of Ω; in this case, we say u has *compact support*. We also set $C_c^k(\Omega) := C_c(\Omega) \cap C^k(\Omega)$, $k \in \mathbb{N} \cup \{\infty\}$.

2.1 Characteristics of initial value problems on $\mathbb{R}$

This section is dedicated to the method of characteristics in the simplest possible case, namely for linear equations in two variables. We will then give an insight into the much more complicated nonlinear situation via the example of Burgers' equation.

2.1.1 Homogeneous problems

We first look at problems on $[0, T] \times \mathbb{R}$, that is, the time variable is in $[0, T]$ and there is one space variable. The initial value $u_0 : \mathbb{R} \to \mathbb{R}$ is thus a function on $\mathbb{R}$.

We start with the first-order homogeneous partial differential equation

$$u_t(t, x) + a(t, x)\, u_x(t, x) = 0, \qquad\qquad x \in \mathbb{R},\ t \in (0, T), \qquad (2.2a)$$
$$u(0, x) = u_0(x), \qquad\qquad x \in \mathbb{R}, \qquad (2.2b)$$

where $T > 0$ is given. This is an *initial value problem*, also known as a *Cauchy problem*. The *initial value* $u_0 : \mathbb{R} \to \mathbb{R}$, a continuous function, is given. The partial differential equation (2.2a) is of first order and has a variable coefficient $a = a(t, x)$. This is a given function $a : \mathbb{R} \times \mathbb{R} \to \mathbb{R}$, which we will assume to be continuously differentiable. We will forget the initial condition (2.2b) for the time being and only consider the partial differential equation (2.2a). The basic idea consists of finding curves in space-time $\mathbb{R} \times \mathbb{R}$ on which every solution of (2.2a) is constant. Such a curve is called a *characteristic* of the equation (2.2a), cf. Figure 2.1.

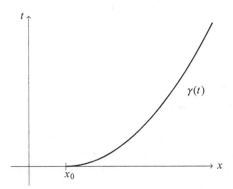

Fig. 2.1 Characteristic: the solution $u(t, \gamma(t))$ is constant and equal to $u_0(x_0)$.

We first make the following *ansatz*. Let $\Gamma : J \to \mathbb{R}^2$ be a curve of the form $\Gamma(s) = (s, \gamma(s))$ for an open interval $J \subset \mathbb{R}$ and $\gamma \in C^1(J)$. Then Γ is a characteristic if and only if

$$0 = \frac{d}{ds} u(s, \gamma(s)) = u_t(s, \gamma(s)) + \dot{\gamma}(s) u_x(s, \gamma(s))$$

for all $s \in J$ and every solution u of the equation. Thus Γ is a characteristic if γ satisfies the ordinary differential equation

$$\dot{\gamma}(s) = a(s, \gamma(s)), \quad s \in J. \qquad (2.3)$$

We now wish to study the initial value problem (2.2) using solutions of this differential equation. Let us suppose that $u \in C^1([0, T] \times \mathbb{R})$ is such a solution of (2.2); we wish to determine $u(t, x)$ for given $0 < t \le T$ and $x \in \mathbb{R}$. By the *Picard–Lindelöf theorem*[1] on existence and uniqueness of solutions to ordinary differential equations [3, (7.6)], the equation (2.3) has a unique maximal solution $\gamma \in C^1(J)$ such that

$$\gamma(t) = x. \tag{2.4}$$

Here $J \subset \mathbb{R}$ is an open interval containing t. The maximality of J means that $\lim_{s \to a+} |\gamma(s)| = \infty$ in the case that $J = (a, b)$ with $a > -\infty$ (that is, if the left endpoint is finite), and analogously in the case of a finite right endpoint b. If $0 \in J$ (which is not always the case; see Example 2.5), then we know that

$$u(t, x) = u(t, \gamma(t)) = u(0, \gamma(0)) = u_0(\gamma(0)).$$

In this case we have determined $u(t, x)$ by solving the initial value problem (2.3), (2.4) (which is actually a final – also known as terminal – value problem). We wish to formulate this as a theorem.

Theorem 2.1 *Let* $u \in C^1([0, T] \times \mathbb{R})$ *be a solution of* (2.2) *and* $\gamma \in C^1([0, t])$ *a solution of* (2.3), (2.4) *for some* $(t, x) \in (0, T] \times \mathbb{R}$. *Then*

$$u(t, x) = u_0(\gamma(0))$$

and u *is constant along the characteristic* Γ.

Thus if we know that a solution of (2.3), (2.4) exists on $[0, T]$ for every x, then we can conclude that (2.2) has at most one solution. This is for example the case if there exists an $L > 0$ such that $|a(s, x)| \le L(1 + |x|)$ for all $x \in \mathbb{R}$ and $s \in [0, T]$ (see Exercise 2.8). In general (2.2) can, however, have multiple solutions for a given initial value (see Example 2.5). At any rate, the initial value problem (2.3), (2.4) gives a strategy for determining solutions; this is the *method of characteristics*. We will now consider a few examples.

Example 2.2 (Linear transport equation) The initial value problem for the linear transport equation is given by

$$u_t + a\, u_x = 0, \qquad x \in \mathbb{R}, \ t > 0,$$
$$u(0, x) = u_0(x), \qquad x \in \mathbb{R},$$

that is, we have (2.2) with constant coefficient $a \equiv \text{const}$. In this case the ordinary differential equation (2.3) reduces to $\dot{\gamma}(s) = a$. Its solutions $\gamma \in C^1(\mathbb{R})$ have the form $\gamma(s) = c + as$, $s \in \mathbb{R}$, where $c \in \mathbb{R}$. Thus in this case the characteristics are straight lines, see Figure 2.2.

Now let $x \in \mathbb{R}$ and $t > 0$. The straight line through the point (t, x) is given by $\gamma(s) = x + a(s - t)$, $s \in \mathbb{R}$, and so by Theorem 2.1 we have $u(t, x) = u_0(\gamma(0)) =$

---

[1] Also known as the *Cauchy–Lipschitz theorem*.

$u_0(x - a\,t)$. If $u_0 \in C^1(\mathbb{R})$, then it is easy to check that this formula defines a solution of the linear transport equation. We have thus proved existence and uniqueness of solutions using the characteristics. $\triangle$

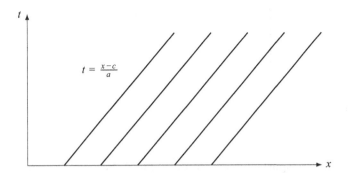

Fig. 2.2 For the linear transport equation the characteristics are straight lines.

Characteristics can also help to explain the role of the choice of initial and boundary values when it comes to existence and uniqueness of solutions of a given partial differential equation. For the meantime we limit ourselves to first-order equations and again consider the transport equation, although this time on an interval in space, which corresponds to a rectangle in space-time.

Example 2.3 We seek the solutions $u \in C^1([0,T] \times [0,1])$, $u = u(t,x)$, of the equation

$$u_t + 2T\,u_x = 0. \tag{2.5}$$

For $(t,x) \in (0,T) \times (0,1)$, a characteristic passing through the point (t,x) is given by $\gamma(s) = x - 2T(t-s)$, $s \in \mathbb{R}$. This line passes through two sides of the rectangle $[0,T] \times [0,1]$, cf. Figure 2.3. If u is a solution of (2.5), then it takes on the same value at both points where the line hits the boundary of the rectangle. With this principle in mind we can see which boundary conditions we can impose on the rectangle in order to obtain existence and uniqueness of solutions. For example, we could impose conditions on the two sides $\{(0,x) : 0 \le x \le 1\}$ and $\{(t,0) : 0 \le t \le T\}$; then every characteristic has exactly one point of intersection with this part of the boundary. At the origin these conditions must be compatible; for example, it is easy to show that if $g \in C^1([0,1])$ and $h \in C^1([0,T])$ with $g(0) = h(0)$ and $h'(0) = -2Tg'(0)$, then there is exactly one solution $u \in C^1([0,T] \times [0,1])$ of (2.5) for which $u(0,x) = g(x)$, $x \in [0,1]$ and $u(t,0) = h(t)$, $t \in [0,T]$, see Exercise 2.6. $\triangle$

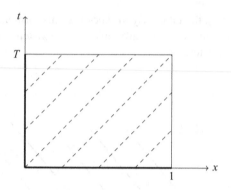

Fig. 2.3 Characteristics $\gamma(s) = 2Ts + (x - 2Tt)$ for (2.5).

Example 2.4 (Transport equation with variable coefficients) Our next example is the problem

$$u_t + x\,u_x = 0, \qquad x \in \mathbb{R},\ t > 0,$$
$$u(0, x) = u_0(x), \qquad x \in \mathbb{R},$$

corresponding to (2.2) with $a(t, x) = x$. The ordinary differential equation for the characteristics reads $\dot{\gamma}(s) = \gamma(s)$, $s \in \mathbb{R}$, which has solutions of the form $\gamma(s) = ce^s$. In order to determine the solution of the partial differential equation at the point (t, x), one can determine the characteristic which passes through this point, that is, $x = ce^t$, and thus $\gamma(s) = xe^{-t}e^s$. This yields $\gamma(0) = xe^{-t}$ and $u(t, x) = u_0(xe^{-t})$; it is easy to check that this is in fact the solution. So in this case the characteristics (s, xe^{s-t}) are not straight lines as they were above, but rather take on the form depicted in Figure 2.4. ∆

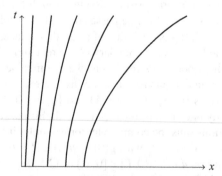

Fig. 2.4 Curved characteristics for the transport equation with variable coefficients.

Here the characteristics also facilitate a physical interpretation of the equation. One can say that the initial data are transported along the characteristics; this means that, at least in the case of pure transport processes (that is, convection and no diffusion) the initial data already completely determine the solution. As mentioned earlier, we will now give an example in which uniqueness of solutions is violated.

Example 2.5 Let $u_0 \in C_c^1(\mathbb{R})$, that is, $u_0 : \mathbb{R} \to \mathbb{R}$ is continuously differentiable and vanishes outside a bounded subset of $\mathbb{R}$. We seek a function $u \in C^1((0, \infty) \times \mathbb{R}) \cap C([0, \infty) \times \mathbb{R})$ satisfying

$$u_t(t, x) = x^2 u_x(t, x), \quad t > 0, \ x \in \mathbb{R}, \tag{2.6a}$$

$$u(0, x) = u_0(x), \qquad x \in \mathbb{R}. \tag{2.6b}$$

We wish to apply the method of characteristics. For $(t, x) \in (0, \infty) \times \mathbb{R}$ we thus need to solve the problem $\dot{\gamma}(s) = -\gamma(s)^2$, $s \in \mathbb{R}$, $\gamma(t) = x$. The solution is given by $\gamma(s) = (s - t + \frac{1}{x})^{-1}$ if $x \neq 0$, and $\gamma \equiv 0$ for $x = 0$, with maximal solution interval

$$J = \begin{cases} \left(t - \frac{1}{x}, \infty\right), & \text{if } x > 0, \\ \left(-\infty, t - \frac{1}{x}\right), & \text{if } x < 0, \\ \mathbb{R}, & \text{if } x = 0. \end{cases}$$

In particular, $0 \in J$ if and only if $x < \frac{1}{t}$, and in this case

$$u(t, x) = u(t, \gamma(t)) = u(0, \gamma(0)) = u_0(\gamma(0)) = u_0\left(\frac{x}{1 - xt}\right).$$

Thus there exists at most one solution in the set $G := \{(t, x) : t > 0, x < \frac{1}{t}\}$, cf. Figure 2.5.

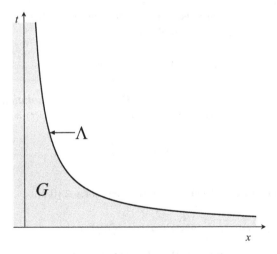

Fig. 2.5 The set $G := \{(t, x) : t > 0, x < \frac{1}{t}\}$ and curve Λ.

Conversely, if we define

$$u(t, x) = u_0\left(\frac{x}{1 - xt}\right) \quad \text{for } x < \frac{1}{t},$$

then $u \in C^1(G)$ and $\lim_{t \downarrow 0} u(t, x) = u_0(x)$ for all $x \in \mathbb{R}$. In addition, $u(t, x) = 0$ in a neighborhood of the curve $\Lambda = \{(t, \frac{1}{t}) : t > 0\}$ since u_0 vanishes outside a bounded set. It is easy to see that this u solves the equation (2.6a) in G. Now, if $u_1 \in C_c^1(\mathbb{R})$ is an arbitrary function and we set

$$w(t, x) := u_1\left(\frac{x}{1 - xt}\right) \quad \text{for } t > 0, \quad x > \frac{1}{t}$$

as well as $w(t, x) := 0$ for $t > 0$ and $x = \frac{1}{t}$, then w solves the initial value problem (2.6a) with $w(0, x) = 0$, so that $u + w$ solves (2.6). Hence uniqueness does not hold here.   △

2.1.2 Inhomogeneous problems

We now consider the inhomogeneous problem

$$u_t + a(t, x) u_x = b(t, x), \ x \in \mathbb{R}, \ t > 0, \quad u(0, x) = u_0(x), \ x \in \mathbb{R}, \tag{2.7}$$

for a given continuous function $b : [0, \infty) \times \mathbb{R} \to \mathbb{R}$. In this example, if $b \neq 0$ it is not clear how to find curves on which all solutions are constant. However, it turns out that there are curves on which the solution of an ordinary differential equation is enough. In this section we will also refer to such curves as *characteristics*. We will again consider the ordinary differential equation

$$\dot{\gamma}(s) = a(s, \gamma(s)), \qquad s \in \mathbb{R}. \tag{2.8}$$

Theorem 2.6 *Let $u \in C^1((0, \infty) \times \mathbb{R}) \cap C([0, \infty) \times \mathbb{R})$ be a solution of (2.7). For $t > 0$ and $x \in \mathbb{R}$ let $\gamma \in C^1([0, t])$ be a solution of (2.8) with $\gamma(t) = x$. Then*

$$u(t, x) = u_0(\gamma(0)) + \int_0^t b(s, \gamma(s)) \, ds.$$

Proof By the chain rule, (2.8) and (2.7), we have, for $s \in [0, t]$,

$$\frac{d}{ds} u(s, \gamma(s)) = u_t(s, \gamma(s)) + u_x(s, \gamma(s)) \dot{\gamma}(s)$$

$$= u_t(s, \gamma(s)) + a(s, \gamma(s)) u_x(s, \gamma(s)) = b(s, \gamma(s)).$$

This shows that the function γ satisfies an inhomogeneous first-order linear differential equation. Physically speaking, one can interpret this as saying that the speed of

u along the characteristic is equal to the external force b. Since $u(0, \gamma(0)) = u_0(\gamma(0))$, it follows from the Fundamental Theorem of Calculus that

$$u(t, x) = u(t, \gamma(t)) = u_0(\gamma(0)) + \int_0^t b(s, \gamma(s)) \, ds,$$

which proves the claim. $\qquad\qquad\square$

In this case u is not constant along the characteristics, but as before $u(t, \gamma(t))$ is uniquely determined by the initial value $u(0, \gamma(0))$ (as the solution of an initial value problem of an ordinary differential equation), if (2.8) is a solution in $[0, t]$.

Example 2.7 (Inhomogeneous linear transport equation) We consider the linear transport equation with the special inhomogeneity $b(t, x) = x$,

$$u_t + u_x = x, \quad x \in \mathbb{R}, t > 0, \qquad u(0, x) = u_0(x), \quad x \in \mathbb{R}.$$

As in the homogeneous case we obtain the characteristics $\gamma(s) = s + c$. Let $(t, x) \in (0, \infty) \times \mathbb{R}$. Then $\gamma(s) = x + s - t$ is the characteristic passing through (t, x), and we obtain the following formula for solutions of the partial differential equation:

$$u(t, x) = u_0(\gamma(0)) + \int_0^t \gamma(s) \, ds = u_0(x - t) + t\left(x - \frac{t}{2}\right).$$

This is indeed a solution if $u_0 \in C^1(\mathbb{R})$. $\qquad\qquad\triangle$

2.1.3* Burgers' equation

Clearly, general quasilinear equations are not covered by the above method, since in the case of quasilinear equations the coefficient a can depend on the solution. However, at least in the special case of Burgers' equation we can find characteristics, as we shall now proceed to do. We recall from Section 1.2.5* that the initial value problem for (the inviscid) Burgers' equation reads

$$u_t + u \, u_x = 0, \quad u(0, x) = u_0(x).$$

Formally, we will use the same *ansatz* as for the linear equations: for a solution $u \in C^1([0, \infty) \times \mathbb{R})$ of Burgers' equation and $(t, x) \in (0, \infty) \times \mathbb{R}$, as above we consider the differential equation

$$\dot{\gamma}(s) = u(s, \gamma(s)), \quad \gamma(t) = x, \tag{2.9}$$

since here the right-hand side of (2.9) is the coefficient of u_x. Now of course we cannot solve the equation in this form, since it contains the unknown solution of Burgers' equation (we are supposing that we wish to find u, and so we do not already know it). We thus require a further condition. This second equation comes from Burgers' equation itself:

$$\ddot{\gamma}(s) = u_t + \dot{\gamma}(s)u_x = u_t + uu_x = 0.$$

This means that in this case the characteristics through (t, x) are again straight lines, and we have $\gamma(s) = \dot{\gamma}(t)(s - t) + x = u(t, x)(s - t) + x$. Hence the solution of (2.9) exists for all $s \geq 0$. Since

$$\frac{d}{ds}u(s, \gamma(s)) = u_t + \dot{\gamma}(s)u_x = u_t + uu_x = 0,$$

we see that u is again constant along the characteristic $(s, \gamma(s))$; hence in particular

$$u(t, x) = u(t, \gamma(t)) = u(0, \gamma(0)) = u_0(\gamma(0)) = u_0(x - tu(t, x)).$$

This is an implicit equation for u.

We now wish to study an important special class of initial functions u_0, namely those u_0 which are linear, or more precisely of the form $u_0(x) = \alpha x$, $\alpha \in \mathbb{R}$, $\alpha \neq 0$. By the above considerations we obtain $u(t, x) = \alpha x - \alpha t u(t, x)$, and so the solution of Burgers' equation is

$$u(t, x) = \frac{\alpha x}{1 + \alpha t}. \tag{2.10}$$

It is easy to check that (2.10) is indeed a solution. Now let $G_c := \{(t, x) : t \geq 0, x \in \mathbb{R}, u(t, x) = c\}$, where u has the form (2.10). We distinguish between two cases.

1st case: $\alpha > 0$. In this case, by (2.10) and since $1 + \alpha t > 0$, $u(t, x)$ has the same sign as x, and the level lines G_c have the form

$$t = \frac{x}{c} - \frac{1}{\alpha}.$$

The characteristics thus have the same form as the one depicted in Figure 2.6. The flow is "thinned out"; we speak of a rarefaction wave. In this case the solution is unique.

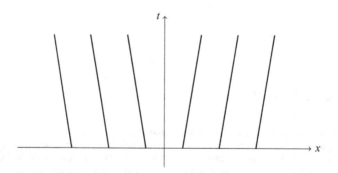

Fig. 2.6 Rarefaction wave for Burgers' equation.

2nd case: $\alpha < 0$. In this case the numerator in (2.10) vanishes, more precisely when $t = |\alpha|^{-1}$. The characteristics G_c thus meet at the point $(0, \frac{1}{|\alpha|})$, as depicted in Figure 2.7. This means that particles with different (initial) velocities collide at time $|\alpha|^{-1}$. The discontinuity this creates is known as a *shock* or sometimes an *implosion*, as the solution breaks down at this point.

These two phenomena, shock and rarefaction, appear in a whole class of *nonlinear* partial differential equations, namely nonlinear *hyperbolic* equations. The simple example considered above should hopefully already give an idea about why the study of such equations can be so challenging. In this book we will not pursue this subject any further.

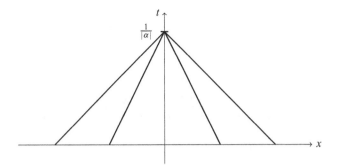

Fig. 2.7 For the Burgers' equation a shock occurs when $\alpha < 0$; the solution becomes discontinuous at $(|\alpha|^{-1}, 0)$ if the initial function is non-constant.

2.2 Equations of second order

We now consider second-order linear partial differential equations, whose general form is

$$\sum_{i,j=1}^{d} a_{i,j}\, u_{x_i, x_j} + \sum_{i=1}^{d} a_i\, u_{x_i} + a_0\, u = f, \qquad (2.11)$$

on a set $\Omega \subseteq \mathbb{R}^d$, where $u_{x_i} := \frac{\partial u}{\partial x_i}$, $u_{x_i, x_j} := \frac{\partial^2 u}{\partial x_i \partial x_j}$, for given $a_{i,j}, a_i, a_0 \in \mathbb{R}$, $i, j = 1, \ldots, d$ and a function $f : \Omega \to \mathbb{R}$. If we consider solutions $u \in C^2(\Omega)$ of (2.11), then $u_{x_i, x_j} = u_{x_j, x_i}$ (this is the statement of Schwarz's theorem). This means that we can always assume that

$$a_{i,j} = a_{j,i}, \quad i, j = 1, \ldots, d, \qquad (2.12)$$

by replacing $a_{i,j}$ by $\frac{1}{2}(a_{i,j} + a_{j,i})$ if necessary. We classify (2.11) according to its *principal part*

$$Au := \sum_{i,j=1}^{d} a_{i,j}\, u_{x_i, x_j} \qquad (2.13)$$

and often imagine terms of lower (first, zeroth) order as perturbations of the principal part.

We seek a coordinate system with respect to which (2.11) takes on as simple a form as possible. This leads to the following classification. Denote the desired coordinate system by $\xi = (\xi_1, \ldots, \xi_d)^T$ and suppose the transformation has the form $\xi = Bx$, $x \in \Omega \subseteq \mathbb{R}^d$, where $B \in \mathbb{R}^{d \times d}$ is orthogonal. If we now replace $u \in C^2(\Omega)$ by $v(x) := u(B^{-1}x)$, that is, $u(x) = v(Bx) = v(\xi)$, then we have

$$u_{x_i} = \sum_{k=1}^{d} v_{\xi_k} b_{k,i}, \quad B = (b_{k,i})_{k,i=1,\dots,d},$$

that is,

$$\frac{\partial}{\partial x_i} = \sum_{k=1}^{d} b_{k,i} \frac{\partial}{\partial \xi_k}, \quad \frac{\partial^2}{\partial x_i \partial x_j} = \sum_{k=1}^{d} \sum_{\ell=1}^{d} b_{k,i} b_{\ell,j} \frac{\partial^2}{\partial \xi_k \partial \xi_\ell}.$$

Hence, for the principal part,

$$Au := \sum_{i,j=1}^{d} a_{i,j} u_{x_i,x_j} = \sum_{k,\ell=1}^{d} \left(\sum_{i,j=1}^{d} b_{k,i} a_{i,j} b_{\ell,j} \right) v_{\xi_k,\xi_\ell} = \tilde{A}v,$$

where $\tilde{A} := BAB^T$. Obviously the transformed principal part $\tilde{A}$ is particularly simple when it is a diagonal matrix. Now since A is symmetric, we can diagonalize it, and so we choose B as the corresponding orthogonal transformation, so that $\tilde{A} = \text{diag}\,(\lambda_1, \dots, \lambda_d)$. Then $\lambda_1, \dots, \lambda_d$ are exactly the eigenvalues of A. The classification is particularly simple when $d = 2$; for the meantime we will limit ourselves to this case. Then A has two real eigenvalues λ_1, λ_2. Assume that $A \neq 0$.

Definition 2.8 When $d = 2$, the differential equation (2.11) is called
 (a) *parabolic* if one eigenvalue of A is zero;
 (b) *elliptic* if both eigenvalues have the same sign; and
 (c) *hyperbolic* if λ_1 and λ_2 have different signs.   △

Recall that $A = (a_{i,j})_{i,j=1,2}$. The following alternative characterization is immediate.

Lemma 2.9 *When $d = 2$, (2.11) is*
(a) parabolic if and only if $\det(A) = 0$*;*
(b) hyperbolic if and only if $\det(A) < 0$*;*
(c) elliptic if and only if $\det(A) > 0$.

Proof For the eigenvalues λ of A we have $0 = \det(A-\lambda I) = \lambda^2 - (a_{1,1}+a_{2,2})\lambda + \det(A)$, that is, $\lambda_{1,2} = \frac{1}{2}(a_{1,1}+a_{2,2}) \pm \sqrt{\frac{1}{4}(a_{1,1} + a_{2,2})^2 - \det(A)}$, from which the claim follows immediately.   □

Next, we wish to give a *normal form* for each of the three categories; each normal form will be representative for its respective category. To simplify notation we write $(x, y) := (x_1, x_2)$; then after applying the above coordinate transformation a general second-order linear partial differential equation with constant coefficients in two variables (x, y) has the form

$$\lambda_1 u_{xx} + \lambda_2 u_{yy} + a_1 u_x + a_2 u_y + a_0 u = f.$$

Here λ_1 and λ_2 are the eigenvalues of A. If $\lambda_1, \lambda_2 \neq 0$, then we may assume that $|\lambda_1| = |\lambda_2| = 1$ by replacing u by $u(x|\lambda_1|^{-1/2}, y|\lambda_2|^{-1/2})$. Now the normal forms are as follows.

Parabolic equations: Since one eigenvalue is zero (without loss of generality λ_2), the normal forms reads

$$u_{xx} + a_1 u_x + a_2 u_y + a_0 u = f.$$

To prevent this equation from reducing to an ordinary differential equation in x parametrized in y, we require that $a_2 \neq 0$. This can also be expressed as requiring that the matrix $(A, (a_1, a_2)^T) \in \mathbb{R}^{2\times3}$ has full rank 2. It is easy to see that the heat equation $u_t - u_{xx} = f$ (with $y = t$) is parabolic.

Hyperbolic equations: In this case the eigenvalues have different signs. The normal form is

$$u_{xx} - u_{yy} + a_1 u_x + a_2 u_y + a_0 u = f.$$

We see that the wave equation $u_{tt} - u_{xx} = f$ is hyperbolic.

Elliptic equations: Since both eigenvalues have the same sign, the normal form is

$$u_{xx} + u_{yy} + a_1 u_x + a_2 u_y + a_0 u = f;$$

for example, Poisson's equation $u_{xx} + u_{yy} = f$ is elliptic.

The names of the three types of equations come from the fact that

- $\dfrac{x^2}{\alpha_1} - \dfrac{y}{\alpha_2} = 1$ describes a parabola,

- $\dfrac{x^2}{\alpha_1} - \dfrac{y^2}{\alpha_2} = 1$ describes an hyperbola, and

- $\dfrac{x^2}{\alpha_1} + \dfrac{y^2}{\alpha_2} = 1$ describes an ellipse

if $\alpha_1 > 0$ and $\alpha_2 > 0$. The association with partial differential equations is, however, purely formal: these three types of curves have no connection to the respective types of partial differential equations.

Of course, we can immediately generalize this approach to second-order linear partial differential equations in $\mathbb{R}^2$ with variable coefficients, by requiring that the above conditions hold at every point $(x, y) \in \mathbb{R}^2$. However, in this case the type may depend on the point (x, y).

It is clear from the above derivation that in dimension $d = 2$ this categorization is complete, that is, a second-order linear partial differential equation of the form (2.11) with real coefficients is either parabolic, hyperbolic or elliptic, as long as $A \neq 0$. It is possible to generalize the above classification to equations of second order in $\mathbb{R}^d$,

$d > 2$, via the use of eigenvalues; however, it will turn out that this classification is no longer complete. We recall that a symmetric matrix is said to be *positive definite* if all its eigenvalues are positive. If they are all non-negative, then we call the matrix *positive semidefinite*.

Definition 2.10 Given the linear partial differential equation

$$-\sum_{i,j=1}^{d} a_{i,j}(x)\, u_{x_i, x_j}(x) + \sum_{i=1}^{d} b_i(x)\, u_{x_i}(x) + c(x)\, u(x) = f(x) \qquad (2.14)$$

with coefficients $A(x) = (a_{i,j}(x))_{i,j=1,\ldots,d} \in \mathbb{R}^{d \times d}$, with $a_{i,j} = a_{j,i}$, as well as $b(x) = (b_i(x))_{i=1,\ldots,d} \in \mathbb{R}^d$ and $c(x) \in \mathbb{R}$, for $x \in \Omega \subseteq \mathbb{R}^d$, we say that (2.14) is
(a) *elliptic* in $x \in \Omega$ if $A(x)$ or $-A(x)$ is positive definite;
(b) *hyperbolic* in $x \in \Omega$ if $A(x)$ or $-A(x)$ has exactly one negative and $(d-1)$ positive eigenvalues;
(c) *parabolic* in $x \in \Omega$ if $A(x)$ or $-A(x)$ is positive semidefinite but not definite *and* the matrix $(A(x), b(x)) \in \mathbb{R}^{d \times (d+1)}$ has maximal rank d. △

Remark 2.11 In Definition 2.10 we can replace part (c) by:
(c') *parabolic* in $x \in \Omega$ if $A(x)$ has exactly one eigenvalue equal to 0 and all other eigenvalues have the same sign.

This is indeed equivalent to (c), since if $A(x)$ or $-A(x)$ is positive semidefinite but not definite, then all nonzero eigenvalues have the same sign. In addition, it follows from the full rank of $(A(x), b(x))$ that only one eigenvalue vanishes. The converse implication is trivial. △

Let us now consider the three examples mentioned above.

Example 2.12 The Laplacian in $\mathbb{R}^d$ has the form

$$\Delta u(x) = \sum_{i=1}^{d} \frac{\partial^2}{\partial x_i^2} u(x), \quad x \in \mathbb{R}^d,$$

so in (2.14) we have $A(x) \equiv \operatorname{diag}(1, \ldots, 1)$. Since this matrix is positive definite, Poisson's equation is elliptic. △

Example 2.13 Now we consider the wave equation $u_{tt} - \Delta u = f$. It is easy to see that in this case the matrix A takes on the form

$$A(x) \equiv A = \begin{pmatrix} -1 & 0 & \cdots & 0 \\ \hline 0 & 1 & & \\ \vdots & & \ddots & \\ 0 & & & 1 \end{pmatrix},$$

and this matrix has -1 as a simple eigenvalue and 1 as an eigenvalue with multiplicity d (note that here $(t, x) \in \mathbb{R}^{d+1}$, that is, we have an equation in $d+1$ dimensions). As such, the wave equation in $\mathbb{R}^{d+1}$ is hyperbolic. △

Remark 2.14 An equation in space and time of the form $u_t + Lu = f$, where L is a second-order elliptic differential operator in the space variables $x \in \Omega \subseteq \mathbb{R}^d$, is parabolic. In particular, if the coefficient matrix of L from (2.14) is positive definite, then we are dealing with the most important case of a parabolic equation.   △

Proof No terms of the form u_{tt} or u_{tx_i}, $i = 1, \ldots, d$ appear. Hence A has the form

$$A = \begin{pmatrix} 0 & 0 \\ 0 & \tilde{A} \end{pmatrix} \in \mathbb{R}^{(d+1)\times(d+1)}$$

with coefficient matrix $\tilde{A} \in \mathbb{R}^{d\times d}$ of L. This is positive definite by assumption, and so A is positive semidefinite but not definite. Hence the equation is parabolic as asserted.   □

Example 2.15 The heat equation $u_t - \Delta u = f$ is parabolic.   △

It is already clear from Definition 2.10 that the three categories introduced there do not give a complete classification in $\mathbb{R}^d$ if $d > 2$, since they do not cover all possibilities for the eigenvalues of the coefficient matrix. One sometimes also comes across the term *ultrahyperbolic* for the situation where the eigenvalues λ_k of A satisfy $\lambda_k \neq 0$ for all $k = 1, \ldots, d$, and at least two eigenvalues are positive and at least two are negative.

At the end of this chapter we will give an overview of the partial differential equations introduced thus far and classify these according to this scheme; see Table 2.1 (page 47).

2.3* Nonlinear equations of second order

Thus far we have only considered linear equations; now we wish to generalize the above classification to nonlinear equations of second order. We first write a general second-order nonlinear partial differential equation in the form

$$F(x, u(x), u_{x_i}(x), u_{x_i, x_j}(x)) = f(x), \tag{2.15}$$

where $x \in \Omega \subseteq \mathbb{R}^d$ is the independent variable (which may involve both space and time), $f : \Omega \to \mathbb{R}$ is a given right-hand side and $u : \Omega \to \mathbb{R}$ is the desired solution. Thus (2.15) describes a scalar equation (that is, (2.15) is an identity involving real numbers, not vectors). The equation itself is described by the (nonlinear) function $F : \Omega \times \mathbb{R} \times \mathbb{R}^d \times \mathbb{R}^{d\times d} \to \mathbb{R}$. Thus we may regard the function F as being applied to the variables $u \in \mathbb{R}$, $q = (q_i)_{i=1,\ldots,d} \in \mathbb{R}^d$ and $p = (p_{ij})_{i,j} \in \mathbb{R}^{d\times d}$, that is, $F(x, u, q, p)$; and we define the matrix

$$A(x) = A(u, x) := F_p\left(x, u(x), u_{x_i}(x), u_{x_i x_j}(x)\right)_{i,j=1,\ldots,d} \tag{2.16}$$

where $x \in \Omega$ and we use the shorthand notation $F_p(x, u, q, p) := \frac{\partial}{\partial p} F(x, u, q, p)$ for the gradient with respect to the last $d \times d$ variables. Now we approximate (2.15) using the linear (first order) Taylor polynomial of F, that is, we write the equation as

$$\nabla u(x)^T A(x) \nabla u(x) + G(x, u, \nabla u) = f(x), \tag{2.17}$$

where the function $G(x, p, q)$ contains the terms of zeroth order of the Taylor series in p. We then say that (2.15) is elliptic (or parabolic, hyperbolic) with respect to u at the point $x \in \Omega$ if (2.17) is of the corresponding type.

Example 2.16 Consider the Monge–Ampère equation in two space variables (call them x and y), $u_{xx}(x, y) u_{yy}(x, y) - u_{xy}^2(x, y) - f(x, y) = 0$. If $u \in C^2(\Omega)$, then its mixed partial derivatives are equal, that is, $u_{xy} u_{yx} = u_{xy} u_{xy} = u_{xy}^2$, and we may write the function F as $F(x, y, u, q, p) = p_{1,1} p_{2,2} - p_{1,2} p_{2,1}$. Hence we obtain

$$F_p(x, y; u) = \begin{pmatrix} u_{yy}(x, y) & -u_{yx}(x, y) \\ -u_{xy}(x, y) & u_{xx}(x, y) \end{pmatrix}.$$

We can use Sylvester's criterion to study when this matrix is positive definite. For this we consider the principal minors, which read (starting from the bottom) u_{xx} for the first and $\det F_p(x, y; u) = u_{xx} u_{yy} - u_{xy} u_{yx}$ for the second. Now by invoking the differential equation we see that $\det F_p(x, y; u) = u_{xx} u_{yy} - u_{xy} u_{yx} = u_{xx} u_{yy} - u_{xy}^2 = f$, and so the matrix $F_p(x, y; u)$ is positive definite if

$$u_{xx}(x, y) > 0, \qquad f(x, y) > 0. \tag{2.18}$$

Hence the Monge–Ampère equation is elliptic in all points (x, y) in which the condition (2.18) is satisfied. If one of the terms in (2.18) is positive or negative and the other is zero, then the equation is parabolic there; if one is positive and one negative, then it is hyperbolic. We thus have an example in which the type of equation depends not just on the point in space but also on the unknown solution u.   △

2.4* Equations of higher order and systems

When treating first-order equations we looked for curves $(t, x(t))$ along which the solutions are constant, or at least determined by an initial value problem for an ordinary differential equation. For equations of second order one looks for local coordinate systems which permit a reduction to some normal form. It is clear that such an approach becomes increasingly complicated as the order of the equation increases.

In some cases, however, it is easy to reduce equations of higher order to lower-order systems. As an example we consider the plate equation $\Delta^2 u(x) = f(x)$, $x \in \Omega \subset \mathbb{R}^d$. Via the change of variables $v = (v_1, v_2) := (u, \Delta u)$ this equation can be reduced to the second-order system

$$\Delta v - \begin{pmatrix} v_2 \\ 0 \end{pmatrix} = \begin{pmatrix} 0 \\ f \end{pmatrix},$$

If we transfer the above classification scheme to systems in the natural way, then it follows that the plate equation is elliptic.

The incompressible Navier–Stokes equations may be written in the form

$$u_t - \Delta u + G(\nabla u, u, p) = f;$$

hence they are parabolic. However, the (quadratic) nonlinearity $(u \cdot \nabla)u$ causes substantial difficulties, both analytically and numerically. For the KdV equation there is no such obvious reduction. Maxwell's equations are in general hyperbolic. However, in certain special cases (e.g.

time-harmonic, two-dimensional, perfect conductors, etc.) they can be reduced to a parabolic or even elliptic problem, see e.g. [24, I.§4]. The Schrödinger equation $i\hbar\psi_t = (-\frac{\hbar^2}{2m}\Delta + V(t, x))\psi$ reads, with $\psi = u + iv$,

$$i\hbar u_t - \hbar v_t = \left(-\frac{\hbar^2}{2m}\Delta + V(t, x)\right)(u + iv)$$

and can thus be reformulated as a system

$$\hbar\begin{pmatrix} u_t \\ v_t \end{pmatrix} = \begin{pmatrix} 0 & -\frac{\hbar^2}{2m}\Delta + V(t, x) \\ \frac{\hbar^2}{2m}\Delta - V(t, x) & 0 \end{pmatrix}\begin{pmatrix} u \\ v \end{pmatrix}.$$

The coefficient matrix of the principal part thus has the form $\mathcal{A} = \begin{pmatrix} 0 & A \\ -A & 0 \end{pmatrix}$, where A is a symmetric positive definite matrix. Hence $\mathcal{A}$ has complex eigenvalues with different signs. If in the above definitions we were to permit complex eigenvalues, then the Schrödinger equation would be hyperbolic. However, the Schrödinger equation is rarely called hyperbolic in the literature, largely owing to the fact that hyperbolic equations have a *finite* speed of propagation, which is not the case for the Schrödinger equation. In this book we will limit ourselves to the study of the standard types: elliptic, parabolic and hyperbolic.

2.5 Exercises

Exercise 2.1 Classify the following equations in $\mathbb{R}^2$:
(a) $(\partial_1 u(x))^2 + e^{x_2}\partial_2 u(x) = \sin(x_1)$, $x = (x_1, x_2)^T$,
(b) $\partial_1^2 u(x) + e^{x_2}\partial_2 u(x) = \sin^2(x_1)$, $x = (x_1, x_2)^T$,
(c) $\partial_1^2 u(x) + \exp(\partial_2 u(x)) = \sin^3(x_1)$, $x = (x_1, x_2)^T$.

Exercise 2.2 Consider the problem

$$\frac{\partial}{\partial x_1}u(x) + \frac{\partial}{\partial x_2}u(x) = u(x)^2, \qquad x = (x_1, x_2)^T \in \mathbb{R}^2,$$

$$u(x_1, -x_1) = x_1, \qquad x_1 \in \mathbb{R}.$$

Solve this equation in a suitable domain using the method of characteristics.

Exercise 2.3 Determine and sketch the characteristics of the partial differential equation $(2x_2 - 3x_1)\partial_{x_1}u - x_2\partial_{x_2}u = x_2^2(2x_2 - 5x_1)$ and determine the solution $u(x_1, x_2)$ for the initial value $u(x_1, 1) = x_1 + 1$.

Exercise 2.4 Determine the type (elliptic, parabolic or hyperbolic) of each of the following partial differential equations, where $(x, y)^T \in \mathbb{R}^2$:
(a) $-2u_{xx} + u_{xy} + u_{yy} = 0$,
(b) $\frac{5}{2}u_{xx} + u_{xy} + u_{yy} = 0$,
(c) $9u_{xx} + 12u_{xy} + 4u_{yy} = 0$.

Exercise 2.5 Determine the type of the differential equation in $\mathbb{R}^2$

$$(x^2 - 1)\frac{\partial^2}{\partial x^2}u + 2xy\frac{\partial^2}{\partial x\partial y}u + (y^2 - 1)\frac{\partial^2}{\partial y^2}u = x\frac{\partial}{\partial x}u + y\frac{\partial}{\partial y}u, (x, y)^T \in \mathbb{R}^2.$$

Exercise 2.6 Let $a > 0$ be a constant and let $g, h \in C^1([0, 1])$ be such that $-ag'(0) = h'(0)$ and $g(0) = h(0)$. Show that there exists a unique $u \in C^1([0, 1] \times [0, 1])$ which satisfies

$$\begin{aligned} u_t(t, x) + au_x(t, x) &= 0, & t, x &\in (0, 1), \\ u(0, x) &= g(x), & x &\in [0, 1], \\ u(t, 0) &= h(t), & t &\in [0, 1]. \end{aligned}$$

Use the method of characteristics.

Exercise 2.7 (Gronwall's lemma) Show that if $y : (\alpha, t] \to \mathbb{R}$ is continuous and $c, \lambda \geq 0$ are such that

$$|y(s)| \leq c + \lambda \int_s^t |y(r)|\, dr, \quad s \in (\alpha, t],$$

then $|y(s)| \leq c\, e^{\lambda(t-s)}, s \in (\alpha, t]$. Also show the following variant of this lemma: if $y \in C([a, b])$ and $c, \lambda \geq 0$ satisfy

$$y(t) \leq c + \lambda \int_a^t y(s)\, ds, \quad t \in [a, b],$$

then $y(t) \leq ce^{\lambda(t-a)}, t \in [a, b]$.
Suggestion: Set $W(t) := c + \lambda \int_a^t y(s)\, ds$, and start by considering $\frac{W'(t)}{W(t)}$ in the case $c > 0$. Prove the claim in this case and then consider the case $c = 0$ afterwards.

Exercise 2.8 Let $a : \mathbb{R} \times \mathbb{R} \to \mathbb{R}$ be continuously differentiable. Assume that for each $T > 0$ there is an $L \geq 0$ such that $|a(t, x)| \leq L(1 + |x|)$ for all $t \in [0, T]$ and $x \in \mathbb{R}$. Show that in this case (2.2) admits at most one solution.
Suggestion: Use Gronwall's lemma from Exercise 2.7.

Exercise 2.9 Show that there is no function $u \in C^1((0, \infty) \times \mathbb{R})$ which satisfies

$$u_t(t, x) = x^2 u_x(t, x), t > 0, \ x \in \mathbb{R}, \qquad \lim_{t \downarrow 0} u(t, x) = \sin x.$$

Exercise 2.10 (Triviality of the characteristics for Laplace's equation) Determine all curves of the form $\Gamma(t) = (\gamma_1(t), \gamma_2(t))$ with $\gamma_1, \gamma_2 \in C^1([0, 1])$ on which all solutions of Laplace's equation $u_{xx} + u_{yy} = 0$ are constant.

Table 2.1 Classification of partial differential equations (alg. type: algebraic type, lin: linear, sl: semilinear, ql: quasilinear, fnl: fully nonlinear, misc.: variable or other type).

equation (name and formula)	alg. type	order	(space, time)	dimension	type
Linear transport equation $u_t + c u_x = f$	lin	1	1	1	——
Burgers' equation $u_t + u u_x = f$	ql	1	1	1	——
Viscous Burgers' equation $u_t + u u_x = \varepsilon u_{xx}$	sl	2	1	1	parabolic
Heat equation $u_t - \Delta u = f$	lin	2	1	d	parabolic
Wave equation $u_{tt} - \Delta u = f$	lin	2	2	d	hyperbolic
Poisson's equation $-\Delta u = f$	lin	2	0	d	elliptic
KdV equation $u_t - 6 u u_x + u_{xxx} = 0$	sl	3	1	1	——
Black–Scholes equation $V_t + \frac{1}{2}\sigma^2 S^2 V_{SS} + (r - \delta)S V_S - rV = 0$	lin	1	2	1	parabolic
Monge-Ampère equation $\det(H(u)) = f$	fnl	0	2	d	misc.
Minimal surface equation $(1 + u_y^2)u_{xx} - 2u_x u_y u_{xy} + (1 + u_x^2)u_{yy} = 0$	ql	2	0	2	misc.
Plate equation $\Delta^2 u = f$	lin	4	0	d	elliptic
Navier–Stokes equations (incompressible) $\mathbf{u}_t - \rho(\mathbf{u} \cdot \nabla)\mathbf{u} - \mu\,\Delta \mathbf{u} + \nabla p = \mathbf{f},\ \mathrm{div}\,\mathbf{u} = 0$	sl	2	1	d	parabolic
Navier–Stokes equations (compressible) $\mathbf{U}_t + \mathrm{div}\,\mathbf{F}_m(\mathbf{U}) + \mathrm{div}\,\mathbf{G}_m(\mathbf{U}) = 0$	ql	2	1	d	hyperbolic
Maxwell's equations $\mathrm{div}\,\boldsymbol{B} = 0,\ \boldsymbol{B}_t + \mathrm{rot}\,\boldsymbol{E} = 0,$ $\mathrm{div}\,\boldsymbol{E} = 4\pi\rho,\ \boldsymbol{E}_t - \mathrm{rot}\,\boldsymbol{B} = -4\pi\boldsymbol{j}$	lin	1	1	3	hyperbolic/ misc.
Schrödinger equation $i\hbar\,\psi_t = \left(-\frac{\hbar^2}{2m}\Delta + V(t, x)\right)\psi$	lin	2	1	$2d$	misc.

Chapter 3
Elementary methods

In this chapter we will derive explicit solutions for a number of partial differential equations. The equations we will be considering here represent a number of important models such as those of the vibrating string, heat diffusion and the pricing of options. But they are additionally prototypes for various partial differential equations which we met in Chapter 1. To be more precise, we will obtain solutions for
- a hyperbolic equation (the wave equation on an interval),
- an elliptic equation (Laplace's equation on rectangles and the disk), and
- a parabolic equation (the heat equation and also the Black–Scholes equation).
These solutions illustrate in particular how differently the three types of equations behave.

The principal method used in this chapter is based on separation of variables; it reduces the equations under consideration to ordinary differential equations and yields special solutions. Fourier series (whose basic properties are derived in Section 3.2) permit us to superimpose these solutions and thus to prove complete existence and uniqueness results for our initial-boundary value problems. In the process, we will also meet the maximum principle as a way of obtaining strong *a priori* estimates on solutions. Finally, we will show how in some cases one can derive explicit formulae for the solution with the help of integral transforms such as the Fourier and Laplace transforms.

Chapter overview

© The Author(s), under exclusive license to Springer Nature Switzerland AG 2023
W. Arendt, K. Urban, *Partial Differential Equations*, Graduate Texts in
Mathematics 294, https://doi.org/10.1007/978-3-031-13379-4_3

3.1 The one-dimensional wave equation

We first consider the one-dimensional wave equation (1.19) with constant wave speed $c > 0$,

$$u_{tt} - c^2 u_{xx} = 0. \tag{3.1}$$

This is a homogeneous equation, cf. Section 1.4.

3.1.1 D'Alembert's formula on $\mathbb{R} \times \mathbb{R}$

Let $u \in C^2(\mathbb{R}^2)$ be a solution of (3.1). Motivated by the characteristics $x \pm ct \equiv$ const we make the change of variables

$$\xi := x + ct, \quad \tau := x - ct, \tag{3.2}$$

and set

$$v(\tau, \xi) := u(t, x) = u\left(\frac{\xi - \tau}{2c}, \frac{\xi + \tau}{2}\right).$$

Since $u \in C^2(\mathbb{R}^2)$, applying the chain rule and (3.2) gives

$$v_\xi = \frac{1}{2c} u_t + \frac{1}{2} u_x,$$

$$v_{\xi\tau} = \frac{1}{2c}\frac{-1}{2c} u_{tt} + \frac{1}{2c}\frac{1}{2} u_{tx} + \frac{1}{2}\frac{-1}{2c} u_{xt} + \frac{1}{2}\frac{1}{2} u_{xx} = \frac{1}{4c^2}(c^2 u_{xx} - u_{tt}) = 0.$$

Hence the function v_ξ is independent of τ, and so there exists a function $F : \mathbb{R} \to \mathbb{R}$ such that $v_\xi(\tau, \xi) = F(\xi)$ for all $\tau \in \mathbb{R}$. Since $v \in C^2(\mathbb{R}^2)$, we must have $F \in C^1(\mathbb{R})$. Let φ be an antiderivative of F, that is, we suppose that $\varphi \in C^2(\mathbb{R})$ satisfies $\varphi' = F$. By the Fundamental Theorem of Calculus, for every $\tau \in \mathbb{R}$ there exists a constant $\psi(\tau)$ (the constant of integration) such that $v(\tau, \xi) = \varphi(\xi) + \psi(\tau)$. Since $\varphi \in C^2(\mathbb{R})$, $v \in C^2(\mathbb{R}^2)$, it follows that $\psi \in C^2(\mathbb{R})$. We have thus shown that

$$u(t, x) = v(\tau, \xi) = v(x - ct, x + ct) = \varphi(x + ct) + \psi(x - ct).$$

Conversely, every such function is a solution of the wave equation (3.1), as is easy to check by explicit calculation. We thus have the following result, which states that the solution is composed of two waves, one moving left and one moving right, both with speed c.

Theorem 3.1 *A function $u \in C^2(\mathbb{R}^2)$ solves the wave equation (3.1) if and only if there exist two functions $\varphi, \psi \in C^2(\mathbb{R})$ such that*

$$u(t, x) = \varphi(x + ct) + \psi(x - ct) \tag{3.3}$$

for all $t, x \in \mathbb{R}$.

We next wish to express φ and ψ in terms of the initial data, in this case the *initial position* $u_0(x) = u(0, x)$ and the *initial velocity* $u_1(x) = u_t(0, x)$. So let u be given by (3.3); then

$$u_0(x) := u(0, x) = \varphi(x) + \psi(x), \quad u_1(x) := u_t(0, x) = c\varphi'(x) - c\psi'(x). \quad (3.4)$$

Differentiating the first equation yields

$$u_0'(x) = \varphi'(x) + \psi'(x). \quad (3.5)$$

Multiplication by the constant c and addition of the second equation in (3.4) leads to $cu_0'(x) + u_1(x) = 2c\varphi'(x)$, that is, $\varphi'(x) = \frac{1}{2}u_0'(x) + \frac{1}{2c}u_1(x)$. Hence there exists a constant (of integration) $c_1 \in \mathbb{R}$ such that

$$\varphi(x) = \frac{1}{2}u_0(x) + \frac{1}{2c}\int_0^x u_1(s)ds + c_1$$

Multiplying (3.5) by c and subtracting the second equation in (3.4) from the result yields $cu_0'(x) - u_1(x) = 2c\psi'(x)$, and so there exists another constant (of integration) $c_2 \in \mathbb{R}$ such that

$$\psi(x) = \frac{1}{2}u_0(x) - \frac{1}{2c}\int_0^x u_1(s)ds + c_2 = \frac{1}{2}u_0(x) + \frac{1}{2c}\int_x^0 u_1(s)ds + c_2$$

for all $x \in \mathbb{R}$. Substituting these expressions into (3.3) finally leads to

$$u(t, x) = \frac{1}{2}\left(u_0(x + ct) + u_0(x - ct)\right) + \frac{1}{2c}\int_{x-ct}^{x+ct} u_1(s)ds + c_1 + c_2.$$

By setting $t = 0$, we see that we must have $c_1 + c_2 = 0$. We have thus proved the following theorem, in which the solution is expressed in terms of the initial position and velocity.

Theorem 3.2 *Given the initial data $u_0 \in C^2(\mathbb{R})$ and $u_1 \in C^1(\mathbb{R})$, there exists a unique function $u \in C^2(\mathbb{R} \times \mathbb{R})$ for which*

$$u_{tt} = c^2 u_{xx} \quad in\ \mathbb{R}^2, \quad (3.6a)$$
$$u(0, x) = u_0(x), \quad x \in \mathbb{R}, \quad (3.6b)$$
$$u_t(0, x) = u_1(x), \quad x \in \mathbb{R}. \quad (3.6c)$$

This function is given by

$$u(t, x) = \frac{1}{2}(u_0(x + ct) + u_0(x - ct)) + \frac{1}{2c}\int_{x-ct}^{x+ct} u_1(s)ds \quad (3.7)$$

for all $x, t \in \mathbb{R}$.

The representation (3.7) is known as *d'Alembert's formula* for the solution of the wave equation (3.6).

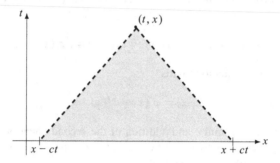

Fig. 3.1 Wave equation: domain of dependence.

Remark 3.3 We already know that c is the speed of propagation of the wave. We see from formula (3.7) that at the point (t, x) the solution only depends on the initial values in the interval $[x - ct, x + ct]$, cf. Figure 3.1. This is a typical property of the wave equation. Another typical property is that the solution u does *not* gain regularity over the course of time: $u(t, \cdot)$ is not infinitely differentiable if u_0 and u_1 are not. We will see that the situation is completely different for the heat equation. △

Remark 3.4 Observe that (3.7) also makes sense if u_0 is only continuous and u_1 even just integrable. In such a case, one could regard the function defined by (3.7) as a *weak solution* of the equation; see Exercise 5.13.   △

3.1.2 The wave equation on an interval

In the model of the vibrating string, as in other physical models, we do not consider the wave equation on the whole real line, but rather on an interval $[a, b] \subset \mathbb{R}$, $-\infty < a < b < \infty$. In order to be consistent with the numerical treatment in Section 9.5 we also consider a finite time interval $[0, T]$, where $0 < T < \infty$. In this case, we wish to find functions $u \in C^2([0, T] \times [a, b])$ for which

$$u_{tt}(t, x) = c^2 u_{xx}(t, x), \quad t \in [0, T], \, x \in (a, b). \tag{3.8}$$

Here $C^2([0, T] \times [a, b])$ is the space of functions which are twice continuously partially differentiable in $(0, T) \times (a, b)$, and all of whose partial derivatives admit a continuous extension to $[0, T] \times [a, b]$. We will again specify an initial position $u_0 \in C([a, b])$ and an initial velocity $u_1 \in C([a, b])$, and require that

$$u(0, x) = u_0(x), \quad u_t(0, x) = u_1(x), \quad x \in [a, b]. \tag{3.9}$$

But on a bounded interval these initial conditions do not yet fully determine a unique solution; we also need to impose boundary conditions in order to recover uniqueness. One possibility is given by the (homogeneous) *Dirichlet boundary conditions*

$$u(t, a) = u(t, b) = 0, \quad t \in [0, T]. \tag{3.10}$$

Substituting these values into (3.9), we obtain the compatibility conditions for the initial values

$$u_0(a) = u_0(b) = 0, \quad u_1(a) = u_1(b) = 0. \tag{3.11}$$

In fact, recall that we are imposing the condition $u \in C^2([0, T] \times [a, b])$; this smoothness requirement also has the following consequence. If $u(t, a) = 0$ for all $t \in [0, T]$, then also $u_t(t, a) = 0$ and in particular $u_1(a) = u_t(0, a) = 0$. A similar statement holds for the right-hand endpoint, that is, $u_1(b) = u_t(0, b) = 0$.

This leads to the following initial-boundary value problem (IBVP): determine the function $u \in C^2([0, T] \times [a, b])$ for which

$$u_{tt} = c^2 u_{xx}, \quad t \in (0, T), \quad x \in (a, b), \tag{3.12a}$$
$$u(t, a) = u(t, b) = 0, \quad t \in (0, T), \tag{3.12b}$$
$$u(0, x) = u_0(x), \quad u_t(0, x) = u_1(x), \quad x \in (a, b). \tag{3.12c}$$

This problem has at most one solution; this will be shown using a theorem on conservation of energy which holds for the solutions of (3.12a). We recall the derivation of Laplace's equation in Section 1.6.4: we saw there that the potential energy is given by the integral of the square of the spatial derivative of u (multiplied by the propagation speed). Here the corresponding expression is

$$E_{\text{pot}}(t) := \int_a^b c^2 \, u_x^2(t, x) \, dx,$$

and we again call this term the *potential energy* (without entering into the question of the physical units of measurement involved). The *kinetic energy*

$$E_{\text{kin}}(t) := \int_a^b u_t^2(t, x) \, dx$$

is correspondingly defined as the (average square) time derivative of u. Using these definitions we can now prove the following theorem for the solutions of the boundary value problem (3.12a), (3.12b).

Theorem 3.5 (Conservation of energy) *Let $u \in C^2([0, T] \times [a, b])$ be a solution of (3.12a), (3.12b). Then the total energy $E(t) := E_{\text{kin}}(t) + E_{\text{pot}}(t)$, $t \in [0, T]$, is constant on $(0, T)$.*

Proof By Schwarz's theorem, $u_{tx} = u_{xt}$ for all $t \in [0, T]$ and $x \in (a, b)$ since $u \in C^2([0, T] \times [a, b])$. Integrating by parts and using (3.12a), we obtain

$$\frac{d}{dt} E_{\text{kin}}(t) = \int_a^b 2u_t(t, x)\, u_{tt}(t, x)\, dx = 2c^2 \int_a^b u_t(t, x)\, u_{xx}(t, x)\, dx$$

$$= 2c^2 u_t(t, x)\, u_x(t, x)\Big|_{x=a}^{x=b} - 2c^2 \int_a^b u_{tx}(t, x)\, u_x(t, x)\, dx$$

$$= -\frac{d}{dt} E_{\text{pot}}(t),$$

where we have also used the boundary condition $u(t, a) = u(t, b) = 0$ and so (as above) $u_t(t, a) = u_t(t, b) = 0$ for $t > 0$, as well.   □

Uniqueness of solutions now follows from conservation of energy as follows. If the initial-boundary value problem (3.12) has two solutions, then their difference u is a solution of (3.12) with the homogeneous initial values $u_0 \equiv u_1 \equiv 0$. For this u, however, conservation of energy implies that for all $t \in (0, T)$

$$E(t) = \int_a^b \left(u_t^2(t, x) + c^2 u_x^2(t, x) \right) dx = E(0)$$

$$= \int_a^b (u_1(x)^2 + c^2 u_0'(x)^2)\, dx = 0,$$

that is, $E(t) = 0$ for all $t \in (0, T)$. It follows that $u_t(t, x) = u_x(t, x) \equiv 0$ and so u is constant. But since $u(0, x) = 0$, we conclude that $u \equiv 0$. We can summarize what we now know in the following theorem.

Theorem 3.6 (Existence and uniqueness of solutions of the IBVP (3.12)) *Let* $u_0 \in C^2([a, b])$ *and* $u_1 \in C^1([a, b])$ *satisfy* $u_0(a) = u_0''(a) = u_0(b) = u_0''(b) = 0$ *and* $u_1(a) = u_1(b) = 0$. *Then* (3.12) *has a unique solution* u.

Proof We may assume that $a = 0$. We extend u_0 and u_1 to odd $2b$-periodic functions $\tilde{u}_0, \tilde{u}_1 : \mathbb{R} \to \mathbb{R}$; then our assumptions imply that $\tilde{u}_0 \in C^2(\mathbb{R})$ and $\tilde{u}_1 \in C^1(\mathbb{R})$, see Exercise 3.21. If we set

$$u(t, x) := \frac{1}{2}(\tilde{u}_0(x + ct) + (\tilde{u}_0(x - ct)) + \frac{1}{2c} \int_{x-ct}^{x+ct} \tilde{u}_1(s)\, ds,$$

then by Theorem 3.2 u is a solution of (3.6) and hence satisfies (3.12a) and (3.12c). Since $\tilde{u}_0$ and $\tilde{u}_1$ are odd and $2b$-periodic, we have $u(t, a) = u(t, b) = 0$ for all $t \geq 0$, as is easy to check. Hence u satisfies (3.12). Uniqueness was proved above.   □

Remark 3.7 Another direct consequence of the conservation of energy is the continuous dependence of the solution on the data; together with Theorem 3.6 this gives us the well-posedness of the IBVP (3.12) for the wave equation. Continuous dependence for the uniform norm can be proved with the help of the Closed Graph Theorem, see Exercise 3.22.   △

We have already seen that it was d'Alembert who showed that a vibrating string satisfies the (one-dimensional) wave equation, see Section 1.4. It was also d'Alembert who showed in 1747 that the general solution is given by (3.3). But it was Euler who discovered one year later that the solution is determined by the initial value and the initial speed. We refer to [38, end of Ch. VI] for a full account of the fascinating history of the problem. What we have not yet seen is the physical nature of the solution, which is hidden in Theorem 3.6. In fact, it turns out that the solution can be expressed as a superposition of simple vibrations, the harmonics, see Exercise 4.11. To see this we will need to study Fourier series, our next subject.

3.2 Fourier series

Many elementary partial differential equations can be solved elegantly using Fourier series. One of the reasons for this is that many such equations model phenomena involving oscillations, as we saw in Chapter 1. Since Fourier series are a superposition of oscillating functions (more precisely sine and cosine functions, or complex exponentials), it is natural to look for solutions in the form of Fourier series. But there are also certain basic questions about Fourier series that are natural to ask, such as when and in what sense the Fourier series of a continuous or differentiable function converges. Here we will introduce Fourier series, collect a few essential facts about them and prove a convergence result (Theorem 3.11).

Remark 3.8 Leonhard Euler (1707–1783), Joseph-Louis Lagrange (1736–1813) and the Swiss brothers Jakob (Jacob) Bernoulli (1655–1705) and Johann Bernoulli (1667–1748) as well as Johann's son Daniel Bernoulli (1700–1782) all worked on series expansions of functions; one of the motivations for doing so was already solving differential equations. But it was Jean Baptiste Joseph Fourier (1768–1830) who claimed in his famous work *Théorie analytique de la chaleur* of 1822 that every function admits a convergent series expansion. These were later named *Fourier series* in his honor. His claim originally met with rejection, in particular from Augustin Louis Cauchy (1789–1857) and Niels Henrik Abel (1802–1829). Johann Peter Gustav Lejeune Dirichlet (1805–1859) proved in 1829 in his work *Sur la convergence des séries trigonométriques qui servent à représenter une fonction arbitraire entre des limites données* that Fourier's claim is at least correct for functions which are piecewise continuously differentiable; see Exercise 5.12.   △

Here we will concentrate on continuous periodic functions and show the convergence of their Fourier series with respect to a weaker notion of convergence, more precisely *convergence in the sense of Abel*. This will have immediate consequences for Laplace's equation and the heat equation.

Definition 3.9 A function $f : \mathbb{R} \to \mathbb{C}$ is called *2π-periodic* if $f(t + 2\pi) = f(t)$ for all $t \in \mathbb{R}$.   △

Typical examples are the trigonometric functions sin and cos. A further example is the complex exponential function e_k, defined for $k \in \mathbb{Z}$ by

$$e_k(t) = e^{ikt} = \cos(kt) + i\sin(kt), \quad t \in \mathbb{R}. \tag{3.13}$$

The functions $\{e_k, k \in \mathbb{Z}\}$ satisfy the orthogonality relation

$$\frac{1}{2\pi} \int_0^{2\pi} e^{ikt} e^{-i\ell t} \, dt = \delta_{k,\ell} = \begin{cases} 1, & \text{if } k = \ell, \\ 0, & \text{if } k \neq \ell, \end{cases} \tag{3.14}$$

as is easy to check. We consider the $\{e_k, k \in \mathbb{Z}\}$ to be "pure oscillations". By superimposing them we can construct more general 2π-periodic functions: let $c_k \in \mathbb{C}$, $k \in \mathbb{Z}$, be complex coefficients for which

$$\sum_{k=-\infty}^{\infty} |c_k| < \infty.$$

Then

$$f(t) = \sum_{k=-\infty}^{\infty} c_k \, e^{ikt}, \quad t \in \mathbb{R} \tag{3.15}$$

defines a 2π-periodic function $f : \mathbb{R} \to \mathbb{C}$. This function is continuous since the series (3.15) converges uniformly on $\mathbb{R}$. We can also recover the coefficients c_k from f,

$$c_k = \frac{1}{2\pi} \int_0^{2\pi} f(t) e^{-ikt} \, dt, \quad k \in \mathbb{Z}, \tag{3.16}$$

as can be seen as follows: due to the uniform convergence of the series (3.15) we may exchange the order of integration and summation. Hence, for $m \in \mathbb{Z}$,

$$\frac{1}{2\pi} \int_0^{2\pi} f(t) e^{-imt} \, dt = \sum_{k=-\infty}^{\infty} c_k \frac{1}{2\pi} \int_0^{2\pi} e^{ikt} e^{-imt} \, dt = c_m,$$

where we have also used the orthogonality relation (3.14). These thoughts lead us to the following definition.

Definition 3.10 (a) The space of continuous 2π-functions is defined to be $C_{2\pi} := \{f : \mathbb{R} \to \mathbb{C} : f \text{ is continuous and } 2\pi\text{-periodic}\}$.
(b) If $f \in C_{2\pi}$ and $k \in \mathbb{Z}$, then we call

$$c_k := \frac{1}{2\pi} \int_0^{2\pi} f(t) e^{-ikt} \, dt \tag{3.17}$$

the kth *Fourier coefficient* of f. The series

$$\sum_{k=-\infty}^{\infty} c_k e^{ikt} \tag{3.18}$$

is called the *Fourier series* of f.   △

In general, the Fourier series (3.18) does not converge pointwise. In fact, for every $t \in \mathbb{R}$ there is a continuous 2π-periodic function $f : \mathbb{R} \to \mathbb{C}$ such that the Fourier series (3.18) of f has unbounded partial sums, and hence diverges, in the point t. A first example was given in 1873 by Paul du Bois-Reymond, see [42, Thm. II.2.1] for more information.

We will now prove the convergence of the Fourier series in a weaker sense (Theorem 3.11). Let $f : \mathbb{R} \to \mathbb{C}$ be continuous and 2π-periodic; then the sequence $(c_k)_{k \in \mathbb{Z}}$ of Fourier coefficients of f is bounded, and the series

$$f_r(t) = \sum_{k=-\infty}^{\infty} r^{|k|} c_k e^{ikt} \tag{3.19}$$

converges uniformly on $\mathbb{R}$ for every $0 < r < 1$ and defines a function $f_r \in C_{2\pi}$. Our goal is to prove the following statement, for which we will need a certain amount of preparation.

Theorem 3.11 (Convergence of Fourier series in the sense of Abel) *We have*

$$\lim_{r \uparrow 1} f_r(t) = f(t)$$

uniformly in $t \in \mathbb{R}$.

Remark 3.12 (Abel convergence) We say that a series $\sum_{k=1}^{\infty} a_k$ *converges to a number c in the sense of Abel if*

$$\lim_{r \uparrow 1} \sum_{k=1}^{\infty} a_k r^k = c.$$

Abel's convergence theorem for power series, cf. [1], states that ordinary convergence of the series, that is, $\lim_{n \to \infty} \sum_{k=1}^{n} a_k = c$, implies convergence to c in the sense of Abel. The notion of convergence in the sense of Abel is, however, strictly weaker than that of convergence of the series, as the above-mentioned example of a divergent Fourier series shows.   △

For the proof of Theorem 3.11 we need to introduce the ∞-norm:

$$\|f\|_{\infty} = \sup_{t \in \mathbb{R}} |f(t)| \,.$$

Equipped with this norm, $C_{2\pi}$ becomes a normed vector space which is complete, that is, a *Banach space*: every Cauchy sequence of functions with respect to this norm has a limit which is also in $C_{2\pi}$. By definition of the norm, a sequence $(f_n)_{n \in \mathbb{N}}$ in $C_{2\pi}$ converges uniformly to some $f \in C_{2\pi}$ if and only if $\lim_{n \to \infty} f_n = f$ with respect to $\| \cdot \|_{\infty}$, that is, if and only if $\lim_{n \to \infty} \|f_n - f\|_{\infty} = 0$.

Definition 3.13 Given functions $f, g \in C_{2\pi}$, we define their *convolution* $f * g$ by

$$(f * g)(t) := \frac{1}{2\pi} \int_0^{2\pi} f(t - s) g(s) \, ds. \qquad \qquad △$$

The following properties of convolutions are easy to see.

Lemma 3.14 *For $f, g, h \in C_{2\pi}$ we have*
(a) commutativity: $f * g = g * f$
(b) associativity: $(f * g) * h = f * (g * h)$
(c) distributivity: $(f + g) * h = f * h + g * h$

These three properties together mean that $C_{2\pi}$ is an *algebra*. Moreover, the norm $\|\cdot\|_\infty$ is *submultiplicative* on $C_{2\pi}$, which means that

$$\|f * g\|_\infty \leq \|f\|_\infty \|g\|_\infty \tag{3.20}$$

for all $f, g \in C_{2\pi}$, as follows immediately from the definition. Hence $C_{2\pi}$ is a *Banach algebra*. The submultiplicativity of the norm implies in particular that the convolution is continuous, in the following sense.

Lemma 3.15 (Continuity of convolution) *Let $f_n, f, g_n, g \in C_{2\pi}$ be such that*

$$\lim_{n \to \infty} \|f_n - f\|_\infty = 0, \qquad \lim_{n \to \infty} \|g_n - g\|_\infty = 0.$$

Then

$$\lim_{n \to \infty} \|f_n * g_n - f * g\|_\infty = 0.$$

In other words: $f_n * g_n \longrightarrow f * g$ *in the norm* $\|\cdot\|_\infty$ *as $n \to \infty$.*

Proof Applying Lemma 3.14, (3.20) and the triangle inequality yields

$$
\begin{aligned}
\|f_n * g_n - f * g\|_\infty &= \|f_n * (g_n - g) + (f_n - f) * g\|_\infty \\
&\leq \|f_n * (g_n - g)\|_\infty + \|(f_n - f) * g\|_\infty \\
&\leq \|f_n\|_\infty \|g_n - g\|_\infty + \|f_n - f\|_\infty \|g\|_\infty \\
&\to 0 \text{ as } n \to \infty,
\end{aligned}
$$

which proves the claim. $\square$

The algebra $C_{2\pi}$ does not possess an identity element, that is, there is no function $u \in C_{2\pi}$ such that $u * f = f$ for all $f \in C_{2\pi}$. Instead, we can give an approximate identity, that is, a family of functions $j_r \in C_{2\pi}$ such that $j_r * f \to f$ for all $f \in C_{2\pi}$. We define these functions, analogously to (3.19), by

$$j_r(t) := \sum_{k=-\infty}^{\infty} r^{|k|} e_k(t) \tag{3.21}$$

for $0 \leq r < 1$. For each such r, this series converges uniformly on $\mathbb{R}$ and thus indeed defines a function $j_r \in C_{2\pi}$. Using the formula for geometric series and the assumption that $0 \leq r < 1$, we see that

$$j_r(t) = \sum_{k=0}^{\infty} r^k e^{ikt} + \sum_{k=1}^{\infty} r^k e^{-ikt} = \frac{1}{1 - re^{it}} + \frac{re^{-it}}{1 - re^{-it}}$$

$$= \frac{1 - re^{-it} + re^{-it}(1 - re^{it})}{(1 - re^{it})(1 - re^{-it})} = \frac{1 - r^2}{1 - 2r\cos t + r^2}. \tag{3.22}$$

Hence, for all $0 \le r < 1$,

$$j_r(t) > 0, \quad t \in \mathbb{R}, \tag{3.23}$$

and for $0 < \delta < 2\pi$

$$\lim_{r \uparrow 1} j_r(t) = 0 \text{ uniformly in } t \in [\delta, 2\pi - \delta]. \tag{3.24}$$

We see from the definition (3.21) of j_r that

$$\frac{1}{2\pi} \int_0^{2\pi} j_r(t)\, dt = 1. \tag{3.25}$$

Note that $\int_0^{2\pi} e_k(t)\, dt = 0$ whenever $k \ne 0$.

We now have all the tools we need in order to show that the family $(j_r)_{0 \le r < 1}$ is an *approximate identity* in $C_{2\pi}$, as claimed above.

Theorem 3.16 (Approximate identity in $C_{2\pi}$) *Given $f \in C_{2\pi}$, we have*

$$\lim_{r \uparrow 1} j_r * f = f,$$

where the convergence is with respect to the norm $\| \cdot \|_\infty$.

Proof Fix $\varepsilon > 0$ arbitrary. Since f is uniformly continuous, there exists a $\delta \in (0, \pi)$ such that for all $|s| \le 2\delta$ we have $|f(t - s) - f(t)| \le \varepsilon$ for all $t \in \mathbb{R}$. Thus, by (3.25) and the periodicity of the function f,

$$(j_r * f)(t) - f(t) = \frac{1}{2\pi} \int_{-\pi}^{\pi} j_r(s)\,(f(t - s) - f(t))\, ds$$

$$= \frac{1}{2\pi} \int_{-\delta}^{\delta} j_r(s)\,(f(t - s) - f(t))\, ds$$

$$+ \frac{1}{2\pi} \int_{\pi \ge |s| \ge \delta} j_r(s)\,(f(t - s) - f(t))\, ds.$$

Hence

$$|(j_r * f)(t) - f(t)| \le \varepsilon \frac{1}{2\pi} \int_{-\delta}^{\delta} j_r(s)\, ds + 2\|f\|_\infty \frac{1}{2\pi} \int_{\pi > |s| > \delta} j_r(s)\, ds.$$

By (3.24) and (3.25) it follows that $\lim\sup_{r\uparrow 1}\|j_r * f - f\|_\infty \leq \varepsilon$. Since $\varepsilon > 0$ was arbitrary, this proves the claim.   □

Proof (of Theorem 3.11) We now merely need to compute $j_r * f$. Firstly, for $f \in C_{2\pi}$ we have

$$(e_k * f)(t) = \frac{1}{2\pi} \int_0^{2\pi} e^{ik(t-s)} f(s)\, ds = \frac{1}{2\pi} e^{ikt} \int_0^{2\pi} e^{-iks} f(s)\, ds = e^{ikt} c_k,$$

where c_k is the kth Fourier coefficient of f. Hence

$$\left(\sum_{k=-n}^{n} r^{|k|} e_k \right) * f = \sum_{k=-n}^{n} r^{|k|}(e_k * f) = \sum_{k=-n}^{n} r^{|k|} c_k e_k.$$

Invoking the continuity of convolutions (see Lemma 3.15) we see that

$$j_r * f = \sum_{k=-\infty}^{\infty} r^{|k|} c_k e_k. \tag{3.26}$$

Abel's convergence theorem for Fourier series, Theorem 3.11, now follows from Theorem 3.16, the theorem about the approximate identity, since the latter states that

$$f_r(t) = \sum_{k=-\infty}^{\infty} r^{|k|} c_k e^{ikt} = j_r * f \longrightarrow f \quad \text{as } r \uparrow 1,$$

where the convergence is with respect to $\|\cdot\|_\infty$.   □

We note a consequence. We denote by

$$\mathcal{T} := \bigcup_{n \in \mathbb{N}} \mathcal{T}_n, \qquad \mathcal{T}_n := \left\{ \sum_{k=-n}^{n} c_k e_k : c_{-n}, \ldots, c_n \in \mathbb{C} \right\}, \; n \in \mathbb{N},$$

the *space of trigonometric polynomials*.

Corollary 3.17 (Density of trigonometric polynomials in $C_{2\pi}$) *The space $\mathcal{T}$ is dense in $C_{2\pi}$ with respect to the norm $\|\cdot\|_\infty$, that is, for every $f \in C_{2\pi}$ there exists a sequence $(f_n)_{n \in \mathbb{N}}$ in $\mathcal{T}$ such that $\lim_{n \to \infty} f_n = f$ with respect to $\|\cdot\|_\infty$.*

Proof If $f \in C_{2\pi}$, then $f = \lim_{r \uparrow 1} j_r * f$ converges uniformly. Hence for every $n \in \mathbb{N}$ there exists an $r_n < 1$ such that $\|f - j_{r_n} * f\|_\infty \leq \frac{1}{2n}$. Due to the uniform convergence of the series (3.26) there exists an $N_n \in \mathbb{N}$ such that

$$\left\| j_r * f - \sum_{|k| \leq N_n} r^{|k|} c_k e_k \right\|_\infty \leq \frac{1}{2n}.$$

Choose $f_n := \sum_{|k| \leq N_n} r^{|k|} c_k e_k$; then $f_n \in \mathcal{T}$, and $\|f - f_n\|_\infty \leq \frac{1}{n}$ by the triangle inequality. $\square$

Finally, we consider expansions of functions in terms of real sine and cosine functions. Let $f \in C_{2\pi}$ and let c_k be its kth Fourier coefficient. Since $e^{-ikt} e^{ikx} + e^{ikt} e^{-ikx} = 2\cos(kt)\cos(kx) + 2\sin(kt)\sin(kx)$ for all $k \in \mathbb{N}$, it follows that $c_k e_k + c_{-k} e_{-k} = a_k \cos(kt) + b_k \sin(kt)$, where

$$a_k = \frac{1}{\pi} \int_0^{2\pi} \cos(kt) f(t)\, dt, \qquad b_k = \frac{1}{\pi} \int_0^{2\pi} \sin(kt) f(t)\, dt. \qquad (3.27)$$

We recall that by Definition 3.16

$$c_0 = \frac{1}{2\pi} \int_0^{2\pi} f(t)\, dt. \qquad (3.28)$$

We thus have

$$(j_r * f)(t) = c_0 + \sum_{k=1}^{\infty} r^k \{ a_k \cos(kt) + b_k \sin(kt) \}. \qquad (3.29)$$

We can now reformulate Abel's convergence theorem (Theorem 3.11).

Corollary 3.18 *For any $f \in C_{2\pi}$, the series*

$$f(t) = \lim_{r \uparrow 1} \left\{ c_0 + \sum_{k=1}^{\infty} r^k \left[a_k \cos(kt) + b_k \sin(kt) \right] \right\} \qquad (3.30)$$

converges uniformly in t, where the Fourier coefficients a_k, b_k, c_0 are defined by (3.27) *and* (3.28).

If f is a real-valued function, then the coefficients c_0, a_k, b_k are also real. Hence the series (3.30) likewise consists exclusively of real terms. We now wish to formulate an important special case explicitly.

Corollary 3.19 *Let $\ell > 0$ and let $g \in C([0, \ell])$ be a real-valued function for which* $g(0) = g(\ell) = 0$. *Set*

$$b_k = \frac{2}{\ell} \int_0^{\ell} g(s) \sin\left(\frac{k}{\ell} \pi s \right) ds.$$

Then

$$g(x) = \lim_{r \uparrow 1} \sum_{k=1}^{\infty} r^k b_k \sin\left(\frac{k\pi}{\ell} x \right)$$

uniformly in $x \in [0, \pi]$.

Proof *1st case:* Let $\ell = \pi$. Since $g(0) = 0$, there exists exactly one function $f \in C_{2\pi}$ such that $f(t) = g(t)$ for all $0 \le t \le \pi$ and $f(t) = -f(-t)$ for all $t \in \mathbb{R}$. Consider the Fourier coefficients a_k, b_k of f. Since f is odd, we have $a_k = 0$ and

$$b_k = \frac{1}{\pi} \int_{-\pi}^{\pi} \sin(kt)\, f(t)\, dt = \frac{2}{\pi} \int_0^{\pi} \sin(kt)\, g(t)\, dt.$$

Now the claim follows from Corollary 3.18.

2nd case: Let $\ell > 0$ be arbitrary. Define $h(t) := g(t\ell/\pi)$ for $0 \le t \le \pi$; then

$$b_k = \frac{1}{\pi} \int_0^{\pi} h(t)\, \sin(kt)\, dt = \frac{1}{\ell} \int_0^{\ell} g(s) \sin\left(\frac{k\pi s}{\ell}\right) ds, \quad k \in \mathbb{N}.$$

By what we have shown in the 1st case,

$$g\left(t\frac{\ell}{\pi}\right) = \lim_{t \uparrow 1} \sum_{k=1}^{\infty} r^k\, b_k\, \sin(kt)$$

uniformly in $t \in [0, \pi]$. Finally, making the substitution $x = t\ell/\pi$ completes the proof.   □

Corollary 3.20 *Let* $g \in C([0, \ell])$ *be such that* $g(0) = g(\ell) = 0$. *Then there exist trigonometric polynomials*

$$g_n(x) = \sum_{k=1}^{\infty} b_k^n \sin\left(\frac{k\pi}{\ell} x\right),$$

where $b_k^n \in \mathbb{R}$ *satisfy* $b_k^n = 0$ *for all* $k \ge N_n$, *for some* $N_n \in \mathbb{N}$, *such that* $\lim_{n\to\infty} g_n(x) = g(x)$ *uniformly in* $[0, \ell]$. *In particular,* $\lim_{n\to\infty} b_k^n = b_k$, $k \in \mathbb{N}$, *where* b_k *is the kth Fourier coefficient of* g *as in Corollary 3.19.*

Proof By Corollary 3.19 there exists an $0 < r_n < 1$ such that

$$\left| g(x) - \sum_{k=1}^{\infty} r_n^k\, b_k \sin\left(\frac{k\pi}{\ell} x\right) \right| \le \frac{1}{2n}, \quad x \in [0, \ell].$$

Choose $N_n \in \mathbb{N}$ such that

$$\left| \sum_{k=N_n+1}^{\infty} r_n^k\, b_k \sin\left(\frac{k\pi}{\ell} x\right) \right| \le \frac{1}{2n}, \quad x \in [0, \ell],$$

and set $b_k^n := r_n^k b_k$ for $k \le N_n$ and $b_k^n = 0$ for $k > N_n$. It now follows from the triangle inequality that $|g(x) - g_n(x)| \le \frac{1}{n}$ for all $x \in [0, \ell]$ and $n \in \mathbb{N}$.   □

3.3 Laplace's equation

In this section we wish to solve Laplace's equation with Dirichlet boundary conditions on certain special domains. To do so, we will use the method of "separation of variables" together with Fourier series. We will start by considering rectangles and then move on to the disk. Afterwards, we will prove what is known as the elliptic maximum principle, which not only yields uniqueness of solutions but also an important *a priori* estimate on them, which we will then use when proving the existence of solutions for arbitrary boundary data.

3.3.1 The Dirichlet problem on the unit square

We first consider Laplace's equation on the unit square

$$\Omega := (0, 1)^2 = \{(x, y) \in \mathbb{R}^2 : 0 < x, y < 1\}$$

with Dirichlet boundary conditions which are inhomogeneous on the upper edge of Ω. The boundary value problem consists of finding a function $u \in C(\overline{\Omega}) \cap C^2(\Omega)$ such that

$$\Delta u = 0, \qquad\qquad 0 < x, y < 1, \qquad (3.31a)$$
$$u(0, y) = u(1, y) = 0 \qquad\qquad 0 \le y \le 1, \qquad (3.31b)$$
$$u(x, 0) = 0, \qquad\qquad 0 \le x < 1, \qquad (3.31c)$$
$$u(x, 1) = g(x), \qquad\qquad 0 < x < 1, \qquad (3.31d)$$

where $g : [0, 1] \to \mathbb{R}$ is a given continuous function which satisfies $g(0) = g(1) = 0$.

We use the method of *separation of variables* to construct special solutions of the equation (3.31). Since $\Omega = (0, 1) \times (0, 1)$ is a Cartesian product of intervals, we may hope to find a solution of (3.31a) of the simple form

$$u(x, y) = X(x) \cdot Y(y) \qquad (3.32)$$

for certain functions $X, Y \in C^2(0, 1) \cap C([0, 1])$.[1] We substitute this form into (3.31a) and obtain $0 = \Delta u(x, y) = X''(x) Y(y) + X(x) Y''(y)$ for all $x, y \in (0, 1)$. If we suppose that $0 \ne u(x, y) = X(x) Y(y)$ for all $x, y \in (0, 1)$, then we can divide the above equation by this term to obtain

---

[1] We will often use spaces of the form $C^k(\Omega) \cap C(\overline{\Omega})$, where $\Omega \subset \mathbb{R}^d$. The functions in this space are by definition k times continuously differentiable in the interior (that is, in Ω) and can be extended continuously to the boundary $\partial\Omega$. In particular, boundary values of such functions on $\partial\Omega$ are well defined. See also the notation at the beginning of Chapter 2 on page 30. Moreover, we just write $C^k(0, 1)$ for $C^k(\Omega)$ with $\Omega = (0, 1)$.

$$-\frac{X''(x)}{X(x)} = \frac{Y''(y)}{Y(y)}. \tag{3.33}$$

We now see that the left-hand side does not depend on y, and the right-hand side does not depend on x. Hence both sides of (3.33) must be equal to the same constant, independent of x and y. Thus there exists some $\lambda \in \mathbb{R}$ such that

$$-X''(x) = \lambda X(x), \quad 0 < x < 1, \tag{3.34a}$$
$$X(0) = X(1) = 0, \tag{3.34b}$$
$$Y''(y) = \lambda Y(y), \quad 0 < y < 1, \tag{3.34c}$$
$$Y(0) = 0. \tag{3.34d}$$

The two ordinary differential equations (3.34a) and (3.34c) follow from the above remarks. The boundary condition (3.34b) follows from (3.31b), since by the separation *ansatz* (3.32) we have $0 = X(0)Y(y) = X(1)Y(y)$, $0 \le y \le 1$, which implies (3.34b). The condition (3.34d) can be shown similarly. We see that (3.34a), (3.34b) is in fact a boundary value problem for a second-order ordinary differential equation. The functions $\sin(k\pi x)$ are obviously solutions of (3.34a), (3.34b) when $\lambda = (k\pi)^2$, $k \in \mathbb{N}$.

Let us now consider (3.34c), (3.34d), another second-order ordinary differential equation, but with one boundary condition missing. We will use this missing condition to satisfy the inhomogeneous condition (3.31d). If we write $\lambda = \beta^2, \beta > 0$, then the general solution of (3.34c) is

$$\alpha_1 e^{\beta y} + \alpha_2 e^{-\beta y}.$$

Substituting this in the boundary condition (3.34d) (that is, when $y = 0$), we obtain $0 = \alpha_1 + \alpha_2$ and thus $Y(y) = \alpha_1(e^{\beta y} - e^{-\beta y}) = 2\alpha_1 \sinh(\beta y)$. Now we saw above that there are solutions of (3.34a), (3.34b) whenever λ is of the form $\lambda_k = (k\pi)^2$; hence for $\beta_k = k\pi$ we obtain the particular solution

$$u_k(x, y) = \sin(k\pi x) \cdot \sinh(k\pi y), \quad k \in \mathbb{N}.$$

Hence under suitable convergence assumptions the series

$$u(x, y) = \sum_{k=1}^{\infty} c_k \, u_k(x, y), \quad c_k \in \mathbb{R}, \tag{3.35}$$

satisfies the conditions (3.31a) – (3.31c). We are still missing the boundary condition (3.31d). Since $g \in C([0, 1])$ and $g(0) = g(1) = 0$, it admits an expansion in sine functions of the form

$$g(x) = \sum_{k=1}^{\infty} b_k \sin(k\pi x)$$

with Fourier coefficients

$$b_k = 2 \int_0^1 g(x) \sin(k\pi x)\, dx. \tag{3.36}$$

The convergence of this series, however, is in general only in the sense of Abel (see Corollary 3.19). We will now proceed with calculations which are, at least for the moment, merely formal. When $y = 1$ the representation (3.35) of $u(x, 1)$ reads

$$u(x, 1) = \sum_{k=1}^{\infty} c_k \sin(k\pi x) \sinh(k\pi).$$

Equating coefficients, we have

$$c_k = \frac{b_k}{\sinh(k\pi)}, \quad k \in \mathbb{N}.$$

We thus obtain the function

$$u(x, y) := \sum_{k=1}^{\infty} \frac{b_k}{\sinh(k\pi)} \sin(k\pi x) \cdot \sinh(k\pi y) \tag{3.37}$$

as a candidate for the solution. This series converges uniformly in every rectangle of the form $[0, 1] \times [0, b]$, $b < 1$. Note that for $0 \le y \le b < 1$

$$\frac{\sinh(k\pi y)}{\sinh(k\pi)} = \frac{e^{k\pi y} - e^{-k\pi y}}{e^{k\pi} - e^{-k\pi}} = \frac{e^{k\pi y}}{e^{k\pi}} \frac{1 - e^{-2k\pi y}}{1 - e^{-2k\pi}} \le \frac{e^{k\pi y}}{e^{k\pi}} = e^{k\pi(y-1)} \le \left(e^{\pi(b-1)} \right)^k.$$

Since the series of partial derivatives of (3.37) of order k also converges uniformly in every such rectangle for every $k \in \mathbb{N}$, (3.37) defines a function $u \in C^{\infty}([0, 1] \times [0, 1))$ and we have

$$\Delta u = 0 \quad \text{in } (0, 1) \times (0, 1).$$

We will see that

$$\lim_{y \uparrow 1} u(x, y) = g(x)$$

uniformly in $x \in [0, 1]$. Thus u solves the Dirichlet problem (3.31a) – (3.31d). To prove this last step we need the maximum principle, which we will prove in Section 3.3.3; this will also yield uniqueness of the solution.

3.3.2 The Dirichlet problem on the disk

We now wish to solve the Dirichlet problem on the disk

$$\mathbb{D} := \{x \in \mathbb{R}^2 : |x| < 1\}$$

using Fourier series; we will even derive an explicit formula for the solutions. Here we denote by $|x| = (x_1^2 + x_2^2)^{1/2}$ the Euclidean norm of $x = (x_1, x_2) \in \mathbb{R}^2$.

To find solutions we will again apply the method of *separation of variables*. Of course this technique only allows us to find very special solutions, namely ones which themselves are characterized by separated variables. However, if we superimpose these, we can construct general solutions. Abel's convergence theorem (Theorem 3.11) gives us exactly the convergence result we will need for this.

The Dirichlet problem reads as follows. Given $g \in C(\partial \mathbb{D})$, we wish to find a function $v \in C^2(\mathbb{D}) \cap C(\overline{\mathbb{D}})$ which solves

$$\Delta v(x) = 0, \qquad\qquad x \in \mathbb{D}, \tag{3.38a}$$

$$v(x) = g(x), \qquad\qquad x \in \partial \mathbb{D}. \tag{3.38b}$$

Here we also introduce the *punctured disk*

$$\dot{\mathbb{D}} := \{x \in \mathbb{R}^2 : 0 < |x| < 1\}.$$

At this stage it makes sense to introduce polar coordinates.

Change of variables: polar coordinates

Lemma 3.21 (Polar coordinates) *Let $v : \dot{\mathbb{D}} \to \mathbb{R}$ and $u : (0, 1) \times \mathbb{R} \to \mathbb{R}$ be such that $u(r, \theta) = v(r \cos \theta, r \sin \theta)$. Then u is twice continuously differentiable if and only if $v \in C^2(\dot{\mathbb{D}})$, and in this case we have*

$$\Delta v(r \cos \theta, r \sin \theta) = u_{rr} + \frac{u_r}{r} + \frac{u_{\theta\theta}}{r^2} .$$

The proof is a simple exercise in calculating the various partial derivatives; see Exercise 3.6.

The boundary of $\mathbb{D}$ is the unit circle

$$\Gamma := \partial \mathbb{D} = \{x \in \mathbb{R}^2 : |x| = 1\}.$$

Let $g \in C(\Gamma)$ and set $f(\theta) := g(\cos \theta, \sin \theta)$; then $f \in C_{2\pi}$. The coordinate transformation of Lemma 3.21 leads us to the following problem: find a function $u \in C^2((0, 1) \times \mathbb{R})$ such that $u(r, \cdot) \in C_{2\pi}$ and

$$u_{rr} + \frac{u_r}{r} + \frac{u_{\theta\theta}}{r^2} = 0, \tag{3.39a}$$

$$\lim_{r \uparrow 1} u(r, \theta) = f(\theta) \qquad \text{uniformly in } \theta \in \mathbb{R}, \tag{3.39b}$$

$$\lim_{r \downarrow 0} u(r, \theta) =: c \qquad \text{exists with uniform convergence in } \theta \in \mathbb{R}. \tag{3.39c}$$

If $u \in C^2((0, 1) \times \mathbb{R})$ is a solution of (3.39) then we set

$$v(r \cos \theta, r \sin \theta) := \begin{cases} f(\theta) & \text{if } r = 1, \\ u(r, \theta) & \text{if } 0 < r < 1, \\ c & \text{if } r = 0. \end{cases}$$

It is to be expected that v is a solution of (3.38). It is clear from Lemma 3.21 that $v \in C(\overline{\mathbb{D}})$, $v_{|\Gamma} = g$, $v \in C^2(\dot{\mathbb{D}})$ and $\Delta v = 0$ on $\dot{\mathbb{D}}$. It remains to show that v is twice continuously differentiable at the origin. This could be derived from (3.39); however, we will first look for a solution of (3.39) and return to the question of differentiability later.

Separation of variables

In order to solve (3.39a) we will again use the method of *separation of variables*. We will first try to find a solution of (3.39) of the form

$$u(r, \theta) = v(r) w(\theta), \tag{3.40}$$

where $v \in C^2(0, 1) \cap C([0, 1])$, and $w \in C^2(\mathbb{R})$ is a 2π-periodic function. In this case,

$$0 = u_{rr} + \frac{u_r}{r} + \frac{u_{\theta\theta}}{r^2} = v''(r)w(\theta) + \frac{v'(r)}{r}w(\theta) + \frac{v(r)}{r^2}w''(\theta).$$

If the functions do not vanish anywhere, then we may deduce that

$$r^2 \frac{v''(r)}{v(r)} + r \frac{v'(r)}{v(r)} = -\frac{w''(\theta)}{w(\theta)}. \tag{3.41}$$

This equation should hold for arbitrary $r \in (0, 1)$ and $\theta \in \mathbb{R}$. Since the left-hand side is independent of θ and the right-hand side is independent of r, both sides of (3.41) must be constant. We conclude that there exists some $\lambda \in \mathbb{R}$ such that

$$r^2 \frac{v''(r)}{v(r)} + r \frac{v'(r)}{v(r)} = \lambda, \tag{3.42}$$

$$\frac{w''(\theta)}{w(\theta)} = -\lambda. \tag{3.43}$$

These are two ordinary differential equations of second order. Since w should be periodic, we require that

$$w(0) = w(2\pi), \quad w'(0) = w'(2\pi).$$

The second equation (3.43) has solutions when $\lambda = n^2$, $n \in \mathbb{N}$, and in this case these are given by

$$w_n(\theta) = a_n \cos(n\theta) + b_n \sin(n\theta). \tag{3.44}$$

The first equation (3.42) reads $r^2 v'' + r v' = n^2 v$, corresponding to $\lambda = n^2$. A solution is $v(r) = r^n$. Hence our *ansatz* (3.40) of separated variables leads us to the special solution

$$u_n(r, \theta) = r^n(a_n \cos(n\theta) + b_n \sin(n\theta)) \tag{3.45}$$

of the partial differential equation (3.39a). It is also easy to check that functions of the form (3.45) are indeed solutions of the equation (3.39a). Here $a_n, b_n \in \mathbb{R}$ are still arbitrary; however, in this case the functions (3.45) will not generally satisfy the boundary condition (3.39b). We now set

$$u(r, \theta) := c_0 + \sum_{n=1}^{\infty} r^n \{a_n \cos(n\theta) + b_n \sin(n\theta)\} \tag{3.46}$$

and choose c_0, a_n, b_n to be the Fourier coefficients of f. Then (3.46) defines a function $u : (0, 1) \times \mathbb{R} \to \mathbb{R}$ which is 2π-periodic in θ. By Abel's convergence theorem in the form of Corollary (3.18) we then have

$$\lim_{r \uparrow 1} u(r, \theta) = f(\theta)$$

uniformly in $\theta \in \mathbb{R}$. Moreover,

$$\lim_{r \downarrow 0} u(r, \theta) = c_0$$

uniformly in θ. Since the series of partial derivatives converges uniformly on $[\delta, 1-\delta]$ for every $\delta \in (0, 1)$, we conclude that the function u solves the problem (3.39).

We wish to express the function u explicitly in terms of f. To this end we first recall that $u(r, \cdot) = j_r * f$; see (3.29). Substituting the expression (3.22) for j_r, we obtain

$$u(r, \theta) = \frac{1}{2\pi} \int_0^{2\pi} f(t) \frac{1 - r^2}{1 - 2r \cos(t - \theta) + r^2} dt.$$

If we switch back to Cartesian coordinates $x = r \cos(\theta)$, $y = r \sin(\theta)$, then $x^2 + y^2 = r^2$ and $(x - \cos t)^2 + (y - \sin t)^2 = 1 - 2r \cos(t - \theta) + r^2$. Hence

$$v(x, y) = u(r, \theta) = \frac{1 - x^2 - y^2}{2\pi} \int_0^{2\pi} \frac{f(t)}{(x - \cos t)^2 + (y - \sin t)^2} dt.$$

Standard theorems on exchanging integral and derivative allow us to conclude that $v \in C^2(\mathbb{D})$, in fact $v \in C^\infty(\mathbb{D})$. We have thus proved the following result.

Theorem 3.22 (Poisson formula for the Dirichlet problem on the disk) *Given a continuous function* $g : \Gamma \to \mathbb{R}$, *the Dirichlet problem* (3.38) *has a solution* $v \in C^2(\mathbb{D}) \cap C(\overline{\mathbb{D}})$ *given by*

$$v(r \cos \theta, r \sin \theta) = \frac{1}{2\pi} \int_0^{2\pi} f(t) \frac{1 - r^2}{1 - 2r \cos(t - \theta) + r^2} \, dt, \qquad (3.47)$$

where $f(t) := g(\cos t, \sin t)$.

Remark 3.23 We will see in the next section that (3.47) is the *unique* solution of (3.38); see Corollary 3.27 (page 71). △

We call (3.47) the *Poisson formula* and

$$R(r, t) = \frac{1 - r^2}{1 - 2r \cos t + r^2}$$

the *Poisson kernel*. The latter is an *integral kernel* for the solution, that is,

$$v(r \cos(\theta), r \sin(\theta)) = \frac{1}{2\pi} \int_0^{2\pi} f(t) \, R(r, t - \theta) \, dt.$$

In the case of the disk we can thus express the solution v explicitly in terms of such a kernel. This also allows us to deduce that v is infinitely differentiable in $\mathbb{D}$. We will see later that this is in fact the case for every solution of the homogeneous Laplace equation, that is, for every harmonic function (Theorem 6.58).

3.3.3 The elliptic maximum principle

We have now managed to construct solutions for the Dirichlet problem in two cases: the unit square and the disk; however, up until now, we have not been able to say whether these are the only possible solutions or whether there are others. In fact, the respective solutions are unique. This is a consequence of the *maximum principle*, which we now wish to formulate and prove in a more general setting, namely for a general domain $\Omega \subset \mathbb{R}^d$. We first require some preparation.

Definition 3.24 A function $u \in C^2(\Omega)$ is called *harmonic (subharmonic, superharmonic)* if

$$\Delta u(x) = 0 \ (-\Delta u(x) \le 0, \ -\Delta u(x) \ge 0)$$

for all $x \in \Omega$. △

Now let Ω be an arbitrary bounded open set in $\mathbb{R}^d$; then its boundary $\partial \Omega$ is a compact set. It is in this general context that we formulate the following problem.

Dirichlet problem: given $g \in C(\partial\Omega)$, find a function $u \in C^2(\Omega) \cap C(\overline{\Omega})$ such that

$$\Delta u(x) = 0, \qquad\qquad x \in \Omega, \qquad\qquad\qquad\qquad (3.48a)$$

$$u(x) = g(x), \qquad\qquad x \in \partial\Omega. \qquad\qquad\qquad\qquad (3.48b)$$

We are thus looking for a harmonic function u in Ω which is continuous up to the boundary $\partial\Omega$ and takes on the prescribed values $g(x)$ there. Uniqueness of the solution, as mentioned above, is a consequence of the following maximum principle.

Theorem 3.25 (Elliptic maximum principle) *Let $\Omega \subset \mathbb{R}^d$ be bounded and open and let $u \in C(\overline{\Omega}) \cap C^2(\Omega)$ be subharmonic. Then*

$$\max_{x \in \overline{\Omega}} u(x) = \max_{x \in \partial\Omega} u(x).$$

Remark 3.26 Note that u has a maximum on $\overline{\Omega}$ since $\overline{\Omega}$ is compact and u is continuous on $\overline{\Omega}$. Theorem 3.25 states that this maximum is attained on the boundary of Ω.   △

Proof Suppose that $u(x) \le M$ for all $x \in \partial\Omega$; it suffices to show that $u(x) \le M$ for all $x \in \Omega$. Without loss of generality we may assume that $M = 0$, since otherwise we simply replace u by $u - M$.

We give a proof by contradiction. Suppose that

$$c := \max_{x \in \overline{\Omega}} u(x) > 0.$$

Set $\varrho := \max_{x \in \overline{\Omega}} |x|^2$ and choose $\varepsilon > 0$ in such a way that $\varepsilon\varrho < c$. Now define

$$v(x) := u(x) + \varepsilon|x|^2, \quad x \in \overline{\Omega};$$

then $\max_{x \in \overline{\Omega}} v(x) \ge c > \varepsilon\varrho$, while on the boundary $\max_{x \in \partial\Omega} v(x) \le \varepsilon\varrho$. Hence there exists some $x_0 \in \Omega$ such that $v(x_0) = \max_{x \in \overline{\Omega}} v(x)$. Since $x_0 \in \Omega$, it follows that

$$\frac{\partial^2}{\partial x_j^2} v(x_0) = \frac{d^2}{dt^2} v(x_0 + te_j)\Big|_{t=0} \le 0$$

for all $j = 1, \ldots, d$, where e_j is the jth unit coordinate vector. Hence $\Delta v(x_0) \le 0$. Now since

$$\Delta(|x|^2) = \Delta\left(\sum_{i=1}^d x_i^2\right) = \sum_{i=1}^d 2 = 2d,$$

it follows that

$$-\Delta u(x_0) = -\Delta v(x_0) + 2d\varepsilon \ge 2d\varepsilon .$$

This is a contradiction to the assumption that u is subharmonic.   □

Corollary 3.27 (Uniqueness) *Given $g \in C(\partial\Omega)$, let u be a solution of the corresponding Dirichlet problem* (3.48). *Then*

$$\min_{s \in \partial\Omega} g(s) \le u(x) \le \max_{s \in \partial\Omega} g(s) \quad \textit{for all } x \in \Omega. \tag{3.49}$$

In particular for each $g \in C(\partial\Omega)$ there exists at most one solution of (3.48).

Proof The second inequality follows from Theorem 3.25. If we apply this theorem to the function $-g$ instead of g, then we also obtain the first inequality. In particular $u \equiv 0$ if $g \equiv 0$.

Uniqueness in the Dirichlet problem now follows by considering the difference of two solutions of (3.48) for the same g. This difference is zero on $\partial\Omega$ due to the identical boundary conditions, and so the above statement implies that the difference is identically zero on the whole of Ω.   □

It follows in particular that the solutions we constructed for the unit square and the disk are indeed unique. But the maximum principle does not just yield the uniqueness of solutions of the Dirichlet problem, it also gives us continuous dependence of the solution (assuming one exists) on the data. More precisely, Corollary 3.27 implies the *a priori* estimate

$$\|u\|_{C(\overline{\Omega})} \le \|g\|_{C(\partial\Omega)}, \tag{3.50}$$

where $g \in C(\partial\Omega)$ is the given boundary function and u is the solution of (3.48). Here we consider the supremum norms

$$\|u\|_{C(\overline{\Omega})} = \sup_{x \in \overline{\Omega}} |u(x)| \quad \text{bzw.} \quad \|g\|_{C(\partial\Omega)} = \sup_{x \in \partial\Omega} |g(x)|$$

on $C(\overline{\Omega})$ and $C(\partial\Omega)$, respectively, which induce uniform convergence on the respective spaces. Thus if $g, g_n \in C(\partial\Omega)$, if $u, u_n \in C(\overline{\Omega})$ are solutions of (3.48) for g and g_n, respectively, and if g_n converges uniformly on $\partial\Omega$ to g, then by linearity

$$\|u_n - u\|_{C(\overline{\Omega})} \le \|g_n - g\|_{C(\partial\Omega)},$$

that is, u_n converges uniformly in $\overline{\Omega}$ to u.

We say that a problem is *well-posed* (in the sense of Hadamard, cf. Section 4.8) if for every input function (in this case $g \in C(\partial\Omega)$) there exists a unique solution, and this solution depends continuously on the input function. Here, the maximum principle yields uniqueness and continuous dependence. We have already proved existence in the cases of the square and the disk; we will now summarize our results.

3.3.4 Well-posedness of the Dirichlet problem for the square and the disk

We start with the square $\Omega = (0, 1) \times (0, 1)$. We first give the missing proof that the function (3.37) really is a solution. For this, we use the *a priori* estimate (3.50).

Theorem 3.28 *The problem* (3.31) *has a unique solution u, which is given by* (3.37). *Moreover,* $u \in C^\infty((0, 1) \times (0, 1))$ *and* $\|u\|_{C([0,1]^2)} \leq \|g\|_{C([0,1])}$.

Proof Uniqueness and continuous dependence on g follow from the maximum principle, as explained above. Let $g \in C([0, 1])$, $g(0) = g(1) = 0$, be the boundary value for $y = 1$ in accordance with (3.31d). Then by Corollary 3.20 there exist trigonometric polynomials of the form

$$g_n(x) := \sum_{k=1}^{N_n} b_k^n \sin(k\pi x)$$

such that $\lim_{n \to \infty} g_n = g$ in $C([0, 1])$. Set

$$u_n(x, y) = \sum_{k=1}^{N_n} b_k^n \frac{1}{\sinh(k\pi)} \sin(k\pi x) \sinh(k\pi y); \qquad (3.51)$$

then u_n is the solution of (3.31) with boundary data g_n in place of g when $y = 1$. It follows from (3.50) that

$$\|u_n - u_m\|_{C(\overline{\Omega})} \leq \|g_n - g_m\|_{C(\partial\Omega)}.$$

Thus $(u_n)_{n \in \mathbb{N}}$ is a Cauchy sequence in $C(\overline{\Omega})$. Let $w := \lim_{n \to \infty} u_n$ be the corresponding limit in $C(\overline{\Omega})$. Since $u_n(x, 1) = g_n(x) \to g(x)$, we have $w(x, 1) = g(x)$ for all $x \in [0, 1]$. Since the coefficients b_k^n are bounded, $\lim_{n \to \infty} b_k^n = b_k$ and

$$0 \leq \frac{\sinh(k\pi y)}{\sinh(k\pi)} \leq \left(\frac{e^{\pi y}}{e^\pi}\right)^k,$$

it follows from (3.51) upon passing to the limit as $n \to \infty$ that

$$w(x, y) = \sum_{k=1}^\infty b_k \frac{1}{\sinh(k\pi)} \sin(k\pi x) \sinh(k\pi y)$$

for all $x \in [0, 1]$ and $y \in [0, 1)$. Thus w is equal to the function u from (3.37) on $[0, 1] \times [0, 1)$. But since $w \in C([0, 1] \times [0, 1])$, it follows that $\lim_{y \uparrow 1} u(x, y) = w(x, 1) = g(x)$ uniformly in $x \in [0, 1]$. Thus the function defined by (3.37) is indeed a (and hence the) solution of (3.31). We already observed in Section 3.3.1 that $u \in C^\infty((0, 1) \times (0, 1))$ and $\Delta u = 0$. This completes the proof.   □

We can now show easily that the Dirichlet problem for the square is well-posed. For this, we consider arbitrary boundary functions (instead of functions which vanish on three sides). We recall that if $\Omega \subset \mathbb{R}^d$ is bounded and open and $g \in C(\partial\Omega)$, then the corresponding Dirichlet problem is as follows: find $u \in C^2(\Omega) \cap C(\overline{\Omega})$ such that

$$\Delta u = 0, \qquad \text{in } \Omega, \tag{3.52a}$$

$$u = g, \qquad \text{on } \partial\Omega. \tag{3.52b}$$

Theorem 3.29 *Let $\Omega = (0, 1) \times (0, 1)$ and $g \in C(\partial\Omega)$. Then there exists a unique solution u of (3.52). Moreover, $u \in C^\infty(\Omega)$ and $\|u\|_{C(\overline{\Omega})} \leq \|g\|_{C(\partial\Omega)}$.*

Proof We only need to prove existence.

1. The problem is linear in g: if $g_1, g_2 \in C(\partial\Omega)$ and $\alpha, \beta \in \mathbb{R}$, and if u_1, u_2 are solutions of (3.52) with boundary data g_1 and g_2, respectively, then $\alpha u_1 + \beta u_2$ is the solution of (3.52) with boundary data $\alpha g_1 + \beta g_2$.
2. Theorem 3.28 yields the existence of a solution for any function $g \in C(\partial\Omega)$ which equals 0 on three sides. Combining this with the observation in 1., we can thus find a unique solution for every function $g \in C(\partial\Omega)$ which vanishes in all four corners.
3. The functions $u_1(x, y) := (1 - x) \cdot (1 - y)$, $u_2(x, y) := x \cdot (1 - y)$, $u_3(x, y) := x \cdot y$, $u_4(x, y) := (1 - x) \cdot y$ are solutions which take on the value 1 at $(0, 0)$, $(1, 0)$, $(1, 1)$, $(0, 1)$, respectively, and in the other three corners take on the value 0. Now let $g \in C(\partial\Omega)$ be arbitrary. Then the function

$$g_0 := g - \left[g(0, 0)u_1 + g(1, 0)u_2 + g(1, 1)u_3 + g(0, 1)u_4 \right]_{|\partial\Omega} \in C(\partial\Omega)$$

is equal to 0 in all four corners. Hence by 1. and 2. there exists a solution v of (3.52) with boundary data g_0. It follows that $u = v + g(0, 0)u_1 + g(1, 0)u_2 + g(1, 1)u_3 + g(0, 1)u_4$ is the solution of (3.52). $\qquad\square$

Small modifications of the above arguments show that the Dirichlet problem on an arbitrary rectangle has a unique solution for any continuous boundary function. We will generalize this statement to a broader class of domains in Chapter 6 (without, however, finding explicit solutions).

For the disk $\mathbb{D}$ we have already found a solution, namely (3.46). We also now know that this solution is unique. We have thus proved the following theorem.

Theorem 3.30 *Let $\Omega = \mathbb{D} = \{x \in \mathbb{R}^2 : |x| < 1\}$ and let $g \in C(\partial\Omega)$. Then (3.52) has exactly one solution, which is given by*

$$u(r \cos\theta, r \sin\theta) = c_0 + \sum_{k=1}^{\infty} r^k \{a_k \cos(k\theta) + b_k \sin(k\theta)\},$$

where $c_0 = \frac{1}{2\pi} \int_0^{2\pi} g(\cos t, \sin t)\, dt$ and

$$a_k = \frac{1}{\pi} \int_0^{2\pi} \cos(kt) \, g(\cos t, \sin t) \, dt, \qquad b_k = \frac{1}{\pi} \int_0^{2\pi} \sin(kt) \, g(\cos t, \sin t) \, dt.$$

In particular, $u \in C^\infty(\Omega)$ and $\|u\|_{C(\overline{\Omega})} \le \|g\|_{C(\partial\Omega)}$.

The fact that

$$\lim_{r \uparrow 1} u(r \cos \theta, r \sin \theta) = g(\cos \theta, \sin \theta)$$

uniformly in $\theta \in \mathbb{R}$ corresponds to the statement of Abel's convergence theorem for Fourier series in this case. This also yields the density of the set of trigonometric polynomials in $C_{2\pi}$. There are however other ways to prove this density statement. For example, it is a direct consequence of the theorem of Stone–Weierstrass (see Theorem A.5). Once we know that the trigonometric polynomials are dense in $C_{2\pi}$, we can directly conclude from the maximum principle that u is the solution of (3.53) using exactly the same arguments as we used in Theorem 3.28 for the square. This leads to another proof of Abel's convergence theorem for Fourier series (Corollary 3.18) based on the elliptic maximum principle; see Exercise 3.4.

3.4 The heat equation

In this section we will consider the heat equation; we will also derive solution formulae for this equation. To this end, we first study the case of one space dimension, for which we use the method of *separation of variables*. Afterwards, we will prove a maximum principle for the heat equation (known as the *parabolic maximum principle*) and infer from it the uniqueness of the obtained solution, among other things. This maximum principle is also valid for the heat equation on an arbitrary domain.

We will then investigate the heat equation on the whole space $\mathbb{R}^d$, where we can give an explicit formula for the solutions, in Section 3.4.4.

3.4.1 Separation of variables

We consider the case of one space dimension, that is, $d = 1$ and $\Omega = (0, \pi)$. Our problem is then as follows. Let $u_0 \in C([0, \pi])$ be a given function such that

$$u_0(0) = u_0(\pi) = 0. \tag{3.53}$$

We seek a continuous function $u : [0, \infty) \times [0, \pi] \to \mathbb{R}$ which is continuously differentiable in $t \in [0, \infty)$ and twice continuously differentiable in $x \in (0, \pi)$, on $(0, \infty) \times (0, \pi)$, such that

$$u_t = u_{xx}, \qquad\qquad t > 0, \ x \in (0, \pi), \qquad\qquad\qquad (3.54a)$$
$$u(t, 0) = u(t, \pi) = 0, \qquad t \geq 0, \qquad\qquad\qquad\qquad (3.54b)$$
$$u(0, x) = u_0(x), \qquad\qquad x \in [0, \pi]. \qquad\qquad\qquad\qquad (3.54c)$$

Here (3.54c) is the initial condition and (3.54b) is the boundary condition (in this case a homogeneous Dirichlet condition). We will prove existence and uniqueness of solutions of the problem (3.54) in Section 3.4.3. We first look for a special solution of (3.54a) via the *ansatz* of separation of variables; that is, we set

$$u(t, x) = w(t) \, v(x)$$

for functions $v : [0, \pi] \to \mathbb{R}$ and $w : [0, \infty) \to \mathbb{R}$ which are to be determined. Now if u is a solution of (3.54a), then

$$0 = u_t(t, x) - u_{xx}(t, x) = v(x) \, \dot{w}(t) - v''(x) \, w(t), \qquad t > 0, \ x \in (0, \pi).$$

If $v(x) \neq 0$ for all $x \in (0, \pi)$ and $w(t) \neq 0$ for all $t > 0$, then we may divide by these terms to obtain

$$\frac{\dot{w}(t)}{w(t)} = \frac{v''(x)}{v(x)} = -\lambda$$

for all $t > 0$ and $x \in (0, \pi)$. The existence of such a constant $\lambda \in \mathbb{R}$ follows, as in the case of Laplace's equation, from the fact that the first term is independent of x and the second is independent of t.

This once again gives us two ordinary differential equations

$$\dot{w}(t) + \lambda \, w(t) = 0, \ t > 0, \qquad v''(x) + \lambda \, v(x) = 0, \ x \in (0, \pi).$$

The solution of the first equation reads $w(t) = c \, e^{-\lambda t}$ for some constant $c \in \mathbb{R}$, and so $u(t, x) = c \, e^{-\lambda t} \, v(x)$.

We only wish to consider bounded solutions[2] and therefore assume that $\lambda \geq 0$. It follows that v, as the solution of the above second-order ordinary differential equation, has the form

$$v(x) = a \cos(\sqrt{\lambda} \, x) + b \sin(\sqrt{\lambda} \, x) \, .$$

The boundary conditions (3.53) imply

$$a = v(0) = 0 \qquad \text{and} \qquad b \sin(\sqrt{\lambda} \, \pi) = v(\pi) = 0.$$

This means that $\sqrt{\lambda} \in \mathbb{N}$, that is, $\lambda = k^2$ for some $k \in \mathbb{N}$, provided that $u \neq 0$. We thus obtain

---

[2] In fact, it is possible to prove using the parabolic maximum principle in the next section that every solution of (3.54) is bounded.

$$u_k(t, x) = b_k e^{-k^2 t} \sin(kx), k \in \mathbb{N},$$

as a family of special solutions.

Linear combinations of these functions are solutions of (3.54a), (3.54b). We now wish to construct series of these special solutions which satisfy the initial condition (3.54c). For this, we apply Corollary 3.19 to the function u_0 and obtain

$$u_0(x) = \lim_{r \uparrow 1} \sum_{k=1}^{\infty} r^k b_k \sin(kx),$$

where

$$b_k = \frac{2}{\pi} \int_0^{\pi} u_0(s) \sin(ks) \, ds.$$

This leads us to look for solutions of (3.54) of the form

$$u(t, x) := \sum_{k=1}^{\infty} b_k e^{-k^2 t} \sin(kx). \tag{3.55}$$

We expect that $\lim_{t \downarrow 0} u(t, x) = u_0(x)$. In any case, u satisfies the conditions (3.54a), (3.54b), as can be seen as follows. We have

$$\sum_{k=1}^{\infty} |b_k| \, k^m \, e^{-k^2 t} < \infty$$

for all $m \in \mathbb{N}_0 := \{0, 1, 2, \dots\}$ and $t > 0$. Hence the function $u : (0, \infty) \times [0, \pi] \to \mathbb{R}$ defined by (3.55) is infinitely differentiable, and for all $t > 0$ and $x \in (0, \pi)$ we have

$$u_t(t, x) = \sum_{k=1}^{\infty} b_k (-k^2) e^{-k^2 t} \sin(kx) = \sum_{k=1}^{\infty} b_k e^{-k^2 t} \frac{d^2}{dx^2} \sin(kx) = u_{xx}(t, x).$$

We have shown that u satisfies the equation (3.54b); moreover, the boundary condition (3.54b) is also satisfied. It remains to show that

$$\lim_{t \downarrow 0} u(t, x) = u_0(x)$$

uniformly in $x \in [0, \pi]$. We will do this in Section 3.4.3 using the parabolic maximum principle.

3.4.2 The parabolic maximum principle

As explained earlier, our goal here is to prove a maximum principle for the heat equation. As done in the case of Laplace's equation we will consider the general case of an arbitrary bounded open set $\Omega \subset \mathbb{R}^d$ and the heat equation in the form $u_t = \Delta u$.

So let $\Omega \subset \mathbb{R}^d$ be a bounded open set with boundary $\partial\Omega$. We consider a time interval $[0, T]$, $T > 0$, and denote by

$$\Omega_T := (0, T) \times \Omega$$

the corresponding cylinder in time-space coordinates. The set

$$\partial^*\Omega_T := ([0, T] \times \partial\Omega) \cup (\{0\} \times \overline{\Omega})$$

is called the *parabolic boundary* of Ω_T. If $d = 2$ and Ω is drawn as a set in the horizontal (that is, xy) plane, then we can imagine the vertical z-axis as the time axis, perpendicular to the space variables. In this case, Ω_T is a piece of cylinder, like a can, with horizontal cross-section Ω, and $\partial^*\Omega_T$ is its boundary without the lid, cf. Figure 3.2.

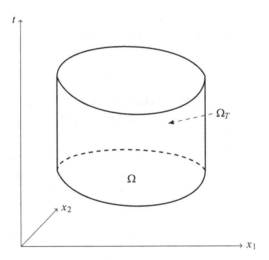

Fig. 3.2 Ω_T together with its parabolic boundary $\partial^*\Omega_T$.

We now define the space $C^{1,2}(\Omega_T)$ as the set of all continuous functions $u :$ $(0, T) \times \Omega \to \mathbb{R}$ whose partial derivatives u_t, $\frac{\partial u}{\partial x_j}$, $\frac{\partial^2 u}{\partial x_i \partial x_j}$, $i, j = 1, \ldots, d$ exist and are continuous in Ω_T. In particular, if $u \in C^{1,2}(\Omega_T)$, then $u_t - \Delta u \in C(\Omega_T)$. The following maximum principle is an easy consequence of simple properties of maxima of differentiable functions of one variable.

Theorem 3.31 (Parabolic maximum principle) *Suppose that the function $u \in$ $C(\overline{\Omega}_T) \cap C^{1,2}(\Omega_T)$ satisfies*

$$u_t(t, x) \leq \Delta u(t, x) \quad \text{for all } (t, x) \in \Omega_T.$$

Then $\max_{(t,x)\in\overline{\Omega}_T} u(t, x) = \max_{(t,x)\in\partial^*\Omega_T} u(t, x)$.

Remark 3.32 Note that the function u is continuous on the compact set $\overline{\Omega}_T = [0, T] \times \overline{\Omega}$ and thus attains a maximum on this set. The parabolic boundary $\partial^* \Omega_T$ is a compact subset of $\overline{\Omega}_T$, and the theorem states that the overall maximum of u on $\overline{\Omega}_T$ is attained on this subset.   △

Proof Fix $0 < T' < T$. We will show that

$$\max_{(t,x) \in \overline{\Omega}_{T'}} u(t, x) = \max_{(t,x) \in \partial^* \Omega_{T'}} u(t, x).$$

This implies the claim since

$$\max_{(t,x) \in \overline{\Omega}_T} u(t, x) = \sup_{T' < T} \max_{(t,x) \in \overline{\Omega}_{T'}} u(t, x)$$

$$= \sup_{T' < T} \max_{(t,x) \in \partial^* \Omega_{T'}} u(t, x) = \max_{(t,x) \in \partial^* \Omega_T} u(t, x).$$

1. We first assume that $u_t < \Delta u$ on $\Omega_{T'}$. Since $\overline{\Omega}_{T'}$ is compact and u is continuous on $\overline{\Omega}_{T'}$, there exists a point $(t_0, x_0) \in \overline{\Omega}_{T'}$ such that

$$u(t_0, x_0) = \max_{(t,x) \in \overline{\Omega}_{T'}} u(t, x).$$

   We claim that $(t_0, x_0) \in \partial^* \Omega_{T'}$. Indeed, if not, then $0 < t_0 \leq T'$ and $x_0 \in \Omega$. Since $u(t_0, x_0) = \max_{0 < s \leq T'} u(s, x_0)$, it follows that $u_t(t_0, x_0) \geq 0$. But since $u(t_0, x_0) = \max_{y \in \Omega} u(t_0, y)$, we obtain that $\frac{\partial^2}{\partial x_j^2} u(t_0, x_0) \leq 0$ for all $j = 1, \ldots, d$, whence $\Delta u(t_0, x_0) \leq 0$. This is a contradiction to our assumption, since then $0 \leq u_t(t_0, x_0) < \Delta u(t_0, x_0) \leq 0$. Hence $(t_0, x_0) \in \partial^* \Omega_{T'}$, as claimed.

2. Set $v : u - \varepsilon t$ for some given $\varepsilon > 0$. Then $v_t = u_t - \varepsilon \leq \Delta u - \varepsilon < \Delta u = \Delta v$. We now apply 1. to v to obtain

$$\max_{(t,x) \in \overline{\Omega}_{T'}} u(t, x) = \max_{(t,x) \in \overline{\Omega}_{T'}} \{v(t, x) + \varepsilon t\} \leq \max_{(t,x) \in \overline{\Omega}_{T'}} \{v(t, x) + \varepsilon T'\}$$

$$= \max_{(t,x) \in \partial^* \Omega_{T'}} \{v(t, x) + \varepsilon T'\} \leq \max_{(t,x) \in \partial^* \Omega_{T'}} \{u(t, x) + \varepsilon T'\}.$$

Since $\varepsilon > 0$ was arbitrary, the claim of the theorem follows.   □

   As a consequence we obtain the following result on uniqueness of solutions of the heat equation.

Corollary 3.33 (Uniqueness) *Suppose that $u, v \in C(\overline{\Omega}_T) \cap C^{1,2}(\Omega_T)$ satisfy*

$$u_t = \Delta u \quad and \quad v_t = \Delta v \quad in \ \Omega_T .$$

If $u = v$ on the parabolic boundary $\partial^* \Omega_T$, then $u = v$ identically on the whole of $\overline{\Omega}_T$.

Proof The difference w of the two solutions vanishes everywhere on the parabolic boundary. It follows from the parabolic maximum principle that $w(t, x) \leq 0$ for all $t \in [0, T]$ and $x \in \overline{\Omega}$. Since the same holds for $-w$ in place of w, we see that $w(t, x) = 0$ for all $t \in [0, T]$ and $x \in \overline{\Omega}$.   □

Thus, if two solutions of the heat equation agree on the parabolic boundary, then they are identical to each other. This leads to a uniqueness result for the following *parabolic initial-boundary value problem* (IBVP).

$$u_t = \Delta u, \qquad\qquad \text{in } \Omega_T, \qquad\qquad\qquad (3.56a)$$

$$u(t, x) = 0, \qquad\qquad x \in \partial\Omega, \ t \in [0, T], \qquad\qquad (3.56b)$$

$$u(0, x) = u_0(x), \qquad\qquad x \in \Omega. \qquad\qquad\qquad (3.56c)$$

We make the natural regularity assumption on u which guarantees that (3.56a) makes sense: the function u should be once continuously differentiable with respect to t and twice with respect to x; this corresponds exactly to the condition $u \in C^{1,2}(\Omega_T)$. Now we only require that the equation hold in the open set $\Omega_T := (0, T) \times \Omega$; however, the function should be continuous up to the boundary, that is, $u \in C(\overline{\Omega_T})$. In (3.56b) we impose the condition that the function vanishes on the boundary, that is, we impose *homogeneous Dirichlet boundary conditions*. The condition (3.56c) is the initial condition for u at $t = 0$. In order for a solution to exist, the boundary and initial conditions must be compatible; for this, we require that the initial value $u_0 \in C_0(\Omega)$, where

$$C_0(\Omega) := \{u \in C(\overline{\Omega}) : u_{|\partial\Omega} = 0\}.$$

This guarantees that the conditions (3.56b) and (3.56c) are compatible with each other. For the initial-boundary value problem (3.56) we have the following maximum principle.

Theorem 3.34 (Parabolic maximum principle) *Let $u_0 \in C_0(\Omega)$ be given. If u is a solution of* (3.56) *with initial condition u_0, then the following estimate holds:*

$$\min_{y \in \overline{\Omega}} u_0(y) \leq u(t, x) \leq \max_{y \in \overline{\Omega}} u_0(y) \qquad\qquad (3.57)$$

for all $t \in [0, T]$ and $x \in \overline{\Omega}$. In particular,

$$|u(t, x)| \leq \sup_{y \in \overline{\Omega}} |u_0(y)| \qquad\qquad (3.58)$$

for all $t \in [0, T]$ and $x \in \overline{\Omega}$.

Proof Set

$$c := \max_{y \in \overline{\Omega}} u_0(y).$$

Then $c \geq 0$ since $u_0 = 0$ on $\partial\Omega$, and thus $u \leq c$ on $\partial^*\Omega_T$. Hence $u \leq c$ on Ω_T by Theorem 3.31. This proves the second inequality in (3.57). Replacing u by $-u$ leads to the first inequality. □

Corollary 3.35 (Uniqueness for the IBVP for the heat equation) *Given an initial value $u_0 \in C_0(\Omega)$, there exists at most one solution of* (3.56).

Proof Theorem 3.34 proves the claim for $u_0 \equiv 0$. The general case follows from considering the difference of two solutions, analogous to the proof of Corollary 3.33. □

3.4.3 Well-posedness of the parabolic initial-boundary value problem on the interval

The parabolic maximum principle yields not only uniqueness of solutions of (3.56), but also the *a priori* estimate (3.58). This allows us to prove well-posedness of the problem in the case where Ω is an interval.

We consider the following *initial-boundary value problem*: given $u_0 \in C_0(0, \pi)$, we seek a solution u of

$$\begin{align}
u_t &= u_{xx}, & \text{in } (0, \infty) \times (0, \pi), & \quad (3.59\text{a}) \\
u(t, 0) &= u(t, \pi) = 0, & t \geq 0, & \quad (3.59\text{b}) \\
u(0, x) &= u_0(x), & x \in [0, \pi]. & \quad (3.59\text{c})
\end{align}$$

Theorem 3.36 *Let $u_0 \in C([0, \pi])$ be such that $u_0(0) = u_0(\pi) = 0$. Then* (3.59) *has a unique solution $u \in C^\infty((0, \infty) \times [0, \pi]) \cap C([0, \infty) \times [0, \pi])$. Moreover, for this solution we have $\|u\|_{C([0,\infty) \times [0,\pi])} \leq \|u_0\|_{C([0,\pi])}$.*

Proof We have already proved uniqueness and continuous dependence on the data; we still need to establish existence. Let $u_{0n} \in C([0, \pi])$ be trigonometric polynomials of the form

$$u_{0n}(x) = \sum_{k=1}^{\infty} b_k^n \sin(kx)$$

such that $u_{0n} \to u_0$ in $C([0, \pi])$ (see Corollary 3.19). Here, for fixed $n \in \mathbb{N}$, $b_k^n = 0$ for all but finitely many $k \in \mathbb{N}$, while $\lim_{n\to\infty} b_k^n = b_k$ (cf. the proof of Theorem 3.28). Now for the initial value u_{0n} the solution of (3.59) is given by

$$u_n(t, x) = \sum_{k=1}^{\infty} b_k^n e^{-k^2 t} \sin(kx). \quad (3.60)$$

We already derived this formula in Section 3.4.2, but it can be easily checked by explicit computation. Now (3.58) implies that

$$\|u_n - u_m\|_{C([0,T]\times[0,\pi])} \le \|u_{n0} - u_{m0}\|_{C[0,\pi]}$$

for any $T > 0$. Consequently, $(u_n)_{n\in\mathbb{N}}$ is a Cauchy sequence in the Banach space $C([0,T] \times [0,\pi])$, and so there exists a function $u \in C([0,\infty) \times [0,\pi])$ such that $u_n(t,x) \to u(t,x)$ uniformly in every rectangle $[0,T] \times [0,\pi]$, $T > 0$. In particular, $u(0,x) := \lim_{n\to\infty} u_n(0,x) = u_0(x)$ for all $x \in [0,\pi]$. Meanwhile, it follows from (3.60) that

$$u(t,x) = \sum_{k=1}^{\infty} b_k e^{-k^2 t} \sin(kx)$$

for all $t > 0$ and $x \in [0,\pi]$. We already saw at the end of Section 3.4.1 that $u \in C^\infty\big((0,\infty) \times [0,\pi]\big)$ and $u_t = u_{xx}$. Hence u is the solution of (3.59).   □

For our numerical treatment of the problem in Chapter 9, it will be convenient to transform the above result to the special case of the unit interval in space and a finite interval in time.

Corollary 3.37 *Let $u_0 \in C([0,1])$ be such that $u_0(0) = u_0(1) = 0$ and let $0 < T < \infty$. Then there exists a unique $u \in C^\infty\big((0,T] \times [0,1]\big) \cap C([0,T] \times [0,1])$ satisfying*

$$\begin{align}
u_t &= u_{xx}, \qquad t \in (0,T], x \in [0,1], \tag{3.61a}\\
u(t,0) &= u(t,1) = 0, t \in [0,T], \tag{3.61b}\\
u(0,x) &= u_0(x), \qquad x \in [0,1]. \tag{3.61c}
\end{align}$$

This function u is given by

$$u(t,x) = \sum_{k=1}^{\infty} e^{-k^2 \pi^2 t} b_k \sin(k\pi x),$$

where $b_k = 2\int_0^1 u_0(x)\sin(k\pi x)\,dx$, $k \in \mathbb{N}$ and $\|u\|_{C([0,T]\times[0,1])} \le \|u_0\|_{C([0,1])}$.

The parabolic maximum principle shows continuous dependence on the data, and that there can be at most one solution $u \in C^{1,2}\big((0,T)\times(0,1)\big) \cap C([0,T]\times[0,1])$. The representation of this solution as a series shows that it is actually in $C^\infty\big((0,T]\times[0,1]\big)$.

For the numerical treatment and in particular for a meaningful error estimate on the approximate solutions we require somewhat more regularity at $t = 0$.

Theorem 3.38 *Let $u_0 \in C^4([0,1])$ be such that $u_0^{(m)}(0) = u^{(m)}(1) = 0$ for $m = 0, 2, 4$. Then the solution u of (3.61) is in $C^{2,4}([0,T]\times[0,1])$, that is, u has partial derivatives at least up to order two in t and four in x in the open rectangle $(0,T) \times (0,1)$, and these all admit continuous extensions to $[0,T] \times [0,1]$.*

Proof We set $s_k(x) := \sin(k\pi x)$, $\lambda_k := k^2\pi^2$, $b_k^{(m)} := 2\int_0^1 u_0^{(m)}(x) s_k(x)\,dx$, $m = 2, 4$. Then $b_k^{(2)} = -\lambda_k b_k$, $b_k^{(4)} = \lambda_k^2 b_k$, as is easy to see via integration by parts. Differentiating componentwise, we obtain

$$u_{xx}(t, x) = \sum_{k=1}^{\infty} e^{-\lambda_k t} b_k s_k''(x) = -\sum_{k=1}^{\infty} e^{-\lambda_k t} b_k \lambda_k s_k(x) = \sum_{k=1}^{\infty} e^{-\lambda_k t} b_k^{(2)} s_k(x)$$

for all $t > 0$ and $x \in [0, 1]$. Thus $\lim_{t \downarrow 0} u_{xx}(t, x) = u_0''(x)$ uniformly in $x \in [0, 1]$, as follows by applying Corollary 3.37 to u_0'' in place of u_0. The same argument shows that

$$u_{xxxx}(t, x) = \sum_{k=1}^{\infty} e^{-\lambda_k t} b_k s_k^{(4)}(x) = \sum_{k=1}^{\infty} e^{-\lambda_k t} b_k^{(4)} s_k(x)$$

converges uniformly to $u_0^{(4)}(x)$. It follows from Exercise 3.20 that $u \in C^{2,4}([0, T] \times [0, 1])$.   □

3.4.4 The heat equation in $\mathbb{R}^d$

We now wish to study the heat equation in the whole space $\mathbb{R}^d$. This will also be useful when we come to the Black–Scholes equation in a later section. We first consider the one-dimensional case $d = 1$. Analogously to what we did above, we denote by $C^{1,2}((0, \infty) \times \mathbb{R})$ the space of those functions $u = u(t, x)$ whose partial derivatives u_t, u_x, u_{xx} exist and are continuous in $(0, \infty) \times \mathbb{R}$.

Our goal is to find solutions $u \in C^{1,2}((0, \infty) \times \mathbb{R})$ of the heat equation

$$u_t = u_{xx}, \quad t > 0, \ x \in \mathbb{R}. \tag{3.62}$$

The following arguments will allow us to construct a solution. Assume that $u \in C^{1,2}((0, \infty) \times \mathbb{R})$ is a solution of (3.62) and let $a > 0$. Then

$$v(t, x) := u(at, \sqrt{a}x) \tag{3.63}$$

also defines a solution of (3.62), as can be checked by direct computation. We will attempt to find a solution u which is invariant under the change of variables (3.63), that is, we want

$$u(t, x) = u(at, \sqrt{a}x), \quad t > 0, \ x \in \mathbb{R}$$

to hold for all $a > 0$. If we make the particular choice $a = \frac{1}{t}$, then we obtain

$$u(t, x) = u\left(1, \frac{x}{\sqrt{t}}\right),$$

whence $u(t, x) = g\left(\frac{x}{\sqrt{t}}\right)$ for $g(y) := u(1, y)$. If such a function u is a solution of (3.62), then $g \in C^2(\mathbb{R})$ and

$$u_t = g'\left(\frac{x}{\sqrt{t}}\right)\left(-\frac{1}{2}\right)\frac{x}{t^{3/2}}, \quad u_x = g'\left(\frac{x}{\sqrt{t}}\right)\frac{1}{\sqrt{t}}, \quad u_{xx} = g''\left(\frac{x}{\sqrt{t}}\right)\frac{1}{t}.$$

This leads to

$$0 = u_t - u_{xx} = -\frac{1}{t}\left(\frac{1}{2}pg'(p) + g''(p)\right),$$

where $p := \frac{x}{\sqrt{t}}$. If we set $h(p) := g'(p)$, then h satisfies the ordinary differential equation

$$\frac{1}{2}p\,h(p) + h'(p) = 0, \quad p \in \mathbb{R}.$$

If $h(p) > 0$ for all $p \in \mathbb{R}$, then

$$\frac{d}{dp}\log\left(h(p)\right) = \frac{h'(p)}{h(p)} = -\frac{p}{2} = \left(-\frac{1}{4}p^2\right)'.$$

This leads to $\log\left(h(p)\right) = -\frac{1}{4}p^2 + c_1$ for some constant c_1, and so $h(p) = c_2\,e^{-p^2/4}$ and

$$g(p) = c_2 \int_0^p e^{-r^2/4}\,dr + c_3,$$

where c_2, c_3 are further constants. Thus

$$u(t, x) = \int_0^{x/\sqrt{t}} e^{-r^2/4}\,dr$$

and so also

$$u_x(t, x) = \frac{1}{\sqrt{t}}e^{-x^2/4t}, \quad t > 0,\ x \in \mathbb{R}$$

are solutions, as can be checked easily by direct calculation. We have thus found a special solution, to which we wish to give a name.

Definition 3.39 The function

$$g(t, x) = \frac{1}{\sqrt{4\pi t}}e^{-x^2/4t}, \quad x \in \mathbb{R},\ t > 0, \tag{3.64}$$

is called the *Gaussian kernel* or the *fundamental solution of the heat equation*.

$\triangle$

The Gaussian kernel has the following properties.

Theorem 3.40 *For the function g defined by (3.64) we have* $g \in C^{\infty}((0, \infty) \times \mathbb{R})$ *and*

$$g_t = g_{xx}, \qquad t > 0, x \in \mathbb{R}, \tag{3.65a}$$

$$\int_{\mathbb{R}} g(t, x) \, dx = 1, \qquad t > 0. \tag{3.65b}$$

Proof See Exercise 3.7. □

Not only is g itself a solution of the heat equation, but so too is the function $(t, x) \mapsto g(t, x - y)$ for any $y \in \mathbb{R}$; the same is thus also true of linear combinations, or more generally superpositions, of such functions. This leads us to the following definition. Let $u_0 : \mathbb{R} \to \mathbb{R}$ be continuous and *exponentially bounded*, that is, we assume the existence of constants $A, a \in \mathbb{R}_+$ such that

$$|u_0(x)| \le A \, e^{a|x|}, \quad x \in \mathbb{R}. \tag{3.66}$$

We define

$$u(t, x) := \int_{\mathbb{R}} g(t, x - y) u_0(y) \, dy = \frac{1}{\sqrt{4\pi t}} \int_{\mathbb{R}} e^{-(x-y)^2/4t} u_0(y) \, dy; \tag{3.67}$$

then we have the following theorem.

Theorem 3.41 *The function u defined by (3.67) has the following properties:*

$$u \in C^{\infty}((0, \infty) \times \mathbb{R}), \tag{3.68a}$$

$$u_t = u_{xx}, \qquad t > 0, \ x \in \mathbb{R}, \tag{3.68b}$$

$$\lim_{t \downarrow 0} u(t, x) = u_0(x), \tag{3.68c}$$

where (3.68c) holds uniformly on bounded sets. Moreover, there exist constants B, $\omega \in \mathbb{R}$ *such that*

$$|u(t, x)| \le B e^{\omega(|x|+t)} \tag{3.69}$$

for all $x \in \mathbb{R}$ *and* $t > 0$.

Thus the function u solves the initial value problem (3.68). We will see later that u is the only solution which is *exponentially bounded*, that is, the only solution for which an estimate of the form (3.69) holds. Before we come to the proof of Theorem 3.41, we first transform the function u.

More precisely, by substitution we can also write u as follows:

$$u(t, x) = \int_{-\infty}^{+\infty} \frac{1}{\sqrt{2\pi}} e^{-z^2/2} u_0(x - \sqrt{2t}z) \, dz. \tag{3.70}$$

To see this, we make a first substitution $w = x - y$ in (3.67) to obtain

$$u(t, x) = \int_{-\infty}^{+\infty} \frac{1}{\sqrt{4\pi t}} e^{-w^2/4t} u_0(x - w) \, dw.$$

If we now set $z := w/\sqrt{2t}$, then we obtain (3.70). It now follows directly from (3.70) that

$$u \equiv 1 \qquad \text{if } u_0 \equiv 1, \tag{3.71}$$

$$u \geq 0 \qquad \text{if } u_0 \geq 0, \tag{3.72}$$

$$|u(t, x)| \leq \|u_0\|_\infty, \tag{3.73}$$

where $\|u_0\|_\infty = \sup_{x \in \mathbb{R}} |u_0(x)|$. In particular, u is bounded if u_0 is.

Proof (of Theorem 3.41) The regularity statement (3.68a) and the equation (3.68b) follow from the regularity of g, namely $g \in C^\infty((0, \infty) \times \mathbb{R})$, and $g_t = g_{xx}$, respectively, after exchanging the order of integration and differentiation.

We will next prove (3.68c). So fix $b > 0$ arbitrary; then we wish to show that $\lim_{t \downarrow 0} u(t, x) = u_0(x)$ uniformly in $x \in [-b, b]$. To this end, choose any sequence $t_n \downarrow 0$; then, since u_0 is uniformly continuous on compact intervals,

$$v_n(z) := \sup_{|x| \leq b} |u_0(x - \sqrt{2t_n}z) - u_0(x)| \longrightarrow 0 \qquad \text{as } n \to \infty, \text{ for all } z \in \mathbb{R}.$$

Hence, by the Dominated Convergence Theorem (Theorem A.9)

$$\sup_{|x| \leq b} |u(t_n, x) - u_0(x)| \leq \int_{-\infty}^{+\infty} \frac{1}{\sqrt{2\pi}} e^{-z^2/2} v_n(z) \, dz$$

converges to 0 as $n \to \infty$. This theorem is applicable as the functions $e^{-z^2/2} v_n(z)$ are easily seen to be dominated by the integrable function $g(z) = 2A e^{a(b+\sqrt{2t_1}z)} e^{-z^2/2}$, where a and A are the constants appearing in the exponential bound on u_0 which we assumed to hold.

It remains to prove (3.69). For this we will use Young's inequality in the form $\sqrt{2t}|z| \leq \varepsilon z^2 + \frac{2t}{4\varepsilon}$ for any $\varepsilon > 0$ (Lemma 5.22, (5.19) on page 173). It thus follows from (3.70) and (3.66) that

$$|u(t, x)| \leq \frac{1}{\sqrt{2\pi}} \int_{-\infty}^{+\infty} e^{-z^2/2} A e^{a|x|} e^{a\sqrt{2t}|z|} \, dz$$

$$\leq \frac{1}{\sqrt{2\pi}} \int_{-\infty}^{+\infty} e^{-(1/2-\varepsilon a)z^2} \, dz \, A e^{a|x|} e^{\frac{a}{2\varepsilon}t}$$

for all $x \in \mathbb{R}$ and $t > 0$. $\qquad\qquad\qquad\qquad\qquad\qquad\qquad\qquad\qquad\qquad\square$

We claimed above, and will see in the next section, that u is the only exponentially bounded function which satisfies (3.68). First, however, we wish to obtain an analogous result in $\mathbb{R}^d$. For this, we introduce *radial functions* in $\mathbb{R}^d$, that is, functions which only depend on the distance to the origin 0. For such functions, the Laplacian reduces to an ordinary differential operator in the radial variable.

Lemma 3.42 (Radial Laplacian) *Let $\Omega = \{x \in \mathbb{R}^d : \rho < |x| < R\}$ be an annulus, where $0 \le \rho < R$, and let $v \in C^2(\rho, R)$ and $u(x) := v(|x|)$. Then*

$$(\Delta u)(x) = \left(v_{rr} + \frac{d-1}{r} v_r \right)(|x|). \tag{3.74}$$

Proof This is an easy calculation; see Exercise 3.8. □

We now define the Gaussian kernel in $\mathbb{R}^d$, analogously to Definition 3.39, by

$$g(t, x) := \frac{1}{(4\pi t)^{d/2}} e^{-x^2/4t}, \quad t > 0, \, x \in \mathbb{R}^d, \tag{3.75}$$

where for $x \in \mathbb{R}^d$ we have written $x^2 := x_1^2 + \cdots + x_d^2 = |x|^2$. This function satisfies

$$g \in C^\infty((0, \infty) \times \mathbb{R}^d), \tag{3.76}$$

$$g_t = \Delta g, \quad t > 0, \, x \in \mathbb{R}^d, \tag{3.77}$$

$$\int_{\mathbb{R}^d} g(t, x) \, dx = 1, \quad t > 0. \tag{3.78}$$

Proof The function $g(t, \cdot)$ is radial. Since $g(t, x) = g_1(t, x_1) \cdots g_d(t, x_d)$, where

$$g_j(t, x_j) = \frac{1}{\sqrt{4\pi t}} e^{-x_j^2/4t},$$

(3.78) follows from (3.65b) by iterated integration. In order to prove (3.77) one can set $v(t, r) = t^{-d/2} e^{-r^2/4t}$, $r > 0$, $t > 0$, and show that $v_t = v_{rr} + \frac{d-1}{r} v_r$. The claim then follows from Lemma 3.42. □

Given $\Omega \subset \mathbb{R}^d$, a function $v : \Omega \to \mathbb{R}$ is said to be *exponentially bounded* if there exists constants $a, A \ge 0$ such that

$$|v(x)| \le A \, e^{a|x|}, \quad x \in \Omega.$$

For such functions we have the following result, analogous to Theorem 3.41.

Theorem 3.43 *Let $u_0 : \mathbb{R}^d \to \mathbb{R}$ be continuous and exponentially bounded, and define*

$$u(t, x) := \frac{1}{(4\pi t)^{d/2}} \int_{\mathbb{R}^d} e^{-(x-y)^2/4t} u_0(y) \, dy. \tag{3.79}$$

Then $u \in C^\infty((0, \infty) \times \mathbb{R}^d)$ and

$$u_t = \Delta u, \qquad t > 0, x \in \mathbb{R}^d, \tag{3.80a}$$

$$\lim_{t \downarrow 0} u(t, x) = u_0(x) \tag{3.80b}$$

uniformly in bounded sets in $\mathbb{R}^d$. Moreover, u is exponentially bounded.

Proof The proof of Theorem 3.43 is entirely analogous to the proof of Theorem 3.41 and is left to the reader; see Exercise 3.9. □

The problem (3.80) is again an initial value problem. Given an initial value u_0, we obtain a corresponding solution via the convolution of u_0 with the Gaussian kernel, cf. (3.79). Since $\mathbb{R}^d$ does not have a boundary, there is no boundary condition; this is replaced by the exponential boundedness. In particular, it is this condition which guarantees uniqueness of the solution. We will now prove this with the help of the following parabolic maximum principle in $\mathbb{R}^d$. This should be compared with the corresponding parabolic maximum principle for domains in Section 3.4.2. Here, the "parabolic boundary" is simply the set $\{0\} \times \mathbb{R}^d$; however, we impose a growth condition (3.81) on the function. We will also use the space $C^{1,2}$ from Section 3.4.2.

Theorem 3.44 (Parabolic maximum principle) *Given $T > 0$, let $u : [0, T] \times \mathbb{R}^d \to \mathbb{R}$ be continuous with $u \in C^{1,2}((0, T) \times \mathbb{R}^d)$ and $u_t(t, x) = \Delta u(t, x)$, $0 < t < T$, $x \in \mathbb{R}^d$. Suppose that there exist constants $a \geq 0$ and $A \geq 0$ such that*

$$u(t, x) \leq A e^{a|x|^2}, \qquad 0 \leq t < T, \ x \in \mathbb{R}^d. \tag{3.81}$$

Then

$$\sup_{(t,x) \in [0,T] \times \mathbb{R}^d} u(t, x) = \sup_{z \in \mathbb{R}^d} u(0, z). \tag{3.82}$$

Proof *1st case:* Suppose that $4aT < 1$. Choose $\varepsilon > 0$ small enough that $4a(T + \varepsilon) < 1$; then $\gamma := \frac{1}{4(T+\varepsilon)} - a > 0$. Now fix $y \in \mathbb{R}^d$ arbitrary; we need to show that

$$u(t, y) \leq \sup_{z \in \mathbb{R}^d} u(0, z) =: c \tag{3.83}$$

for all $t \in [0, T]$. Let $\mu > 0$ be given. Since for $A', B' > 0$ we have

$$\lim_{r \to \infty} \{A' e^{a(|y|+r)^2} - B' e^{(a+\gamma)r^2}\} = \lim_{r \to \infty} e^{a(|y|+r)^2} \{A' - B' e^{-a|y|^2} e^{\gamma r^2 - 2a|y|r}\}$$

$$= -\infty,$$

we may choose $r > 0$ so large that

$$A e^{a(|y|+r)^2} - \mu (4(a + \gamma))^{d/2} e^{(a+\gamma)r^2} \leq c.$$

Now set $v(t, x) := u(t, x) - \frac{\mu}{(T+\varepsilon-t)^{d/2}} e^{(x-y)^2/4(T+\varepsilon-t)}$; then $v_t - \Delta v = 0$ (use Lemma 3.42). Set $\Omega = B(y, r)$ and $\Omega_T = (0, T) \times \Omega$. The parabolic maximum principle of Theorem 3.31 applied to v yields

$$\sup_{(t,x)\in\Omega_T} v(t, x) \leq \sup_{(t,x)\in\partial^*\Omega_T} v(t, x).$$

Now the parabolic boundary $\partial^*\Omega_T$ consists of two parts:

(a) the part where $x \in \Omega$ and $t = 0$. Here, $v(0, x) \leq u(0, x) \leq c$;
(b) the part where $t \in [0, T]$ and $x \in \partial\Omega$. Here, $|x - y| = r$, and by (3.81),

$$\begin{aligned}
v(t, x) &= u(t, x) - \frac{\mu}{(T + \varepsilon - t)^{d/2}} e^{r^2/4(T+\varepsilon-t)} \\
&\leq A e^{a(|y|+r)^2} - \frac{\mu}{(T + \varepsilon - t)^{d/2}} e^{r^2/4(T+\varepsilon-t)} \\
&\leq A e^{a(|y|+r)^2} - \frac{\mu}{(T + \varepsilon)^{d/2}} e^{r^2/4(T+\varepsilon)} \\
&= A e^{a(|y|+r)^2} - \mu(4(a + \gamma))^{d/2} e^{r^2(a+\gamma)} \quad \leq c.
\end{aligned}$$

Hence $\sup_{(t,x)\in\partial^*\Omega_T} v(t, x) \leq c$ and so by the maximum principle

$$u(t, y) - \frac{\mu}{(T + \varepsilon - t)^{d/2}} = v(t, y) \leq c$$

for all $0 \leq t \leq T$. We now obtain (3.83) by letting μ converge to 0.
2nd case: If $4aT \geq 1$, then we apply the 1st case successively to the time intervals $[0, T_1], [T_1, 2T_1], \ldots$, where $T_1 = \frac{1}{8a}$. $\square$

The parabolic maximum principle gives us uniqueness of solutions on every time interval. We may thus summarize our results in the form of the following statement on well-posedness of a suitable initial value problem.

Theorem 3.45 (Heat equation in $\mathbb{R}^d$) *Let $u_0 : \mathbb{R}^d \to \mathbb{R}$ be continuous and exponentially bounded. Then the initial value problem*

$$u_t = \Delta u, \qquad 0 < t, x \in \mathbb{R}^d, \tag{3.84a}$$

$$u(0, x) = u_0(x), \qquad x \in \mathbb{R}^d. \tag{3.84b}$$

admits a unique exponentially bounded solution $u \in C^{1,2}((0, \infty) \times \mathbb{R}^d) \cap C([0, \infty) \times \mathbb{R}^d)$. This solution u is given by (3.79), and in particular $u \in C^\infty((0, \infty) \times \mathbb{R}^d)$ and $\|u\|_{C([0,\infty)\times\mathbb{R}^d)} \leq \|u_0\|_{C(\mathbb{R}^d)}$.

Proof We only have to prove uniqueness. If u_1, u_2 are two solutions, then $u = u_1 - u_2$ satisfies (3.84) with $u_0 \equiv 0$. It follows from (3.82) that $u(t, x) \leq 0$, $t \in [0, T]$ for all $x \in \mathbb{R}^d$, for any given $T > 0$. Exchanging u_1 and u_2 shows that $u \equiv 0$. $\square$

The condition that $u \in C^{1,2}((0, \infty) \times \mathbb{R}^d)$ is the minimal possible regularity assumption we can make on the solution u in order that the equation (3.84a) should make sense. In particular, we have the result that the solution is automatically infinitely differentiable even if the initial value u_0 is nowhere differentiable. Here this is a consequence of the representation (3.79) of the solution in terms of the Gaussian kernel. However, we will see later that solutions of the heat equation in a domain (rather than $\mathbb{R}^d$) are also automatically C^∞ (see Chapter 8).

As noted above, since $\mathbb{R}^d$ does not have a boundary, there is no boundary condition; this is replaced by the assumption that u is exponentially bounded. There are however other solutions which grow extremely quickly as $|x| \to \infty$ (see [28, § 2.3]).

Physically, we interpret the solutions of the heat equation as follows: if u_0 is a given distribution of heat in $\mathbb{R}^d$, then the solution $u(t, x)$ of (3.84) describes the temperature at the point $x \in \mathbb{R}^d$ at time t. Another model described by the same equation is the diffusion of, say, ink in water. Then u_0 is the initial concentration, that is, for any measurable set $B \subset \mathbb{R}^d$, $\int_B u_0(x)\, dx$ gives the quantity of ink in B. The solution $u(t, x)$ gives the concentration at time t. We refer to Chapter 1 for the derivation of the heat equation.

The representation (3.79) of the solution in terms of the Gaussian kernel also allows us to see that

$$\int_{\mathbb{R}^d} u(t, x)\, dx = \int_{\mathbb{R}^d} u_0(x)\, dx \tag{3.85}$$

for all $t > 0$, that is, the total mass is conserved, as is to be expected. Nevertheless, we are dealing here with an idealized model. Suppose that u_0 is an initial concentration such that $\int_{\mathbb{R}^d} u_0(x)\, dx = 1$ and $u_0 \geq 0$ but $u_0(x) = 0$ for all $|x| \geq \varepsilon$, where $\varepsilon > 0$ is given; that is, u_0 is concentrated in a small ball. Let x be a point a long way from the origin. Then no matter how small we take $t > 0$, we still have $u(t, x) > 0$, as can be seen from (3.79). This is a contradiction to the theory of relativity; in fact, including relativistic effects in the model leads to a nonlinear equation.

Finally, we wish to say something about the normalization of the Gaussian kernel. The function $g(t, \cdot)$ given in (3.75) is the density function of a normal distribution with variance $\sigma = 2t$ and expected value $\mu = 0$. For this reason, the renormalized function $g(t/2, x)$ is generally preferred by probability theorists; this is the kernel of the solutions of the equation

$$u_t = \frac{1}{2}\Delta u.$$

That is, in place of the Laplacian Δ one works with $\frac{1}{2}\Delta$.

3.5 The Black–Scholes equation

In the previous section we determined all exponentially bounded solutions of the heat equation

$$u_t = u_{xx}, \quad t > 0, \ x \in \mathbb{R} \tag{3.86}$$

explicitly. Here, taking these solutions as our point of departure, via a suitable change of variables and scaling we will obtain the polynomially bounded solutions of the Black–Scholes equation, and finish by deriving a formula for option pricing.

We start by considering the general parabolic equation

$$w_t = \alpha w_{xx} + \beta w_x + \gamma w, \quad t > 0, \ x \in \mathbb{R}. \tag{3.87}$$

Here $\alpha, \beta, \gamma \in \mathbb{R}$ are constants, with $\alpha > 0$. Now let $T > 0$ and suppose that $u \in C^{1,2}((0, T) \times \mathbb{R})$ is a solution of $u_t = u_{xx}$. We will modify u successively in order to obtain a solution of the more general equation (3.87). To this end, let $a, \lambda \in \mathbb{R}$ be two constants, yet to be determined.

1st Step: Define $v(t, x) := e^{ax} u(t, x)$, then

$$v_t = e^{ax} u_t,$$
$$v_x = e^{ax}(au + u_x),$$
$$v_{xx} = e^{ax}(u_{xx} + 2au_x + a^2 u) = e^{ax} u_t + 2ae^{ax}(au + u_x) - a^2 u e^{ax}$$
$$= v_t + 2av_x - a^2 v$$

since $u_t = u_{xx}$. Thus the function v solves the equation $v_t = v_{xx} - 2av_x + a^2 v$.

2nd Step: We now adjust the velocity by setting $w(t, x) := e^{ax} u(\alpha t, x) = v(\alpha t, x)$. Then $w_t = \alpha w_{xx} - 2a\alpha w_x + \alpha a^2 w$.

3rd Step: We next insert the factor $e^{-\lambda t}$; more precisely, we set

$$w(t, x) := e^{-\lambda t} e^{ax} u(\alpha t, x); \tag{3.88}$$

then we have

$$w_t(t, x) = \alpha w_{xx} - 2a\alpha w_x + (\alpha a^2 - \lambda)w. \tag{3.89}$$

Hence w solves the equation (3.87) if

$$a = -\frac{\beta}{2\alpha}, \quad \gamma = \alpha a^2 - \lambda. \tag{3.90}$$

Conversely, if w is a solution of (3.89), then $u(t, x) := e^{\lambda t/\alpha} e^{-ax} w(\frac{t}{\alpha}, x)$ solves the equation $u_t = u_{xx}$, as is easy to calculate just as above. Finally, $w \in C([0, T] \times \mathbb{R})$ if and only if $u \in C([0, T] \times \mathbb{R})$. Combining this with Theorem 3.45, we obtain the following result.

Theorem 3.46 *Let $\alpha, \beta, \gamma \in \mathbb{R}$, $\alpha > 0$ and let $0 < T < \infty$. Suppose that $w_0 : \mathbb{R} \to \mathbb{R}$ is continuous and exponentially bounded. Then there exists a unique exponentially bounded function $w \in C^{1,2}((0,T) \times \mathbb{R}) \cap C([0,T] \times \mathbb{R})$ such that*

$$w_t = \alpha w_{xx} + \beta w_x + \gamma w, \qquad t \in (0,T], x \in \mathbb{R}, \tag{3.91a}$$

$$w(0,x) = w_0(x), \qquad\qquad\qquad x \in \mathbb{R}. \tag{3.91b}$$

This function w is given by

$$w(t,x) = e^{-\lambda t} \frac{1}{\sqrt{2\pi}} \int_{\mathbb{R}} e^{-z^2/2} e^{a\sqrt{2\alpha t}z} w_0(x - \sqrt{2\alpha t}z) \, dz, \tag{3.92}$$

where a and λ are defined by (3.90). In particular, we have $w \in C^\infty((0,T] \times \mathbb{R})$ and $\|w\|_{C([0,T]\times\mathbb{R})} \le \|w_0\|_{C(\mathbb{R})}$.

Proof We only have to prove (3.92). Since $w(t,x) = e^{-\lambda t} e^{ax} u(\alpha t, x)$, in particular $w(0,x) = e^{ax} u(0,x) = e^{ax} w_0(x)$. Hence, by Theorem 3.45,

$$w(t,x) = e^{-\lambda t} e^{ax} \frac{1}{\sqrt{2\pi}} \int_{\mathbb{R}} e^{-z^2/2} e^{-a(x-\sqrt{2\alpha t}z)} w_0(x - \sqrt{2\alpha t}z) \, dz,$$

$$= e^{-\lambda t} \frac{1}{\sqrt{2\pi}} \int_{\mathbb{R}} e^{-z^2/2} e^{a\sqrt{2\alpha t}z} w_0(x - \sqrt{2\alpha t}z) \, dz,$$

as claimed. $\qquad\qquad\qquad\qquad\qquad\qquad\qquad\qquad\qquad\qquad\qquad\qquad\qquad\qquad\square$

We now turn to the Black–Scholes equation. We first recall the model for a European call option described in Section 1.5 (see also Appendix A.4). We take the strike K, the timepoint of maturity $T > 0$, the volatility $\sigma > 0$ and the interest rate $r > 0$ as given. As in (1.21) we will use the letter $y \ge 0$ for the variable giving the current asset price, not S as in the financial mathematics literature. We again consider the case of a call option (that is, an option to buy), so that the payoff function in (1.20) in the new variable y reads

$$y \mapsto (y - K)^+.$$

We wish to determine the fair price $V(t,y)$ of this option at time $t \in [0,T]$. The function V satisfies the following final value problem for all $y > 0$:

$$V_t(t,y) + \frac{\sigma^2}{2} y^2 V_{yy}(t,y) + ry V_y(t,y) - rV(t,y) = 0, \qquad 0 < t < T, \tag{3.93a}$$

$$V(T,y) = (y - K)^+. \tag{3.93b}$$

This problem is well-posed, and its solutions can be given explicitly. In order to guarantee uniqueness of solutions, however, we need to assume that they grow at most polynomially. We say that a function $f : (0,\infty) \to \mathbb{R}$ is *polynomially bounded* if there exist constants $m \in \mathbb{N}$ and $c \ge 0$ such that

$$|f(y)| \le c \begin{cases} y^m & \text{if } y \ge 1, \\ y^{-m} & \text{if } 0 < y < 1. \end{cases} \tag{3.94}$$

Thus f is polynomially bounded if and only if the function $g : \mathbb{R} \to \mathbb{R}$ given by $g(x) := f(e^x)$ is exponentially bounded. We next extend these terms to apply to functions of time and space: we say that a function $u : [0,T] \times (0,\infty) \to \mathbb{R}$ is *polynomially bounded* if there exist constants $m \in \mathbb{N}$ and $c \ge 0$ such that (3.94) holds for $f = u(t, \cdot)$, for all $t \in [0,T]$, where m and c are independent of t. Finally, we will use the notation

$$\mathcal{N}(x) := \mathcal{N}_{0,1}(x) = \frac{1}{\sqrt{2\pi}} \int_{-\infty}^{x} e^{-z^2/2} \, dz, \quad x \in \mathbb{R},$$

that is, $\mathcal{N}$ is the (cumulative) distribution function of the standard normal distribution.

Theorem 3.47 *There exists a unique polynomially bounded function*

$$V \in C([0,T] \times (0,\infty)) \cap C^{1,2}((0,T) \times (0,\infty))$$

satisfying (3.93). *Moreover,* $V \in C^{\infty}([0,T) \times (0,\infty))$, *and* V *is given by the explicit formula*

$$V(t, y) = y\mathcal{N}(d_1) - Ke^{-r(T-t)}\mathcal{N}(d_2), \tag{3.95}$$

where

$$d_1 := \frac{\log\left(\frac{y}{K}\right) + \left(r + \frac{\sigma^2}{2}\right)(T-t)}{\sigma\sqrt{T-t}}, \quad d_2 := d_1 - \sigma\sqrt{T-t}, \quad 0 \le t < T. \tag{3.96}$$

Proof We first consider the transformation $t \mapsto T - t$ of the time variable, which will convert the equation into one which runs "forward in time". Setting $W(t, y) := V(T - t, y)$ we obtain a function $W \in C([0,T] \times (0,\infty)) \cap C^{1,2}((0,T) \times (0,\infty))$ which solves the initial value problem

$$W_t = \frac{\sigma^2}{2} y^2 W_{yy} + ry W_y - rW, \quad y > 0, \, 0 < t < T, \tag{3.97a}$$

$$W(0, y) = (y - K)^+ \tag{3.97b}$$

if and only if V is a solution of (3.93). We next transform the problem (3.97) from one on $(0,\infty)$ to one on $\mathbb{R}$. For any continuous function $w : [0,T] \times \mathbb{R} \to \mathbb{R}$ we define a new function by

$$W(t, y) := w(t, \log y), \quad y > 0, \, 0 < t < T.$$

Let us suppose that $w \in C^{1,2}((0,T) \times \mathbb{R})$, which is equivalent to $W \in C^{1,2}((0,T) \times (0,\infty))$. Then we have

$$W_y(t, y) = \frac{1}{y} w_x(t, \log y), \text{ that is, } y W_y(t, y) = w_x(t, \log y), \text{ and}$$

$$W_{yy}(t, y) = -\frac{1}{y^2} w_x(t, \log y) + \frac{1}{y^2} w_{xx}(t, \log y)$$

$$= -\frac{1}{y} W_y(t, y) + \frac{1}{y^2} w_{xx}(t, \log y).$$

This implies that $y^2 W_{yy}(t, y) + y W_y(t, y) = w_{xx}(t, \log y)$, and so W solves the equation (3.97a) if and only if

$$\frac{\sigma^2}{2} w_{xx}(t, x) + \left(r - \frac{\sigma^2}{2}\right) w_x(t, x) - r\, w(t, x) = w_t(t, x) \tag{3.98}$$

for all $0 < t < T$ and $x \in \mathbb{R}$. Moreover, $W(0, y) = (y - K)^+$ if and only if

$$w(0, x) = (e^x - K)^+ =: w_0(x), \quad x \in \mathbb{R}. \tag{3.99}$$

Finally, W is polynomially bounded if and only if w is exponentially bounded. We may thus apply Theorem 3.46 to infer that (3.97) has exactly one such solution W. This function is in $C^\infty((0, T] \times (0, \infty))$, and is given explicitly by

$$W(t, y) = w(t, \log y) = e^{-\lambda t} \frac{1}{\sqrt{2\pi}} \int_{\mathbb{R}} e^{-z^2/2} e^{a\sigma\sqrt{t}z} w_0(\log y - \sigma\sqrt{t}z)\, dz$$

$$= e^{-\lambda t} \frac{1}{\sqrt{2\pi}} \int_{-\infty}^{+\infty} e^{-z^2/2} e^{a\sigma\sqrt{t}z} (y e^{-\sigma\sqrt{t}z} - K)^+ \, dz =: I_1 - I_2,$$

where $a := \frac{1}{2} - \frac{r}{\sigma^2}$, $\lambda := \frac{\sigma^2}{2} a^2 + r$ and

$$I_1 := y e^{-\lambda t} \frac{1}{\sqrt{2\pi}} \int_{-\infty}^{\frac{1}{\sigma\sqrt{t}} \log(y/K)} e^{-z^2/2} e^{(a-1)\sigma\sqrt{t}z} \, dz,$$

$$I_2 := K e^{-\lambda t} \frac{1}{\sqrt{2\pi}} \int_{-\infty}^{\frac{1}{\sigma\sqrt{t}} \log(y/K)} e^{-z^2/2} e^{a\sigma\sqrt{t}z} \, dz.$$

In order to calculate I_1 we complete the square in the exponent of the integrand:

$$-z^2/2 + (a-1)\sigma\sqrt{t}z = -\frac{1}{2}\{z^2 - 2(a-1)\sigma\sqrt{t}z\}$$

$$= -\frac{1}{2}\{(z - (a-1)\sigma\sqrt{t})^2 - (a-1)^2\sigma^2 t\}.$$

Making the substitution $z' := z - (a-1)\sigma\sqrt{t}$ leads to

$$I_1 = y e^{-\lambda t} N\left(\frac{1}{\sigma\sqrt{t}} \log \frac{y}{K} - (a-1)\sigma\sqrt{t}\right) e^{\frac{1}{2}(a-1)^2\sigma^2 t} = y N(d_1'),$$

where

$$d_1' = \frac{\log \frac{y}{K} - (a-1)\sigma^2 t}{\sigma\sqrt{t}} = \frac{\log \frac{y}{K} + \left(\frac{\sigma^2}{2} + r\right)t}{\sigma\sqrt{t}}$$

since

$$-\lambda + \frac{1}{2}(a-1)^2\sigma^2 = -\frac{\sigma^2}{2}a^2 - r + \frac{1}{2}(a^2 - 2a + 1)\sigma^2 = -r - a\sigma^2 + \frac{\sigma^2}{2} = 0.$$

In order to calculate I_2 we again consider the exponent of the integrand:

$$-\frac{1}{2}z^2 + a\sigma\sqrt{t}z = -\frac{1}{2}\{z^2 - 2a\sigma\sqrt{t}z\} = -\frac{1}{2}\{(z - a\sigma\sqrt{t})^2 - a^2\sigma^2 t\}.$$

Making the substitution $z' = z - a\sigma\sqrt{t}$, we obtain

$$I_2 = Ke^{-\lambda t}N\left(\frac{1}{\sigma\sqrt{t}}\log\frac{y}{K} - a\sigma\sqrt{t}\right)e^{\frac{1}{2}a^2\sigma^2 t} = KN(d_2')e^{-rt},$$

where

$$d_2' = \frac{\log\frac{y}{K} - a\sigma^2 t}{\sigma\sqrt{t}} = d_1' - \sigma\sqrt{t},$$

where we have used that $-\lambda + \frac{1}{2}a^2\sigma^2 = -r$. Since $V(t, y) = W(T - t, y)$, this gives us the formula (3.95). $\qquad\square$

Of particular interest is the solution $V(0, y)$ at time $t = 0$. This is the price of the option: if y is the share price at time $t = 0$, then the price for the option to buy the share at time T for the price K is given by

$$V(0, y) = yN(d_1) - Ke^{-rT}N(d_2), \tag{3.100}$$

where

$$d_1 := \frac{\log\left(\frac{y}{K}\right) + \left(r + \frac{\sigma^2}{2}\right)T}{\sigma\sqrt{T}}, \qquad\qquad d_2 := d_1 - \sigma\sqrt{T}.$$

This is the famous *Black–Scholes formula*, which to this day is in widespread use in the finance industry for calculating prices. It can be evaluated for concrete data using, for example, Maple® (cf. Section 10.1.5). Our derivation here was based on explicitly solving the partial differential equation of Black and Scholes (3.93), which we had in turn derived from the no arbitrage principle using stochastic techniques in Appendix A.4. This was the approach taken in the groundbreaking work of Black, Scholes and Merton in 1973. There is, however, another way to obtain the Black–

Scholes formula, which was developed by Cox, Ross and Rubinstein in 1979 and which is presented in many textbooks (see for example [44] or [15]).

We shall finish this section by considering a certain question of consistency. The solution V given by (3.95) depends, of course, on $\sigma > 0$. It will be interesting to look at the limit of $V(t, y)$ as $\sigma \to 0$. Here we distinguish between two cases.

1st Case: $\log\left(\frac{y}{k}\right) + rT > 0$. In this case, $\lim_{\sigma \to 0} d_1 = \lim_{\sigma \to 0} d_2 = \infty$, whence $\lim_{\sigma \to 0} \mathcal{N}(d_1) = \lim_{\sigma \to 0} \mathcal{N}(d_2) = 1$. We thus have

$$u(t, y) := \lim_{\sigma \to 0} V(t, y) = y - Ke^{-r(T-t)}.$$

In particular, the price of the option in the limit $\sigma = 0$ is given by $u(0, y) = y - Ke^{-rT}$. This is consistent with our model: if one invests the amount $y - Ke^{-rT}$ at the fixed interest rate r, then after time T its value is

$$e^{rT}(y - Ke^{-rT}) = e^{rT}y - K.$$

If the volatility is $\sigma = 0$, that is, if the increase in value is $r\%$ with complete certainty, then the share is worth $e^{rT}y$ at time T (cf. the following Remark 3.48). If the option to buy the share at time T for the price K is used, then the corresponding profit is $e^{tT}y - K$, exactly equal to the profit from depositing the amount $y - Ke^{-rT}$ at the fixed interest rate r. Hence this is exactly the price which should logically be paid for this risk-free option.

2nd Case: If $\log\left(\frac{y}{K}\right) + rT \leq 0$, then $u(0, y) := \lim_{\sigma \to 0} V(0, y) = 0$, that is, the purchase price of the option is 0. Indeed, in this case, the deterministic evolution of the share price leads to a value $e^{rT}y$ at time T which is under the strike price K. In this case, the option to buy the share for the price K is worthless.

* Remark 3.48 (Continuously compounded interest) Suppose that the fixed amount z_0 is deposited in a bank for a fixed interest rate $r\%$, which refers to the return on the investment over the course of a year. The bank pays interest after a certain period of time; suppose that this amount is added to the deposit and hence itself attracts interest. Let us assume that the deposit earns interest for a period of t years. If the bank pays interest only at the end of this period, then the total capital value becomes $z(t) = z_0 + trz_0 = (1 + tr)z_0$. If the bank pays out interest at the halfway point $\frac{t}{2}$, then the value of the capital at this time is

$$z\left(\frac{t}{2}\right) = z_0 + \frac{t}{2}rz_0 = \left(1 + \frac{t}{2}r\right)z_0.$$

This larger value $z\left(\frac{t}{2}\right)$ then earns interest for the remaining $\frac{t}{2}$ years, and so has the value

$$z(t) = z\left(\frac{t}{2}\right) + z\left(\frac{t}{2}\right)\frac{t}{2}r = \left(1 + \frac{t}{2}r\right)^2 z_0$$

at time t. If the bank pays interest at the end of each time interval of length $\frac{t}{n}$, $n \in \mathbb{N}$ then the total value at time $\frac{t}{n}$ is now $z(\frac{t}{n}) = z_0(1 + \frac{t}{n}r)$. This new amount earns interest in the period $[\frac{t}{n}, \frac{2t}{n}]$ and results in the value

$$z\left(\frac{2t}{n}\right) = z\left(\frac{t}{n}\right) + z\left(\frac{t}{n}\right)\frac{t}{n}r = z_0\left(1 + \frac{t}{n}r\right)^2$$

at time $\frac{2t}{n}$. At time $\frac{3t}{n}$ the deposit then has the value

$$z\left(\frac{3t}{n}\right) = z\left(\frac{2t}{n}\right) + z\left(\frac{2t}{n}\right)\frac{t}{n}r = z_0\left(1 + \frac{t}{n}r\right)^3.$$

Proceeding in this fashion, we end up with

$$z(t) = z_0\left(1 + \frac{t}{n}r\right)^n \tag{3.101}$$

at time t. This is the value of the capital after t years if interest is payed out every $\frac{t}{n}$ years. But this does not reflect the investment of the extra interest in time intervals smaller than $\frac{t}{n}$. To be fair the bank should pay out interest continuously. In this case, the value $z(t)$ of the capital after t years given an initial deposit of z_0, a yearly rate of interest of $r\%$ and continuous interest payments is

$$z(t) = \lim_{n\to\infty} z_0\left(1 + \frac{t}{n}r\right)^n = z_0 e^{rt}.$$

The limit of $\left(1+\frac{t}{n}r\right)^n$ as a formula for interest payments is due to Jakob Bernoulli (Acta Eruditorum, 1690), long before Euler introduced the letter e for the limit $\lim_{n\to\infty}\left(1+\frac{1}{n}\right)^n$ in a letter to Goldbach. This is an example of how natural constants can not only play a role in the laws of physics but also in the laws of finance. If money is deposited with compound interest, then regardless of the planet on which it is deposited, Euler's number $e \approx 2.71828$ is the value of the capital which has accumulated after one year (measured on that planet) if one unit of money was deposited at the rate of 100% with continuous interest payments. ◻

3.6 Integral transforms

It is very common, especially when dealing with partial differential equations arising from problems in engineering, to transform the partial differential equation using an integral transform. This often permits the reduction of the equation to an ordinary differential equation, which in turn can often be reduced to an algebraic equation, and thus solved, via the use of such integral transforms. These transforms often have a physical interpretation, for example the transformation of a wave in space-time variables into phase space (the frequency spectrum). In phase space the partial differential equation often becomes an ordinary differential equation. If one can solve the resulting equation in phase space, then the solution of the original differential equation can be obtained by applying the inverse transform.

3.6.1 The Fourier transform

We start with the Fourier transform, which was developed by Jean Baptiste Joseph Fourier in 1822 in his work *Théorie analytique de la chaleur*, already mentioned earlier. Here we will take a practical approach to Fourier transforms: we will show how they can be used to solve equations explicitly. Later, in Chapter 6, they will be used for a more systematic investigation in the context of Sobolev spaces.

Unlike in the previous sections, here we consider complex-valued functions. For $1 \le p < \infty$ we define

$$L_p(\mathbb{R}, \mathbb{C}) := \left\{ f : \mathbb{R} \to \mathbb{C} \text{ measurable} : \int_{\mathbb{R}} |f(t)|^p \, dt < \infty \right\}.$$

If we identify functions which coincide almost everywhere, then $L_p(\mathbb{R}, \mathbb{C})$ becomes a Banach space when equipped with the norm

$$\|f\|_p := \left(\int_{\mathbb{R}} |f(t)|^p \, dt \right)^{1/p}.$$

The space $L_2(\mathbb{R}, \mathbb{C})$ is a Hilbert space.

Remark 3.49 A remark about our notation is in order: the above spaces are often denoted by $L^p(\Omega, \mathbb{C})$ in the literature. We will write p (the power appearing in the integrand) as a subscript to distinguish it from the order of differentiation as appears in spaces like C^k and H^k. In dimension one we also avoid double brackets, that is, we write $L_2(0, 1)$ instead of $L_2((0, 1))$ (for example), even though the latter would be more consistent. $\triangle$

Definition 3.50 For $f \in L_1(\mathbb{R}, \mathbb{C})$ the function defined by

$$\mathcal{F}f(\omega) := \hat{f}(\omega) := \frac{1}{\sqrt{2\pi}} \int_{-\infty}^{\infty} f(t) e^{-i\omega t} \, dt, \ \omega \in \mathbb{R},$$

is called the *Fourier transform* (also the *spectral function*) of f. $\triangle$

In the context of physics, the Fourier transform can be interpreted as the transformation of a time-amplitude signal $(t, f(t))$ into its frequency spectrum $(\omega, \hat{f}(\omega))$, in other words, $\hat{f}(\omega)$ gives the proportion of the waves in the signal which have frequency ω.

Examples

Before we turn to the properties of $\mathcal{F}$, we wish to start with a few simple examples.

Example 3.51 The *rectangular pulse* is defined by

$$f(t) = \begin{cases} 1, & \text{if } |t| \le 1, \\ 0, & \text{otherwise}, \end{cases}$$

cf. Figure 3.3 (left). Clearly $f \in L_1(\mathbb{R}, \mathbb{C})$, and so the Fourier transformation of f when $\omega \ne 0$ is given by

$$\hat{f}(\omega) = \frac{1}{\sqrt{2\pi}} \int_{-1}^{1} e^{-i\omega t} \, dt = \frac{1}{\sqrt{2\pi}} \frac{-1}{i\omega} e^{-i\omega t} \Big|_{t=-1}^{t=1} = \sqrt{\frac{2}{\pi}} \frac{\sin \omega}{\omega},$$

while $\hat{f}(0) = \sqrt{2/\pi}$. Thus $\hat{f}$ is the *sinc function* (from the Latin *sinus cardinalis*), cf. Figure 3.3 (right).   △

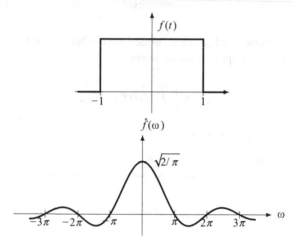

Fig. 3.3 Rectangular pulse (left) and its Fourier transform (right).

Example 3.52 The *exponentially decaying pulse* has the form $f(t) = e^{-\alpha|t|}$ for some $\alpha > 0$. Since for any $R > 0$

$$\int_{-R}^{R} |f(t)|\, dt = \int_{-R}^{R} e^{-\alpha|t|}\, dt = 2\int_{0}^{R} e^{-\alpha t}\, dt = 2\left(-\frac{1}{\alpha}\right) e^{-\alpha t}\Big|_{t=0}^{t=R}$$

$$= \left(-\frac{2}{\alpha}\right)(e^{-\alpha R} - 1) \xrightarrow{R\to\infty} \frac{2}{\alpha},$$

we see that $f \in L_1(\mathbb{R})$. The Fourier transform is easy to calculate:

$$\hat{f}(\omega) = \frac{1}{\sqrt{2\pi}}\left\{\int_{0}^{\infty} e^{-\alpha t} e^{-i\omega t}\, dt + \int_{0}^{\infty} e^{-\alpha t} e^{i\omega t}\, dt\right\}$$

$$= \frac{1}{\sqrt{2\pi}}\left\{\frac{(-1)}{\alpha + i\omega} e^{-(\alpha+i\omega)t}\Big|_{t=0}^{\infty} + \frac{1}{-\alpha + i\omega} e^{(-\alpha+i\omega)t}\Big|_{t=0}^{\infty}\right\}$$

$$= \frac{1}{\sqrt{2\pi}}\left\{\frac{1}{\alpha + i\omega} + \frac{1}{\alpha - i\omega}\right\} = \frac{1}{\sqrt{2\pi}}\frac{\alpha - i\omega + \alpha + i\omega}{\alpha^2 + \omega^2} = \sqrt{\frac{2}{\pi}}\frac{\alpha}{\alpha^2 + \omega^2}.$$

In particular, $\hat{f}$ is a rational function; see Figure 3.4.   △

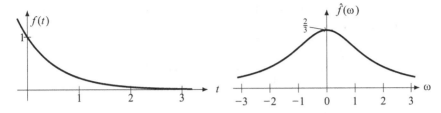

Fig. 3.4 Exponentially decaying pulse ($\alpha = 1.5$, left) and its Fourier transform (right).

Properties of the Fourier transform

We now wish to collect a few essential properties of Fourier transforms; we will focus on those we will need when solving partial differential equations.

Theorem 3.53 (Properties and rules of calculation) *Let $f \in L_1(\mathbb{R}, \mathbb{C})$. Then:*

(i) The Fourier transform $\hat{f} : \mathbb{R} \to \mathbb{C}$ is continuous and $\lim_{|\omega| \to \infty} \hat{f}(\omega) = 0$.

(ii) Linearity: if $f_1, \ldots, f_n \in L_1(\mathbb{R}, \mathbb{C})$ and $c_1, \ldots, c_n \in \mathbb{C}$, then

$$\mathcal{F}\left(\sum_{k=1}^{n} c_k f_k\right) = \sum_{k=1}^{n} c_k \, \mathcal{F} f_k.$$

(iii) If f is continuously differentiable with $f' \in L_1(\mathbb{R}, \mathbb{C})$ and $\lim_{|t| \to \infty} f(t) = 0$, then

$$\mathcal{F}(f')(\omega) = i\omega \, \mathcal{F} f(\omega). \tag{3.102}$$

(iv) If $\int_{-\infty}^{\infty} |t f(t)| \, dt < \infty$, then

$$\frac{d}{d\omega} \mathcal{F} f(\omega) = (-i)\mathcal{F}(\cdot \, f(\cdot))(\omega). \tag{3.103}$$

(v) For any $\alpha \in \mathbb{R}$ we have $\mathcal{F}(f(\cdot - \alpha))(\omega) = e^{-i\alpha\omega} \mathcal{F} f(\omega)$.

(vi) For any $\alpha \in \mathbb{R} \setminus \{0\}$ we have $\mathcal{F}(f(\alpha \cdot))(\omega) = \frac{1}{|\alpha|} \mathcal{F} f\left(\frac{\omega}{\alpha}\right)$.

Proof Here we will only prove (iii) and (iv), as these two properties will play a central role in what follows. The other statements are left to the reader as an exercise (see Exercise 3.10).

(iii) For any $R \geq 0$, if we integrate by parts and use the assumptions of the theorem, then we have

$$\frac{1}{\sqrt{2\pi}} \int_{-R}^{R} f'(t) e^{-i\omega t} \, dt = \frac{1}{\sqrt{2\pi}} f(t) e^{-i\omega t} \Big|_{t=-R}^{R} + \frac{i\omega}{\sqrt{2\pi}} \int_{-R}^{R} f(t) e^{-i\omega t} \, dt.$$

By assumption, the first term converges to 0 as $R \to \infty$, while the second converges to $i\omega \hat{f}(\omega)$.

(iv) Since we may exchange the order of differentiation and integration, the derivative is given by

$$\frac{d}{d\omega}\left[\frac{1}{\sqrt{2\pi}}\int\limits_{-\infty}^{\infty} f(t)e^{-i\omega t}\,dt\right] = (-i)\frac{1}{\sqrt{2\pi}}\int\limits_{-\infty}^{\infty} tf(t)e^{-i\omega t}\,dt = (-i)\,\mathcal{F}(\cdot f(\cdot))(\omega).$$

This proves the claim. $\square$

One of the fundamental reasons for the importance of the Fourier transform has to do with the following theorems. Given $f, g \in L_1(\mathbb{R}, \mathbb{C})$, by Fubini's theorem $f(\cdot)g(t - \cdot)$ is in $L_1(\mathbb{R}, \mathbb{C})$ for almost every $t \in \mathbb{R}$ and the *convolution* $f * g$ defined by

$$(f * g)(t) := \int_{-\infty}^{\infty} f(\tau)g(t-\tau)\,d\tau, \quad t \in \mathbb{R}, \tag{3.104}$$

is likewise in $L_1(\mathbb{R}, \mathbb{C})$. But we can say more.

A function $g : \mathbb{R} \to \mathbb{C}$ is said to have *compact support* if it vanishes identically outside a compact set (cf. the notation introduced at the beginning of Chapter 2 and in particular equation (2.1)). We set

$$C_c(\mathbb{R}, \mathbb{C}) := \{g \in C(\mathbb{R}, \mathbb{C}) : g \text{ has compact support}\} \tag{3.105}$$

and

$$C_c^m(\mathbb{R}, \mathbb{C}) := C_c(\mathbb{R}, \mathbb{C}) \cap C^m(\mathbb{R}, \mathbb{C}) \quad \text{for all } m \in \mathbb{N}.$$

Theorem 3.54 *Let $g \in C_c^m(\mathbb{R}, \mathbb{C})$ for some $m \in \mathbb{N}$ and let $f \in L_1(\mathbb{R}, \mathbb{C})$. Then the convolution $f * g$ is m times continuously differentiable and*

$$\frac{d^m}{dt^m}(f * g) = f * \left(\frac{d^m}{dt^m}g\right).$$

Proof Since g has compact support, $\sup_{\tau \in \mathbb{R}} |g(\tau)| < \infty$, and so for the function $h(\tau, t) := f(\tau)g(t - \tau)$ we have the bound

$$|h(\tau, t)| \le |f(\tau)| \sup_{\sigma \in \mathbb{R}} |g(\sigma)| < \infty;$$

in particular, the convolution is well defined and continuous. This proves the claim when $m = 0$. If $m \ge 1$, then we have $\frac{d}{dt}h(\tau, t) = f(\tau)\frac{d}{dt}g(t - \tau)$, and so

$$\left|\frac{d}{dt}h(\tau, t)\right| \le |f(\tau)| \sup_{\sigma \in \mathbb{R}} \left|\frac{d}{dt}g(\sigma)\right|.$$

Hence $f * g \in C^1(\mathbb{R})$ and

$$\frac{d}{dt}(f * g)(t) = \frac{d}{dt} \int_{-\infty}^{\infty} h(\tau, t) \, d\tau = \int_{-\infty}^{\infty} f(\tau) \frac{d}{dt} g(t - \tau) \, d\tau = \left(f * \frac{d}{dt} g \right)(t).$$

The rest follows inductively.   □

We can now formulate and prove the above-mentioned essential property of Fourier transforms.

Theorem 3.55 (Convolution theorem) *If $f, g \in L_1(\mathbb{R}, \mathbb{C})$, then $f * g \in L_1(\mathbb{R}, \mathbb{C})$ and $\widehat{f * g} = \sqrt{2\pi} \, \hat{f} \hat{g}$.*

Proof It follows from Fubini's theorem that $f * g \in L_1(\mathbb{R}, \mathbb{C})$. By definition

$$\mathcal{F}(f * g)(\omega) = \frac{1}{\sqrt{2\pi}} \int_{-\infty}^{\infty} (f * g)(t) \, e^{-i\omega t} \, dt$$

$$= \frac{1}{\sqrt{2\pi}} \int_{-\infty}^{\infty} \int_{-\infty}^{\infty} f(\tau) g(t - \tau) \, d\tau e^{-i\omega t} \, dt.$$

We now introduce the change of variables $t \mapsto s := t - \tau$, which leads to

$$\mathcal{F}(f * g)(\omega) = \frac{1}{\sqrt{2\pi}} \int_{-\infty}^{\infty} \int_{-\infty}^{\infty} f(\tau) g(s) \, d\tau \, e^{-i\omega(s+\tau)} \, ds,$$

which proves the claim.   □

We next have two theorems which state that the energy (here the norm in L_2) is preserved by the Fourier transform. We will not give the proof here.

Theorem 3.56 (Plancherel's theorem) *Let $f, g \in L_1(\mathbb{R}, \mathbb{C}) \cap L_2(\mathbb{R}, \mathbb{C})$. Then $\hat{f}, \hat{g} \in L_2(\mathbb{R}, \mathbb{C})$ and we have*

$$\int_{-\infty}^{\infty} \hat{f}(\omega) \overline{\hat{g}(\omega)} \, d\omega = \int_{-\infty}^{\infty} f(t) \overline{g(t)} \, dt \, .$$

Proof The proof can be found, for example, in [54, Thm. 9.13].   □

An immediate consequence of Plancherel's theorem 3.56 is an important relation known as *Parseval's identity*, which, as mentioned above, states that energy is preserved under the Fourier transform.

Theorem 3.57 (Parseval's identity) *If $f \in L_1(\mathbb{R}, \mathbb{C}) \cap L_2(\mathbb{R}, \mathbb{C})$, then $\hat{f} \in L_2(\mathbb{R}, \mathbb{C})$ and*

$$\int_{-\infty}^{\infty} |\hat{f}(\omega)|^2 \, d\omega = \int_{-\infty}^{\infty} |f(t)|^2 \, dt,$$

that is, $\|\hat{f}\|_{L_2(\mathbb{R})} = \|f\|_{L_2(\mathbb{R})}$.

Proof This follows from applying Plancherel's theorem to $f = g$.   □

Inverse Fourier transforms

As mentioned in the introduction to this section, we will use Fourier transforms to convert certain partial differential equations into ordinary differential equations. This is done principally using the formulae (3.102) and (3.103). Once the resulting ordinary differential equation has been solved, we need to invert the Fourier transformation in order to obtain a concrete solution of the original equation.

Theorem 3.58 (Fourier inversion formula) *If $f \in L_1(\mathbb{R}, \mathbb{C})$ and $\hat{f} \in L_1(\mathbb{R}, \mathbb{C})$, then*

$$f(x) = \frac{1}{\sqrt{2\pi}} \int_{-\infty}^{\infty} \hat{f}(\omega) e^{i\omega x} d\omega =: \mathcal{F}^{-1}(\hat{f})(x)$$

for almost every $x \in \mathbb{R}$.

Proof Compare for example [54, Thm. 9.11].   □

Solving differential equations using Fourier transforms

As announced, the properties of the Fourier transform established in Theorem 3.53 can be used to solve certain partial differential equations. We will illustrate this here using a selection of examples. It is to be emphasized that this method serves to determine a *formula* for a solution of a concrete differential equation. This does not in general imply any statements about uniqueness or continuous dependence on the data.

In each case the strategy consists of four steps:
1. Transform the original partial differential equation into an ordinary differential equation.
2. Transform the initial conditions (if applicable).
3. Solve the initial value problem for the ordinary differential equation.
4. Applying the inverse transformation yields the desired formula for the solution.

In what follows, we will assume that all the functions we are dealing with satisfy the conditions of the relevant theorems from the preceding sections, so that the rules of calculation we stated there can be applied. Our goal is to show how explicit solutions can be obtained.

Example 3.59 (IVP for the wave equation on $\mathbb{R}$) Consider the initial value problem (IVP) for the wave equation,

$$\begin{align}
u_{tt} &= u_{xx}, & t &\geq 0,\ x \in \mathbb{R}, & \text{(3.106a)} \\
u(0, x) &= f(x), & x &\in \mathbb{R}, & \text{(3.106b)} \\
u_t(0, x) &= g(x), & x &\in \mathbb{R}, & \text{(3.106c)} \\
u(t, x) &\to 0, & x &\to \pm\infty,\ t \geq 0, & \text{(3.106d)}
\end{align}$$

where $f, g : \mathbb{R} \to \mathbb{R}$ are given functions assumed to be sufficiently smooth, and consistent with the *asymptotic boundary conditions* (3.106d). In this case we already know the solution via d'Alembert's formula (3.7). Here we wish to show using this example how Fourier transforms can be used to obtain such formulae for the solution. To this end we will perform the four steps sketched above.

1. Transformation of the differential equation

We define the Fourier transform of the solution to be determined with respect to the space variable x,

$$\mathcal{U}(t, \omega) := \mathcal{F}[u(t, \cdot)](\omega) = \frac{1}{\sqrt{2\pi}} \int_{-\infty}^{\infty} u(t, x)\, e^{-ix\omega} dx.$$

Now we transform the equation (3.106a), which is possible due to the asymptotic boundary conditions (3.106d). By the rules of calculation for Fourier transforms and (3.106a) we have

$$\mathcal{F}[u_{xx}(t, \cdot)](\omega) = \mathcal{F}[u_{tt}(t, \cdot)](\omega) = -\omega^2 \mathcal{U}(t, \omega).$$

2. Transformation of the initial conditions

Since we are considering the Fourier transform with respect to the space variable x, the initial conditions are easy to transform.

$$\mathcal{U}_{tt}(t, \omega) + \omega^2 \mathcal{U}(t, \omega) = 0, \qquad\qquad t \geq 0, \qquad\qquad\qquad (3.107a)$$

$$\mathcal{U}(0, \omega) = F(\omega), \qquad\qquad\qquad\qquad\qquad (3.107b)$$

$$\mathcal{U}_t(0, \omega) = G(\omega). \qquad\qquad\qquad\qquad\qquad (3.107c)$$

This is clearly a (linear) ordinary differential equation in the time variable t, for each fixed $\omega \in \mathbb{R}$: more precisely, it is an initial value problem for a second-order ordinary differential equation.

3. Solution of the initial value problem

Using known methods from the solution theory for ordinary differential equations we can solve (3.107). For any $\omega \neq 0$ the solution is

$$\mathcal{U}(t, \omega) = F(\omega)\cos(\omega t) + G(\omega)\frac{\sin(\omega t)}{\omega}.$$

If $\omega = 0$, then we obtain $\mathcal{U}(t, 0) = F(0) + t\, G(0)$, that is, the continuous extension.

4. Inverse transformation

Finally, in order to obtain an expression for the solution $u(t, x)$ of (3.106), we apply the inversion formula of Theorem 3.58 to $\mathcal{U}$:

$$u(t, x) = \mathcal{F}^{-1}[\mathcal{U}(t, \cdot)](x)$$
$$= \frac{1}{\sqrt{2\pi}} \int_{-\infty}^{\infty} \mathcal{U}(t, \omega)\, e^{i\omega x}\, d\omega$$

$$= \frac{1}{\sqrt{2\pi}} \left\{ \int_{-\infty}^{\infty} F(\omega) \cos(\omega t) e^{i\omega x} \, d\omega + \int_{-\infty}^{\infty} G(\omega) \frac{\sin(\omega t)}{\omega} e^{i\omega x} \, d\omega \right\}$$

$$=: u_1(t, x) + u_2(t, x).$$

For the first term, we apply the identities $\cos \alpha = \frac{1}{2}(e^{i\alpha} + e^{-i\alpha})$ and $\sin \alpha = \frac{1}{2i}(e^{i\alpha} - e^{-i\alpha})$ to obtain

$$u_1(t, x) = \frac{1}{\sqrt{2\pi}} \int_{-\infty}^{\infty} F(\omega) \frac{1}{2}(e^{i\omega t} + e^{-i\omega t}) e^{i\omega x} \, d\omega$$

$$= \frac{1}{2} \frac{1}{\sqrt{2\pi}} \int_{-\infty}^{\infty} F(\omega) e^{i\omega(t+x)} \, d\omega + \frac{1}{2} \frac{1}{\sqrt{2\pi}} \int_{-\infty}^{\infty} F(\omega) e^{-i\omega(t-x)} \, d\omega$$

$$= \frac{1}{2}(\mathcal{F}^{-1}F)(x + t) + \frac{1}{2}(\mathcal{F}^{-1}F)(x - t)$$

$$= \frac{1}{2}(f(x + t) + f(x - t)).$$

For the second term we have

$$u_2(t, x) = \frac{1}{\sqrt{2\pi}} \int_{-\infty}^{\infty} G(\omega) \frac{1}{\omega} \frac{1}{2i}(e^{i\omega t} - e^{-i\omega t}) e^{i\omega x} \, d\omega$$

$$= \frac{1}{2} \frac{1}{\sqrt{2\pi}} \int_{-\infty}^{\infty} \frac{1}{i\omega} G(\omega)(e^{i\omega(x+t)} - e^{-i\omega(x-t)}) \, d\omega.$$

Now we apply (3.102) in the form

$$G(\omega) = \hat{g}(\omega) = (i\omega)\mathcal{F}\left[\int_{-\infty}^{\bullet} g(\xi) \, d\xi\right]$$

and obtain

$$u_2(t, x) = \frac{1}{2} \frac{1}{\sqrt{2\pi}} \int_{-\infty}^{\infty} \mathcal{F}\left[\int_{-\infty}^{\bullet} g(\xi) \, d\xi\right] (e^{i\omega(x+t)} - e^{-i\omega(x-t)}) \, d\omega$$

$$= \frac{1}{2} \left\{ \int_{-\infty}^{x+t} g(\xi) \, d\xi - \int_{-\infty}^{x-t} g(\xi) \, d\xi \right\} = \frac{1}{2} \int_{x-t}^{x+t} g(\xi) \, d\xi.$$

Thus we have derived d'Alembert's formula (3.7)

$$u(t, x) = \frac{1}{2}(f(x + t) + f(x - t)) + \frac{1}{2} \int_{x-t}^{x+t} g(\xi) \, d\xi.$$

We already obtained this formula in Theorem 3.2 via other methods; there we also proved uniqueness of solutions.   △

For our next example we will consider the heat equation on the real line. In Section 3.4.4 we found a solution via an invariance argument; here we will show that it can also be obtained using Fourier transforms. For this we assume that the functions appearing in the arguments are such that the rules of calculation can be applied.

Example 3.60 (IVP for the heat equation on $\mathbb{R}$) We consider the initial value problem for the heat equation on $\mathbb{R}$; here we will again consider asymptotic boundary conditions. The corresponding system reads

$$u_t = u_{xx}, \qquad x \in \mathbb{R}, \, t \geq 0, \tag{3.108a}$$

$$u(0, x) = f(x), \qquad x \in \mathbb{R}, \tag{3.108b}$$

$$u(t, x) \to 0, \qquad x \to \pm\infty. \tag{3.108c}$$

As in Example 3.59 we will consider the Fourier transform of the desired solution with respect to the space variable $x \in \mathbb{R}$, that is,

$$\mathcal{U}(t, \omega) := \mathcal{F}[u(t, \cdot)](\omega).$$

1. Transformation of the differential equation

The equation (3.108a) becomes $\mathcal{U}_t(t, \omega) = -\omega^2\,\mathcal{U}(t, \omega)$ for all $t \geq 0$, $\omega \in \mathbb{R}$.

2. Transformation of the initial conditions

The initial conditions take the form $\mathcal{U}(0, \omega) = \hat{f}(\omega)$, $\omega \in \mathbb{R}$, in phase space; hence we obtain an initial value problem for a homogeneous linear ordinary differential equation, which can be solved by standard means.

3. Solution of the initial value problem

The Fourier transform of the solution is thus $\mathcal{U}(t, \omega) = \hat{f}(\omega)e^{-\omega^2 t}$.

4. Inverse transformation

Applying the inversion formula we are led to the following formula for the solution of (3.108)

$$u(t, x) = \frac{1}{\sqrt{2\pi}} \int_{-\infty}^{\infty} \hat{f}(\omega)e^{-\omega^2 t}e^{i\omega x}\,d\omega.$$

This formula can be further simplified using properties of Fourier transforms. If, as in (3.64), we define the function $g(t, x) := \frac{1}{\sqrt{4\pi t}}e^{-x^2/4t}$, which by Exercise 3.5 satisfies $\hat{g}(t, \omega) = \mathcal{F}(g(t, \cdot))(\omega) = (2\pi)^{-1/2}\,e^{-t\omega^2}$, then by the convolution theorem we obtain

$$u(t, x) = \int_{-\infty}^{\infty} \hat{f}(\omega)\,\hat{g}(t, \omega)\,e^{i\omega x}\,d\omega = \frac{1}{\sqrt{2\pi}} \int_{-\infty}^{\infty} \mathcal{F}(g(t, \cdot) * f)(\omega)\,e^{i\omega x}\,d\omega$$

$$= (g(t, \cdot) * f)(x) = \int_{-\infty}^{\infty} g(t, x - y)\,f(y)\,dy.$$

For this reason we call the function g the *fundamental solution of the heat equation*. We will treat this example in Section 10.1 with the help of Maple®.   △

3.6.2* The Laplace transform

We have seen that in order to apply the method of Fourier transforms to solve partial differential equations it is necessary to assume suitable decay of the solutions, that is, asymptotic boundary conditions. If, however, other boundary conditions are given, then we need a different integral transform. For the sake of completeness we will give a brief introduction to the Laplace transform here, without however going into much detail. This transform is also highly significant in many parts of the theory of partial differential equations; a systematic treatment can be found in [6], for example. We denote by

$$L_{1,\mathrm{loc}}(\mathbb{R}_+, \mathbb{C}) := \left\{ f : [0, \infty) \to \mathbb{C} \text{ measurable} : \int_0^c |f(t)| \, dt < \infty \text{ for all } c > 0 \right\}$$

the *space of locally integrable functions* on $\mathbb{R}_+ := [0, \infty)$.

Definition 3.61 For a function $f \in L_{1,\mathrm{loc}}(\mathbb{R}_+, \mathbb{C})$ we set

$$\mathcal{L}f(s) = F(s) := \int_0^\infty f(t)e^{-st} \, dt = \lim_{c \to \infty} \int_0^c f(t)e^{-st} \, dt, \quad s \in \mathbb{C},$$

if the indefinite integral exists, and call the resulting function the *Laplace transform* of f. △

Theorem 3.62 (Existence of the Laplace transform) *Let* $f \in L_1(\mathbb{R}_+, \mathbb{C})$ *be* exponentially bounded, *that is, we have the bound* $|f(t)| \leq M e^{\gamma t}$, $t \geq 0$, *for some constants* $M \geq 0$ *and* $\gamma \in \mathbb{R}$. *Then* $\mathcal{L}f(s)$ *exists for all* $s \in \mathbb{C}$ *for which* $\mathrm{Re}\, s > \gamma$.

Proof See Exercise 3.11. □

We call the number γ in Theorem 3.62 an *exponential bound* for the function f.

Remark 3.63 The pair $f(t)$, $F(s) = \mathcal{L}f(s)$ is sometimes known as a *Laplace correspondence*, especially in the engineering literature, written $f(t) \circ\!\!-\!\!-\!\!\bullet F(s)$. △

For functions $f, g \in L_{1,\mathrm{loc}}(\mathbb{R}_+, \mathbb{C})$, we define their *convolution* via

$$(f * g)(t) := \int_0^t f(t - s) \, g(s) \, ds.$$

This is consistent with (3.104) when $f, g \in L_{1,\mathrm{loc}}(\mathbb{R}_+, \mathbb{C})$ and the functions are extended to $\mathbb{R}$ by 0. In any case $f * g \in L_{1,\mathrm{loc}}(\mathbb{R}_+, \mathbb{C})$; moreover, $f * g$ is exponentially bounded if f and g are. We will now summarize some essential properties and rules of calculation of Laplace transforms.

Theorem 3.64 (Properties of the Laplace transform) *Let* $f, g, f_1, \ldots, f_n \in L_{1,\mathrm{loc}}(\mathbb{R}_+, \mathbb{C})$ *all be exponentially bounded for the common exponential bound* γ. *Then we have:*
  (i) *Decay for large s:* $\lim_{\mathrm{Re}\, s \to \infty} \mathcal{L}f(s) = 0$.
  (ii) *Linearity: we have* $\mathcal{L}\left(\sum_{k=1}^n c_k f_k \right) = \sum_{k=1}^n c_k \mathcal{L}f_k$ *for any constants* $c_1, \ldots, c_n \in \mathbb{C}$.
  (iii) *If* $f \in C^n(\mathbb{R}_+, \mathbb{C})$ *is such that* $f^{(k)}$, $1 \leq k \leq n$, *is exponentially bounded with bound* γ, *then*

$$\mathcal{L}f^{(n)}(s) = s^n \mathcal{L}f(s) - \sum_{k=0}^{n-1} f^{(k)}(0) s^{n-1-k}, \quad \mathrm{Re}\, s > \gamma.$$

  (iv) *We have* $\left(\frac{d}{ds} \right)^n \mathcal{L}f(s) = (-1)^n \mathcal{L}((\cdot)^n f)(s)$, $\mathrm{Re}\, s > \gamma$.
  (v) *Translation: For all* $\alpha > 0$ *we have* $\mathcal{L}(f(\cdot - \alpha))(s) = e^{-\alpha s} \mathcal{L}f(s)$, $\mathrm{Re}\, s > \gamma$.
  (vi) *Convolution theorem: For f, g as above we have* $\mathcal{L}(f * g) = (\mathcal{L}f)(\mathcal{L}g)$.

Proof Here we will limit ourselves to proving (iii); for the other statements we refer to Exercise 3.17.
By assumption and integration by parts we have

$$\int_0^\infty f'(t)e^{-st}\,dt = f(t)e^{-st}\Big|_{t=0}^{t=\infty} + s\int_0^\infty f(t)e^{-st}\,dt = -f(0) + s\,\mathcal{L}f(s),$$

which is the statement for $n = 1$. For $n > 1$ the statement may be proved inductively.   □

As with the Fourier transform we also require an inversion formula for the Laplace transform
in order to be able to solve (partial) differential equations.

For this we will require the following terminology. We say that a function $f : [0, \ell] \to \mathbb{C}$ is
piecewise continuously differentiable (or more loosely *piecewise smooth*) if there exist $0 = t_0 < t_1 <
\cdots < t_n = \ell$ and $g_j \in C^1([t_{j-1}, t_j])$, $j = 1, \ldots, n$, such that $g_j(t) = f(t)$ for all $t \in (t_{j-1}, t_j)$,
$j = 1, \ldots, n$. A function $f : \mathbb{R}_+ \to \mathbb{C}$ is then called *piecewise continuously differentiable* (or
piecewise smooth) if $f|_{[0,\ell]}$ is piecewise continuously differentiable for every $\ell > 0$. In this case
the one-sided limits

$$f(t+) := \lim_{\varepsilon \downarrow 0} f(t + \varepsilon), \qquad f(t-) := \lim_{\varepsilon \downarrow 0} f(t - \varepsilon).$$

always exist, for every $t > 0$.

Theorem 3.65 (Inversion formula for the Laplace transformation) *Let $f : [0, \infty) \to \mathbb{C}$ be
piecewise smooth and exponentially bounded with exponential bound γ. Then for all $x > \gamma$*

$$\frac{1}{2\pi} \lim_{R \to \infty} \int_{-R}^R \mathcal{L}f(x + is)e^{(x+is)t}\,ds = \begin{cases} \frac{1}{2}\big(f(t+) + f(t-)\big), & \text{if } t > 0, \\ \frac{1}{2}f(0+), & \text{if } t = 0. \end{cases}$$

Proof For the proof we refer to the literature, for example [6].   □

The inverse transform can be calculated, for example, using the residue theorem, especially
if the functions involved are rational functions. There are also extensive tables of both Laplace
transforms and inverse Laplace transforms of functions available, which can be consulted when
solving differential equations. Maple® is another useful tool, as we will demonstrate in Section
10.1. Here we wish to describe an example which shows how one can derive a formula for solutions
of a differential equation using Laplace transforms.

Example 3.66 (The heat equation on an interval) We will consider the following initial-boundary
value problem for the heat equation on the interval $(0, 1)$,

$$\begin{aligned}
u_t - u_{xx} &= f(t, x), \quad f(t, x) := -(t^2 + x)e^{-tx}, \quad x \in (0, 1),\ t \geq 0, \\
u(0, x) &= 1, & x \in (0, 1), \qquad (3.109) \\
u(t, 0) &= a(t) := 1, \quad u(t, 1) = b(t) := e^{-t}, \quad t \geq 0.
\end{aligned}$$

The boundary conditions $a(t) := 1$ and $b(t) := e^{-t}$ are clearly exponentially bounded functions.
We will calculate the Laplace transform in the time variable t, and so we define

$$\mathcal{U}(s, x) := \mathcal{L}[u(\cdot, x)](s).$$

As when working with Fourier transforms, we proceed in four steps.

1. Transformation of the differential equation
Using the rules for calculating Laplace transforms we can transform the equation as follows:
$\mathcal{L}[u_t(\cdot, x)](s) = s\,\mathcal{U}(s, x) - u(0, x) = s\,\mathcal{U}(s, x) - 1$.

2. Transformation of the boundary conditions

Unlike the above examples involving Fourier transforms, here we need to transform not the initial condition but the boundary conditions $a(t) \equiv 1$ and $v(t) = e^{-t}$:

$$\mathcal{U}(s, 0) = \mathcal{L}[a](s) = s^{-1}, \qquad \mathcal{U}(s, 1) = \mathcal{L}[b](s) = (1 + s)^{-1}, \quad \mathrm{Re}\, s > 0.$$

This transforms the partial differential equation (3.109) into a boundary value problem for an ordinary differential equation of the form

$$-\mathcal{U}_{xx} + s\, \mathcal{U} = 1 + \mathcal{F}(s, x), \ \text{for}\, \mathrm{Re}\, s > 0, \ x \in (0, 1),$$

$$\mathcal{U}(s, 0) = s^{-1}, \qquad\qquad \mathrm{Re}\, s > 0, \qquad\qquad (3.110)$$

$$\mathcal{U}(s, 1) = (1 + s)^{-1}, \qquad \mathrm{Re}\, s > 0,$$

where $\mathcal{F}(s, x) := \mathcal{L}[f(\cdot, x)](s) = -2(s + x)^{-3} - \frac{x}{s+x}$ is the Laplace transform of the right-hand side, cf. Exercise 3.18.

3. Solution of the boundary value problem

The above homogeneous linear boundary value problem for a second-order ordinary differential equation can be solved by standard methods; the solution is $\mathcal{U}(s, x) = (s + x)^{-1}$.

4. Inverse transformation

With the help of a partial fraction expansion one can check that $u(t, x) = e^{-tx}$ is a solution of the problem (3.109), cf. Section 10.1   △

We finish by giving an example in which Fourier and Laplace transforms can be combined.

Example 3.67 (The inhomogeneous heat equation on $\mathbb{R}$) We consider the following initial value problem for the inhomogeneous heat equation

$$u_t = u_{xx} + f, \quad x \in \mathbb{R}, \ t \geq 0,$$

$$u(0, x) = u_0(x), \ x \in \mathbb{R},$$

$$u(t, x) \to 0, \qquad \text{as}\ |x| \to \infty, \ t \geq 0,$$

where $u_0 \in C_c(\mathbb{R})$ and $f \in C([0, \infty) \times \mathbb{R})$ is bounded. We begin by taking the Fourier transform in the space variable x and using the rules in Theorem 3.53 obtain the following initial value problem for $\hat{u}(t, \omega) := \mathcal{F}(u(t, \cdot))(\omega), \omega \in \mathbb{R}$:

$$\hat{u}_t(t, \omega) = -|\omega|^2\, \hat{u}(t, \omega) + \hat{f}(t, \omega), \ t \geq 0, \ \omega \in \mathbb{R},$$

$$\hat{u}(0, \omega) = \hat{u}_0(\omega), \qquad\qquad\qquad \omega \in \mathbb{R}.$$

The second step is to calculate the Laplace transform with respect to the time variable t; using the rules of calculation for Laplace's equation from Theorem 3.64 we are led to an algebraic equation for $\hat{U}(s, \omega) := \mathcal{L}(\hat{u}(\cdot, \omega))(s)$, namely $s\, \hat{U}(s, \omega) - \hat{u}_0(\omega) = -|\omega|^2\, \hat{U}(s, \omega) + \hat{F}(s, \omega)$ for all $s \in \mathbb{C}$ such that $\mathrm{Re}\, s > 0$, where $\hat{F}(s, \omega) := \mathcal{L}(\hat{f}(\cdot, \omega))(s)$. This equation is easy to solve; we obtain

$$\hat{U}(s, \omega) = \frac{1}{s + |\omega|^2}\left(\hat{u}_0(\omega) + \hat{F}(s, \omega)\right).$$

Applying the inverse Laplace transform (cf. Exercise 3.11) and using the convolution theorem (Theorem 3.64(iv)) gives

$$\hat{u}(t, \omega) = \hat{u}_0(\omega)\, e^{-t|\omega|^2} + \int_0^t e^{-|\omega|^2(t-s)} \hat{f}(s, \omega)\, ds.$$

Finally we use that (cf. Exercise 3.5)

$$g(t, x) := \frac{1}{\sqrt{4\pi t}} e^{-x^2/4t} \text{ satisfies } \hat{g}(\omega) = \frac{1}{\sqrt{2\pi}} e^{-t\omega^2}.$$

An application of the convolution theorem (Theorem 3.55) now leads to the formula

$$u(t, x) = \int_{\mathbb{R}} g(t, x - y) u_0(y) \, dy + \int_0^t \int_{\mathbb{R}} g(t - s, x - y) f(s, y) \, dy \, ds$$

for the solution, which expresses it in terms of convolutions with the source term f and the initial function u_0. $\qquad\qquad\qquad\qquad\qquad\qquad\qquad\qquad\qquad\qquad\qquad\qquad\qquad\qquad\qquad\quad$ △

3.7 Outlook

In this chapter we investigated the following three partial differential equations using elementary methods:

$$\begin{aligned} u_{xx} + u_{yy} &= 0 && \text{(Laplace's equation)} \\ u_t &= u_{xx} && \text{(heat equation)} \\ u_{tt} &= u_{xx} && \text{(wave equation)} \end{aligned}$$

As we saw when classifying equations in Chapter 2, the first is an elliptic, the second a parabolic and the third a hyperbolic equation. It should already be clear from our elementary study in this chapter that these three kinds of equations (elliptic, parabolic and hyperbolic) evince very different properties. We wish to summarize the most important of these.

Well-posedness: In all three cases we can prove existence and uniqueness of solutions, as long as the equations are paired with suitable boundary conditions (and initial conditions, as the case may be). Uniqueness (and continuous dependence on the data) for Laplace's equation is due to the elliptic maximum principle (Section 3.3.3), in the case of the heat equation it is a consequence of the parabolic maximum principle (Section 3.4.2), and for the wave equation it follows from the principle of conservation of energy (Section 3.1.2).

Smoothing: Both Laplace's equation and the heat equation exhibit a quite strong smoothing effect: the solution of the Dirichlet problem is always in $C^\infty(\Omega)$ (Theorems 3.28 and 3.30); the solution of the heat equation is in $C^\infty((0, \infty) \times \Omega)$, that is, smooth in both time and space variables for any initial values (Theorem 3.36, Corollary 3.37 and Theorem 3.41). The solutions of the wave equation, on the other hand, are no more regular than the data: for example, d'Alembert's formula from Theorem 3.2 shows that the solution is exactly as regular as the initial data.

Conservation of information: The "information" contained in the initial data is preserved in the solution of the wave equation (here we are thinking, for example, of the form of a signal); this can again be seen via d'Alembert's formula (3.7). If the initial data have compact support, then so too does the solution at every point in time. The speed of propagation of solutions of the wave equation is finite. The situation is completely different in the case of the heat equation. The explicit solution (3.79) in terms of the Gaussian kernel shows that $u(t, x) > 0$ for all $x \in \mathbb{R}^d$ and $t > 0$ as long as $u_0 \geq 0$ and $u_0 \not\equiv 0$. This means that the speed of propagation is infinite.

Asymptotic behavior as $t \to \infty$: The solutions of the heat equation in $\mathbb{R}^d$ converge to 0 as $t \to \infty$. The same is true on an interval with Dirichlet boundary conditions. The wave equation is completely different: the energy is constant in time, see Theorem 3.5.

Both the heat equation and the wave equation are evolution equations: they describe the evolution in time of a certain state. We summarize the properties of the solutions of these equations in the following table.

Table 3.1 Properties of different kinds of equations.

property	wave/transport	diffusion
well-posedness	yes	yes
maximum principle	no	yes
smoothing	no	C^∞
conservation of info	yes	no
propagation speed	$\leq C$	∞
behavior for $t \to \infty$	energy constant	$u(t, x) \to 0$

Over the course of the book we will study these three prototypical equations (Laplace, heat and wave equations) in higher dimensions. Properties of parabolic equations can generally be easily deduced from those of the corresponding elliptic equations. For this reason, our principal focus in Chapters 5, 6 and 7 is on elliptic equations. In Chapter 8 we turn to heat equations on domains. We will see that the properties listed in the above table remain valid in higher dimensions and when the equations are defined on more complicated domains.

3.8 Exercises

Exercise 3.1 Solve the following initial value problem for the *telegraph equation* using Laplace transforms

$$u_{xx} - a u_{tt} - b u_t - c u = 0, \ x \in \mathbb{R}, \ t > 0,$$

where $u(0, x) = u_t(0, x) = 0$, $u(t, 0) = g(t)$ and $\lim_{x \to \infty} u(t, x) = 0$, and under the assumption that $b^2 - 4ac = 0$. Which conditions does g need to satisfy?

Exercise 3.2 Solve using Laplace transforms

$$u_t = u_{xx}, \quad (t, x) \in [0, \infty) \times [0, \infty),$$

where $u(0, x) = 1$ for all $x > 0$, $u(t, 0) = 0$ for all $t > 0$, and $\lim_{x \to \infty} u(t, x) = 1$.

Exercise 3.3 Using Fourier transforms, derive a formula for the solution of the Dirichlet problem $u_{xx} + u_{yy} = 0$, $(x, y) \in \mathbb{R} \times [0, \infty)$, with boundary values $u(x, 0) = f(x)$, $x \in \mathbb{R}$. What conditions do we need to assume on f?

Exercise 3.4 (Abelian convergence of Fourier series)
(a) Show that the space $\mathcal{T}$ of trigonometric polynomials is dense in $C_{2\pi}$.
   Suggestion: Use the theorem of Stone–Weierstrass (Theorem A.5).
(b) Let $f \in C_{2\pi}$ have Fourier series

$$f(t) = c_0 + \sum_{k=1}^{\infty} \left(a_k \cos(kt) + b_k \sin(kt) \right).$$

In accordance with (a), let $f_n(t) = c_0^{(n)} + \sum_{k=1}^{N_n} \left(a_k^{(n)} \cos(kt) + b_k^{(n)} \sin(kt) \right)$ be such that $\| f_n - f \|_\infty \to 0$ as $n \to \infty$. Show that then $c_0^{(n)} \to c_0$, $a_k^{(n)} \to a_k$ and $b_k^{(n)} \to b_k$ as $n \to \infty$.
(c) Let $u_n(r, t) = c_0^{(n)} + \sum_{k=1}^{N_n} r^k \left(a_k^{(n)} \cos(kt) + b_k^{(n)} \sin(kt) \right)$. Show that u_n converges uniformly in $[0, 1] \times \mathbb{R}$ to a function $u \in C([0, 1] \times \mathbb{R})$.
(d) Show that $u(r, t) = c_0 + \sum_{k=1}^{\infty} r^k \left(a_k \cos(kt) + b_k \sin(kt) \right)$ for all $0 \le r < 1$.
(e) Conclude that $\lim_{r \uparrow 1} u(r, t) = f(r)$ uniformly in $t \in \mathbb{R}$. This is exactly the convergence of the Fourier series of f in the sense of Abel (Theorem 3.11).

Exercise 3.5 For the function $g_a(t) := \frac{1}{\sqrt{a\pi}} e^{-x^2/a}$, where $a \in \mathbb{R}_+$, show that $\hat{g}_a(\omega) = (2\pi)^{-1/2} e^{-a\omega^2/4}$. Use that $\int_{-\infty}^{\infty} e^{-x^2} dx = \sqrt{\pi}$.

Exercise 3.6 Prove Lemma 3.21.

Exercise 3.7 Prove Theorem 3.40.

Exercise 3.8 Prove Lemma 3.42.

Exercise 3.9 Prove Theorem 3.43.

Exercise 3.10 Prove Theorem 3.53 parts (i), (ii), (v) and (vi).

Exercise 3.11 Prove Theorem 3.62 (existence of the Laplace transform).

Exercise 3.12 Let $u_0 \in C^2(\mathbb{R})$, $u_1 \in C^1(\mathbb{R})$ be initial values for the initial value problem for the wave equation

$$\begin{aligned}
u_{tt} - c^2 u_{xx} &= 0 \quad \text{in } \mathbb{R}^+ \times \mathbb{R}, \\
u(0, \cdot) &= u_0 \quad \text{in } \mathbb{R}, \\
u_t(0, \cdot) &= u_1 \quad \text{in } \mathbb{R}.
\end{aligned}$$

Show using d'Alembert's formula that the solution $u \in C^2(\mathbb{R} \times \mathbb{R})$ at the point (t, x) only depends on the initial values $u_0(y), u_1(y)$ in the *domain of dependence* $y \in A(t, x) := [x - c|t|, \ x + c|t|]$, cf. Figure 3.1 on page 52.

Exercise 3.13 Consider the *Heaviside function* $H : \mathbb{R} \to \mathbb{R}$, defined by

$$H(x) := \begin{cases} 1, & x \geq 0, \\ 0, & \text{otherwise.} \end{cases}$$

Show that $(H * \varphi)(x) = \int_{-\infty}^{x} \varphi(s) \, ds$ for all $\varphi \in \mathcal{D}(\mathbb{R})$.

Exercise 3.14 Solve the partial differential equation

$$x^2 u_{xx} - y^2 u_{yy} + x u_x - y u_y = 0$$

in $\mathbb{R}^2$ via a suitable variable transformation. How can uniqueness of solutions be attained?

Exercise 3.15 Determine the solution $u(t, x)$ of the initial-boundary value problem

$$\begin{aligned}
u_{xx} &= 4 u_{tt}, & t &> 0, \ x \in (0, 1), \\
u(t, 0) = u(t, 1) &= 0, & t &> 0, \\
u(0, x) &= \sin(2\pi x), & x &\in [0, 1], \\
u_t(0, x) &= x(x - 1), & x &\in [0, 1].
\end{aligned}$$

Exercise 3.16 Use a Fourier series *ansatz* to solve the problem

$$\begin{aligned}
u_{tt} + a^2 u_{xxxx} &= 0, & t &> 0, \ x \in (0, \ell), \\
u(t, 0) = u(t, \ell) &= 0, & t &> 0, \\
u_{xx}(t, 0) = u_{xx}(t, \ell) &= 0, & t &> 0, \\
u(0, x) &= f(x), & x &\in (0, \ell), \\
u_t(0, x) &= g(x), & x &\in (0, \ell),
\end{aligned}$$

where f, g are odd 2ℓ-periodic functions and $a \in \mathbb{R}$.

Exercise 3.17 Prove Theorem 3.64 parts (i), (ii), (iv), (v) and (vi).

Exercise 3.18 Show that:

(a) $a(t) := 1$, $\mathcal{L}[a](s) = s^{-1}$

(b) $b(t) := e^{-t}$, $\mathcal{L}[b](s) = (1+s)^{-1}$

(c) $f(t, x) := -(t^2 + x)e^{-tx}$, $\mathcal{L}[f(\cdot, x)](s) = -2(s+x)^{-3} - \frac{x}{s+x}$

(d) $\mathcal{U}(s, x) := (s+x)^{-1}$, $\mathcal{L}^{-1}[\mathcal{U}(\cdot, x)](t) = e^{-tx}$

Exercise 3.19 (Harmonic functions do not see individual points) Let $\Omega \subset \mathbb{R}^2$ be an open set, $x_0 \in \Omega$ and $\Omega^* := \Omega \setminus \{x_0\}$. We denote by $\mathcal{H}(\Omega) := \{u \in C^2(\Omega) : \Delta u = 0\}$ the space of harmonic functions on Ω. We wish to show that the sets $\mathcal{H}(\Omega^*) \cap C(\Omega)$ and $\mathcal{H}(\Omega)$ agree. We let $B := B(x_0, R)$ be a ball in $\mathbb{R}^d$ and $B^* := B \setminus \{x_0\}$.

(a) Determine all functions $f \in C^2(0, R)$ such that $rf''(r) + f'(r) = 0$ for all $r \in (0, R)$.

(b) Let $u \in \mathcal{H}(B^*) \cap C(\overline{B})$ and $u_{|\partial B} = 0$. Show that u is invariant under rotations about x_0: more precisely, $u(x_0 + y_1) = u(x_0 + y_2)$ for all $y_1, y_2 \in \mathbb{R}^2$ such that $|y_1| = |y_2| \leq R$.

(c) Prove that if $w \in C^2(B^*)$ is invariant under rotations about x_0, then there exists a function $\tilde{w} \in C^2(0, R)$ such that $\tilde{w}(|x - x_0|) = w(x)$ for all $x \in B^*$. Moreover, this function satisfies $\Delta w(x) = \tilde{w}''(r) + \frac{1}{r}\tilde{w}'(r)$, where $r := r(x) := |x - x_0|$.

(d) Let u and v be functions in $\mathcal{H}(B^*) \cap C(\overline{B})$ such that $u_{|\partial B} = v_{|\partial B}$. Show that $u = v$.

   Suggestion: It can be shown using part (c) that $w := u - v$ vanishes identically in B.

(e) Let $u \in \mathcal{H}(B^*) \cap C(\overline{B})$. Show that $u \in \mathcal{H}(B)$. Thus Ω^* is not Dirichlet regular.

(f) Let $u \in \mathcal{H}(\Omega^*)$. Show that there is a unique extension $v \in \mathcal{H}(\Omega)$ of v to Ω if and only if the limit $\lim_{x \to x_0} u(x)$ exists.

(g) Does the assertion of part (f) also hold in dimension $d = 1$? In other words, is every function which is in $\mathcal{H}((-1, 0) \cup (0, 1)) \cap C([-1, 1])$ also in $\mathcal{H}(-1, 1)$?

Exercise 3.20 Let $f_n \in C^2([0, 1])$ and $f, g \in C([0, 1])$ be functions such that $\lim_{n \to \infty} f_n = f$ and $\lim_{n \to \infty} f_n'' = g$ uniformly in $[0, 1]$. Show that $(f_n')_{n \in \mathbb{N}}$ converges uniformly in $[0, 1]$ to a function $h \in C([0, 1])$, and deduce that $f \in C^2([0, 1])$, $f' = h$, $f'' = g$.

Exercise 3.21 Let $g \in C^1([0, b])$.

(a) Assume that $g(0) = g(b) = 0$. Show that the $2b$-periodic odd extension G of g is in $C^1(\mathbb{R})$.

(b) Assume that $g'(0) = g'(b) = 0$. Show that the $2b$-periodic even extension of g is in $C^1(\mathbb{R})$.

(c) Let $G \in C^1(\mathbb{R})$. Show that if G is odd (i.e., $G(x) = -G(-x)$ for all $x \in \mathbb{R}$) then G' is even (i.e., $G(x) = G(-x)$ for all $x \in \mathbb{R}$).

(d) Let $g \in C^2([0, b])$ satisfy $g(0) = g(b) = g''(0) = g''(b) = 0$. Show that in this case the $2b$-periodic odd extension of g is in $C^2(\mathbb{R})$.

Exercise 3.22 (Well-posedness of the IBVP for the wave equation) Background:
For $f \in C([a, b])$ let $\|f\|_{C([a,b])} := \sup_{x \in [a,b]} |f(x)|$ denote the supremum norm.
Then $C^1([a, b])$ is a Banach space for

$$\|f\|_{C^1([a,b])} := \|f\|_{C([a,b])} + \|f'\|_{C([a,b])}.$$

Similarly, $C^2([a, b])$ is a Banach space for the norm

$$\|f\|_{C^2([a,b])} := \|f\|_{C^1([a,b])} + \|f''\|_{C([a,b])}.$$

Moreover, $X := \{u_0 \in C^1([a, b]) : u_0(a) = u_0(b) = 0\}$ is a closed subspace of
$C^1([a, b])$ and $Y := \{u_1 \in C^2([a, b]) : u_1(a) = u_1(b) = u_1''(a) = u_1''(b) = 0\}$ a closed
subspace of $C^2([a, b])$. Finally, $C^2([0, T] \times [a, b])$ is a Banach space for the norm

$$\|u\|_{C^2} := \|u\|_C + \|u_t\|_C + \|u_x\|_C + \|u_{tx}\|_C + \|u_{tt}\|_C + \|u_{xx}\|_C,$$

where for $v \in C^2([0, T] \times [a, b])$, we set

$$\|v\|_C := \sup\{|v(t, x)| : t \in [0, T], x \in [a, b]\}.$$

The space

$$Z := \{u \in C^2([0, T] \times [a, b]) : u \text{ satisfies } (3.12a) \text{ and } (3.12b)\}$$

is a closed subspace of $C^2([0, T] \times [a, b])$. All these assertions are routine arguments
based on the completeness of $C([a, b])$.

The exercise consists in showing the following: Let $u_0 \in X$, $u_1 \in Y$ and let u
be the solution of (3.12). Given $n \in \mathbb{N}$, $u_0^n \in X$, $u_1^n \in Y$, let u^n be the solution of
(3.12) with u_0, u_1 replaced by u_0^n, u_1^n. Use the Closed Graph Theorem to show that
$\lim_{n \to \infty} \|u^n - u\|_{C^2} = 0$.

Chapter 4
Hilbert spaces

The goal of this chapter is to give an elementary but comprehensive introduction to the theory of Hilbert spaces, where we wish to highlight the many-faceted interplay of their geometric and analytic properties. This culminates in the theorem of Riesz–Fréchet, which describes continuous linear forms in a Hilbert space. Important for us will be that this theorem can be interpreted as an existence and uniqueness result. A generalization known as the Lax–Milgram theorem, which we will give in Section 4.5, is at the heart of the solution theory of elliptic equations which we will present in Chapters 5, 6 and 7. For the reader who is principally interested in these equations, the part of the current chapter up to and including Section 4.5 provides sufficient background.

The rest of the chapter, from Section 4.6 on, is devoted to spectral theory. The key result here, the spectral theorem of Section 4.8, asserts that, for certain operators, the ambient Hilbert space has an orthonormal basis consisting of eigenvectors of the operator. This result is the most important tool for studying evolution equations in Chapter 8. The spectral theorem will permit us to apply the method of separation of variables (more precisely, space and time variables will be separated) to far more general situations than we could manage in Chapter 2 using just Fourier series.

Chapter overview

© The Author(s), under exclusive license to Springer Nature Switzerland AG 2023
W. Arendt, K. Urban, *Partial Differential Equations*, Graduate Texts in
Mathematics 294, https://doi.org/10.1007/978-3-031-13379-4_4

By a Hilbert space we mean a vector space equipped with an inner product, such that the vector space is complete with respect to the norm induced by the inner product. We will now proceed to introduce the necessary terms step-by-step.

4.1 Inner product spaces

Since all Hilbert spaces are *inner product spaces*, sometimes also called *pre-Hilbert spaces*, we will naturally start with these. Let E be a vector space over the field $\mathbb{K} = \mathbb{R}$ or $\mathbb{K} = \mathbb{C}$. A mapping

$$(\cdot, \cdot) : E \times E \to \mathbb{K} \quad f, g \mapsto (f, g)$$

is called an *inner product* or *scalar product* if the following conditions are satisfied:
(a) $(f + g, h) = (f, h) + (g, h), \quad f, g, h \in E$;
(b) $(\lambda f, g) = \lambda(f, g), \quad f, g \in E, \lambda \in \mathbb{K}$;
(c) $(f, g) = \overline{(g, f)}, \quad f, g \in E$;
(d) $(f, f) > 0 \quad (f \neq 0), \quad f \in E$.

Notice that (c) implies that $(f, f) = \overline{(f, f)} \in \mathbb{R}$, for all $f \in E$. Thus (d) does in fact make sense when $\mathbb{K} = \mathbb{C}$. We call (c) *symmetry* and (d) *positive definiteness*. The symmetry property also implies
(a') $(f, g + h) = (f, g) + (f, h), \quad f, g, h \in E$;
(b') $(f, \lambda g) = \bar{\lambda}(f, g), \quad f, g \in E$.
Here and in what follows, $\bar{\lambda}$ denotes the complex conjugate of the number $\lambda \in \mathbb{C}$. Inner products are thus *linear* in the first variable (that is, (a) and (b) hold), while they are *antilinear* in the second (that is, (a') and (b') hold). We shall now consider a few examples.

Example 4.1 (a) Let $E = \mathbb{R}^d$, then $(x, y) := \sum_{j=1}^{d} x_j y_j = x^T y$ defines the natural inner product on $\mathbb{R}^d$.
(b) Let $E = \mathbb{C}^d$, then $(x, y) := \sum_{j=1}^{d} x_j \overline{y_j}$ is the natural inner product on $\mathbb{C}^d$.
(c) Let $a < b$ and set $C([a, b]) := \{f : [a, b] \to \mathbb{K} : f \text{ continuous}\}$ to be the space of continuous functions on $[a, b]$. Then

$$(f, g) := \int_a^b f(t)\overline{g(t)}dt$$

defines an inner product on $C([a, b])$. Observe that $C([a, b])$ is infinite dimensional, while $\mathbb{R}^d$ and $\mathbb{C}^d$ are finite dimensional.   △

We call a vector space E equipped with an inner product, or more precisely the pair $(E, (\cdot, \cdot))$, an *inner product space* (or sometimes *pre-Hilbert space*). We now wish to establish a number of geometric properties of inner products. To this end, we need to introduce a further real-valued quantity. The number $\|x\| := \sqrt{(x, x)}$ is called the *norm* of the vector $x \in E$; we also speak of the norm *induced* by the inner product $(\cdot, \cdot)$. It has the following properties:

(N1) $\|f\| = 0 \iff f = 0, \quad f \in E$;

(N2) $\|\lambda f\| = |\lambda| \|f\|, \quad \lambda \in \mathbb{K}, f \in E$;

(N3) $\|f + g\| \le \|f\| + \|g\|, \quad f, g \in E$.

Properties (N1) and (N2) follow directly from the axioms of inner products. The *triangle inequality* (N3), however, is not obvious, and we will need a couple of preliminaries in order to prove it. To this end, we first consider the instructive example $E = \mathbb{R}^2$ with the natural inner product

$$(x, y) := x_1 y_1 + x_2 y_2 \quad \text{for } x = \begin{pmatrix} x_1 \\ x_2 \end{pmatrix}, \ y = \begin{pmatrix} y_1 \\ y_2 \end{pmatrix} \in \mathbb{R}^2.$$

By Pythagoras' theorem in $\mathbb{R}^2$, the norm $\|x\| = \sqrt{x_1^2 + x_2^2}$ of $x \in \mathbb{R}^2$ is the distance between x and the origin. If $x, y \in E$ are on the unit sphere, that is, $\|x\| = \|y\| = 1$, then we can write them as $x = (\cos\theta_1, \sin\theta_1)$, $y = (\cos\theta_2, \sin\theta_2)$. Hence the inner product

$$(x, y) = \cos\theta_1 \cdot \cos\theta_2 + \sin\theta_1 \cdot \sin\theta_2 = \cos(\theta_1 - \theta_2)$$

is the cosine of the angle between the two vectors x and y. In particular, x and y are orthogonal (that is, perpendicular) to each other if and only if $(x, y) = 0$. If $x, y \in \mathbb{R}^2$, $x \ne 0$, $y \ne 0$, are now two vectors with arbitrary lengths, then the points $\frac{x}{\|x\|}$, $\frac{y}{\|y\|}$ lie on the unit sphere. Since

$$(x, y) = \|x\| \cdot \|y\| \cdot \left(\frac{x}{\|x\|}, \frac{y}{\|y\|} \right),$$

we see that in this more general case, x and y are still orthogonal to each other if and only if $(x, y) = 0$, cf. Figure 4.1.

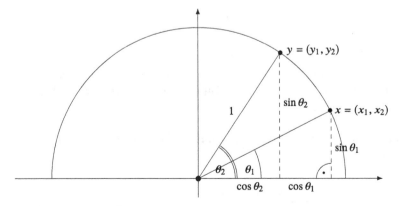

Fig. 4.1 Two points $x, y \in \mathbb{R}^2$ on the unit sphere and the angle between them.

This leads us to the following definition. Let E be a pre-Hilbert space over $\mathbb{K}$. Two vectors $x, y \in E$ are said to be *orthogonal* (and we write $x \perp y$) if $(x, y) = 0$. Using the axioms for inner products, we can obtain the following characterization of orthogonality via the norm. We emphasize that here $(E, (\cdot, \cdot))$ is an arbitrary inner product space.

Theorem 4.2 (Pythagoras' theorem for inner product spaces) *Let E be an inner product space. If $f, g \in E$ are orthogonal, then $\|f + g\|^2 = \|f\|^2 + \|g\|^2$. In the case where $\mathbb{K} = \mathbb{R}$ also the converse holds.*

Proof We have, using that $(f, g) = \overline{(g, f)}$,

$$\|f + g\|^2 = (f + g, f + g) = (f, f) + (f, g) + (g, f) + (g, g)$$
$$= (f, f) + (g, g) + 2\operatorname{Re}(f, g) = \|f\|^2 + \|g\|^2 + 2\operatorname{Re}(f, g).$$

This shows that $\operatorname{Re}(f, g) = 0$ if and only if $\|f + g\|^2 = \|f\|^2 + \|g\|^2$. $\square$

For an arbitrary subset $M \subset E$ of E, we denote by

$$M^\perp := \{f \in E : f \perp g \text{ for all } g \in M\}$$

the *orthogonal complement* of M. One can see easily that $M^\perp$ is always a (linear) subspace of E, for any set $M \subset E$. Moreover, due to the definiteness of the inner product, we always have $M \cap M^\perp = \{0\}$. We next consider straight lines in inner product spaces E.

Theorem 4.3 *Let $g \in E$ with $\|g\| = 1$. Consider the straight line $G := \{\lambda g : \lambda \in \mathbb{K}\}$ passing through 0 and g. Then, for any $f \in E$, we have $f - (f, g)g \in G^\perp$.*

Proof Using the properties of the inner product, we obtain

$$\left(f - (f, g)g, \lambda g\right) = (f, \lambda g) - (f, g)(g, \lambda g) = \bar{\lambda}(f, g) - (f, g)\bar{\lambda}(g, g) = 0$$

for all $f \in E$, since by assumption $(g, g) = 1$. $\square$

We call the vector $Pf := (f, g)\, g$ the *orthogonal projection* of f onto the line G, since $Pf \in G$ and $f - Pf \in G^\perp$, the latter holding by Theorem 4.3. We can now prove a central inequality.

Theorem 4.4 (Cauchy–Schwarz inequality[1]) *We have $|(f, g)| \le \|f\| \, \|g\|$ for all $f, g \in E$.*

---

[1] Sometimes the name Cauchy–Bunyakovsky–Schwarz inequality is used, after the Russian mathematician Viktor Yakovlevich Bunyakovsky (1804–1889).

Proof We proceed in two steps:

Step 1: $\|g\| = 1$. Then $f = Pf + f - Pf = (f, g)g + (f - Pf)$. Since $(f, g)g \perp (f - Pf)$, by Pythagoras' theorem and (N2),

$$\|f\|^2 = \|(f, g)g\|^2 + \|f - Pf\|^2 = |(f, g)|^2 + \|f - Pf\|^2 \geq |(f, g)|^2$$

due to the positivity of the norm.

Step 2: Now let $g \in E$ be arbitrary. The inequality is trivial if $g = 0$, so we may suppose that $g \neq 0$. We may then apply Step 1 to $g_1 := \frac{1}{\|g\|} g$ to obtain the claim. $\quad\square$

The triangle inequality (that is, (N3)) is a direct consequence of the Cauchy–Schwarz inequality. Let $f, g \in E$; then

$$\|f + g\|^2 = (f + g, f + g) = (f, f) + (f, g) + (g, f) + (g, g)$$
$$= \|f\|^2 + (f, g) + \overline{(f, g)} + \|g\|^2 = \|f\|^2 + 2\,\mathrm{Re}\,(f, g) + \|g\|^2$$
$$\leq \|f\|^2 + 2|(f, g)| + \|g\|^2 \leq \|f\|^2 + 2\|f\|\,\|g\| + \|g\|^2$$
$$= (\|f\| + \|g\|)^2,$$

which establishes (N3). Thus $\|\cdot\|$ is a norm on E. This permits us to define convergence: if $f_n, f \in E$, then we say that the sequence $(f_n)_{n \in \mathbb{N}}$ *converges to* f (and write $\lim_{n \to \infty} f_n = f$ or $f_n \to f$), if $\lim_{n \to \infty} \|f_n - f\| = 0$.

Example 4.5 Let $-\infty < a < b < \infty$ and let $E = C([a, b])$ be equipped with the corresponding norm:

$$(f, g) := \int_a^b f(t)\,\overline{g(t)}\,dt, \qquad \|f\| = \left(\int_a^b |f(t)|^2\,dt \right)^{\frac{1}{2}} = (f, f)^{\frac{1}{2}}$$

We also call this norm the 2-norm and write $\|f\|_{L_2(a,b)} := \|f\|$. If there is no danger of confusion, then we also write $\|\cdot\|_{L_2}$. For $f_n, f \in E$, we thus have $f_n \to f$ in E if and only if

$$\left(\int_a^b |f_n(t) - f(t)|^2\,dt \right)^{\frac{1}{2}} \to 0,$$

that is, the 2-norm describes *convergence in quadratic mean*. $\quad\triangle$

Remark 4.6 In addition to the 2-norm, we will also sometimes consider the supremum norm (∞-norm), which is given by

$$\|f\|_\infty := \sup_{t \in [a,b]} |f(t)| \equiv \|f\|_{C([a,b])}.$$

This leads to the notion of *uniform convergence*: $f_n \to f$ with respect to $\|\cdot\|_\infty$ if and only if for all $\varepsilon > 0$ there is some $n_0 \in \mathbb{N}$ such that $|f_n(t) - f(t)| < \varepsilon$ for all $n \geq n_0$ and all $t \in [a, b]$. $\quad\triangle$

4.2 Orthonormal bases

The goal of this section is to introduce orthonormal bases. To this end let $(E, (\cdot, \cdot))$ be an inner product space over $\mathbb{K} = \mathbb{C}$ or $\mathbb{R}$. Let $I = \mathbb{N}, \mathbb{Z}$, or $\{1, \ldots, d\}$. A family $\{e_n : n \in I\}$ in E is called *orthonormal* if

$$(e_n, e_m) = \begin{cases} 1 \text{ if } n = m, \\ 0 \text{ if } n \neq m. \end{cases}$$

Let us first consider the finite case $I = \{1, \ldots, d\}$. Let

$$F := \text{span}\{e_1, \ldots, e_d\} = \left\{ \sum_{n=1}^{d} \lambda_n e_n : \lambda_1, \ldots, \lambda_d \in \mathbb{K} \right\}$$

be the span (also known as the linear span or the linear hull) of $\{e_1, \ldots, e_d\}$. For $f = \sum_{n=1}^{d} \lambda_n e_n \in F$ and $m \in \{1, \ldots, d\}$, we have

$$(f, e_m) = \left(\sum_{n=1}^{d} \lambda_n e_n, e_m \right) = \lambda_m.$$

Hence $f = \sum_{n=1}^{d}(f, e_n)e_n$ is the unique representation of $f \in F$ as a linear combination of the vectors $\{e_1, \ldots, e_d\}$. If we consider infinite index sets I, then we need infinite series. To this end, we will again consider convergence with respect to the norm associated with the inner product $\|f\| := (f, f)^{\frac{1}{2}}$.

Definition 4.7 A subset M of E is called *dense* in E if for every $g \in E$ there is a sequence $(f_n)_{n \in \mathbb{N}}$ in M such that $\lim_{n \to \infty} f_n = g$ in $(E, (\cdot, \cdot))$. The set M is called *total* if

$$\text{span}M := \left\{ \sum_{j=1}^{m} \lambda_j g_j : m \in \mathbb{N}, \lambda_1, \ldots, \lambda_m \in \mathbb{K}, g_1, \ldots, g_m \in M \right\}$$

is dense in E. △

Definition 4.8 Let $I = \mathbb{N}$ or $\mathbb{Z}$. A family $\{e_n : n \in I\}$ is called an *orthonormal basis* of E if it is orthonormal and total in E. △

We next define when a series converges in E. Let $f_n \in E$, $n \in I$, and let $f \in E$. We say that the series $\sum_{n \in I} f_n$ *converges to* f and write

$$\sum_{n \in I} f_n = f$$

if $\lim_{n \to \infty} \sum_{m=1}^{n} f_m = f$ for $I = \mathbb{N}$; and if $f = \lim_{n \to \infty} \sum_{m=-n}^{n} f_n$ in the case $I = \mathbb{Z}$. An (infinite) series $\sum_{n \in I} f_n$ is said to be *convergent* if there exists a vector $f \in E$ such that $\sum_{n \in I} f_n = f$.

Theorem 4.9 (Orthonormal expansion) *Let $I = \mathbb{N}$ or $\mathbb{Z}$ and let $\{e_n : n \in I\}$ be an orthonormal basis of E, as well as $f \in E$. Then:*
(a) Orthonormal expansion: $\sum_{n \in I}(f, e_n)e_n = f$;
(b) Parseval's identity: $\sum_{n \in I}|(f, e_n)|^2 = \|f\|^2$.

Proof Let $E_n := \text{span}\{e_m : m \in I, |m| \leq n\}$. For $f \in E$ define

$$P_n f := \sum_{m \in I, |m| \leq n}(f, e_m)\, e_m.$$

We need to show that $\lim_{n \to \infty} P_n f = f$.

1. We claim that $P_n f \in E_n$ and $(f - P_n f) \in E_n^\perp$.
   The first statement is clear. To show the second, choose $|k| \leq n$. Then

$$(f - P_n f, e_k) = (f, e_k) - \sum_{|m| \leq n}(f, e_m)(e_m, e_k) = (f, e_k) - (f, e_k) = 0.$$

   This shows that $f - P_n f$ is orthogonal to all e_k for which $|k| \leq n$; and thus $(f - P_n f) \in E_n^\perp$ by definition.

2. By Pythagoras' theorem, $\|f\|^2 = \|P_n f\|^2 + \|f - P_n f\|^2$. In particular, $\|P_n f\| \leq \|f\|$ for all $f \in E$ and all $n \in \mathbb{N}$. If now $f \in E_n$ for some $n \in \mathbb{N}$, then $P_m f = f$ for all $m \geq n$. Hence $\lim_{m \to \infty} P_m f = f$ trivially in this case.

Now let $f \in E$ and $\varepsilon > 0$ be arbitrary. By assumption $\{e_n : n \in I\}$ is an orthonormal basis of E, hence there exist an $n \in \mathbb{N}$ and a vector $g_n \in E_n$ such that $\|f - g_n\| < \varepsilon/2$. Thus $P_m g_n = g_n$ for all $m \geq n$ and so, for all $m \geq n$,

$$\|f - P_m f\| = \|f - g_n + P_m g_n - P_m f\| \leq \|f - g_n\| + \|P_m(g_n - f)\|$$
$$\leq 2\|f - g_n\| < \varepsilon.$$

Since $\varepsilon > 0$ was arbitrary, we have shown that $\lim_{m \to \infty} P_m f = f$ for all $f \in E$. This proves statement (a). But it also establishes (b), since, by Pythagoras' theorem,

$$\|P_n f\|^2 = \left\| \sum_{|k| \leq n}(f, e_k)\, e_k \right\|^2 = \sum_{|k| \leq n}\|(f, e_k)\, e_k\|^2 = \sum_{|k| \leq n}|(f, e_k)|^2.$$

By (a), $\|f\|^2 = \lim_{n \to \infty} \|P_n f\|^2$ for all $f \in E$. This proves (b).   □

We already know one important example of an orthonormal basis. Consider

$$C_{2\pi} := \{f : \mathbb{R} \to \mathbb{C} \text{ continuous, } 2\pi\text{-periodic}\}$$

from Chapter 3, which is an inner product space with respect to

$$(f, g) := \frac{1}{2\pi} \int_{-\pi}^{\pi} f(t)\, \overline{g(t)}\, dt.$$

Let $e_k(t) := e^{ikt}$, $t \in \mathbb{R}$. Then $\{e_k : k \in \mathbb{Z}\}$ is an orthonormal basis of $C_{2\pi}$. One can establish the orthonormality simply by integrating. But we also saw that the space $\mathcal{J} = \mathrm{span}\{e_k : k \in \mathbb{Z}\}$ is dense in $C_{2\pi}$ with respect to the supremum norm $\| \cdot \|_\infty$ (Corollary 3.17). But since

$$\|f\|_{L_2} = \sqrt{(f, f)} = \left(\frac{1}{2\pi} \int_{-\pi}^{\pi} |f(t)|^2 \, dt \right)^{\frac{1}{2}} \leq \|f\|_\infty,$$

it follows immediately that $\mathcal{J}$ is also dense in $C_{2\pi}$ with respect to the norm $\| \cdot \|_{L_2}$ of the inner product space. We have thus shown that $\{e_k : k \in \mathbb{Z}\}$ is an orthonormal basis of $C_{2\pi}$. Applying Theorem 4.9, we obtain the following classical result about the Fourier series expansion of a function $f \in C_{2\pi}$.

Theorem 4.10 *Let $f \in C_{2\pi}$, let $c_k = (f, e_k) = \frac{1}{2\pi} \int_{-\pi}^{\pi} f(t) e^{-ikt} \, dt$ be the kth Fourier coefficient of f, for $k \in \mathbb{Z}$, and let $s_n := \sum_{k=-n}^{n} c_k e_k$ be the nth Fourier polynomial of f. Then $\lim_{n \to \infty} \|s_n - f\|_{L_2} = 0$.*

The theorem says that the *Fourier series* of a function $f \in C_{2\pi}$ converges to f in quadratic mean. In this case we also write

$$f = \sum_{k=-\infty}^{\infty} c_k e_k \quad \text{with respect to } \| \cdot \|_{L_2}.$$

While in general the series does not converge pointwise, we nevertheless have two positive results: the quadratic convergence of Theorem 4.10 and convergence in the sense of Abel from Theorem 3.11 (page 57).

The question arises as to whether every inner product space has an orthonormal basis; this is indeed the case if the space is "not too big". We wish to describe this more precisely. We say that an inner product space E is *separable* if there is a (countable) sequence $(x_n)_{n \in \mathbb{N}}$ in E such that the set $\{x_n : n \in \mathbb{N}\}$ is dense in E. This is always the case if there is a total sequence $(u_n)_{n \in \mathbb{N}}$ in E, since we may then set (if, say, $\mathbb{K} = \mathbb{R}$)

$$F_n := \left\{ \sum_{k=1}^{n} q_k u_k : n \in \mathbb{N}, \, q_k \in \mathbb{Q} \right\};$$

if the sequence $(u_n)_{n \in \mathbb{N}}$ is total, then $F := \bigcup_{n=1}^{\infty} F_n$ is dense in E. The set F is countable as the countable union of countable sets. If $\mathbb{K} = \mathbb{C}$ then we may replace $\mathbb{Q}$ by $\mathbb{Q} + i\mathbb{Q}$.

Theorem 4.11 (Orthonormalization) *Every separable inner product space E admits an orthonormal basis.*

Proof Let $(u_n)_{n \in \mathbb{N}}$ be a total sequence in E. We will assume that the space E is infinite dimensional. We can then assume that the set $\{u_1, \ldots, u_n\}$ is linearly independent for each $n \in \mathbb{N}$ (otherwise we simply remove unnecessary vectors). Using the Gram–Schmidt orthogonalization process, we will now construct an orthonormal sequence $(e_n)_{n \in \mathbb{N}}$ such that

$$\text{span}\{u_1, \ldots, u_n\} = \text{span}\{e_1, \ldots, e_n\} \tag{4.1}$$

for all $n \in \mathbb{N}$.

To this end, first set $e_1 := \|u_1\|^{-1} u_1$ and for $n \in \mathbb{N}$ assume that $e_1, \ldots, e_n$ have already been found. Now define

$$w_{n+1} := u_{n+1} - \sum_{j=1}^{n} (u_{n+1}, e_j) e_j.$$

Then $(w_{n+1}, e_i) = 0$ for all $i = 1, \ldots, n$. Moreover, $w_{n+1} \neq 0$ due to our assumption that the u_k are linearly independent. Now set $e_{n+1} := \|w_{n+1}\|^{-1} w_{n+1}$. Then (4.1) is satisfied for $n + 1$ and the construction has been completed. The sequence $(e_n)_{n \in \mathbb{N}}$ is orthonormal, and total in E by (4.1). Hence $(e_n)_{n \in \mathbb{N}}$ is an orthonormal basis.

If $\dim E < \infty$, then we choose an arbitrary basis $\{u_1, \ldots, u_n\}$ of E and orthonormalize it in accordance with the above scheme. □

The above proof clearly yields a constructive procedure to orthonormalize n linearly independent vectors. The assumption of separability is satisfied in practically all applications. In particular all inner product spaces which are useful for partial differential equations, are separable.

4.3 Completeness

Let E be an inner product space over $\mathbb{K} = \mathbb{R}$ or $\mathbb{C}$, with inner product $(\cdot, \cdot)$. We consider convergence in E with respect to the induced norm $\|u\| := \sqrt{(u, u)}$, as well as Cauchy sequences with respect to this norm. The space E is said to be *complete* if for every Cauchy sequence $(u_n)_{n \in \mathbb{N}}$ in E there is a vector $u \in E$ such that $\lim_{n \to \infty} u_n = u$. A *Hilbert space* is a complete inner product space. In what follows, we will always denote Hilbert spaces by H or V. We start by giving an example.

Example 4.12 (a) Each finite-dimensional inner product space is complete (see [2, §2.6]).
(b) Let $I = \mathbb{N}$ or $\mathbb{Z}$ and

$$\ell_2(I) := \left\{ (x_n)_{n \in I} \subset \mathbb{K} : \|(x_n)_{n \in I}\|^2_{\ell_2(I)} := \sum_{n \in I} |x_n|^2 < \infty \right\}.$$

This space is a Hilbert space with respect to the inner product

$$(x, y) := \sum_{n \in I} x_n \overline{y_n}$$

and the norm $\|x\|_{\ell_2(I)} := \sqrt{(x, x)}$ induced by it. Moreover, if we define $e_n := (0, 0, \ldots, 1, 0, \ldots)$ to be the sequence with exactly one 1, in position n, and 0 elsewhere, then $\{e_n : n \in I\}$ is an orthonormal basis. We refer to Exercise 4.1 for the proof.   △

Under the assumption of completeness, we can say more about the representation of vectors with respect to an orthonormal basis.

Theorem 4.13 *Let H be a separable Hilbert space and let $\{e_n : n \in I\}$ be an orthonormal basis, where $I = \mathbb{N}$ or $\mathbb{Z}$. Let $x = (x_n)_{n \in I} \in \ell_2(I)$. Then the series $u = \sum_{n \in I} x_n e_n$ converges in H and $(u, e_n) = x_n$ for all $n \in I$.*

Proof We will give the proof for $I = \mathbb{N}$; for $I = \mathbb{Z}$ one argues analogously. For every $\varepsilon > 0$ there is a number $n_0 \in \mathbb{N}$ such that

$$\sum_{n \geq n_0} |x_n|^2 < \varepsilon.$$

Set $u_n := \sum_{k=1}^n x_k e_k$. Then, by Pythagoras' theorem, for $n, m \geq n_0$ we have

$$\|u_n - u_m\|^2 = \left\| \sum_{k=m+1}^n x_k e_k \right\|^2 = \sum_{k=m+1}^n |x_k|^2 < \varepsilon,$$

where we have assumed that $n > m$. This shows that $(u_n)_{n \in \mathbb{N}}$ is a Cauchy sequence; hence it has a limit $u := \lim_{n \in \mathbb{N}} u_n$, and

$$(u, e_m) = \lim_{n \to \infty} (u_n, e_m) = x_m,$$

which proves the claim.   □

In the other direction, we saw in Theorem 4.9 that

$$\sum_{n \in I} |(u, e_n)|^2 = \|u\|_H^2 < \infty$$

for every $u \in H$. It follows that the mapping $U : H \to \ell_2(I)$ given by $u \mapsto \left((u, e_n)\right)_{n \in I}$ is linear and bijective with $(Uu, Uv) = (u, v)$ for all $u, v \in H$. Such a mapping is called *unitary*. This means that the two Hilbert spaces H and $\ell_2(I)$ are essentially the same object: the mapping U preserves addition and scalar multiplication (that is, the vector space structure of the spaces) as well as the inner product. Nevertheless, Hilbert spaces often have a very different form from $\ell_2(I)$; two such examples follow. Further examples are Sobolev spaces, to be introduced in Chapter 5.

Example 4.14 The space

$$L_2((0, 2\pi), \mathbb{C}) := \left\{ f : (0, 2\pi) \to \mathbb{C} : f \text{ is measurable and } \int_0^{2\pi} |f(t)|^2 dt < \infty \right\}$$

is a complex Hilbert space with respect to the inner product

$$(f, g) := \frac{1}{2\pi} \int_0^{2\pi} f(t) \overline{g(t)} \, dt,$$

where the integration is with respect to Lebesgue measure. The functions $\{e_k : k \in \mathbb{Z}\}$, $e_k(t) = e^{ikt}$, form an orthonormal basis. Hence the mapping defined as $U: L_2((0, 2\pi), \mathbb{C}) \to \ell_2(\mathbb{Z})$, $f \mapsto ((f, e_k))_{k \in \mathbb{Z}}$ is unitary.   △

Proof We have already seen that $\{e_k : k \in \mathbb{Z}\}$ is total in $C_{2\pi}$. We can identify $C_{2\pi}$ with the space $F := \{f : [0, 2\pi] \to \mathbb{C} : f \text{ is continuous and } f(0) = f(2\pi)\}$. From measure theory we know that F is dense in $L_2((0, 2\pi), \mathbb{C})$, see e.g. [54, Ch. III]. Hence $\{e_k : k \in \mathbb{Z}\}$ is total in $L_2((0, 2\pi), \mathbb{C})$.   □

The following Hilbert space will be used frequently throughout the book.

Example 4.15 Let $\Omega \subset \mathbb{R}^d$ be an open set and $\mathbb{K} = \mathbb{R}$ or $\mathbb{C}$. We set

$$L_2(\Omega, \mathbb{K}) := \left\{ f : \Omega \to \mathbb{K} : f \text{ is measurable and } \int_\Omega |f(x)|^2 dx < \infty \right\}.$$

Then the space $L_2(\Omega, \mathbb{K})$ is a Hilbert space over $\mathbb{K}$ with respect to the inner product

$$(f, g) := (f, g)_{L_2} := \int_\Omega f(x) \overline{g(x)} \, dx.$$

Here dx is the Lebesgue measure on Ω. We set $L_2(\Omega) := L_2(\Omega, \mathbb{R})$.   △

4.4 Orthogonal projections

Let H be a Hilbert space over $\mathbb{K} = \mathbb{R}$ or $\mathbb{C}$. If M is a subset of H, then, analogously to Section 4.1, we may define the *orthogonal complement* $M^\perp$ of M by

$$M^\perp := \{u \in H : (u, v) = 0 \text{ for all } v \in M\}.$$

The set $M^\perp$ is a closed subspace of H. In what follows, we will need the *parallelogram identity*

$$\|u + v\|^2 + \|u - v\|^2 = 2\|u\|^2 + 2\|v\|^2 \tag{4.2}$$

for all $u, v \in H$, which is an immediate consequence of the definition of the norm via the inner product. The following theorem shows that every closed subspace of H, that is, every subspace of H which is also closed as a set, has an orthogonal complement.

Theorem 4.16 (Projection theorem) *Let F be a closed subspace of a Hilbert space H. Then $H = F \oplus F^\perp$. In words, every $u \in H$ admits a unique decomposition $u = v + w$, where $v \in F$ and $w \in F^\perp$.*

The mapping P which maps $u \in H$ to the vector $v \in F$ for which $u - v \in F^\perp$, is called the *orthogonal projection* of H onto F, cf. Figure 4.2.

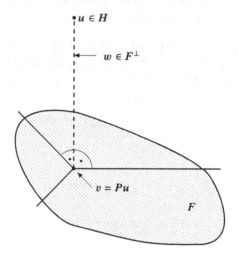

Fig. 4.2 The orthogonal projection of $u \in H$ onto the subspace $F \subset H$.

Proof By the definiteness property (d) of the inner product, we have $F \cap F^\perp = \{0\}$; the essential part of the projection theorem is the assertion that $F + F^\perp = H$. Given $u \in H$, we need to find $u_0 \in F$ such that $u - u_0 \in F^\perp$. Consider the distance d of u to F,

$$d := \mathrm{dist}(u, F) := \inf\{\|u - v\| : v \in F\}.$$

We will show that this distance is realized, that is, that there exists a vector $v_0 \in F$ for which $d = \|u - v_0\|$. For such a v_0, we necessarily have $u - v_0 \in F^\perp$, that is, $(u - v_0, z) = 0$ for all $z \in F$. In particular, there is at most one minimizer. To show this, choose an arbitrary $z \in F$ with $\|z\| = 1$ (recall F is a vector space). Since also $v_0 \in F$, we have

$$
\begin{aligned}
\|u - v_0\|^2 = d^2 &\leq \|u - v_0 - (u - v_0, z)z\|^2 \\
&= \|u - v_0\|^2 - (u - v_0, (u - v_0, z)z) \\
&\quad - (u - v_0, z)(z, u - v_0) + |(u - v_0, z)|^2 \\
&= \|u - v_0\|^2 - \overline{(u - v_0, z)}(u - v_0, z) \\
&\quad - (u - v_0, z)\overline{(u - v_0, z)} + |(u - v_0, z)|^2
\end{aligned}
$$

$$= \|u - v_0\|^2 - |(u - v_0, z)|^2,$$

whence $(u - v_0, z) = 0$.

It remains to show that the distance is in fact realized. Choose a minimizing sequence, that is, choose $(v_n)_{n \in \mathbb{N}} \subset F$ such that $\|u - v_n\| \to d$ as $n \to \infty$. The parallelogram identity (4.2) shows that

$$
\begin{aligned}
\|v_n - v_m\|^2 &= \|v_n - u - (v_m - u)\|^2 \\
&= 2\|v_n - u\|^2 + 2\|v_m - u\|^2 - \|(v_n - u) + (v_m - u)\|^2 \\
&= 2\|v_n - u\|^2 + 2\|v_m - u\|^2 - 4\left\|\frac{v_n + v_m}{2} - u\right\|^2 \\
&\leq 2\|v_n - u\|^2 + 2\|v_m - u\|^2 - 4d^2,
\end{aligned}
$$

since $\frac{v_n + v_m}{2} \in F$ and hence $\|\frac{v_n + v_m}{2} - u\|^2 \geq d^2$. This establishes that $(v_n)_{n \in \mathbb{N}}$ is a Cauchy sequence, and hence it has a limit $v_0 := \lim_{n \to \infty} v_n$. Since F is closed, $v_0 \in F$; moreover, $\|u - v_0\| = \lim_{n \to \infty} \|u - v_n\| = d$. $\qquad\square$

Remark 4.17 The above proof also shows that for any given $u \in H$, the orthogonal projection Pu is the unique element of F with minimal distance to u, that is, $\|u - Pu\| = \min\{\|u - f\| : f \in F\}$. In other words, the orthogonal projection is the (unique) best approximation. $\qquad\triangle$

As a consequence, we note the following very useful criterion for proving the density of a subspace in a given Hilbert space.

Corollary 4.18 *A subspace F of a Hilbert space H is dense in H if and only if $F^\perp = \{0\}$.*

Proof 1. Let F be dense in H and suppose $x \in F^\perp$. Then by density of F there is a sequence $(x_n)_{n \in \mathbb{N}} \subset F$ such that $\lim_{n \to \infty} x_n = x$. But since $x \in F^\perp$, we have $\|x\|^2 = (x, x) = \lim_{n \to \infty}(x, x_n) = 0$ and hence $x = 0$.

2. The closure $\overline{F}$ of F is a closed subspace of H. Suppose that $\overline{F} \neq H$, then by the projection theorem, $\overline{F}^\perp \neq \{0\}$. But since $(\overline{F})^\perp = F^\perp$, we also have $F^\perp \neq \{0\}$, which completes the proof. $\qquad\square$

In particular, we obtain a simple criterion for a sequence to be an orthonormal basis of a complete space.

Corollary 4.19 *Let $\{e_n : n \in I\}$ be an orthonormal family in a Hilbert space H, where $I = \mathbb{N}$ or $\mathbb{Z}$. Then $\{e_n : n \in I\}$ is an orthonormal basis of H if and only if, for all $v \in H$, the condition $(e_n, v) = 0$ for all $n \in I$ implies that $v = 0$.*

Proof By definition, the orthonormal family $\{e_n : n \in I\}$ is an orthonormal basis if and only if $\mathrm{span}\{e_n : n \in I\}$ is dense in H. By Corollary 4.18, this is equivalent to $(\mathrm{span}\{e_n : n \in I\})^\perp = \{0\}$. The claim now follows from the easily verified statement that for every subset M of H, $M^\perp = (\mathrm{span}\,M)^\perp$. $\qquad\square$

4.5 Linear and bilinear forms

In this section we will introduce two efficient tools for the mathematical analysis of partial differential equations: the theorem of Riesz–Fréchet and the Lax–Milgram theorem. To simplify matters, and since we will almost exclusively be interested in real partial differential equations, we will always take our underlying field to be $\mathbb{K} = \mathbb{R}$, even though the results continue to hold with minimal modifications for $\mathbb{K} = \mathbb{C}$.

Let H be a real Hilbert space. A *linear form*, also known as a *(linear) functional*, is a linear mapping $\varphi : H \to \mathbb{R}$. It can be shown that it is continuous if and only if it is bounded in the sense that there is a constant $c \geq 0$ such that

$$|\varphi(v)| \leq c\|v\| \quad \text{for all } v \in H. \tag{4.3}$$

In this case we set

$$\|\varphi\| := \inf\{c \geq 0 : |\varphi(v)| \leq c\|v\| \text{ for all } v \in H\};$$

then clearly $|\varphi(v)| \leq \|\varphi\| \, \|v\|$ for all $v \in H$. We denote the set of all continuous linear forms on H by H'. This set is a vector space with respect to pointwise addition and scalar multiplication: for example, if $\varphi, \psi \in H'$, then $\varphi + \psi$ is the mapping from H to $\mathbb{R}$ defined by $(\varphi + \psi)(v) = \varphi(v) + \psi(v)$, and so on. We refer to H' as the *dual space* of H.

Example 4.20 Let $u \in H$ be fixed. Then $\varphi(v) := (u, v)$, $v \in H$, defines a continuous linear form φ on H, and moreover $\|\varphi\| = \|u\|$.   △

Proof The linearity of φ follows from the bilinearity of the inner product. The Cauchy–Schwarz inequality $|\varphi(v)| = |(u, v)| \leq \|u\| \, \|v\|$, $v \in H$, implies that φ is continuous and $\|\varphi\| \leq \|u\|$. But $\|u\|^2 = (u, u) = \varphi(u)$ implies that $\|\varphi\| \geq \|u\|$.   □

The following theorem shows that every continuous linear form on H can be represented by taking the inner product with a given element of H, as in Example 4.20.

Theorem 4.21 (Riesz–Fréchet) *Let $\varphi : H \to \mathbb{R}$ be a continuous linear form. Then there exists a unique $u \in H$ such that $\varphi(v) = (u, v)$ for all $v \in H$.*

Proof We establish existence and uniqueness of such a $u \in H$ separately.

Uniqueness: This is a consequence of the positive definiteness of the inner product: let $u_1, u_2 \in H$ be such that $(u_1, v) = (u_2, v)$ for all $v \in H$. With the choice $v = u_1 - u_2$, this means that $\|u_1 - u_2\|^2 = (u_1 - u_2, v) = 0$, whence $u_1 = u_2$.

Existence: If $\varphi = 0$ (that is, $\varphi(v) = 0$ for all $v \in H$), then we choose $u = 0$. We may thus assume that $\varphi \neq 0$. Then its kernel $\ker \varphi := \{v \in H : \varphi(v) = 0\}$ is not equal to H. Since $\ker \varphi$ is a closed subspace of H, it follows from the projection theorem that $(\ker \varphi)^\perp \neq \{0\}$. We may thus choose $u_1 \in (\ker \varphi)^\perp$ with $u_1 \neq 0$; without loss of generality, we suppose that $\varphi(u_1) = 1$ (otherwise we multiply u_1 by a suitable scalar). For $v \in H$ arbitrary, by definition of $(\ker \varphi)^\perp$ we have $v - \varphi(v)u_1 \in \ker \varphi$. Hence $v - \varphi(v)u_1$ is orthogonal to u_1, that is, $(v - \varphi(v)u_1, u_1) = 0$. It follows that $(v, u_1) = \varphi(v)\|u_1\|^2$. If we choose $u := \|u_1\|^{-2}u_1$, then we conclude that $\varphi(v) = \|u_1\|^{-2}(v, u_1) = (v, u) = (u, v)$ for all $v \in H$.   □

Remark 4.22 This theorem is also sometimes known as the *Riesz representation theorem*; the unique element $u = u(\varphi)$ is sometimes called the *Riesz representative* of φ. There are however other representation theorems named after Riesz; see Theorem 7.10. To avoid ambiguity, we will always use the name *theorem of Riesz–Fréchet* for Theorem 4.21 (which is also historically justified).   △

Next we wish to generalize the theorem of Riesz–Fréchet to non-symmetric bilinear forms; this will greatly expand the number of partial differential equations to which the theorem can be applied to establish existence and uniqueness of solutions. For the generalization, we need to replace the inner product by a more general bilinear form. We will usually denote the domain of definition of such a bilinear form by V; in what follows, we will thus take V to be a real Hilbert space. A *bilinear form* is thus a mapping $a : V \times V \to \mathbb{R}$ such that

$$a(\alpha u_1 + \beta u_2, v) = \alpha a(u_1, v) + \beta a(u_2, v),$$
$$a(v, \alpha u_1 + \beta u_2) = \alpha a(v, u_1) + \beta a(v, u_2)$$

for all $u_1, u_2, v \in V$ and all $\alpha, \beta \in \mathbb{R}$. We thus demand that the mappings $a(\cdot, v)$ and $a(v, \cdot) : V \to \mathbb{R}$ be linear for each $v \in V$. A bilinear form is said to be *continuous* if $\lim_{n\to\infty} a(u_n, v_n) = a(u, v)$ whenever $\lim_{n\to\infty} u_n = u$ and $\lim_{n\to\infty} v_n = v$. One sees, much as in Theorem A.1 (see Exercise 4.4), that $a(\cdot, \cdot)$ is continuous if and only if there exists a constant $C \geq 0$ such that

$$|a(u, v)| \leq C\|u\|\,\|v\| \quad \text{for all } u, v \in V. \tag{4.4}$$

We shall use the abbreviation

$$a(u) := a(u, u), \quad u \in V. \tag{4.5}$$

A bilinear form is called *coercive* if there exists an $\alpha > 0$ such that

$$a(u) \geq \alpha \|u\|^2 \quad \text{for all } u \in V, \tag{4.6}$$

and *symmetric* if

$$a(u, v) = a(v, u) \quad \text{for all } u, v \in V.$$

The following theorem will play a central role in the study of partial differential equations in the sequel.

Theorem 4.23 (Lax–Milgram theorem) *Let $a : V \times V \to \mathbb{R}$ be a continuous and coercive bilinear form, and let $\varphi \in V'$. Then there is a unique $u \in V$ such that*

$$a(u, v) = \varphi(v) \quad \text{for all } v \in V. \tag{4.7}$$

Proof Once again we treat existence and uniqueness separately.

Uniqueness: This is a consequence of the coercivity. Let $u_1, u_2 \in V$ be such that $a(u_1, v) = a(u_2, v)$ for all $v \in V$. Then $a(u_1 - u_2, v) = 0$ for all $v \in V$. If we choose $v = u_1 - u_2$, then we arrive at $a(u_1 - u_2) = 0$. Since a is coercive, it follows that $u_1 - u_2 = 0$.

Existence: Let $u \in V$ be fixed. Then $g(v) := a(u, v)$, $v \in V$ defines a continuous linear form on V. By the theorem of Riesz–Fréchet (Theorem 4.21) there is exactly one $Tu \in V$ such that $a(u, v) = (Tu, v)$ for all $v \in V$. By the uniqueness of Tu and the bilinearity of $a(\cdot, \cdot)$, the mapping $T : V \to V$ is linear. Since $a(\cdot, \cdot)$ is continuous, it follows that $|(Tu, v)| = |a(u, v)| \le C\|u\| \|v\|$ for all $v \in V$, where C is as in (4.4). Hence $\|Tu\| \le C\|u\|$ (see Theorem A.1). The operator $T : V \to V$ is thus continuous, with $\|T\| \le C$.

The coercivity of $a(\cdot, \cdot)$ now implies that $(Tu, u) \ge \alpha\|u\|^2$ for all $u \in V$, for some $\alpha > 0$. The Cauchy–Schwarz inequality implies that

$$\alpha\|u\|^2 \le a(u) = (Tu, u) \le \|Tu\| \|u\|$$

and hence

$$\alpha\|u\| \le \|Tu\| \text{ for all } u \in V. \tag{4.8}$$

We now wish to show that T is surjective. To this end, we consider the *range* of T, defined as

$$T(V) := \{Tu : u \in V\}. \tag{4.9}$$

Since T is linear, $T(V)$ is a subspace of V. We claim that $T(V)$ is closed. Given $u_n \in V$, $n \in \mathbb{N}$, assume that $v := \lim_{n\to\infty} Tu_n$ exists; we have to show that $v \in T(V)$. By (4.8) and the linearity of T, we have $\alpha\|u_n - u_m\| \le \|Tu_n - Tu_m\|$. It follows that $(u_n)_{n\in\mathbb{N}}$ is a Cauchy sequence in the Hilbert space V, and hence the limit $u := \lim_{n\to\infty} u_n$ exists. Moreover,

$$Tu = \lim_{n\to\infty} Tu_n = v,$$

so that $v \in T(V)$. Now by the projection theorem (Theorem 4.16), it follows that $T(V) \oplus (T(V))^\perp = V$. Suppose that $u \in (T(V))^\perp$, so that $(Tv, u) = 0$ for all $v \in V$ and in particular for $v = u$. Then $\alpha\|u\|^2 \le (Tu, u) = 0$, whence $u = 0$. We have shown that $T(V)^\perp = \{0\}$ and hence that $T(V) = V$, that is, T is surjective.

Now let $\varphi \in V'$. By the theorem of Riesz–Fréchet (Theorem 4.21) there is a $w \in V$ such that $\varphi(v) = (w, v)$ for all $v \in V$. Since T is surjective, there exists $u \in V$ such that $Tu = w$. Hence, by definition of T, $a(u, v) = (Tu, v) = (w, v) = \varphi(v)$ for all $v \in V$. $\qquad\qquad\square$

If the bilinear form in the Lax–Milgram theorem is also symmetric, then (4.7) can be solved as a minimization problem. In such cases we speak of a *variational problem*.

Theorem 4.24 (Variational Lax–Milgram) *Let $a : V \times V \to \mathbb{R}$ be a continuous, coercive and symmetric bilinear form, and let $\varphi \in V'$, $u \in V$. Then the following are equivalent:*
(i) We have $a(u, v) = \varphi(v)$ for all $v \in V$.
(ii) For the functional $J(v) := \frac{1}{2}a(v) - \varphi(v)$, we have $J(u) \leq J(v)$ for all $v \in V$.

Since the vector u is unique by Theorem 4.23, we may write $u = \arg\min_{v \in V} J(v)$.

Proof Statement (ii) is equivalent to $J(u) \leq J(u + w)$ for all $w \in V$ (just choose $v = u + w \in V$). Since

$$J(u + w) = \frac{1}{2}a(u + w) - \varphi(u + w)$$
$$= \frac{1}{2}a(u) + a(u, w) + \frac{1}{2}a(w) - \varphi(u) - \varphi(w),$$

is (ii) thus equivalent to

$$0 \leq a(u, w) + \frac{1}{2}a(w) - \varphi(w) \tag{4.10}$$

for all $w \in V$. By replacing w by tw for $t > 0$ in (4.10) we obtain

$$0 \leq t\,a(u, w) + \frac{1}{2}t^2 a(w) - t\varphi(w).$$

Dividing by t yields $0 \leq a(u, w) + \frac{1}{2}ta(w) - \varphi(w)$ for all $t > 0$ and $w \in V$. If we now let t tend to 0, then we arrive at

$$0 \leq a(u, w) - \varphi(w) \text{ for all } w \in V.$$

If we now replace w by $-w$, then we obtain the reverse inequality $0 \geq a(u, w) - \varphi(w)$ for all $w \in V$. Thus $a(u, w) = \varphi(w)$ for all $w \in V$. We have shown that (ii) implies statement (i). Conversely, inequality (4.10) follows from (i) since $a(w) \geq 0$ for all $w \in V$. But (4.10) is equivalent to (ii), as we have seen. □

The inner product on V is an important special case of a continuous, coercive and symmetric bilinear form on V. As such, the Lax–Milgram theorem contains the theorem of Riesz–Fréchet as a special case. Theorem 4.24 is, among other things, the basis for a number of efficient numerical methods for solving linear systems of equations with symmetric positive definite matrices such as the conjugate gradient method.

Remark 4.25 The coercivity of a and the Lax–Milgram theorem also immediately yield continuous dependence on the data. Indeed, let $u \neq 0$ be a solution of (4.7), then $\alpha\|u\|^2 \leq a(u, u) = \varphi(u) \leq \|\varphi\| \|u\|$, that is,

$$\|u\| \leq \frac{1}{\alpha} \|\varphi\|,$$

and this estimate also holds trivially for $u = 0$. For this reason, when it comes to the well-posedness of variational problems, we will restrict ourselves to investigating existence and uniqueness of solutions. $\triangle$

We have already seen that the Cauchy–Schwarz inequality is an essential property of the inner product. We will now give a second, less geometric proof of this inequality in a general situation, which will be of use to us later.

Theorem 4.26 (Generalized Cauchy–Schwarz inequality) *Let* $a : V \times V \to \mathbb{R}$ *be bilinear and symmetric, and suppose* $a(u) \geq 0$ *for all* $u \in V$. *Then for all* $u, v \in H$ *we have*

$$|a(u, v)| \leq a(u)^{\frac{1}{2}} a(v)^{\frac{1}{2}}.$$

Proof Let $u, v \in H$, then $0 \leq a(u - tv) = a(u) - 2ta(u, v) + t^2 a(v)$ for all $t \in \mathbb{R}$. If $a(v) = 0$, then it follows that $2ta(u, v) \leq a(u)$ for all $t \in \mathbb{R}$ and so $a(u, v) = 0$. Hence this claim is true in this case. If $a(v) \neq 0$, then we make the special choice $t = \frac{a(u,v)}{a(v)}$ and obtain

$$0 \leq a(u) - 2\frac{a(u, v)^2}{a(v)} + \frac{a(u, v)^2}{a(v)} = a(u) - \frac{a(u, v)^2}{a(v)}.$$

Hence $a(u, v)^2 \leq a(u) a(v)$. $\square$

4.5.1* Extensions and generalizations

So far, we have seen that certain variational problems of the form (4.7), associated with a continuous and coercive bilinear form $a : V \times V \to \mathbb{R}$, are well-posed. The converse of this statement is, however, not true; that is, there are well-posed variational problems with more general bilinear forms, as we will now see.

Let V and W be two real Hilbert spaces with norms $\| \cdot \|_V$ and $\| \cdot \|_W$, respectively. We will call V the *trial space* and W the *test space*. Analogously to what we did above, we will call $b : V \times W \to \mathbb{R}$ a *bilinear form* if

$$b(\alpha v_1 + \beta v_2, w) = \alpha b(v_1, w) + \beta b(v_2, w), \quad b(v, \alpha w_1 + \beta w_2) = \alpha b(v, w_1) + \beta b(v, w_2),$$

for all $v_1, v_2, v \in V$, $w_1, w_2, w \in W$ and $\alpha, \beta \in \mathbb{R}$. Analogously to (4.4), we will say the bilinear form b is *continuous* (or *bounded*) if there exists a $C > 0$ such that

$$|b(v, w)| \le C \|v\|_V \|w\|_W \quad \text{for all } v \in V \text{ and } w \in W.$$

Obviously, coercivity in the sense of (4.6) can have no direct equivalent here if V and W are different. However, we can give a more general condition which is actually weaker than coercivity in the case $V = W$. We will say that the bilinear form b satisfies an *inf-sup condition* if there exists a constant $\beta > 0$ (called an *inf-sup constant*) such that

$$\sup_{w \in W} \frac{b(v, w)}{\|w\|_W} \ge \beta \|v\|_V \quad \text{for all } v \in V. \tag{4.11}$$

Obviously, if $V = W$, then every coercive bilinear form satisfies (4.11) with $\alpha = \beta$.

Theorem 4.27 (Banach–Nečas) *Let $b : V \times W \to \mathbb{R}$ be a continuous bilinear form. Then the following statements are equivalent:*
(i) For each $\varphi \in W'$ there exists a unique $u \in V$ such that

$$b(u, w) = \varphi(w) \text{ for all } w \in W. \tag{4.12}$$

(ii) (a) The condition (4.11) holds, and

$$\text{(b) for all } 0 \ne w \in W \text{ there exists } v \in V \text{ with } b(v, w) \ne 0. \tag{4.13}$$

Proof We first prove that (ii) implies (i). Analogously to the proof of Theorem 4.23 we obtain a continuous linear mapping $T : V \to W$ such that $(Tv, w)_W = b(v, w)$ and

$$\|Tv\|_W \le C \|v\|_V \tag{4.14}$$

for all $v \in V$. The inf-sup condition (4.11) implies that

$$\beta \|v\|_V \le \sup_{w \in W} \frac{b(v, w)}{\|w\|_W} = \sup_{w \in W} \frac{(Tv, w)_W}{\|w\|_W} = \|Tv\|_W \quad \text{for all } v \in V,$$

whence T is injective and has closed image $T(V) = \{Tv : v \in V\}$ (cf. the proof of Theorem 4.23). To show that $T(V) = W$ we apply Corollary 4.18. Let $w_0 \in (T(V))^\perp$, then $b(v, w_0) = (Tv, w_0) = 0$ for all $v \in V$. By (4.13) we have that $w_0 = 0$; from Corollary 4.18 we obtain that $T(V) = W$. Hence T is surjective and (i) is proved. Conversely, suppose that (i) holds. Then T is invertible and T^{-1} is continuous, that is, there exists some $\beta > 0$ such that $\|T^{-1}w\|_V \le \beta^{-1} \|w\|_W$ for all $w \in W$. It follows that, for all $v \in V$,

$$\sup_{w \in W} \frac{b(v, w)}{\|w\|_W} = \sup_{w \in W} \frac{(Tv, w)_W}{\|w\|_W} = \|Tv\|_W \ge \beta \|v\|_V,$$

that is, (4.11) holds. Since T is, by assumption, an isomorphism, for each $0 \ne w \in W$ there is a $0 \ne v \in V$ such that $Tv = w$ and hence $b(v, w) = (Tv, w)_W = (w, w)_W = \|w\|_W^2 \ne 0$. Thus (4.13) holds. $\qquad\square$

Remark 4.28 (a) The above theorem establishes that the well-posedness of variational problems for continuous bilinear forms is *equivalent* to the conditions (4.11) and (4.13). In this sense, the Lax–Milgram theorem may be regarded as a special case of the theorem of Banach–Nečas.
(b) The name "inf-sup condition" comes from the following equivalent form of (4.11): there exists a $\beta > 0$ such that

$$\inf_{v \in V} \sup_{w \in W} \frac{b(v, w)}{\|w\|_W \|v\|_V} \ge \beta. \tag{4.11'}$$

(c) Let $u \in V$ be the unique solution of (4.12); then the inf-sup condition implies that $\beta \|u\|_V \le \sup_{w \in W} \frac{b(u,w)}{\|w\|_W} = \sup_{w \in W} \frac{\varphi(w)}{\|w\|_W} = \|\varphi\|_{W'}$, that is, it implies the continuous dependence on the data.

(d) Let $v \in V$. Then by the theorem of Riesz–Fréchet (Theorem 4.21) there exists a unique $s_v \in W$ such that $(s_v, w)_W = b(v, w)$ for all $w \in W$, where we have written $(\cdot, \cdot)_W$ for the inner product of the space W. Then

$$\|s_v\|_W = \sup_{w \in W} \frac{(s_v, w)_W}{\|w\|_W} = \sup_{w \in W} \frac{b(v, w)_W}{\|w\|_W},$$

that is, $s_v = \arg\sup_{w \in W} \frac{b(v,w)_W}{\|w\|_W}$. Hence we also call this s_v the *supremizer* of v with respect to b. It plays an important role in the numerics of variational problems (see also Remark 8.39).

(e) The above proof also shows that if the inf-sup condition (4.11) holds, then the mapping T is surjective if and only if (4.13) holds.   △

We now wish to introduce a second generalization of the Lax–Milgram theorem in the form of Theorem 4.24. The following theorem due to J.-L. Lions may be considered "dual" to the theorem of Banach–Nečas, Theorem 4.27. It has the advantage that one of the two spaces need not be complete.

Theorem 4.29 (J.-L. Lions) *Let V be a Hilbert space and let W be an inner product space. Let $b : V \times W \to \mathbb{R}$ be bilinear, and assume $b(\cdot, w) \in V'$ for all $w \in W$. Then the following statements are equivalent:*

(i) For each $\varphi \in W'$ there exists a $u \in V$ such that $b(u, w) = \varphi(w)$ for all $w \in W$.

(ii) There exists a $\beta > 0$ such that

$$\sup_{v \in V} \frac{|b(v, w)|}{\|v\|_V} \ge \beta \|w\|_W \tag{4.15}$$

for all $w \in W$.

Proof We first show that (ii) implies (i). Let $w \in W$. Since $b(\cdot, w) \in V'$, the theorem of Riesz–Fréchet (Theorem 4.21) yields the existence of a unique $Tw \in V$ such that $b(v, w) = (v, Tw)_V$ for all $v \in V$. Hence $T : W \to V$ is linear, and the assumption (ii) says exactly that $\|Tw\|_V \ge \beta \|w\|_W$ for all $w \in W$. This means that T is injective and the inverse mapping $T^{-1} : T(W) \to W$ satisfies the estimate $\|T^{-1}v\|_W \le \frac{1}{\beta} \|v\|_V$ for all $v \in T(W)$, the image of W under T. By Theorem A.2, T^{-1} has a unique continuous extension $R : \overline{T(W)} \to \tilde{W}$, where $\tilde{W}$ denotes the completion of W (see Exercise 4.4).

Let $u \in W$. Then the desired relation $\varphi(w) = b(u, w) = (u, Tw)_V$ for all $w \in W$ holds if and only if $\varphi(T^{-1}v) = (u, v)_V$ for all $v \in T(W)$. This, in turn, is equivalent to $\varphi(Rv) = (u, v)_V$ for all $v \in \overline{T(W)}$. Since $\varphi \circ R \in (\overline{T(W)})'$, the theorem of Riesz–Fréchet yields a unique $u \in \overline{T(W)}$ for which this relation is satisfied.

We now show that (i) implies (ii). For this direction we need the *uniform boundedness principle* in the following form: if $w_n \in W$, $n \in \mathbb{N}$ such that $\sup_{n \in \mathbb{N}} |f(w_n)| < \infty$ for all $f \in W'$, then $\sup_{n \in \mathbb{N}} \|w_n\|_W < \infty$ (see, for example, [2, Thm. 7.2]).

Suppose now that (i) holds and assume that (ii) is false. Then there exist $w_n \in W$ such that $\|w_n\|_W = 1$ but $|b(\cdot, w_n)\|_{V'} < \frac{1}{n}$. Let $f \in W'$. By (i) there exists a $u_f \in V$ such that $f(w_n) = b(u_f, w_n)$, and hence $|f(n w_n)| = n |b(u_f, w_n)| < n \|u_f\|_V \frac{1}{n} = \|u_f\|_V$ for all $n \in \mathbb{N}$. This means that $\sup_{n \in \mathbb{N}} |f(n w_n)| < \infty$ for all $f \in W'$ and so, by the uniform boundedness principle, the sequence $(n w_n)_{n \in \mathbb{N}}$ is bounded, a contradiction to $\|w_n\|_W = 1$.   □

Remark 4.30 (a) In general, for each $\varphi \in W$ there can be multiple $u \in V$ which satisfy (i). Uniqueness is equivalent to b *separating* the space (that is, for each $0 \neq v \in V$ there is a

$w \in W$ such that $b(v, w) \neq 0$; or in other words, W is not in the orthogonal complement of V with respect to b). This is exactly dual to the separation property (4.13) which appears in the theorem of Banach–Nečas. The condition (4.15) means that

$$\inf_{w \in W} \sup_{v \in V} \frac{|b(v, w)|}{\|v\|_V \|w\|_W} \geq \beta > 0,$$

and so it is dual to (4.11).

(b) In the theorem of Lions we only require continuity in the first variable: $b(\cdot, w) \in V'$ for all $w \in W$; the roles of V and W with respect to inf and sup are swapped.

(c) If $W \hookrightarrow V$, that is, if $W \subset V$ and there exists a constant $c > 0$ such that $\|w\|_V \leq c \|w\|_W$ for all $w \in W$, then condition (ii) in Theorem 4.29 is satisfied if there exists an $\alpha > 0$ such that $b(w, w) \geq \alpha \|w\|_W^2$ for all $w \in W$, that is, (ii) is satisfied if b is coercive on W.

Proof. Let $w \in W$ and set $v := \frac{1}{c\|w\|_W} w$. Then $\|v\|_V \leq c \|v\|_W = 1$, and so $\sup_{v \in V} \frac{|b(w,v)|}{\|v\|_V} \geq \frac{1}{c\|w\|_W} b(w, w) \geq \frac{1}{c\|w\|_W} \alpha \|w\|_W^2 = \frac{\alpha}{c} \|w\|_W$. △

4.6 Weak convergence

One of the most important results of elementary analysis states that in any finite-dimensional Hilbert space, every bounded sequence has a convergent subsequence. In infinite-dimensional spaces, however, this statement is not true (see Example 4.32 below). Fortunately, it continues to hold if we use a weaker notion of convergence. This is the content of Theorem 4.35, the main theorem of this section. Here we will always take H to be a real Hilbert space.

Definition 4.31 (Weak convergence) Let $(u_n)_{n \in \mathbb{N}}$ be a sequence in H.

(a) Let $u \in H$ be given. We say that $(u_n)_{n \in \mathbb{N}}$ *converges weakly to u*, and write

$$\mathrm{w} - \lim_{n \to \infty} u_n = u \text{ or } u_n \rightharpoonup u, \quad n \to \infty,$$

if $\lim_{n \to \infty} (u_n, v) = (u, v)$ for all $v \in H$.

(b) We say that the sequence $(u_n)_{n \in \mathbb{N}}$ is *weakly convergent* if there exists some $u \in H$ such that $\mathrm{w} - \lim_{n \to \infty} u_n = u$. △

We first observe that weak limits are always unique: if for a further $w \in H$,

$$w = \mathrm{w} - \lim_{n \to \infty} u_n = u,$$

then $(u, v) = (w, v)$ for all $v \in H$. If we choose $v = u - w$, then we see that $\|u - w\|^2 = (u - w, v) = 0$ and so $u - w = 0$. It is immediate that every convergent sequence also converges weakly, and to the same limit. The converse, however, is false in general, as the following example shows.

Example 4.32 Let $\{e_n : n \in \mathbb{N}\}$ be an orthonormal basis of H; then

$$\sum_{n=1}^{\infty} |(v, e_n)|^2 = \|v\|^2 \text{ for all } v \in H.$$

Since the terms of a convergent series tend to zero, it follows from this identity that $e_n \to 0$ as $n \to \infty$. But since $\|e_n - e_m\|^2 = \|e_n\|^2 + \|e_m\|^2 = 2$ for $n \neq m$ by the orthogonality of the sequence, $(e_n)_{n \in \mathbb{N}}$ cannot have a convergent subsequence, even though it is weakly convergent.   △

Convergence in norm does however follow from weak convergence under an additional condition; we have the following useful criterion:

Theorem 4.33 *Let* $(u_n)_{n \in \mathbb{N}}$ *be a sequence in H and* $u \in H$. *Then the following statements are equivalent:*
(i) $\lim_{n \to \infty} u_n = u$.
(ii) $u_n \rightharpoonup u$ *and* $\|u_n\| \to \|u\|$ *as* $n \to \infty$.

Proof We will only show that (ii) implies (i), since the other implication is obvious. We thus suppose that $u_n \rightharpoonup u$ and $\|u_n\| \to \|u\|$ as $n \to \infty$. Then

$$\|u_n - u\|^2 = (u_n - u, u_n - u) = \|u_n\|^2 - (u, u_n) - (u_n, u) + \|u\|^2 \to 0,$$

that is, (i) holds.   □

The following lemma is often useful when one wishes to prove weak convergence.

Lemma 4.34 *Let* $(u_n)_{n \in \mathbb{N}}$ *be a bounded sequence in H, and assume that there exists a total subset M of H, such that* $\big((u_n, w)\big)_{n \in \mathbb{N}}$ *converges for all* $w \in M$. *Then* $(u_n)_{n \in \mathbb{N}}$ *is weakly convergent.*

Proof It follows directly from our assumptions that

$$\varphi_0(w) := \lim_{n \to \infty} (u_n, w)$$

exists for all $w \in F := \operatorname{span} M$. Let $c \geq 0$ be such that $\|u_n\| \leq c$ for all $n \in \mathbb{N}$; then $|\varphi_0(w)| \leq c\|w\|$ for all $w \in F$. The mapping $\varphi_0 : F \to \mathbb{R}$ is thus linear and continuous. Since F is dense in H, φ_0 has an extension $\varphi \in H'$ (see Theorem A.2). By the theorem of Riesz–Fréchet (Theorem 4.21) there exists a $u \in H$ such that $\varphi(v) = (u, v)$ for all $v \in H$. If $v \in F$, then

$$(u, v) = \varphi(v) = \varphi_0(v) = \lim_{n \to \infty} (u_n, v).$$

Now let $v \in H$ and $\varepsilon > 0$ be arbitrary; then there exists $w \in F$ such that $\|v - w\| \leq \varepsilon$. Since $\lim_{n \to \infty} (u_n, w) = (u, w)$ it follows that

$$\begin{aligned}
\limsup_{n \to \infty} |(u_n, v) - (u, v)| &= \limsup_{n \to \infty} |(u_n - u, v - w) + (u_n - u, w)| \\
&\leq \limsup_{n \to \infty} |(u_n - u, v - w)| \\
&\leq \lim_{n \to \infty} \|u_n - u\| \, \|v - w\| \\
&\leq (c + \|u\|)\varepsilon.
\end{aligned}$$

Since $\varepsilon > 0$ was arbitrary, $\lim_{n\to\infty} |(u_n, v) - (u, v)| = 0$, that is,

$$\lim_{n\to\infty} (u_n, v) = (u, v),$$

which proves the assertion. □

The proof of the following theorem rests on a *diagonal argument*, an important and frequently used method of proof in analysis.

Theorem 4.35 *Let H be a real, separable Hilbert space. Then every bounded sequence in H admits a weakly convergent subsequence.*

Proof We start with three preliminary remarks, to which we will refer during the later course of the proof:
1. Every subsequence of a convergent sequence is itself convergent; moreover, it converges to the same limit as the original sequence.
2. We may always alter finitely many terms in a sequence without affecting its limit.
3. Every bounded sequence in $\mathbb{R}$ admits a convergent subsequence.

Now let $u_n \in H$ be such that $\|u_n\| \le c$ for all $n \in \mathbb{N}$. We wish to find a weakly convergent subsequence of $(u_n)_{n\in\mathbb{N}}$. Let $\{w_p : p \in \mathbb{N}\}$ be a total sequence in H. By Lemma 4.34, it is sufficient to find a subsequence $(u_{n_k})_{k\in\mathbb{N}}$ of $(u_n)_{n\in\mathbb{N}}$ such that

$$\lim_{k\to\infty} (u_{n_k}, w_p)$$

exists for each $p \in \mathbb{N}$. Now since $|(u_n, w_p)| \le c\|w_p\|$ for all $n \in \mathbb{N}$, the sequence $\big((u_n, w_p)\big)_{n\in\mathbb{N}}$ admits a convergent subsequence, for each $p \in \mathbb{N}$.

We will construct a subsequence which works for all $p \in \mathbb{N}$ simultaneously. Let $p = 1$, then there is a subsequence $(u_{n(1,k)})_{k\in\mathbb{N}}$ of $(u_n)_{n\in\mathbb{N}}$ for which $(u_{n(1,k)}, w_1)$ converges as $k \to \infty$. Here $n(1, k) \in \mathbb{N}$ is an index such that $n(1, k) < n(1, k + 1)$ for all $k \in \mathbb{N}$. Now there is a subsequence $(u_{n(2,k)})_{k\in\mathbb{N}}$ of $(u_{n(1,k)})_{k\in\mathbb{N}}$ such that

$$\lim_{k\to\infty} (u_{n(2,k)}, w_2)$$

exists. We proceed in this fashion, finding indices $n(\ell, k) \in \mathbb{N}$ for $\ell, k \in \mathbb{N}$ such that
(a) $n(\ell, k) < n(\ell, k + 1)$;
(b) $(n(\ell + 1, k))_{k\in\mathbb{N}}$ is a subsequence of $(n(\ell, k))_{k\in\mathbb{N}}$, that is, $n(\ell + 1, k) \in \{n(\ell, m) : m \in \mathbb{N}\}$ for all $k \in \mathbb{N}$; in particular
(c) $n(\ell + 1, k) \ge n(\ell, k)$ for all $k \in \mathbb{N}$;
(d) $\lim_{k\to\infty}(u_{n(p,k)}, w_p) =: c_p$ exists for all $p \in \mathbb{N}$. □

Set $n_k := n(k, k)$, then by (a) and (c) we have $n_k < n(k, k+1) \le n(k+1, k+1) = n_{k+1}$. Now let $p \in \mathbb{N}$. Since also $n_k \in \{n(p, m) : m \in \mathbb{N}\}$ for all $k \ge p$ by (b), it finally follows from (d) that $\lim_{k\to\infty}(u_{n_k}, w_p) = c_p$. □

Theorem 4.35 remains true if H is not separable (since each sequence is contained in a separable, closed subspace). We shall finish by mentioning the following theorem, whose proof we will however omit.

Theorem 4.36 *Every weakly convergent sequence is bounded.*

Proof The proof is based on the Baire category theorem and for it we refer to books on functional analysis, such as [2, Remark 8.3(5)].   □

4.7 Continuous and compact operators

A linear mapping T is continuous if and only if the image of the unit sphere under T is bounded (see Appendix A.1). If this image is even relatively compact (that is, its closure is compact), then we call T *compact*. We have thus made the following definition:

Definition 4.37 Let E and F be normed spaces. A linear mapping $T : E \to F$ is called *compact* if the following condition is satisfied: for any bounded sequence $(u_n)_{n \in \mathbb{N}} \subset E$ there exists a subsequence $(u_{n_k})_{k \in \mathbb{N}}$ such that $(Tu_{n_k})_{k \in \mathbb{N}}$ converges in F.   △

Here we wish to consider such mappings in the special situation of Hilbert spaces, where we have both convergence in norm and weak convergence. We will assume throughout that H_1 and H_2 are real or complex separable Hilbert spaces.

Theorem 4.38 *Let $T : H_1 \to H_2$ be linear and continuous. Then T is weakly continuous, that is, $u_n \rightharpoonup u$ in H_1 implies $Tu_n \rightharpoonup Tu$ in H_2.*

Proof Let $w \in H_2$, then $\varphi(v) := (Tv, w)_{H_2}$ defines a continuous linear form on H_1. By the theorem of Riesz–Fréchet, there is thus a unique $T^*w \in H_1$ such that $(Tv, w)_{H_2} = (v, T^*w)_{H_1}$ for all $v \in H_1$. If $u_n \rightharpoonup u$ in H_1, then

$$(Tu_n, w)_{H_2} = (u_n, T^*w)_{H_1} \to (u, T^*w)_{H_1} = (Tu, w)_{H_2}.$$

Since $w \in H_2$ was arbitrary, it follows that $Tu_n \rightharpoonup Tu$ in H_2.   □

The converse of Theorem 4.38 is also true: every weakly continuous operator is also continuous; see Exercise 4.2. Now we can describe compact operators as being exactly those operators which "improve" convergence in the following sense.

Theorem 4.39 *Let $T : H_1 \to H_2$ be linear. Then the following statements are equivalent:*

(i) T is compact.
(ii) $u_n \rightharpoonup u$ in H_1 implies $Tu_n \to Tu$ in H_2.

For the proof we will use the following lemma which, despite looking like splitting hairs, turns out to be extremely useful.

Lemma 4.40 *Let $(u_n)_{n \in \mathbb{N}}$ be a sequence in a normed space E and let $u \in E$. Then the sequence converges to u if and only if every subsequence of $(u_n)_{n \in \mathbb{N}}$ itself has a subsequence which converges to u.*

Proof If the sequence does not converge to u, then there exists an $\varepsilon > 0$ and a subsequence $(u_{n_k})_{k \in \mathbb{N}}$ such that $\|u_{n_k} - u\| \geq \varepsilon$ for all $k \in \mathbb{N}$. Hence no subsequence of this subsequence can converge to u. The converse implication is clear.   □

Proof (of Theorem 4.39) That (i) implies (ii) is seen as follows. Assume that $u_n \rightharpoonup u$ in H_1, then $(u_n)_{n \in \mathbb{N}}$ is bounded by Theorem 4.36. By assumption there is a subsequence $(u_{n_k})_{k \in \mathbb{N}}$ such that the limit (in norm) $v := \lim_{k \to \infty} T u_{n_k}$ exists. Since T is also weakly continuous, $v = Tu$. Lemma 4.40 now implies that $\lim_{n \to \infty} T u_n = Tu$.

Now assume (ii). Let $(u_n)_{n \in \mathbb{N}}$ be a bounded sequence in H_1; then by Theorem 4.35 it has a weakly convergence subsequence $(u_{n_k})_{k \in \mathbb{N}}$. By assumption, the sequence $(T u_{n_k})_{k \in \mathbb{N}}$ converges in norm, meaning that T is compact.   □

4.8 The spectral theorem

We will often be able to write partial differential equations as operator equations of the form $Au = f$, where A is a linear operator on some Hilbert space H and $f \in H$ is given. The desired unknown is $u \in H$. Most commonly $H = L_2(\Omega)$ for an open set $\Omega \subset \mathbb{R}^d$, and A is a differential operator. In this case, A cannot be defined for all functions in H, but only for those in a certain subspace $D(A)$ of H, the domain of definition of A. More precisely, one should choose a set of functions in H which are differentiable sufficiently many times to allow the application of A. We thus wish to speak of linear mappings $A : D(A) \to H$, where $D(A)$ is a subspace of H; we will capture this notion in the following definition.

Definition 4.41 Let H be a real Hilbert space.
(a) An *operator* on H is a pair $(A, D(A))$, where $D(A) \subseteq H$ is a subspace of H and $A : D(A) \to H$ is a linear mapping. We call $D(A)$ the *domain* of A. To simplify notation, in this case we will also speak of A as being an operator on H; to this operator belong the domain $D(A)$ and the linear mapping $A : D(A) \to H$.
(b) Two operators A and B on H are *equal* (and we write $A = B$) if $D(A) = D(B)$ and $Au = Bu$ for all $u \in D(A)$.
(c) If A and B are operators on H, then we write $A \subset B$ and say that A is a *restriction* of B, or equivalently that B is an *extension* of A, if $D(A) \subset D(B)$ and $Au = Bu$ for all $u \in D(A)$.   △

What we are calling an operator here, is often known as an *unbounded operator*, although the word "unbounded" is to be interpreted as "not necessarily bounded". Thus an unbounded operator on H is simply a linear mapping whose domain is not (necessarily) the whole of H. Often, the domain of an operator A is dense in H: we then say that A is *densely defined*.

Example 4.42 Let $H = L_2(0, 1)$. We can consider the second derivative on $(0, 1)$ with various domains, for example:

(a) Let $D(A_0) := \{u \in C^2([0,1]) : u(0) = u(1) = 0\}$ and $A_0 u := -u''$ for all $u \in D(A_0)$.

(b) Let $D(A_1) := \{u \in C^2([0,1]) : u'(0) = u'(1) = 0\}$ and $A_1 u := -u''$ for all $u \in D(A_1)$.

(c) Let $D(B) := C^2([0,1])$ and $Bu := -u''$ for all $u \in D(B)$.

Then $A_0 \subset B$, $A_1 \subset B$, and $A_0 \neq A_1$. The rule defining the linear mapping is always the same; nevertheless, all three operators are different. Dirichlet boundary conditions are encoded into the domain of A_0, while Neumann boundary conditions are encoded into the one of A_1. We will return to this subject in the following chapter.

$\triangle$

Let A be an operator on H. Our goal will be to analyze equations of the form

$$Au = f. \tag{4.16}$$

In terms of the solvability of (4.16), there are three properties for which one may reasonably hope:

1. **Existence:** for each $f \in H$ there is a solution $u \in D(A)$ of (4.16).
2. **Uniqueness:** for each $f \in H$ there is at most one solution $u \in D(A)$.
3. **Continuity of the solution with respect to the given data:** if $f_n, f \in H$ and $u_n, u \in D(A)$ are such that $Au_n = f_n$ and $Au = f$, and $f_n \to f$ in H, then we must have $u_n \to u$ in H.

The existence statement is equivalent to the surjectivity of the mapping

$$A : D(A) \to H;$$

uniqueness means that A is injective. Now since A is linear, the kernel (or null space) of A, defined by

$$\ker A := \{u \in D(A) : Au = 0\},$$

is a subspace of $D(A)$, and A is injective if and only if $\ker A = \{0\}$. Existence and uniqueness together are thus equivalent to the bijectivity of the mapping $A : D(A) \to H$. In this case $A^{-1} : H \to D(A)$ is also linear, and since $D(A) \subset H$, we may consider A^{-1} as a mapping from H to H. The requirement of continuity in 3. corresponds exactly to the continuity of this mapping. If these three conditions are satisfied, then we shall call the operator A *invertible*: to summarize, this means that $A : D(A) \to H$ is bijective and $A^{-1} : H \to H$ is continuous. Following Hadamard, we say that the problem (4.16) is *well-posed* if all three conditions listed above, namely existence, uniqueness and continuous dependence on the data, are satisfied. Thus the problem (4.16) is well-posed if and only if the operator A is invertible.

Remark 4.43 Put slightly differently, an operator A on H is invertible if the following two conditions hold:

(a) for each $f \in H$ there is exactly one $u \in D(A)$ such that $Au = f$, and

(b) there exists a $c \geq 0$ such that the *a priori estimate* $\|u\| \leq c\|f\|$ holds. $\triangle$

The following observation is useful for helping to understand the role of the domain.

Remark 4.44 Let A be an invertible operator and let B be a restriction or an extension of A. If B is invertible, then $A = B$. △

Proof 1. Let $B \subset A$. If B is surjective, then we claim that $B = A$. In fact, if $u \in D(A)$, then by surjectivity of B there exists a $v \in D(B)$ such that $Bv = Au$. Since A is injective and $Av = Bv = Au$, it follows that $u = v$.

2. Let $A \subset B$. If B is injective, then we claim that $A = B$. This time, let $u \in D(B)$, then by surjectivity of A there exists a $v \in D(A)$ such that $Av = Bu$. Since B is injective, as before we get $u = v \in D(A)$. □

If $\lambda \in \mathbb{R}$, then we may define the operator $A - \lambda I$ by

$$(A - \lambda I)u := Au - \lambda u, \quad D(A - \lambda I) := D(A).$$

Here I stands for the identity mapping. We say that $\lambda \in \mathbb{R}$ is an *eigenvalue* of A if $A - \lambda I$ is not injective, that is, if there exists some $u \in D(A)$ such that $u \neq 0$ and $Au = \lambda u$.

Of particular interest are operators associated with a bilinear form. Here we will continue to assume that H is a real Hilbert space, and we will denote its inner product by $(\cdot, \cdot)_H$ and its norm by $\| \cdot \|_H$. Let V be another Hilbert space with inner product $(\cdot, \cdot)_V$ and norm $\| \cdot \|_V$. We say that V *is continuously imbedded in H* and write $V \hookrightarrow H$ if $V \subset H$ and there exists a constant $c \geq 0$ such that

$$\|u\|_H \leq c\|u\|_V, \quad u \in V. \tag{4.17}$$

This means exactly that the identity mapping of V into H is continuous. We also wish to assume that V is dense in H.

Now let $a : V \times V \to \mathbb{R}$ be a continuous bilinear form; thus there exists a $C > 0$ such that

$$|a(u, v)| \leq C\|u\|_V\|v\|_V \quad \text{for all } u, v \in V. \tag{4.18}$$

We wish to associate an operator A on H with the form $a(\cdot, \cdot)$. To this end we first define the *domain* of A by

$$D(A) := \{u \in V : \exists f \in H \text{ such that } a(u, v) = (f, v)_H \text{ for all } v \in V\}, \tag{4.19}$$

a subspace of H. If $u \in D(A)$, then there exists an $f \in H$ such that

$$a(u, v) = (f, v)_H \quad \text{for all } v \in H. \tag{4.20}$$

This f is unique. Indeed, if $\tilde{f} \in H$ is some other vector for which $a(u, v) = (\tilde{f}, v)_H$ for all $v \in V$, then $(f - \tilde{f}, v)_H = 0$ for all $v \in V$. Since V is dense in H, this continues to hold for all $v \in H$. In particular, $\|f - \tilde{f}\|_H^2 = (f - \tilde{f}, f - \tilde{f})_H = 0$ and thus $f = \tilde{f}$.

If $u \in D(A)$ and if $f \in H$ is the unique element of H for which (4.20) holds, then we set $Au = f$. This defines a linear mapping $A : D(A) \to H$; we call A *the operator associated with* $a(\cdot, \cdot)$ *on* H. We shall now consider the equation

$$Au = f. \tag{4.21}$$

Given $f \in H$, we seek $u \in D(A)$ for which (4.21) holds. By the Lax–Milgram theorem we have the following result:

Theorem 4.45 *If* $a(\cdot, \cdot)$ *is continuous and coercive, then the problem* (4.21) *is well-posed, that is, the operator* A *is invertible.*

Proof Let $\alpha > 0$ be such that $a(u) \geq \alpha \|u\|_V^2$ for all $u \in V$, and let $f \in H$. Then $F(v) := (f, v)_H$ defines a continuous linear form on V, as follows from the continuity of the imbedding of V in H.

By the Lax–Milgram theorem there exists exactly one $u \in V$ such that $a(u, v) = (f, v)_H$ for all $v \in V$. This means exactly that $u \in D(A)$ and $Au = f$. We have thus shown that A is bijective. Now (4.17) also implies that

$$\frac{\alpha}{c} \|u\|_H^2 \leq \alpha \|u\|_V^2 \leq a(u, u) = (f, u)_H \leq \|f\|_H \|u\|_H,$$

and hence $\|u\|_H \leq \frac{c}{\alpha} \|f\|_H$. Since $u = A^{-1} f$, this shows that $\|A^{-1}\| \leq \frac{c}{\alpha}$. $\square$

In place of A we can also consider the operator $A + \lambda I$ for $\lambda \in \mathbb{R}$. It is an immediate consequence of the definition that $A + \lambda I$ is associated with the bilinear form

$$a_\lambda : V \times V \to \mathbb{R}, \qquad a_\lambda(u, v) := a(u, v) + \lambda(u, v)_H.$$

This form is continuous since V is continuously imbedded in H. We say that the bilinear form $a(\cdot, \cdot)$ is *H-elliptic* if there exist $\omega \in R$ and $\alpha > 0$ such that

$$a(u) + \omega \|u\|_H^2 \geq \alpha \|u\|_V^2 \quad \text{for all } u \in V. \tag{4.22}$$

We see that the form $a(\cdot, \cdot)$ is H-elliptic if and only if there is an $\omega \in \mathbb{R}$ such that $a_\omega(\cdot, \cdot)$ is coercive. In this case, $a_\lambda(\cdot, \cdot)$ is also coercive, and $A + \lambda I$ is invertible, for all $\lambda \geq \omega$. The inequality (4.22) is also known as *Gårding's inequality*, especially in numerical mathematics.[2]

Now we make the additional assumption that the form is symmetric. One sees easily that the associated operator A is then also *symmetric*, that is,

$$(Au, v)_H = (u, Av)_H \quad \text{for all } u, v \in D(A).$$

If $V = H$ and $\dim H < \infty$, then we know from linear algebra that H admits an orthonormal basis consisting of eigenvectors of A. We now wish to generalize this

---

[2] This appellation is somewhat misleading since (4.22) is not a proved inequality but rather an assumption.

result to our infinite-dimensional setting. To do so, we need a stronger relationship between V and H than just continuity of the imbedding. We say that V *is compactly imbedded in* H, and write

$$V \overset{c}{\hookrightarrow} H,$$

if $V \subset H$ and the imbedding $u \mapsto u$ of V into H is compact. By Theorem 4.39 this means exactly that

$$u_n \rightharpoonup u \text{ in } V \text{ implies } u_n \to u \text{ in } H.$$

Under this additional assumption H admits an orthonormal basis of eigenvectors of A. This is the statement of the following theorem, which also gives a precise description of the domain of A in terms of the orthonormal basis.

Theorem 4.46 (Operator version of the spectral theorem)
Let $\dim H = \infty$ *and let V be compactly and densely imbedded in H. Let $a : V \times V \to \mathbb{R}$ be a symmetric, continuous and H-elliptic bilinear form. Then there exist an orthonormal basis $\{e_n : n \in \mathbb{N}\}$ of H and numbers $\lambda_n \in \mathbb{R}$ with*

$$\lambda_n \leq \lambda_{n+1}, \, n \in \mathbb{N} \quad and \quad \lim_{n \to \infty} \lambda_n = \infty,$$

such that the operator A associated with $a(\cdot, \cdot)$ is given by

$$D(A) = \left\{ u \in H : \sum_{n=1}^{\infty} \lambda_n^2 (u, e_n)_H^2 < \infty \right\}, \tag{4.23}$$

$$Au = \sum_{n=1}^{\infty} \lambda_n (u, e_n)_H \, e_n, \quad u \in D(A). \tag{4.24}$$

Before we prove the theorem, we first wish to state a few consequences.

Remark 4.47 Under the assumptions of the spectral theorem, we have:
(a) $e_n \in D(A)$ and $Ae_n = \lambda_n e_n$. In particular, every λ_n is an eigenvalue of A. This follows from (4.23) and (4.24).
(b) The λ_n are *exactly* the eigenvalues of A: indeed, if $\lambda \in \mathbb{R}$ is an eigenvalue, then there exists some $u \in D(A)$ such that $Au = \lambda u$, with $u \neq 0$. Hence there exists an $m \in \mathbb{N}$ such that $(u, e_m)_H \neq 0$. By (4.24) it then follows that $\lambda(u, e_m)_H = (Au, e_m)_H = \lambda_m (u, e_m)_H$. Thus $\lambda = \lambda_m$.
(c) By Theorem 4.13 we know that, for any sequence $(x_n)_{n \in \mathbb{N}} \subset \mathbb{R}$, the series $\sum_{n=1}^{\infty} x_n e_n$ converges in H if and only if $\sum_{n=1}^{\infty} x_n^2 < \infty$. The domain of A thus consists of exactly those $u \in H$ for which $\sum_{n=1}^{\infty} \lambda_n (u, e_n)_H \, e_n$ converges in H. △

The spectral theorem has a number of notable consequences for the equation

$$Au - \lambda u = f. \tag{4.25}$$

The uniqueness of solutions of (4.25) corresponds exactly to λ not being an eigenvalue of A. The following theorem states that the equation is already well-posed, that is, $A - \lambda I$ is invertible, as soon as $A - \lambda I$ is injective. In other words, what is known as the *Fredholm alternative* holds for the equation (4.25): either the equation has multiple solutions for $f = 0$, or it is well-posed.

Corollary 4.48 *If $\lambda \in \mathbb{R} \setminus \{\lambda_n : n \in \mathbb{N}\}$, then $A - \lambda I$ is invertible. Moreover, given $f \in H$, the solution of (4.25) is given by the series*

$$u = \sum_{n=1}^{\infty} (\lambda_n - \lambda)^{-1} (f, e_n)_H \, e_n,$$

which converges in H.

Proof First observe that $\delta := \inf\{|\lambda - \lambda_n| : n \in \mathbb{N}\} > 0$ since $\lambda_n \to \infty$. Hence $|\lambda - \lambda_n|^{-1} \leq 1/\delta$ for all $n \in \mathbb{N}$. This means that

$$Rf := \sum_{n=1}^{\infty} (\lambda_n - \lambda)^{-1} (f, e_n)_H \, e_n$$

defines a continuous linear operator $R : H \to H$. By Theorem 4.9,

$$\|Rf\|_H^2 = \sum_{n=1}^{\infty} |\lambda - \lambda_n|^{-2} (f, e_n)_H^2 \leq \delta^{-2} \|f\|_H^2$$

for all $f \in H$. Now for $u \in D(A)$ we have

$$(A - \lambda I)u = \sum_{n=1}^{\infty} (\lambda_n - \lambda)(u, e_n)_H \, e_n;$$

hence $R(A - \lambda I)u = u$ for all $u \in D(A)$. In the other direction, since the sequence

$$\left(\frac{\lambda_n}{\lambda_n - \lambda} \right)_{n \in \mathbb{N}}$$

is bounded, we have $Rf \in D(A)$ and $(A - \lambda I)Rf = f$ for all $f \in H$. $\square$

Remark 4.49 (Self-adjoint operators) a) Let B be an operator on H with dense domain $D(B)$. Then the adjoint B^* of B is defined as follows

$$D(B^*) := \{v \in H : \exists w_v \in H \text{ s.t. } (Bu, v)_H = (u, w_v)_H \text{ for all } u \in D(B)\},$$
$$B^* v := w_v.$$

More precisely, for $v \in D(B^*)$, $B^* v$ is the unique element in H such that

$$(Bu, v)_H = (u, B^* v)_H \quad \text{for all } v \in D(B).$$

Uniqueness follows from the density of the domain.

The operator B is called *self-adjoint* if $B = B^*$ (with identical domains). Note that B is symmetric (see the definition given above) if and only if $B \subset B^*$. For self-adjointness we require in addition that $D(B) = D(B^*)$.

b) The operator A in Theorem 4.46 is self-adjoint (as it is easy to see using the theorem). In addition, A is *bounded from below*, i.e., $(Au, u)_H + \omega\|u\|_H^2 \geq 0$ for all $u \in D(A)$ and some $\omega \geq 0$. Finally, A has compact *resolvent*, i.e., $(\lambda - A)^{-1} : H \to H$ is compact for all $\lambda \in \mathbb{R} \setminus \{\lambda_n : n \in \mathbb{N}\}$. Conversely, every adjoint operator which is bounded and has compact resolvent is induced by a form exactly in the way described in Theorem 4.46. $\triangle$

We now turn to the proof of the spectral theorem; to begin with, we diagonalize the form.

Theorem 4.50 (Form version of the spectral theorem) *Suppose that* $\dim H = \infty$ *and that V is compactly and densely imbedded in H. Let $a : V \times V \to \mathbb{R}$ be a continuous, symmetric, H-elliptic bilinear form. Then there exist an orthonormal basis $\{e_n : n \in \mathbb{N}\}$ of H and $\lambda_n \in \mathbb{R}$ with $\lambda_n \leq \lambda_{n+1}$, $n \in \mathbb{N}$, $\lim_{n \to \infty} \lambda_n = \infty$, such that*

$$V = \left\{ u \in H : \sum_{n=1}^{\infty} |\lambda_n(u, e_n)|_H^2 < \infty \right\}, \tag{4.26}$$

$$a(u, v) = \sum_{n=1}^{\infty} \lambda_n(u, e_n)_H (e_n, v)_H \quad \text{for all } u, v \in V. \tag{4.27}$$

Proof We may assume that $a(\cdot, \cdot)$ is coercive; if not, then we consider $a_\omega(\cdot, \cdot)$ in place of $a(\cdot, \cdot)$, where ω is the constant from (4.22). So we suppose that $a(\cdot, \cdot)$ is coercive, that is, we assume that (4.6) holds for all $u \in V$. Then it follows that

$$\sqrt{\alpha}\|u\|_V \leq a(u)^{1/2} \leq \sqrt{C}\|u\|_V \quad \text{for all } u \in V.$$

Since $a(\cdot, \cdot)$ is symmetric, $a(u)^{1/2}$ thus defines an equivalent norm on V (called the *energy norm* $\| \cdot \|_a$). Since the inner product of V does not appear explicitly in the formulation of the theorem, we may as well suppose that in fact

$$(u, v)_V = a(u, v), \quad u, v \in V.$$

We now define $\lambda_1 := \inf\{a(u) : u \in V, \|u\|_H = 1\}$. It follows from the coercivity of a, see (4.6), that $\lambda_1 > 0$.

1. We claim that there exists an $e_1 \in V$ such that $\|e_1\|_H = 1$, $\lambda_1 = a(e_1)$ and

$$a(e_1, v) = \lambda_1(e_1, v)_H \quad \text{for all } v \in V. \tag{4.28}$$

To this end we consider the form $a_1(u, v) := a(u, v) - \lambda_1(u, v)_H$, $u, v \in V$. Then $\inf\{a_1(u) : u \in V, \|u\|_H = 1\} = 0$. We can thus find vectors $u_n \in V$ with $\|u_n\|_H = 1$, such that $\lim_{n\to\infty} a_1(u_n) = 0$, and in particular the sequence $(u_n)_{n\in\mathbb{N}}$ is bounded in V. Theorem 4.35 allows us to assume without loss of generality that $u_n \rightharpoonup e_1$ in V, for some $e_1 \in V$ (otherwise we pass to a subsequence). Since V is compactly imbedded in H, it follows that $u_n \to e_1$ in H, and so $\|e_1\|_H = 1$. The Cauchy–Schwarz inequality now yields

$$a_1(e_1) = a_1(e_1 - u_n, e_1) + a_1(u_n, e_1)$$
$$\leq a_1(e_1 - u_n, e_1) + a_1(u_n)^{1/2} a_1(e_1)^{1/2}.$$

Since $f(v) := a_1(v, e_1)$ defines a continuous linear form on V and $u_n \rightharpoonup e_1$ in V, we have $\lim_{n\to\infty} a_1(e_1 - u_n, e_1) = 0$. Since $a_1(u_n) \to 0$, the above inequality implies that $a_1(e_1) = 0$. Hence $a(e_1) = \lambda_1$. Furthermore, by Theorem 4.26, for $v \in V$ we have $a_1(e_1, v) \leq a_1(e_1)^{1/2} a_1(v)^{1/2} = 0$. By replacing v by $-v$, we see that the reverse inequality also holds. Thus $a_1(e_1, v) = 0$, that is, we have $a(e_1, v) = \lambda_1(e_1, v)_H$ for all $v \in V$. This proves (4.28).

2. We now consider the space $V_1 := \{u \in V : (u, e_1)_H = 0\}$, which is a closed subspace of V. It is not equal to $\{0\}$ since V is dense in H and $\dim H = \infty$. If we apply 1. to the restriction of $a(\cdot, \cdot)$ to $V_1 \times V_1$ and to $H_1 := \{u \in H : (u, e_1)_H = 0\}$, then we obtain an $e_2 \in V_1$ such that $\|e_2\|_H = 1$ and $\lambda_2 := a(e_2) = \min\{a(u) : u \in V_1, \|u\|_H = 1\}$, as well as $a(e_2, v) = \lambda_2(e_2, v)_H$ for all $v \in V_1$. Since by 1. we have $a(e_2, e_1) = a(e_1, e_2) = \lambda_1(e_1, e_2)_H = 0 = \lambda_2(e_2, e_1)_H$, it follows that $a(e_2, v) = \lambda_2(e_2, v)_H$ for all $v \in V$ (by the projection theorem).

3. If we continue in this fashion, we find a sequence $(e_n)_{n\in\mathbb{N}}$ in V such that $\|e_n\|_H = 1$ and a monotonically increasing sequence $(\lambda_n)_{n\in\mathbb{N}}$ in $\mathbb{R}$, such that for all $n \geq 1$,

$$\lambda_{n+1} = a(e_{n+1}) = \min\{a(u) : u \in V_n, \|u\|_H = 1\},$$

where $V_n := \{u \in V : (u, e_k)_H = 0, k = 1, \ldots, n\}$, $e_{n+1} \in V_n$ and $a(e_n, v) = \lambda_n(e_n, v)_H$ for all $v \in V$ and all $n \in \mathbb{N}$. In particular, the sequence $(e_n)_{n\in\mathbb{N}}$ is orthonormal.

4. We will now show that $\lim_{n\to\infty} \lambda_n = \infty$. Suppose not; then after passing to a subsequence if necessary we may assume that the sequence $a(e_n) = \lambda_n$ is bounded. Then $(e_n)_{n\in\mathbb{N}}$ is bounded in V, and so there exists a subsequence $(e_{n_k})_{k\in\mathbb{N}}$ which converges weakly in V, say to $w \in V$. Then, by compactness of the imbedding of V in H, in fact $\lim_{k\to\infty} e_{n_k} = w$ in H. Since $e_n \rightharpoonup 0$ in H as an orthonormal sequence in H (see Example 4.32), it follows that $w = 0$. But this yields the desired contradiction to $\|w\|_H = \lim_{k\to\infty} \|e_{n_k}\|_H = 1$. We conclude that $\lim_{n\to\infty} \lambda_n = \infty$.

5. We will next show that $\{e_n : n \in \mathbb{N}\}$ is total in H. To this end, suppose that $v \in H$ is such that $(e_n, v)_H = 0$ for all $n \in \mathbb{N}$; then we wish to show that $v = 0$ (see Corollary 4.19). Suppose that $v \neq 0$. Let $v_1 := \|v\|_H^{-1} v$; then $\|v_1\|_H = 1$ and $(e_n, v_1)_H = 0$ for all $n \in \mathbb{N}$. Thus $v_1 \in V_n$ (as defined in 3.) for all $n \in \mathbb{N}$, and so

$$a(v_1) \geq \inf\{a(u) : u \in V_n, \|u\|_H = 1\} = \lambda_n$$

for all $n \in \mathbb{N}$. This contradicts the fact that $\lim_{n \to \infty} \lambda_n = \infty$.

6. Finally, set $\tilde{e}_n := \frac{1}{\sqrt{\lambda_n}} e_n$. Then $a(\tilde{e}_n, \tilde{e}_m) = \frac{1}{\sqrt{\lambda_n}} \frac{1}{\sqrt{\lambda_m}} a(e_n, e_m)$ is equal to 1 if $n = m$, and 0 otherwise. Since for $u \in V$ we have that

$$(u, \tilde{e}_n)_V = a(u, \tilde{e}_n) = \frac{1}{\sqrt{\lambda_n}} a(u, e_n) = \sqrt{\lambda_n} \, (u, e_n)_H, \qquad (4.29)$$

it follows from $(u, \tilde{e}_n)_V = 0$ for all $n \in \mathbb{N}$ that $u = 0$. Hence $\{\tilde{e}_n : n \in \mathbb{N}\}$ is an orthonormal basis of V. By Theorem 4.9 we thus obtain, for all $u, v \in V$,

$$a(u, v) = (u, v)_V = \sum_{n=1}^{\infty} (u, \tilde{e}_n)_V \, (\tilde{e}_n, v)_V = \sum_{n=1}^{\infty} \lambda_n (u, e_n)_H \, (e_n, v)_H.$$

In particular, $\sum_{n=1}^{\infty} \lambda_n (u, e_n)_H^2 = a(u) < \infty$. In the other direction, let $u \in H$ be such that $\sum_{n=1}^{\infty} \lambda_n (u, e_n)_H^2 < \infty$. Set $x_n := \sqrt{\lambda_n} \, (u, e_n)_H$. Then by Theorem 4.13 the series $\sum_{n=1}^{\infty} x_n \tilde{e}_n$ converges to an element w of V. Since $x_n \tilde{e}_n = (u, e_n)_H \, e_n$, we have $w = \sum_{n=1}^{\infty} (u, e_n)_H \, e_n = u$ and thus $u = w \in V$. This completes the proof of Theorem 4.50. □

It is remarkable that the assumptions of Theorem 4.50 already imply that the space H is separable.

Proof (of Theorem 4.46 (the spectral theorem)) By Theorem 4.50, V is given by (4.26), and $a(\cdot, \cdot)$ by (4.27). Let A be the operator associated with $a(\cdot, \cdot)$. Let $u \in D(A)$ with $Au = f$; then by definition, $a(u, v) = (f, v)_H$ for all $v \in V$. In particular, for $v = e_n$, the equation $(f, e_n)_H = a(u, e_n) = \lambda_n (u, e_n)_H$ holds, whence

$$f = \sum_{n=1}^{\infty} (f, e_n)_H e_n = \sum_{n=1}^{\infty} \lambda_n (u, e_n)_H e_n.$$

In the other direction, let $u \in H$ such that $\sum_{n=1}^{\infty} \lambda_n^2 \, (u, e_n)_H^2 < \infty$. Then since $\lambda_n \to \infty$ as $n \to \infty$ we also have $\sum_{n=1}^{\infty} \lambda_n \, (u, e_n)_H^2 < \infty$ and so $u \in V$. Moreover,

$$f := \sum_{n=1}^{\infty} \lambda_n \, (u, e_n)_H \, e_n$$

in H. For $n \in \mathbb{N}$, we have $(f, e_n)_H = \lambda_n (u, e_n)_H = a(u, e_n)$ and thus $(f, v)_H = a(u, v)$ for all $v \in \text{span}\{e_n : n \in \mathbb{N}\}$. Since we already showed in the proof of Theorem 4.50 that $\{\lambda_n^{-1/2} e_n : n \in \mathbb{N}\}$ is an orthonormal basis of V, it follows that $(f, v)_H = a(u, v)$ for all $v \in V$. Hence $u \in D(A)$ and $Au = f$. □

Remark 4.51 (Gelfand triple) Let V and H be real Hilbert spaces such that $V \hookrightarrow H$ (i.e., V is a subspace of H and $\|v\|_H \leq c \, \|v\|_V$ for all $v \in V$ and some $c \geq 0$). Also assume that V is dense in H.

Then, for every fixed $f \in H$, $(f, \cdot)_H$ defines a continuous linear form on V. Moreover, since V is dense in H, the mapping $f \mapsto (f, \cdot)_H : H \to V'$ is injective.

Identifying $f \in H$ with $(f, \cdot)_H$ we thus obtain a continuous imbedding $H \hookrightarrow V'$. Indeed,

$$\|(f, \cdot)_H\|_{V'} = \sup_{v \in V} \frac{|(f, v)_H|}{\|v\|_V} \leq \sup_{v \in V} \frac{\|f\|_H \|v\|_H}{\|v\|_V} \leq c \|f\|_H,$$

which establishes the claimed continuity of the imbedding. We thus have two imbeddings

$$V \hookrightarrow H \hookrightarrow V'.$$

These three spaces, together with the corresponding imbeddings, are sometimes called a *Gelfand triple*. For $\varphi \in V'$ one frequently uses the notation

$$\langle \varphi, v \rangle := \varphi(v) \qquad (v \in V),$$

which is also known as a *duality pairing*. Thus, if $\varphi = f \in H \subset V'$, then

$$\langle f, v \rangle = (f, v)_H \qquad (v \in V).$$

The bracket $\langle \cdot, \cdot \rangle$ extends the scalar product of H. It is important to realize that identifying H with a subspace of V' prevents us from identifying V with V' at the same time: V becomes a proper subspace of V' (unless $V = H$).

Now we assume that the imbedding of V in H is, in addition, compact. Then we choose the scalar product of V as bilinear form and apply the form version of the spectral theorem, Theorem 4.50. It yields an orthonormal basis $(e_n)_{n \in \mathbb{N}}$ of H and a sequence $1 \leq \lambda_1 \leq \lambda_2 \leq \cdots \leq \lambda_n \leq \lambda_{n+1}$ with $\lim_{n \to \infty} = \infty$ such that

$$V = \left\{ v \in H : \sum_{n=1}^{\infty} \lambda_n (v, e_n)_H^2 < \infty \right\}, \qquad (u, v)_V = \sum_{n=1}^{\infty} \lambda_n (u, e_n)_H (e_n, v)_H$$

for all $u, v \in V$. Now the imbedding of H in V' can be described very elegantly. The space

$$W := \left\{ w = (w_n)_{n \in \mathbb{N}} \subset \mathbb{R} : \sum_{n=1}^{\infty} \frac{w_n^2}{\lambda_n} < \infty \right\}$$

is a Hilbert space when equipped with the scalar product

$$(w, v)_W = \sum_{n=1}^{\infty} \frac{1}{\lambda_n} w_n v_n, \qquad w, v \in W.$$

It is isometrically isomorphic to V' if we define, for $w \in W$ and $v \in V$,

$$\langle w, v \rangle := \sum_{n=1}^{\infty} w_n (e_n, v)_H.$$

Proof Let $w = (w_n)_{n \in \mathbb{N}} \in W$. By the Cauchy–Schwarz inequality we have for $v \in V$,

$$\sum_{n=1}^{\infty} |w_n (e_n, v)_H| = \sum_{n=1}^{\infty} \frac{1}{\sqrt{\lambda_n}} |w_n| \sqrt{\lambda_n} |(e_n, v)_H|$$

$$\leq \left(\sum_{n=1}^{\infty} \frac{1}{\lambda_n} w_n^2 \right)^{1/2} \left(\sum_{n=1}^{\infty} \lambda_n (e_n, v)_H^2 \right)^{1/2} = \|w\|_W \|v\|_V.$$

Thus $\langle w, \cdot \rangle \in V'$.

Conversely, let $\varphi \in V'$. By the theorem of Riesz–Fréchet (Theorem 4.21) there exists a unique $u \in V$ such that

$$\langle \varphi, v \rangle = (u, v)_V = \sum_{n=1}^{\infty} \lambda_n (u, e_n)_H (e_n, v)_H$$

for all $v \in V$. Moreover, $\|\varphi\|_{V'}^2 = \|u\|_V^2 = \sum_{n=1}^{\infty} \lambda_n (u, e_n)_H^2$ (see Example 4.20). Let $w_n := \lambda_n (u, e_n)_H = \varphi(e_n)$. Then $\sum_{n=1}^{\infty} \frac{1}{\lambda_n} w_n^2 = \sum_{n=1}^{\infty} \lambda_n (u, e_n)_H^2$. Thus, we have $w \in W$, $\|w\|_W = \|\varphi\|_{V'}$ and

$$\langle \varphi, v \rangle = \sum_{n=1}^{\infty} \lambda_n (u, e_n)_H (e_n, v)_H = \sum_{n=1}^{\infty} w_n (e_n, w)_H = \langle w, v \rangle$$

for all $v \in V$. $\qquad \square$

Notice that the desired isomorphism is given by $V' \to W : \varphi \mapsto (\varphi(e_n))_{n \in \mathbb{N}}$. $\quad \triangle$

Let us again consider the situation described in the spectral theorem, Theorem 4.46. Here, λ_n is the nth eigenvalue of A, since we are assuming that the eigenvalues are ordered as an increasing sequence. From the proof of Theorem 4.50 we can extract the following formula for the first eigenvalue

$$\lambda_1 = \min\{a(u) : u \in V, \|u\|_H = 1\}. \tag{4.30}$$

The following max-min formula allows us more generally to calculate the nth eigenvalue of A from the form $a(\cdot, \cdot)$.

Theorem 4.52 (Max-min formula) *Under the assumptions of the spectral theorem, the nth eigenvalue of A is given by*

$$\lambda_n = \max_{\substack{W \subset V \\ \operatorname{codim} W \leq n-1}} \min_{\substack{u \in W \\ \|u\|_H = 1}} a(u). \tag{4.31}$$

Remark 4.53 Let W be a subspace of V. (a) We set $\operatorname{codim} W := \dim W^{\perp}$, where $W^{\perp} := \{v \in V : (w, v)_V = 0 \text{ for all } w \in W\}$.

(b) If U is a subspace of V such that $U \cap W = \{0\}$, then $\dim U \leq \operatorname{codim} W$. In fact, denote by $P : V \to W^{\perp}$ the orthogonal projection. Then $P_{|U}$ is injective (because for $u \in U$, $Pu = 0$ implies that $u \in W \cap U$). Thus $\dim U \leq \dim W^{\perp}$. $\triangle$

Proof (of Theorem 4.52) Let $n > 1$ and let W be a subspace of V with $\operatorname{codim} W \leq n - 1$. Then by Remark 4.53 (b) there exists $u \in \operatorname{span}\{e_1, \ldots, e_n\} \cap W$ with $\|u\|_H = 1$. For such a u, by (4.27),

$$a(u) = \sum_{k=1}^{n} \lambda_k (u, e_n)_H^2 \leq \lambda_n \sum_{k=1}^{n} (u, e_k)_H^2 = \lambda_n.$$

Hence $\inf\{a(u) : u \in W, \|u\|_H = 1\} \leq \lambda_n$ for every subspace W of V with $\operatorname{codim} W \leq n - 1$. It remains to show that such a W exists for which $\inf\{a(u) : u \in W, \|u\|_H = 1\} = \lambda_n$. We claim that $W_{n-1} := \{e_1, \ldots, e_{n-1}\}^{\perp}$ will work. In fact, let $u \in W_{n-1}$ such that $\|u\|_H = 1$; then

$$a(u) = \sum_{k \geq n} \lambda_k (u, e_k)_H^2 \geq \lambda_n \sum_{k \geq n} (u, e_k)_H^2 = \lambda_n.$$

Since $e_n \in W_{n-1}$ and $a(e_n) = \lambda_n$, we finally have $\min\{a(u) : u \in W_{n-1}, \|u\|_H = 1\} = \lambda_n$. $\square$

If $a(\cdot, \cdot)$ is not symmetric, then the associated operator A is not in general a diagonal operator. Nevertheless, it still satisfies the Fredholm alternative. For the proof, we refer to [2, Thm. 12.8].

Theorem 4.54 (Fredholm alternative) *Let H be a real separable Hilbert space and let H be a Hilbert space which is compactly and densely imbedded in H. Suppose that $a : V \times V \to \mathbb{R}$ is a continuous, H-elliptic bilinear form. Then for the operator A associated with a, if $\lambda \in \mathbb{R}$ is not an eigenvalue of A, then $A - \lambda I$ is invertible.*

4.9* Comments on Chapter 4

David Hilbert (1862–1943) was one of the pre-eminent mathematicians of the first half of last century. At the International Congress of Mathematicians (ICM) in Paris in 1900, Hilbert presented his famous 23 mathematical problems, which continue to exert considerable influence on mathematical research to this day. He was the central figure of the Göttingen school, which up until the Nazis seized power in Germany in 1933 was one of the foremost mathematical institutions worldwide. Between 1904 and 1910, Hilbert published a series of articles on integral equations, in which bilinear forms on ℓ_2 appeared as an essential tool.

The definition of Hilbert spaces was only given in 1929, by John von Neumann. Decisive for the success of Hilbert space theory was the invention of the Lebesgue integral, in the doctoral thesis of Henri Lebesgue completed in 1902. Building on Lebesgue's theory, Frigyes Riesz showed in 1907 that every element of $\ell_2(\mathbb{Z})$ is the sequence of Fourier coefficients of some function in $L_2(0, 2\pi)$. In the same year, Ernst Fischer (1875–1954) proved that $L_2(0, 2\pi)$ is complete, which is equivalent to Riesz' result. This is why the name "theorem of Riesz–Fischer" is often used for the fact that $L_2(\Omega)$

is complete (see the appendix). Riesz and Maurice Fréchet (1878–1973) determined the set of all continuous linear forms on $L_2(0, 1)$ independently of each other in 1907, thus proving Theorem 4.21.

Fréchet obtained his doctorate in Paris in 1906 under Hadamard; in his dissertation, metric spaces are introduced for the first time. The theorem of Riesz–Fréchet is often known as the Riesz representation theorem; however, there is another Riesz representation theorem for measures, which will play an important role in Section 7.2.

It was Hilbert who proved a first version of the spectral theorem, Theorem 4.50. But subsequent to his works on integral equations, it was operators and not bilinear forms which took centre stage. Bounded operators on Hilbert spaces became a central concept in mathematics, and to this day operator theory remains an important area of mathematical research. In the 1930s, quantum theory was a key driving force in the development of functional analysis. The decisive breakthrough in the mathematical formulation of quantum physics was made by John von Neumann, who had visited Göttingen in 1926/27 and to whom the concept of unbounded operators is due (cf. Mathematische Annalen, No. 33, 1932). An unbounded self-adjoint operator (as was defined in Section 4.8) models an observable in quantum theory. The book *Mathematische Grundlagen der Quantenmechanik* (Mathematical Foundations of Quantum Mechanics) by John von Neumann, which appeared in 1932 in German, describes the mathematical modeling of quantum theory as is still used today, and indeed with great success. Although operators are ideal for describing equations, it is interesting that bilinear forms reached their pinnacle as a method for solving partial differential equations around 50 years after Hilbert's work.

The Lax–Milgram theorem was published in 1954; since then it has become a standard tool for every analyst and numericist. Peter David Lax counts among the most important mathematicians of the second half of last century. He has made material contributions to the theory and the numerics of partial differential equations, as well as mathematical physics and functional analysis. But he was equally responsible for the proof of fundamental theorems in numerical mathematics. Lax was born in 1926 in Budapest and emigrated in 1941 to the United States with his parents. He was a professor at New York University and director of the Courant Institute of Mathematical Sciences for many years. Among the numerous awards and distinctions he received was the Abel Prize in 2005.

The key notion of a well-posed problem, which we met in Section 4.8, was introduced by Jacques Hadamard (1865–1963) in 1898, in an article on boundary value problems (which will be the topic of Chapters 5, 6 and 7). In 1910, a year after Hadamard took up a position as professor for mechanics at the Collège de France, he published his *Leçons sur le calcul des variations*, in which the notion of a functional is introduced for the first time. Hadamard had an enormous influence on analysis; we will meet him again when we see his contributions to the dispute about the variational formulation of the Dirichlet problem in Chapter 6. The Fredholm alternative is particularly attractive when taken together with Hadamard's concept of well-posedness. In the setting of Theorem 4.54, for example, it shows that well-posedness already follows from uniqueness of solutions alone. It was proved by Erik Fredholm (1866–1927) in 1903; Fredholm had obtained his doctorate in Uppsala in 1893 and became well known for his work on integral equations and spectral theory.

4.10 Exercises

Exercise 4.1 Show that the space $\ell_2(I)$ is a Hilbert space with respect to the inner product $(x, y) := \sum_{n \in I} x_n \overline{y_n}$, where I is either a finite index set, $\mathbb{N}$, or $\mathbb{Z}$.

Exercise 4.2 Let H_1 and H_2 be Hilbert spaces and $T : H_1 \to H_2$ a linear mapping. If on both H_1 and H_2 we consider two kinds of convergence, namely convergence in norm and weak convergence, then there are four possibilities in total to define sequential continuity of T. Which of these four are the same?
Suggestion: Use the closed graph theorem (see Theorem A.2).

Exercise 4.3 (Polarization identity)
(a) Let E be a real vector space and let $a : E \times E \to \mathbb{R}$ be bilinear and symmetric. Show that $a(u, v) = \frac{1}{4}(a(u + v) - a(u - v))$ for all $u, v \in E$. Here $a(u) := a(u, u)$, $u \in E$.
(b) Let E_1 and E_2 be inner product spaces and let $T : E_1 \to E_2$ be linear. Show that the following statements are equivalent:
   (i) $(Tu, Tv)_{E_2} = (u, v)_{E_1}$ for all $u, v \in E_1$.
   (ii) T is isometric, that is, $\|Tu\| = \|u\|$ for all $u \in E_1$.
   Thus T is isometric if and only if it preserves the inner product.

Exercise 4.4 (Completion) Let E be a separable real inner product space and let $\{e_n : n \in \mathbb{N}\}$ be an orthonormal basis of E.
(a) Prove that the mapping $J : E \to \ell_2$, $J(u) = ((u, e_n))_{n \in \mathbb{N}}$ is linear and isometric, and has dense range.
(b) Deduce that there exists a Hilbert space H such that E is a dense subspace of H.
   Suggestion: Identify E and $J(E)$.
(c) Uniqueness: Let H_1 and H_2 be Hilbert spaces and $J_k : E \to H_k$ linear and isometric with dense range, $k = 1, 2$. Show that there exists a unitary operator $U : H_1 \to H_2$ such that $UJ_1x = J_2x$ for all $x \in H_1$.

Exercise 4.5 Let E be a real normed vector space and suppose $a : E \times E \to \mathbb{R}$ is bilinear. We say that $a(\cdot, \cdot)$ is continuous if $u_n \to u, v_n \to v \Rightarrow a(u_n, v_n) \to a(u, v)$.
(a) Show that a is continuous if and only if there exists a constant $C \geq 0$ such that $|a(u, v)| \leq C\|u\| \, \|v\|$, $u, v \in E$.
(b) Let a be symmetric and $a(u) \geq 0$, $u \in E$. Show that a is continuous if and only if $u_n \to u \Rightarrow a(u_n) \to a(u)$.
   Suggestion: Use Theorem 4.26.
(c) Let H be a Hilbert space and $T : H \to H$ linear and symmetric, that is, $(Tu, v) = (u, Tv)$, $u, v \in H$. Show that T is continuous.
   Suggestion: Closed graph theorem.
(d) Let H be a Hilbert space, $a : H \times H \to \mathbb{R}$ bilinear, symmetric and such that $u_n \to u \Rightarrow a(u_n, v) \to a(u, v)$ for all $v \in H$. Show using (c) that a is continuous.

Exercise 4.6 Consider the operators from Example 4.42. Determine $\ker A_0$, $\ker A_1$ and $\ker B$. Which of the operators is invertible?

Exercise 4.7 Let A be an operator on the Hilbert space H. We say that A is *bounded* if there exists $c > 0$ such that $\|Au\| \leq c\|u\|$ for all $u \in D(A)$, cf. Appendix A.1.

(a) Let A be invertible. Show that A is bounded if and only if $D(A)$ is closed in H.

(b) Let A be invertible and suppose that $A^{-1} : H \to H$ is compact. Show that A is bounded if and only if $\dim H < \infty$.

(c) We say that A is *closed* if the graph $G(A) := \{(u, Au) : u \in D(A)\}$ of A is closed in $H \times H$. Show that if A is invertible, then A is closed.

(d) Let $D(A) = H$. Show that A is bounded if and only if A is closed.

(e) Show that A is closed if and only if $D(A)$ is complete with respect to the norm $\|u\|_A^2 := \|u\|_H^2 + \|Au\|_H^2$.

Exercise 4.8 Let H be a Hilbert space and let $\{e_n : n \in \mathbb{N}\}$ be an orthonormal basis of H. Let $\lambda_n \in (0, \infty)$, $\lambda_n \leq \lambda_{n+1}$, $\lim_{n \to \infty} \lambda_n = \infty$.

(a) Set $V := \{u \in H : \sum_{n=1}^{\infty} \lambda_n (u, e_n)_H^2 < \infty\}$. Show that

$$a(u, v) := \sum_{n=1}^{\infty} \lambda_n (u, e_n)_H (e_n, v)_H$$

defines an inner product on V with respect to which V is a Hilbert space.

(b) Show that V is compactly imbedded in H.

(c) Choose $H = \ell^2$ and $e_n = (0, \ldots, 0, 1, 0 \ldots)$, where the 1 is the nth entry. Let A be the operator associated with $a(\cdot, \cdot)$. Show that $D(A) = \{x \in \ell^2 : (\lambda_n x_n)_{n \in \mathbb{N}} \in \ell^2\}$ and $Ax = (\lambda_n x_n)_{n \in \mathbb{N}}$. Thus A is a *diagonal operator*.

Exercise 4.9 We consider the setup described in Theorem 4.46. Let λ be an eigenvalue of A. Show:

(a) $\ker(A - \lambda I)$ is finite dimensional;

(b) given $f \in H$, (4.25) has a solution if and only if $(f, e_m)_H = 0$ for all $m \in \mathbb{N}$ such that $\lambda_m = \lambda$.

Exercise 4.10 Let $g \in C([0, \pi])$ be real valued, and set

$$a_k = \frac{2}{k} \int_0^{\pi} \cos(ks)\, g(s)\, ds, \ k \in \mathbb{N}_0, \qquad b_k = \frac{2}{k} \int_0^{\pi} \sin(ks)\, g(s)\, ds, \ k \in \mathbb{N},$$

cf. Corollary 3.19.

(a) Show that $\sum_{k=1}^{\infty} (a_k^2 + b_k^2) < \infty$.

Suggestion: Use Theorem 4.9.

(b) Let $g \in C^1([0, \pi])$ with $g(0) = g(\pi) = 0$. Show that $\sum_{k=1}^{\infty} |b_k| < \infty$, and conclude that $g(x) = \sum_{k=1}^{\infty} b_k \sin(kx)$, where the series converges uniformly in x.

Suggestion: Apply (a) to $a_k' = \frac{2}{k} \int_0^{\pi} \cos(ks)\, g'(s)\, ds$ and show that $b_k = \frac{a_k'}{k}$.

Exercise 4.11 Consider the initial-boundary value problem (3.12) (page 53) with $a = 0$, $b = \pi$ and assume that u_0, u_1 satisfy the hypotheses of Theorem 3.6 (page 54). The aim of this exercise is to show that the solution u of (3.12) is given by

$$u(t, x) = \sum_{k=1}^{\infty} \sin(kx)\left(a_k \cos(ckt) + b_k \sin(ckt)\right), \tag{4.32}$$

where $a_k = \dfrac{2}{\pi}\displaystyle\int_0^{\pi} u_0(x) \sin(kx)\,dx$ and $b_k = \dfrac{2}{ck\pi}\displaystyle\int_0^{\pi} u_1(x) \sin(kx)\,dx$.

This shows that the solution u is a superposition of the harmonics $u_k(t, x) = \sin(kx)\left(a_k \cos(ckt) + b_k \sin(ckt)\right)$. Observe that in the expression for $u_k(t, x)$ only the amplitude of the wave $\sin(k\cdot)$ is time dependent.

(a) Use Exercise 4.10 to show that $\sum_{k=1}^{\infty} \left(|a_k| + |b_k|\right) < \infty$, so that (4.32) converges uniformly to the continuous function u.

(b) Assume that $u_1 = 0$ and let $\tilde{u}_0(x) = \sum_{k=1}^{\infty} a_k \sin(kx)$, $x \in \mathbb{R}$.

Show that $\tilde{u}_0 \in C^2(\mathbb{R})$ and $\tilde{u}_0|_{[0,\pi]} = u_0$.
Suggestion: Use Exercise 3.21.

(c) Assume that $u_1 = 0$. We know from Theorem 3.6 and its proof that $u(t, x) = \frac{1}{2}\left(\tilde{u}_0(x + ct) + \tilde{u}_0(x - ct)\right)$. Use this to prove (4.32).

(d) Assume that $u_0 = 0$. Let $c_k = \frac{2}{\pi}\int_0^{\pi} u_1(x) \sin(kx)\,dx$. Define $H : \mathbb{R} \to \mathbb{R}$ by

$$H(x) := -\sum_{k=1}^{\infty} \frac{c_k}{k} \cos(kx).$$ Show that $H \in C^1(\mathbb{R})$ and $H' = \tilde{u}_1$. By Theorem 3.6 and its proof the solution u of (3.12) is given by $u(t, x) = \frac{1}{2c}\left(H(x+ct) - H(x-ct)\right)$. Use this to show (4.32) in the case $u_0 = 0$.

(e) Put (c) and (d) together to prove the claim.

Exercise 4.12 Let $V \hookrightarrow H \hookrightarrow V'$ be a Gelfand triple as in Remark 4.51; in particular, V is taken to be dense in H. Show that H is dense in V'.
Suggestion: Use Corollary 4.18.

Chapter 5
Sobolev spaces and boundary value problems in dimension one

Sobolev spaces on an interval have special appeal: they consist of functions which can be written as indefinite integrals of integrable functions. Numerous properties and rules of calculation, such as the integration by parts formula, still hold, and give us a version of calculus which is almost as practicable as the classical version taught in first year. But what one gains by basing everything on integrable rather than continuous functions, is the structure of a Hilbert space, namely the Hilbert space of (weakly) differentiable functions. As such, the theorem of Riesz–Fréchet, and more generally the Lax–Milgram theorem, can be applied to obtain existence and uniqueness of solutions of differential equations with boundary conditions.

In this chapter we carry out the entire program in the simple situation of dimension one, which we then repeat for higher dimensions in Chapters 6 and 7. But even in dimension one we can see how efficiently and elegantly Hilbert space methods can be applied. The contents of this chapter are essentially from the area of ordinary differential equations: we consider problems of Sturm–Liouville type or – expressed in the language of our model examples – stationary reaction and diffusion equations.

The various boundary conditions (Dirichlet, Neumann, mixed and Robin) do not cause much trouble: Sobolev spaces on intervals consist of functions which are continuous up to the boundary.

The chapter consists of two parts: the introduction to Sobolev spaces, and the boundary value problems as applications. Some of the exercises at the end of the chapter also give additional information about the theory, others lead to the solution of further boundary value problems.

Chapter overview

© The Author(s), under exclusive license to Springer Nature Switzerland AG 2023
W. Arendt, K. Urban, *Partial Differential Equations*, Graduate Texts in
Mathematics 294, https://doi.org/10.1007/978-3-031-13379-4_5

5.1 Sobolev spaces in one variable

Let $-\infty < a < b < \infty$. We consider the space $L_2(a, b)$ of real-valued square integrable functions on the interval (a, b). This is a real Hilbert space with respect to the inner product

$$(f, g)_{L_2} = \int_a^b f(x) g(x) \, dx,$$

which induces the norm

$$\|f\|_{L_2} := \sqrt{(f, f)} = \left(\int_a^b (f(x))^2 \, dx \right)^{\frac{1}{2}}.$$

In this section we wish to study *weak derivatives*. For this, we will need the following class of C^1-functions with compact support.

Let $C_c^1(a, b)$ denote the space of the functions $v \in C^1(a, b)$ with *compact support*, that is,

$$C_c^1(a, b) := \{v \in C^1(a, b) : \exists \varepsilon > 0 \text{ such that}$$
$$v(x) = 0 \text{ for } x \in [a, a + \varepsilon] \cup [b - \varepsilon, b]\}.$$

For $f \in C_c^1(a, b)$ one calls $\text{supp} f := \overline{\{x \in [a, b] : f(x) \neq 0\}}$ the *support* of f, cf. (2.1). We start by noting that every function $f \in L_2(a, b)$ can be approximated by functions in $C_c^1(a, b)$.

Lemma 5.1 *The space $C_c^1(a, b)$ is dense in $L_2(a, b)$.*

Proof The statement follows, for example, from the definition of $L_2(a, b)$ (since the step functions, which can be easily smoothed, are dense in $L_2(a, b)$, see Theorem A.11). We will, however, also give a proof in higher dimensions (Corollary 6.9). □

If $f \in C^1([a, b])$, that is, if f is continuously differentiable in the usual sense, then by integration by parts, for all $v \in C_c^1(a, b)$ we obtain

$$-(f, v')_{L_2} = -\int_a^b f(x) v'(x) \, dx = \int_a^b f'(x) v(x) \, dx = (f', v)_{L_2}, \qquad (5.1)$$

where the boundary terms disappear since $v(a) = v(b) = 0$.

The integrals which appear here are clearly still well defined if f' is only Lebesgue integrable. This observation leads us to the following definition.

Definition 5.2 Let $f \in L_2(a, b)$. A function $g \in L_2(a, b)$ is called a *weak derivative* of f if

$$-\int_a^b f(x) v'(x) \, dx = \int_a^b g(x) v(x) \, dx \text{ for all } v \in C_c^1(a, b),$$

or, written differently, $-(f, v')_{L_2} = (g, v)_{L_2}$. △

We next prove that weak derivatives are unique.

Lemma 5.3 *A function* $f \in L_2(a, b)$ *has at most one weak derivative* $g \in L_2(a, b)$.

Proof Suppose that $h \in L_2(a, b)$ is a second weak derivative of f. Then it follows from the definition that

$$\int_a^b g(x) v(x) \, dx = \int_a^b h(x) v(x) \, dx,$$

that is, $(g - h, v)_{L_2} = 0$ for all $v \in C_c^1(a, b)$. Since $C_c^1(a, b)$ is dense in $L_2(a, b)$, it follows that $(g - h, v)_{L_2} = 0$ for all $v \in L_2(a, b)$. In particular, $(g - h, g - h)_{L_2} = 0$ and thus $g = h$ in $L_2(a, b)$. □

Remark 5.4 If $f \in L_2(a, b) \cap C^1(a, b)$ and $f' \in L_2(a, b)$, then f' is also the weak derivative of f, as follows directly from (5.1). That is, if a derivative exists in the classical sense, then it coincides with the weak derivative. We therefore denote the weak derivative of a function $f \in L_2(a, b)$, if it exists, by f'. △

We now define the Sobolev space $H^1(a, b)$ of order one as the space of those functions in $L_2(a, b)$ which have a weak derivative in $L_2(a, b)$, that is,

$$H^1(a, b) := \left\{ f \in L_2(a, b) : \exists f' \in L_2(a, b) \text{ such that} \right.$$

$$\left. -\int_a^b f(x) v'(x) \, dx = \int_a^b f'(x) v(x) \, dx \ \forall v \in C_c^1(a, b) \right\}.$$

We next give an example of a function which is weakly but not classically differentiable.

Example 5.5 (Absolute value function) Let $f(x) := |x|$. Then $f \in H^1(-1, 1)$ and

$$f'(x) = \text{sign}(x) := \begin{cases} 1, & \text{if } x > 0, \\ 0, & \text{if } x = 0, \\ -1, & \text{if } x < 0; \end{cases}$$

that is, the weak derivative is the piecewise derivative; see Exercise 5.2 for the proof. △

It is immediate that $H^1(a, b)$ is a vector subspace of $L_2(a, b)$ and the mapping $f \mapsto f' : H^1(a, b) \to L_2(a, b)$ is linear.

Theorem 5.6 *The space* $H^1(a, b)$ *is a Hilbert space with respect to the inner product*

$$(f, g)_{H^1} := (f, g)_{L_2} + (f', g')_{L_2}.$$

If necessary, we may also write $(\cdot, \cdot)_{H^1(a,b)}$. We denote the associated norm on $H^1(a, b)$ by $\|u\|_{H^1(a,b)} := \sqrt{(u, u)_{H^1}}$.

Proof We need to show that the space $H^1(a, b)$ is complete with respect to the norm

$$\|f\|_{H^1}^2 := \int_a^b |f(x)|^2\, dx + \int_a^b |f'(x)|^2\, dx = \|f\|_{L_2}^2 + \|f'\|_{L_2}^2.$$

Now the Cartesian product $H := L_2(a, b) \times L_2(a, b)$ is a Hilbert space with respect to the norm

$$\|[f, g]\|_H^2 := \|f\|_{L_2}^2 + \|g\|_{L_2}^2, \quad \text{where } [f, g] \in H \text{ is any pair, i.e., element of } H.$$

The mapping $j : H^1(a, b) \to H$ given by $j(f) := [f, f']$ is linear, and isometric since $\|j(f)\|_H = (\|f\|_{L_2}^2 + \|f'\|_{L_2}^2)^{1/2} = \|f\|_{H^1}$. Thus $H^1(a, b)$ is isometrically isomorphic to its image $j(H^1(a, b)) =: F$. We therefore need to show that $F = \{[f, f'] : f \in H^1(a, b)\}$ is closed in H. This will prove that F is complete, since H is. So let $[f, g] \in \overline{F}$ be given. Then there exist $f_n \in H^1(a, b)$ such that $f_n \to f$ and $f_n' \to g$ in $L_2(a, b)$. Hence, for all $v \in C_c^1(a, b)$,

$$-\int_a^b f(x)\, v'(x)\, dx = (-1) \lim_{n\to\infty} \int_a^b f_n(x)\, v'(x)\, dx$$

$$= \lim_{n\to\infty} \int_a^b f_n'(x)\, v(x)\, dx = \int_a^b g(x)\, v(x)\, dx.$$

This means that $f' = g$ according to our definition, and thus $[f, g] \in F$.   □

The following results show that many properties of classical derivatives continue to hold for weak derivatives. As always, we will identify functions which agree almost everywhere.

Lemma 5.7 *Let $f \in H^1(a, b)$ be such that $f' = 0$. Then f is constant.*

Proof Fix $\psi \in C_c^1(a, b)$ such that $\int_a^b \psi(x)\, dx = 1$. Let $w \in C_c^1(a, b)$; we claim that there exists $v \in C_c^1(a, b)$ such that

$$v'(x) = w(x) - \psi(x) \int_a^b w(y)\, dy.$$

To see this, we define $g(x) := w(x) - \psi(x) \int_a^b w(y)\, dy$. Then $g \in C_c^1(a, b)$ and $\int_a^b g(x)\, dx = 0$. We can then set $v(x) := \int_a^x g(y)\, dy$. Now since $f' = 0$ weakly, with v defined as above, we have

$$0 = -\int_a^b f'(x)\, v(x)\, dx = \int_a^b f(x)\, v'(x)\, dx$$

$$= \int_a^b f(x)\,w(x)\,dx - \left(\int_a^b w(x)\,dx \right)\left(\int_a^b f(x)\,\psi(x)\,dx \right)$$

$$= \int_a^b \left(f(x) - \left(\int_a^b f(y)\,\psi(y)\,dy \right) \right) w(x)\,dx.$$

Since $w \in C_c^1(a,b)$ was arbitrary and $C_c^1(a,b)$ is dense in $L_2(a,b)$, it follows that $f(x) - \int_a^b f(y)\,\psi(y)\,dy = 0$ in $L_2(a,b)$, that is,

$$f(x) = \int_a^b f(y)\,\psi(y)\,dy \equiv \text{const}$$

for almost all $x \in (a,b)$. □

In what follows, we will require Fubini's theorem in the following form.

Lemma 5.8 (Fubini's theorem) *Let* $f, g \in L_1(a,b)$. *Then for* $a \le x \le b$ *we have*

$$\int_a^b \int_a^x g(y)\,dy\, f(x)\,dx = \int_a^b \int_y^b f(x)\,dx\, g(y)\,dy.$$

Proof Let

$$\mathbb{1}_{[a,x]}(y) = \begin{cases} 1 & \text{if } a \le y \le x, \\ 0 & \text{otherwise.} \end{cases}$$

The theorems of Fubini and Tonelli on exchanging the order of integration show that

$$\int_a^b \int_a^b \mathbb{1}_{[a,x]}(y)g(y)f(x)\,dy\, dx = \int_a^b \int_a^b \mathbb{1}_{[a,x]}(y)g(y)f(x)\,dx\, dy.$$

This is exactly the desired identity. □

The following theorem is a weak version of the Fundamental Theorem of Calculus.

Theorem 5.9 (Lebesgue version of the fundamental theorem of calculus)
(a) *Let* $g \in L_2(a,b)$, $c \in \mathbb{R}$ *and* $f(x) := c + \int_a^x g(y)\,dy$ *for* $x \in (a,b)$. *Then* $f \in H^1(a,b)$ *and* $f' = g$.
(b) *In the opposite direction, let* $f \in H^1(a,b)$. *Then there exists* $c \in \mathbb{R}$ *such that* $f(x) = c + \int_a^x f'(y)\,dy$ *for almost every* $x \in (a,b)$.

Proof (a) Let $v \in C_c^1(a,b)$. Since $\int_y^b v'(x)\,dx = v(b) - v(y) = -v(y)$, by Lemma 5.8 we have

$$-\int_a^b f(x)v'(x)\,dx = -\int_a^b \int_a^x g(y)\,dy\, v'(x)\,dx - c\int_a^b v'(x)\,dx$$

$$= -\int_a^b \int_y^b v'(x)\,dx\,g(y)\,dy = \int_a^b v(y)\,g(y)\,dy.$$

Thus g is the weak derivative of f.

(b) Let $f \in H^1(a, b)$ be given. Define the function w by $w(x) := f(x) - \int_a^x f'(y)\,dy$. Then $w \in H^1(a, b)$ and $w' = 0$ by (a). Hence w is constant by Lemma 5.7.   □

Corollary 5.10 *We have* $H^1(a, b) \subset C([a, b])$.

The statement of the above Corollary 5.10 requires an explanation. In accordance with the definition of $L_2(a, b)$, we identify functions which agree almost everywhere. If two continuous functions agree almost everywhere, then they are identically equal (since if they were to differ in a single point, then they would differ in a neighborhood of that point). Theorem 5.9 permits us to identify each function $f \in H^1(a, b)$ with a (or rather its) *continuous representative*,

$$f(x) = f(a) + \int_a^x f'(y)\,dy \tag{5.2}$$

for all $x \in [a, b]$. In the sequel we will always choose the continuous representative of $f \in H^1(a, b)$. The following result follows directly from (5.2).

Corollary 5.11 *A function* $f \in H^1(a, b)$ *is in* $C^1([a, b])$ *if and only if* $f' \in C([a, b])$.

Now $C([a, b])$ is a Banach space, that is, a complete normed space, with respect to the norm

$$\|f\|_\infty := \sup_{x \in [a,b]} |f(x)|.$$

In this norm, the imbedding of $H^1(a, b)$ in the space $C([a, b])$ is continuous, that is, $H^1(a, b) \hookrightarrow C([a, b])$, cf. (4.17). But more is true.

Theorem 5.12 *The imbeddings of* $H^1(a, b)$ *in* $C([a, b])$ *and* $L_2(a, b)$ *are compact.*

Proof 1. The imbedding of $H^1(a, b)$ in $C([a, b])$ is continuous. This follows directly from the closed graph theorem (Theorem A.2); see Exercise 5.11. However, we will give a direct proof. We have to show that there exists a constant $c \geq 0$ such that $\|f\|_\infty \leq c\|f\|_{H^1(a,b)}$ for all $f \in H^1(a, b)$. Assume that this statement is false. Then there exists a sequence $(f_n)_{n \in \mathbb{N}}$ in $H^1(a, b)$ such that $\|f_n\|_{H^1(a,b)} \leq 1$, but $\|f_n\|_\infty \geq n$. By Theorem 5.9,

$$f_n(x) = f_n(a) + \int_a^x f_n'(y)\,dy. \tag{5.3}$$

By the Cauchy–Schwarz inequality,

$$\left| \int_a^x f_n'(y)\, dy \right| \le (b-a)^{\frac{1}{2}} \left(\int_a^b |f_n'(y)|^2\, dy \right)^{\frac{1}{2}} \le (b-a)^{\frac{1}{2}} \|f_n\|_{H^1(a,b)}$$

$$\le (b-a)^{\frac{1}{2}}.$$

If $(f_n(a))_{n \in \mathbb{N}}$ were bounded, then by (5.3), the sequence $(f_n(x))_{n \in \mathbb{N}}$ would be too. Thus $\lim_{n \to \infty} |f_n(a)| = \infty$. Suppose without loss of generality that in fact $\lim_{n \to \infty} f_n(a) = \infty$. By what we have shown above, $f_n(x) \ge f_n(a) - (b-a)^{\frac{1}{2}}$ and thus we get $\lim_{n \to \infty} \|f_n\|_{L_2} = \infty$. But since on the other hand $\|f_n\|_{L_2} \le \|f_n\|_{H^1} \le 1$, we have found a contradiction.

2. We will now show that the imbedding of $H^1(a,b)$ in $C([a,b])$ is compact. Denote by $B := \{f \in H^1(a,b) : \|f\|_{H^1} \le 1\}$ the unit sphere in $H^1(a,b)$. We wish to show that this set is relatively compact in $C([a,b])$. We already know from (a) that B is bounded in $C([a,b])$. We estimate $f \in B$ with the help of Hölder's inequality as follows:

$$|f(x) - f(y)| = \left| \int_y^x f'(t)\, dt \right| \le \left(\int_y^x |f'(t)|^2\, dt \right)^{\frac{1}{2}} |y - x|^{\frac{1}{2}}$$

$$\le \|f\|_{H^1} |y - x|^{\frac{1}{2}} \le |y - x|^{\frac{1}{2}}.$$

This means that B is equicontinuous, and the claim follows from the Arzelà–Ascoli theorem (Theorem A.6).

3. Since the imbedding of $C([a,b])$ in $L_2(a,b)$ is continuous, the second claim of the theorem follows from the first one. □

Next we shall provide a weak version of the product rule. Since weak derivatives are defined via integrals, this rule can be proved via Fubini's theorem in the form of Lemma 5.8.

Theorem 5.13 (Product rule and integration by parts) *Let $f, g \in H^1(a,b)$. Then*
(a) $f \cdot g \in H^1(a,b)$ and $(f \cdot g)' = f'g + fg'$;
(b) integration by parts:

$$\int_a^b f(x)\, g'(x)\, dx = f(b)\, g(b) - f(a)\, g(a) - \int_a^b f'(x)\, g(x)\, dx.$$

Proof 1. By Lemma 5.8 we have

$$\int_a^b f(x)\, g'(x)\, dx = \int_a^b \left(f(a) + \int_a^x f'(y)\, dy \right) g'(x)\, dx$$

$$= f(a)\, g(b) - f(a)\, g(a) + \int_a^b \int_y^b g'(x)\, dx\, f'(y)\, dy$$

$$= f(a)\, g(b) - f(a)\, g(a) + \int_a^b (g(b)\, f'(y) - g(y)\, f'(y))\, dy$$

$$= f(a) g(b) - f(a) g(a) + g(b) f(b) - g(b) f(a)$$

$$- \int_a^b g(y) f'(y) \, dy.$$

This proves part (b).

2. If in 1. we replace the upper limit of integration b by $x \in [a, b)$, then we obtain

$$\int_a^x f(y) g'(y) \, dy = f(x) g(x) - f(a) g(a) - \int_a^x f'(y) g(y) \, dy,$$

and thus

$$f(x) g(x) = f(a) g(a) + \int_a^x \{ f(y) g'(y) + f'(y) g(y) \} \, dy.$$

It now follows from Theorem 5.9 that $f \cdot g \in H^1(a, b)$ and $(f \cdot g)' = f'g + fg'$.
□

We next show using integration by parts how one can glue together H^1-functions defined piecewise.

Theorem 5.14 *Let* $-\infty < a = t_0 < t_1 < \ldots < t_n = b$ *be a partition of the interval* $[a, b]$. *If* $f \in C([a, b])$ *is such that* $f_{|(t_{i-1}, t_i)} \in H^1(t_{i-1}, t_i)$ *for all* $i = 1, \ldots, N$, *then* $f \in H^1(a, b)$.

Proof Denote by $f_i' \in L_2(t_{i-1}, t_i)$ the weak derivative of $f_{|(t_{i-1}, t_i)}$ and define a function $g \in L_2(a, b)$ by $g(t) := f_i'(t)$ for $t \in (t_{i-1}, t_i)$, $g(t) = 0$ for $t \in \{t_0, t_1, \ldots, t_N\}$. We will show that g is the weak derivative of f, which in particular implies that $f \in H^1(a, b)$. To this end, we let $v \in C_c^1(a, b)$ be arbitrary and obtain, upon integrating by parts,

$$- \int_a^b f v' \, dt = - \sum_{i=1}^N \int_{t_{i-1}}^{t_i} f v' \, dt$$

$$= \sum_{i=1}^N \left\{ \int_{t_{i-1}}^{t_i} f_i' v \, dt - (f(t_i) v(t_i) - f(t_{i-1}) v(t_{i-1})) \right\}$$

$$= \int_a^b g v \, dt - f(t_N) v(t_N) + f(t_0) v(t_0) = \int_a^b g v \, dt,$$

since $v(t_N) = v(t_0) = 0$. Hence $f' = g$.
□

We will need the following special case of Theorem 5.14 in Chapter 9 on numerical methods. A function $f : I \to \mathbb{R}$ is called *affine* if there exist $\alpha, \beta \in \mathbb{R}$ such that $f(x) = \alpha x + \beta$ for all $x \in I$.

Remark 5.15 Let $a = t_0 < t_1 < \ldots < t_N = b$ be a partition of the interval $[a, b]$. Then the space $A := \{ f \in C([a, b]) : f_{|(t_{i-1}, t_i)} \text{ is affine, } i = 1, \ldots, N \}$ of *piecewise affine functions* is contained in $H^1(a, b)$. The weak derivative of a function f in A is constant on each of the intervals (t_{i-1}, t_i), that is, f' is a step function.   △

We next wish to define Sobolev spaces of higher order. We set

$$H^2(a, b) := \{f \in H^1(a, b) : f' \in H^1(a, b)\}.$$

Then, for $f \in H^2(a, b)$, $f'' := (f')'$ is in $L_2(a, b)$. Since $f' \in H^1(a, b) \hookrightarrow C([a, b])$ and $f(x) = f(a) + \int_a^x f'(y)\, dy$, it follows that $H^2(a, b) \subset C^1([a, b])$. Similarly to Theorem 5.6 we see that $H^2(a, b)$ is a Hilbert space with respect to the inner product

$$(f, g)_{H^2} := (f, g)_{L_2} + (f', g')_{L_2} + (f'', g'')_{L_2}.$$

More generally, for $k \in \mathbb{N}$ we define inductively

$$H^{k+1}(a, b) := \{f \in H^1(a, b) : f' \in H^k(a, b)\},$$

and for $f \in H^{k+1}(a, b)$ we set $f^{(k+1)} := (f')^{(k)}$. Then $H^k(a, b)$ is a Hilbert space with respect to the inner product

$$(f, g)_{H^k} := \sum_{m=0}^k (f^{(m)}, g^{(m)})_{L_2},$$

where $f^{(0)} := f$. We then have

$$H^{k+1}(a, b) \subset C^k([a, b]) \quad \text{for all } k \in \mathbb{N}, \tag{5.4}$$

as one can see easily by induction (see Exercise 5.11).

Finally, we wish to consider those Sobolev functions which are zero on the boundary. Since $H^1(a, b) \hookrightarrow C([a, b])$, the spaces

$$H^1_{(a)}(a, b) := \{f \in H^1(a, b) : f(a) = 0\},$$
$$H^1_{(b)}(a, b) := \{f \in H^1(a, b) : f(b) = 0\} \quad \text{and}$$
$$H^1_0(a, b) := \{f \in H^1(a, b) : f(a) = f(b) = 0\}$$

are closed subspaces of $H^1(a, b)$. The following estimate plays a central role in the study of boundary value problems.

Theorem 5.16 (Poincaré inequality) *We have*

$$\int_a^b u(x)^2\, dx \leq \frac{1}{2}(b - a)^2 \int_a^b \left(u'(x)\right)^2 dx$$

for all $u \in H^1_{(a)}(a, b) \cup H^1_{(b)}(a, b)$.

Proof For $u \in H_{(a)}^1(a, b)$, (5.2) yields the representation $u(x) = \int_a^x u'(y)\, dy$. Hence we may apply Hölder's inequality to obtain the estimate

$$|u(x)| \leq (x - a)^{\frac{1}{2}} \left(\int_a^b (u'(y))^2 \, dy \right)^{\frac{1}{2}}.$$

It follows that

$$\int_a^b u(x)^2 \, dx \leq \int_a^b (x - a) \, dx \int_a^b (u'(y))^2 \, dy = \frac{1}{2}(b - a)^2 \int_a^b (u'(y))^2 \, dy,$$

as claimed. Given $u \in H_{(b)}^1(a, b)$, apply the first case to v defined by $v(x) := u(a+b-x)$ to prove the estimate in the second case.   □

The Poincaré inequality shows in particular that

$$|u|_{H^1(a,b)} := \left(\int_a^b (u'(x))^2 \, dx \right)^{\frac{1}{2}} = \|u'\|_{L_2}$$

defines an equivalent norm on $H_{(a)}^1(a, b)$ and on $H_0^1(a, b)$, since it implies that

$$\|u\|_{H^1(a,b)} \leq c \, |u|_{H^1(a,b)}, \quad u \in H_{(a)}^1(a, b), \tag{5.5}$$

for some constant $c > 0$. One should, however, be aware that $|\cdot|_{H^1(a,b)}$ only defines a seminorm on $H^1(a, b)$, since $|f|_{H^1(a,b)} = 0$ for any constant function f.

5.2 Boundary value problems on the interval

The problems we will consider in this section involve linear differential equations with various boundary conditions; we will use them to demonstrate the effectiveness of Hilbert space methods. Everything is based on the theorem of Riesz–Fréchet or, more generally, the Lax–Milgram theorem. The same arguments also carry over to domains in $\mathbb{R}^d$, that is, where we replace the second derivative by the Laplacian in two or more dimensions. For this, however, we require Sobolev spaces in higher dimensions, among other tools, to which we will turn in the next chapter. Here, in this section, we will always work on a bounded interval $(a, b) \subset \mathbb{R}$.

5.2.1 Dirichlet boundary conditions

We will start by studying Dirichlet boundary conditions. Let the constant λ be nonnegative and the function $f \in L_2(a, b)$ be given. We wish to consider the problem of determining $u \in H^2(a, b)$ such that

$$\lambda u - u'' = f \quad \text{in } (a, b), \tag{5.6a}$$
$$u(a) = u(b) = 0. \tag{5.6b}$$

Suppose that u is a solution of (5.6). If we multiply (5.6a) by a function $v \in H_0^1(a, b)$ and integrate, then by integration by parts (Theorem 5.13) we obtain

$$\int_a^b f(x)\, v(x)\, dx = \lambda \int_a^b u(x)\, v(x)\, dx - \int_a^b u''(x)\, v(x)\, dx$$
$$= \lambda \int_a^b u(x)\, v(x)\, dx + \int_a^b u'(x)\, v'(x)\, dx$$
$$\quad - u'(b)\, v(b) + u'(a)\, v(a)$$
$$= \lambda \int_a^b u(x)\, v(x)\, dx + \int_a^b u'(x)\, v'(x)\, dx$$

since $v(a) = v(b) = 0$. This motivates us to consider the bilinear form

$$a(u, v) := \lambda \int_a^b u(x)\, v(x)\, dx + \int_a^b u'(x)\, v'(x)\, dx$$

on $H_0^1(a, b) \times H_0^1(a, b)$. It is easy to check that this form is continuous and symmetric; moreover, it is coercive since $a(u) \geq |u|_{H^1(a,b)}^2 \geq \frac{1}{c} \|u\|_{H^1(a,b)}^2$ by (5.5), recall (4.5). We also define a continuous linear form F on $H_0^1(a, b)$ by

$$F(v) := \int_a^b v(x)\, f(x)\, dx.$$

By the Lax–Milgram theorem (Theorem 4.23) there exists a unique $u \in H_0^1(a, b)$ such that

$$a(u, v) = F(v) \quad \text{for all } v \in H_0^1(a, b). \tag{5.7}$$

We will now show that this u is a solution of (5.6). The identity (5.7) says that

$$\int_a^b u'(x)\, v'(x)\, dx = \int_a^b (f(x) - \lambda u(x))v(x)\, dx$$

for all $v \in H_0^1(a, b)$ and in particular for all $v \in C_c^1(a, b)$. Since $u' \in L_2(a, b)$ and $f - \lambda u \in L_2(a, b)$, by Definition 5.2 this means that $u' \in H^1(a, b)$, so that $u \in H^2(a, b)$ and $-u'' = f - \lambda u$. Since also $u \in H_0^1(a, b)$, it is thus a solution of (5.6).

We next prove uniqueness. Let u be a solution of (5.6), that is, let $u \in H_0^1(a, b) \cap H^2(a, b)$ satisfy $\lambda u - u'' = f$. We already showed above that in this case $a(u, v) = F(v)$ for all $v \in H_0^1(a, b)$. But then uniqueness in the Lax–Milgram theorem means that only one such u can exist.

Now we wish to investigate the regularity of the solution u of (5.6). It follows directly from the representation of a function $w \in H^1(a, b)$ as an indefinite integral (Theorem 5.9) that $w \in H^1(a, b)$ is in $C^1([a, b])$ if and only if $w' \in C([a, b])$. More generally, one sees by induction that, for $k \in \mathbb{N}$ and $w \in H^k(a, b)$,

$$w \in C^k([a, b]) \iff w^{(k)} \in C([a, b]). \tag{5.8}$$

Now let u be the solution of (5.6). Since $u'' = \lambda u - f$, it follows from (5.8) with $k = 2$ that $u \in C^2([a, b])$ if and only if $f \in C([a, b])$. We thus obtain inductively that, for any $k \in \mathbb{N}$, $u \in C^{k+2}([a, b])$ if (and only if) $f \in C^k([a, b])$. Such a statement is known as a *shift theorem*, since the regularity of the solution is increased (shifted) by exactly the order of the differential operator (here 2), compared with the orther of the right-hand side. Summarizing, we have proved the following statement about solvability and regularity of problem (5.6). Here we set $C^0([a, b]) := C([a, b])$.

Theorem 5.17 (Dirichlet boundary conditions) *Let* $\lambda \geq 0$ *and* $f \in L_2(a, b)$ *be given. Then problem (5.6) has exactly one solution* $u \in H^2(a, b) \cap H_0^1(a, b)$. *Moreover,* $u \in C^{k+2}([a, b])$ *if* $f \in C^k([a, b])$ *for some* $k \in \mathbb{N}_0$.

The case of inhomogeneous Dirichlet boundary conditions can be reduced to problem (5.6). Suppose that we wish to find $u \in H^2(a, b)$ such that

$$\lambda u - u'' = f \qquad \text{in } (a, b), \tag{5.9a}$$

$$u(a) = A, \quad u(b) = B, \tag{5.9b}$$

for given $f \in L_2(a, b)$ and $A, B \in \mathbb{R}$. We may then choose any function $g \in C^\infty([a, b])$ such that $g(a) = A$ and $g(b) = B$, and solve the homogeneous problem (5.6) with right-hand side $\tilde{f} := f - \lambda g + g''$; let us call this solution $u_0 \in H^2(a, b) \cap H_0^1(a, b)$. Then $u := u_0 + g$ solves the inhomogeneous problem (5.9) since $\lambda u - u'' = \lambda u_0 - u_0'' + \lambda g - g'' = \tilde{f} + \lambda g - g'' = f$ and $u(a) = u_0(a) + g(a) = A$, as well as $u(b) = u_0(b) + g(b) = B$. We will also apply this method of reduction to homogeneous Dirichlet boundary conditions to domains in $\mathbb{R}^2$, in Chapter 7. Here we have shown that for all $A, B \in \mathbb{R}$ and $f \in L_2(a, b)$ there is exactly one function $u \in H^2(a, b)$ which solves problem (5.9).

5.2.2 Neumann boundary conditions

Now we turn to the case of Neumann boundary conditions. Let $\lambda > 0$ and $f \in L_2(a, b)$ be given. Here we seek a solution $u \in H^2(a, b)$ of the problem

$$\lambda u - u'' = f \qquad \text{in } (a, b), \tag{5.10a}$$

$$u'(a) = u'(b) = 0. \tag{5.10b}$$

One should note that $H^2(a, b) \subset C^1([a, b])$, and thus the *Neumann boundary condition* (5.10b) is well defined.

Theorem 5.18 (Neumann boundary conditions) *Let $f \in L_2(a, b)$ and $\lambda > 0$. Then there exists a unique function $u \in H^2(a, b)$ such that (5.10) holds. Moreover, $u \in C^{k+2}([a, b])$ if $f \in C^k([a, b])$ for some $k \in \mathbb{N}_0$.*

Proof Suppose that $u \in H^2(a, b)$ is a solution of (5.10) and $v \in H^1(a, b)$. Integrating by parts (Theorem 5.13), we obtain

$$
\int_a^b f(x) v(x) \, dx = \lambda \int_a^b u(x) v(x) \, dx - \int_a^b u''(x) v(x) \, dx
$$

$$
= \lambda \int_a^b u(x) v(x) \, dx + \int_a^b u'(x) v'(x) \, dx \qquad (5.11)
$$
$$
- (u'(b) v(b) - u'(a) v(a))
$$

$$
= \lambda \int_a^b u(x) v(x) \, dx + \int_a^b u'(x) v'(x) \, dx, \qquad (5.12)
$$

where we used (5.10b) in the last step. We now consider the Hilbert space $V :=$ $H^1(a, b)$ and the bilinear form $a : V \times V \to \mathbb{R}$ given by

$$
a(u, v) := \lambda \int_a^b u(x) v(x) \, dx + \int_a^b u'(x) v'(x) \, dx.
$$

Clearly, $a(\cdot, \cdot)$ is continuous and coercive, since it corresponds exactly to the inner product on V (cf. Theorem 5.6). We also define a continuous linear form $F \in V'$ by $F(v) := \int_a^b f(x) v(x) \, dx$. We have seen that if $u \in H^2(a, b)$ is a solution of (5.10), then $a(u, v) = F(v)$ for all $v \in V$. But by the Lax–Milgram theorem we know that there exists a unique function $u \in V$ such that $a(u, v) = F(v)$ for all $v \in H^1(a, b)$; only this function u can be a solution. This already proves uniqueness.

We will now show that u is actually a solution. To this end, we note that by definition of a and F, we have

$$
\int_a^b u'(x) v'(x) \, dx = \int_a^b (f(x) - \lambda u(x)) v(x) \, dx \quad \text{for all } v \in H^1(a, b). \qquad (5.13)
$$

If we take $v \in C_c^1(a, b)$ in (5.13), then we see that $u \in H^2(a, b)$ and $u'' = \lambda u - f$ by definition of weak derivatives. We only need to check the boundary condition. For this, we substitute $f - \lambda u = -u''$ into (5.13) and obtain

$$
\int_a^b u'(x) v'(x) \, dx = - \int_a^b u''(x) v(x) \, dx \quad \text{for all } v \in H^1(a, b).
$$

Integration by parts (Theorem 5.13) now yields

$$
- \int_a^b u''(x) v(x) \, dx = \int_a^b u'(x) v'(x) \, dx - u'(b) v(b) + u'(a) v(a)
$$

and thus

$$-u'(b)\,v(b) + u'(a)\,v(a) = 0 \qquad (5.14)$$

for all $v \in H^1(a, b)$. By choosing $v \in H^1(a, b)$ such that $v(a) = 1$ and $v(b) = 0$, we see that $u'(a) = 0$. We now set $v \equiv 1 \in H^1(a, b)$ in (5.14) and conclude that $u'(b) = 0$. Thus u is a solution of (5.10). The statement about regularity follows from (5.8). □

One can treat inhomogeneous Neumann boundary conditions analogously to inhomogeneous Dirichlet boundary conditions, namely by reduction to homogeneous conditions.

5.2.3 Robin boundary conditions

Next, we wish to consider Robin boundary conditions. Let $\beta_0, \beta_1 \geq 0$, $\lambda > 0$ and $f \in L_2(a, b)$ be given.

Theorem 5.19 (Robin boundary conditions) *There exists a unique $u \in H^2(a, b)$ such that*

$$\lambda u - u'' = f \quad \text{in } (a, b), \qquad (5.15a)$$
$$-u'(a) + \beta_0 u(a) = 0, \quad u'(b) + \beta_1 u(b) = 0. \qquad (5.15b)$$

The condition (5.15b) is called a *Robin boundary condition*, or sometimes a *boundary condition of the third kind*. Since $H^2(a, b) \subset C^1([a, b])$, this condition makes sense for $u \in H^2(a, b)$. In the special case $\beta_0 = \beta_1 = 0$ we recover Neumann boundary conditions.

Proof (of Theorem 5.19) Suppose that $u \in H^2(a, b)$ is a solution of (5.15). We then have, for $v \in H^1(a, b)$,

$$\int_a^b f(x)v(x)\,dx = \int_a^b \lambda u(x)\,v(x)\,dx - \int_a^b u''(x)\,v(x)\,dx$$

$$= \int_a^b \lambda u(x)\,v(x)\,dx + \int_a^b u'(x)\,v'(x)\,dx - u'(b)\,v(b) + u'(a)\,v(a)$$

$$= \lambda \int_a^b u(x)\,v(x)\,dx + \int_a^b u'(x)\,v'(x)\,dx + \beta_1\,u(b)\,v(b) + \beta_0\,u(a)\,v(a).$$

This observation leads us to set $V := H^1(a, b)$ and define $a : V \times V \to \mathbb{R}$ by

$$a(u, v) := \lambda \int_a^b u(x)\,v(x)\,dx + \int_a^b u'(x)\,v'(x)\,dx + \beta_1\,u(b)\,v(b) + \beta_0\,u(a)\,v(a).$$

Then $a(\cdot, \cdot)$ is bilinear, continuous and coercive. We again set

$$F(v) := \int_a^b f(x)\, v(x)\, dx,$$

so that $F \in V'$ is a continuous linear form. Again by the Lax–Milgram theorem there exists a unique $u \in V$ such that $a(u, v) = F(v)$ for all $v \in H^1(a, b)$. Hence in the special case of a function $v \in C_c^1(a, b)$ we have

$$\int_a^b u'(x)\, v'(x)\, dx = \int_a^b (f(x) - \lambda u(x)) v(x)\, dx,$$

and thus $u \in H^2(a, b)$ with $-u'' = f - \lambda u$, by definition of weak derivatives. We then substitute this identity into the relation $a(u, v) = F(v)$, $v \in H^1(a, b)$ and obtain

$$-\int_a^b u''(x)\, v(x)\, dx = \int_a^b (f(x) - \lambda u(x)) v(x)\, dx$$

$$= a(u, v) - \lambda \int_a^b u(x)\, v(x)\, dx$$

$$= \int_a^b u'(x)\, v'(x)\, dx + \beta_1\, u(b)\, v(b) + \beta_0\, u(a)\, v(a)$$

for all $v \in H^1(a, b)$. On the other hand, integrating by parts we see that

$$-\int_a^b u''(x)\, v(x)\, dx = \int_a^b u'(x)\, v'(x)\, dx - u'(b)\, v(b) + u'(a)\, v(a)$$

and thus $-u'(b)\, v(b) + u'(a)\, v(a) = \beta_1\, u(b)\, v(b) + \beta_0\, u(a)\, v(a)$ for all $v \in H^1(a, b)$. It follows that $-u'(b) = \beta_1\, u(b)$ and $u'(a) = \beta_0\, u(a)$. Thus u is a solution of (5.15). The Lax–Milgram theorem guarantees uniqueness of solutions.   □

5.2.4 Mixed and periodic boundary conditions

We will now consider a somewhat more complicated differential operator. In Chapter 1 we saw how one can interpret spatial derivatives physically for problems involving time and space variables. More precisely, we know that second derivatives describe diffusion, first derivatives transport processes (convection) and terms of zeroth order (i.e., no derivatives) reactive processes. The study of solutions which are constant in time is an essential step towards understanding equations in space and time. We thus wish to consider stationary equations in which terms of zeroth, first and second order appear. Such general elliptic differential equations are referred to as stationary *diffusion-convection-reaction equations*. As an example we will study *mixed boundary conditions*, that is, Dirichlet boundary conditions at one endpoint and Neumann boundary conditions at the other endpoint of the interval.

Theorem 5.20 (Mixed boundary conditions) *Let $p \in C^1([a, b])$ and $r \in C([a, b])$ with $r \geq 0$. Suppose that there exists $\alpha > 0$ such that $p(x) \geq \alpha$ for all $x \in [a, b]$, and let $f \in L_2(a, b)$. Then there exists a unique solution $u \in H^2(a, b)$ of*

$$-(p\,u')' + ru = f \quad \text{almost everywhere in } (a, b), \tag{5.16a}$$
$$u(a) = 0, \ u'(b) = 0. \tag{5.16b}$$

If $f \in C([a, b])$, then $u \in C^2([a, b])$.

One should observe that, by Theorem 5.13 (the product rule), $pu' \in H^1(a, b)$ if $u \in H^2(a, b)$, meaning that (5.16a) is well defined.

Proof (of Theorem 5.20) Suppose that $u \in H^2(a, b)$ is a solution of the boundary value problem (5.16) and let $v \in H^1_{(a)}(a, b) := \{u \in H^1(a, b) : u(a) = 0\}$. Then by integrating by parts we obtain

$$\int_a^b f(x)\,v(x)\,dx = -\int_a^b (p(x)\,u'(x))'\,v(x)\,dx + \int_a^b r(x)\,u(x)\,v(x)\,dx$$

$$= \int_a^b p(x)\,u'(x)\,v'(x)\,dx - p(b)\,u'(b)\,v(b) + p(a)\,u'(a)\,v(a)$$

$$\qquad + \int_a^b r(x)\,u(x)\,v(x)\,dx$$

$$= \int_a^b p(x)\,u'(x)\,v'(x)\,dx + \int_a^b r(x)\,u(x)\,v(x)\,dx$$

since $u'(b) = 0$ and $v(a) = 0$. It follows from Theorem 5.12 that the space $H^1_{(a)}(a, b)$ is closed in $H^1(a, b)$ and thus a Hilbert space. The bilinear form

$$a(u, v) := \int_a^b p(x)\,u'(x)\,v'(x)\,dx + \int_a^b r(x)\,u(x)\,v(x)\,dx$$

is continuous on $H^1_{(a)}(a, b) \times H^1_{(a)}(a, b)$, since we have

$$|a(u, v)| \leq c\{\|u'\|_{L_2}\|v'\|_{L_2} + \|u\|_{L_2}\|v\|_{L_2}\}$$

$$\leq 2c(\|u'\|_{L_2}^2 + \|u\|_{L_2}^2)^{\frac{1}{2}}(\|v'\|_{L_2}^2 + \|v\|_{L_2}^2)^{\frac{1}{2}} = 2c\,\|u\|_{H^1}\,\|v\|_{H^1}$$

for $c := \max\{\|p\|_{C([a,b])}, \|r\|_{C([a,b])}\}$. Using the Poincaré inequality (5.5) and the assumptions on the functions p and r, we can prove the coercivity of $a(\cdot, \cdot)$ as follows:

$$a(u) \geq \alpha \int_a^b (u'(x))^2\,dx = \alpha |u|^2_{H^1(a,b)}\,dx \geq \alpha \frac{1}{c^2}\|u\|^2_{H^1}$$

for all $u \in H^1_{(a)}(a, b)$. Now let $f \in L_2(a, b)$, then $F(v) := \int_a^b f(x) v(x) \, dx$ defines a continuous linear form F on $H^1_{(a)}(a, b)$. By the Lax–Milgram theorem there exists a unique $u \in H^1_{(a)}(a, b)$ such that $a(u, v) = \int_a^b f(x) v(x) \, dx$ for all $v \in H^1_{(a)}(a, b)$, that is,

$$\int_a^b p(x) u'(x) v'(x) \, dx + \int_a^b r(x) u(x) v(x) \, dx = \int_a^b f(x) v(x) \, dx \qquad (5.17)$$

for all $v \in H^1_{(a)}(a, b)$. In particular, (5.17) holds for all $v \in C^1_c(a, b)$, whence $pu' \in H^1(a, b)$ and $-(pu')' = (f - ru)$. If we substitute these identities into (5.17), then we obtain by integration by parts that

$$\int_a^b p(x) u'(x) v'(x) \, dx + \int_a^b r(x) u(x) v(x) \, dx = \int_a^b f(x) v(x) \, dx$$

$$= -\int_a^b (p(x) u'(x))' v(x) + \int_a^b r(x) u(x) v(x) \, dx$$

$$= \int_a^b p(x) u'(x) v'(x) \, dx - p(b) u'(b) v(b) + p(a) u'(a) v(a)$$

$$+ \int_a^b r(x) u(x) v(x) \, dx$$

for all $v \in H^1_{(a)}(a, b)$. Since $v(a) = 0$, it follows that $-p(b) u'(b) v(b) = 0$ for all $v \in H^1_{(a)}(a, b)$, and so $u'(b) = 0$. Since $pu' \in H^1(a, b)$ and $\frac{1}{p} \in C^1([a, b]) \subset H^1(a, b)$, it follows from Theorem 5.13 that $u' = \frac{1}{p}(pu') \in H^1(a, b)$, that is, $u \in H^2(a, b)$. Hence u is a solution of (5.16).

If u is a solution of (5.16), then $u \in H^1_{(a)}(a, b)$, and the above argument shows that $a(u, v) = \int_a^b f(x) v(x) \, dx$ for all $v \in H^1_{(a)}(a, b)$. Uniqueness thus follows from the Lax–Milgram theorem.

Finally, if $f \in C([a, b])$, then $(pu')' = ru - f \in C([a, b])$. Thus $pu' \in C^1([a, b])$ and so $u' \in C^1(a, b]$. It follows that $u \in C^2([a, b])$. $\qquad \square$

Periodic boundary conditions can also be treated using the Lax–Milgram theorem.

Theorem 5.21 (Periodic boundary conditions) *Let $p \in C^1([a, b])$ and $r \in C([a, b])$ be such that $p(a) = p(b)$ and $p(x) \geq \alpha > 0$, as well as $r(x) \geq \alpha > 0$, for some $\alpha > 0$ and all $x \in [a, b]$. Let $f \in L_2(a, b)$ be given. Then there exists a unique solution $u \in H^2(a, b)$ of*

$$-(pu')' + ru = f \qquad \text{almost everywhere in } (a, b), \qquad (5.18a)$$
$$u(a) = u(b), \ u'(a) = u'(b), \qquad\qquad\qquad\qquad\qquad (5.18b)$$

If $f \in C([a, b])$, then $u \in C^2([a, b])$.

Proof We consider the space $H^1_{\text{per}}(a, b) := \{f \in H^1(a, b) : f(a) = f(b)\}$. Since $H^1(a, b)$ is continuously imbedded in $C([a, b])$, the space $H^1_{\text{per}}(a, b)$ is a closed subspace of $H^1(a, b)$ and thus a Hilbert space. The mapping

$$a(u, v) := \int_a^b \{p(x)\, u'(x)\, v'(x) + r(x)\, u(x)\, v(x)\}\, dx$$

defines a continuous bilinear form on $H^1_{\text{per}}(a, b)$; since

$$a(u) \geq \alpha \int_a^b \left\{ (u'(x))^2 + u(x)^2 \right\} dx,$$

it is coercive. Now let $f \in L_2(a, b)$. By the Lax–Milgram theorem there exists a unique $u \in H^1_{\text{per}}(a, b)$ such that $a(u, v) = \int_a^b f(x)\, v(x)\, dx$ for all $v \in H^1_{\text{per}}(a, b)$. If we choose $v \in C^1_c(a, b)$, then it follows from the definition of weak derivatives that $-(pu') \in H^1(a, b)$ and $-(pu')' + ru = f$. Since $pu' \in H^1(a, b)$, it follows from Theorem 5.13 that $u' = \frac{1}{p}(pu') \in H^1(a, b)$, and so $u \in H^2(a, b)$. If we substitute this expression for f, then we obtain

$$\int_a^b \{p(x)\, u'(x)\, v'(x) + r(x)\, u(x)\, v(x)\}\, dx = a(u, v) = \int_a^b f(x)\, v(x)\, dx$$

$$= \int_a^b \{-(p(x)\, u'(x))'\, v(x) + r(x)\, u(x)\, v(x)\}\, dx$$

$$= \int_a^b \{p(x)\, u'(x)\, v'(x) + r(x)\, u(x)\, v(x)\}\, dx$$

$$\quad - p(b)\, u'(b)\, v(b) + p(a)\, u'(a)\, v(a)$$

for all $v \in H^1_{\text{per}}(a, b)$. Since $v(a) = v(b)$ and $p(a) = p(b)$, it follows that

$$0 = -p(b)\, u'(b)\, v(b) + p(a)\, u'(a)\, v(a) = (-u'(b) + u'(a))\, p(a)\, v(a)$$

for all $v \in H^1_{\text{per}}(a, b)$. This leads to $u'(a) = u'(b)$, and thus u is a solution of (5.18).

By inverting the above argument, one sees that, for any solution u of (5.18), the equation $a(u, v) = \int_a^b f(x)\, v(x)\, dx$ holds for all $v \in H^1_{\text{per}}(a, b)$. Thus the Lax–Milgram theorem also yields uniqueness. □

5.2.5 Non-symmetric differential operators

All problems we have considered thus far lead to symmetric bilinear forms. As such, we could equally have used the theorem of Riesz–Fréchet in place of the Lax–Milgram theorem. Now we will add a term to the differential operator of Theorem 5.20 which will make the associated form in fact non-symmetric. To keep things simple, we will limit ourselves to Dirichlet boundary conditions. In the proof of coercivity we will use a simple but extremely useful inequality:

Lemma 5.22 (Young's inequality) *Let* $\alpha, \beta \geq 0$ *and* $\varepsilon > 0$. *Then*

$$\alpha\beta \leq \varepsilon\alpha^2 + \frac{1}{4\varepsilon}\beta^2. \tag{5.19}$$

Proof Since $0 \leq (\alpha - \beta)^2 = \alpha^2 - 2\alpha\beta + \beta^2$, we have $\alpha\beta \leq \frac{1}{2}\alpha^2 + \frac{1}{2}\beta^2$. If, in this inequality, we replace α by $\sqrt{2\varepsilon}\alpha$ and β by $\frac{1}{\sqrt{2\varepsilon}}\beta$, then we obtain (5.19).   □

Theorem 5.23 *Let* $p \in C^1([a, b])$ *and* $r, q \in C([a, b])$. *Suppose that there exist* $0 < \beta < \alpha$ *such that* $p(x) \geq \alpha$ *and* $q(x)^2 \leq 4\beta r(x)$ *for all* $x \in [a, b]$. *Let* $f \in L_2(a, b)$ *be given. Then there exists a unique function* $u \in H^2(a, b)$ *such that*

$$-(pu')' + qu' + ru = f \qquad \text{almost everywhere in } (a, b), \tag{5.20a}$$
$$u(a) = u(b) = 0. \tag{5.20b}$$

If $f \in C([a, b])$, *then* $u \in C^2([a, b])$.

Proof This time, we define a continuous bilinear form on $H_0^1(a, b)$ by

$$a(u, v) := \int_a^b \{p(x)\, u'(x)\, v'(x) + q(x)\, u'(x)\, v(x) + r(x)\, u(x)\, v(x)\}\, dx.$$

Now

$$\int_a^b p(x)\left(u'(x)\right)^2 dx \geq \alpha \int_a^b \left(u'(x)\right)^2 dx,$$

and from Young's inequality it follows that

$$|q(x)\, u'(x)\, u(x)| \leq \beta\left(u'(x)\right)^2 + \frac{1}{4\beta}q(x)^2\, u(x)^2$$

for $u \in H_0^1(a, b)$. Hence

$$q(x)\, u'(x)\, u(x) \geq -|q(x)\, u'(x)\, u(x)| \geq -\beta\left(u'(x)\right)^2 - \frac{1}{4\beta}q(x)^2 u(x)^2$$

and so, for all $u \in H_0^1(a, b)$,

$$a(u) \geq \alpha \int_a^b \left(u'(x)\right)^2 dx - \beta \int_a^b \left(u'(x)\right)^2 dx + \int_a^b \left(r(x) - \frac{q(x)^2}{4\beta}\right)u(x)^2 dx$$

$$\geq (\alpha - \beta) \int_a^b \left(u'(x)\right)^2 dx,$$

since $(r - q^2/4\beta) \geq 0$ by assumption. It now follows from the Poincaré inequality (5.5) that $a(\cdot, \cdot)$ is coercive.

Now let $f \in L_2(a, b)$; then by the Lax–Milgram theorem there exists a unique $u \in H_0^1(a, b)$ such that $a(u, v) = \int_a^b f(x) v(x) dx$ for all $v \in H_0^1(a, b)$. By choosing $v \in C_c^1(a, b)$ in Section 5.2.1 or the proof of Theorem 5.18, we deduce that $pu' \in H^1(a, b)$ and $-(pu')' + qu' + ru = f$. It follows that $u' = \frac{1}{p}(pu') \in H^1(a, b)$; hence $u \in H^2(a, b)$, and u is a solution of (5.20). This proves existence.

Uniqueness follows once again from the Lax–Milgram theorem: one sees easily that for each solution $u \in H^2(a, b)$ of (5.20) the equation $a(u, v) = \int_a^b f(x) v(x) dx$ holds for all $v \in H_0^1(a, b)$.

It remains to prove the regularity statement. Since $u \in H^2(a, b) \subset C^1([a, b])$, we have $qu' + ru \in C([a, b])$. Now if $f \in C([a, b])$, then it follows that $(pu')' \in C([a, b])$. By Corollary 5.11, we have $pu' \in C^1([a, b])$; since $\frac{1}{p} \in C^1([a, b])$, we conclude that $u \in C^2([a, b])$.   □

We say that equation (5.20a) is given in *divergence form* (since the leading term is $(pu')'$). Equations which are not in divergence form can easily be reduced to (5.20a):

Corollary 5.24 *Given* $p, q, r \in C([a, b])$, *let* $\alpha > 0$ *and* $0 < \beta < 1$ *be constants such that* $p(x) \geq \alpha > 0$ *and* $q(x)^2 \leq 4\beta p(x) r(x)$ *for all* $x \in [a, b]$. *Then for each* $f \in L_2(a, b)$ *there exists a unique* $u \in H^2(a, b)$ *such that*

$$-pu'' + qu' + ru = f \qquad \text{almost everywhere in } (a, b), \qquad (5.21a)$$
$$u(a) = u(b) = 0. \qquad (5.21b)$$

If $f \in C([a, b])$, *then* $u \in C^2([a, b])$.

Proof By Theorem 5.23 there exists a unique $u \in H^2(a, b) \cap H_0^1(a, b)$ such that $-u'' + \frac{q}{p} u' + \frac{r}{p} = \frac{f}{p}$. This is equivalent to (5.21). If $f \in C([a, b])$, then also $\frac{f}{p} \in C([a, b])$, and thus $u \in C^2([a, b])$ by Theorem 5.23.   □

5.2.6* A variational approach to singularly perturbed problems and the transport equation

In this subsection we discuss another case where we have coercivity (in which, however, the coercivity constant can be arbitrarily small), and then an example in which the Banach–Nečas theorem can be used in the absence of coercivity.

Example 5.25 Let $0 < \varepsilon < 1$ be "small" and let $f \in L_2(0, 1)$. We consider the problem

$$-\varepsilon u''(x) + u'(x) + u(x) = f(x) \quad \text{for all } x \in (0, 1),$$
$$u(0) = u(1) = 0.$$

The (non-symmetric) bilinear form a for the weak formulation of this boundary value problem on $V := H_0^1(0, 1)$ reads $a(u, v) := \varepsilon(u', v')_{L_2} + (u' + u, v)_{L_2}$. It is easy to verify that a is continuous. Regarding the coercivity, for all $u \in H_0^1(0, 1)$ we have by Theorem 5.13 (b) that $(u', u)_{L_2} = -(u, u')_{L_2}$, hence

$$(u', u)_{L_2} = \int_0^1 u'(x) u(x) \, dx = 0,$$

and so $a(u, u) = \varepsilon \|u'\|_{L_2}^2 + \|u\|_{L_2}^2 \geq \varepsilon \|u\|_{H^1}^2$. Hence the bilinear form is coercive with constant of coercivity $\alpha = \varepsilon$, and the above problem is well-posed. In fact, as in Theorem 5.23 one sees that for each $f \in L_2(0, 1)$, there exists a unique solution $u \in H^2(0, 1)$. We will see later (Céa's lemma, Theorem 9.14) that the reciprocal of the coercivity constant has substantial influence on the error analysis of finite element methods in numerics. Clearly, in this regard $\alpha = \varepsilon$ is an issue, which is also the reason why such problems are called *singular perturbed*. △

In the above example the limit as $\varepsilon \to 0$ is of particular interest. As the diffusion part $-\varepsilon u''(x)$ vanishes when $\varepsilon \to 0$, the second order problem is replaced by a first order problem. This also implies that in the limit we have "too many" boundary conditions. One option for the limit is the initial value transport problem:

$$u'(x) + u(x) = f(x) \quad \text{for all } x \in (0, 1), \tag{5.22a}$$
$$u(0) = 0. \tag{5.22b}$$

We can also derive a variational formulation of this problem; here, however, we require different trial and test spaces: we take $V := H_{(0)}^1(0, 1) := \{v \in H^1(0, 1) : v(0) = 0\}$ and $W := L_2(0, 1)$ and set

$$b : V \times W \to \mathbb{R}, \qquad b(v, w) := (v' + v, w)_{L_2} = \int_0^1 (v'(x) + v(x)) \, w(x) \, dx.$$

It is easy to show using the Cauchy–Schwarz inequality that b is continuous. We now show that the theorem of Banach–Nečas (Theorem 4.27) guarantees well-posedness. To this end, we first check the inf-sup condition (4.11): let $0 \neq v \in V$; then $v' + v \in W$ and

$$\sup_{w \in W} \frac{b(v, w)}{\|w\|_W} \geq \frac{b(v, v' + v)}{\|v' + v\|_W} = \frac{(v' + v, v' + v)_{L_2}}{\|v' + v\|_{L_2}} = \|v' + v\|_{L_2}.$$

Moreover, by Theorem 5.13,

$$\|v' + v\|_{L_2}^2 = (v' + v, v' + v)_{L_2} = \|v'\|_{L_2}^2 + \|v\|_{L_2}^2 + 2(v', v)_{L_2}$$
$$= \|v'\|_{L_2}^2 + \|v\|_{L_2}^2 + v(1)^2 \geq \|v\|_{H^1}^2,$$

and thus

$$\sup_{w \in W} \frac{b(v, w)}{\|w\|_W} \geq \|v' + v\|_{L_2} \geq \|v\|_{H^1}.$$

This estimate obviously also holds for $v = 0$; we have thus established (4.11) with $\beta = 1$. To prove (4.13), we let $0 \neq w \in W$ and define $v(x) := \int_0^x w(s) \, ds$. Then $v \in V$, $v'(x) = w(x)$ and by Theorem 5.13 (b)

$$b(v, w) = (v' + v, w)_{L_2} = \|w\|_{L_2}^2 + (v, v')_{L_2} = \|w\|_{L_2}^2 + \frac{1}{2}v(1)^2 > 0.$$

Thus the hypotheses of Theorem 4.27 are satisfied. Consequently, for all $f \in L_2(0, 1)$, there is a unique $v \in \mathbb{V}$ such that $b(u, v) = (f, w)_{L_2}$ for all $w \in \mathbb{W}$. This means that for all $f \in L_2(0, 1)$ there exists a unique solution $u \in H^1(0, 1)$ of (5.22).

5.3* Comments on Chapter 5

The Fundamental Theorem of Calculus was discovered separately by Newton and Leibniz in 1680. Of course, before the development of the Lebesgue integral it could not be stated in the form of Theorem 5.9. And indeed, Lebesgue proved in 1904 that, for $f \in L^1(a, b)$ and its Lebesgue integral $F(t) := \int_a^t f(s)\,ds$, we have

$$\lim_{h \to 0} \frac{F(t + h) - F(t)}{h} = f(t)$$

for almost every $t \in (a, b)$ (see [54, Ch. 7]). Since the notion of weak derivatives, which are defined via integrals, is inherently simpler and more useful for treating partial differential equations, we will not dwell further on this "almost everywhere" derivative. Further results on Sobolev spaces in one variable can be found in the highly recommendable book [18] by H. Brezis.

5.4 Exercises

Exercise 5.1 Prove the statement of Example 5.5.

Exercise 5.2 Determine the weak derivative of the "hat function"

$$h(x) := \begin{cases} x, & 0 \le x \le 1, \\ 2 - x, & 1 < x \le 2. \end{cases}$$

Exercise 5.3 Let $f \in L_1(a, b)$ be given, and define the function $g : [a, b] \to \mathbb{R}$ by $g(x) := \int_a^x f(y)\,dy$. Show that g is continuous.
Suggestion: Write $g(x) = \int_a^b \mathbb{1}_{[a,x]}(y)f(y)\,dy$. Let $x_n \in [a, b]$, $\lim_{n \to \infty} x_n = x$. Show that $\mathbb{1}_{[a,x_n]}(y)f(y) \to \mathbb{1}_{[a,x]}(y)f(y)$ and apply the Dominated Convergence Theorem.

Exercise 5.4 Let $\lambda > 0$, $A, B \in \mathbb{R}$, $f \in L_2(a, b)$. Show that there exists a unique function $u \in H^2(a, b)$ such that $\lambda u - u'' = f$ in (a, b) and $u'(a) = A$, $u'(b) = B$.
Suggestion: Proceed as with the inhomogeneous Dirichlet boundary conditions just after Theorem 5.17.

Exercise 5.5 (a) Show that $\int_0^1 u(x)^2\,dx \le 2\big(u(0)^2 + \int_0^1 (u'(x))^2\,dx\big)$ for all $u \in H^1(0, 1)$.

(b) Let $\alpha > 0$. Show that $a(u, v) := \alpha u(0) v(0) + \int_0^1 u'(x) v'(x) \, dx$ defines a coercive
form on $H^1(0, 1)$.
(c) Let $\alpha > 0$ and $f \in L_2(0, 1)$. Show that there exists a unique $u \in H^2(0, 1)$ such
that

$$-u''(x) = f(x), \quad x \in (0, 1),$$
$$-u'(0) + \alpha u(0) = 0, \quad u'(1) = 0.$$

Exercise 5.6 Suppose that the assumptions of Theorem 5.23 are satisfied, and in
addition $r(x) > 0$ for all $x \in [a, b]$. Show that, given $f \in L_2(a, b)$, there exists a
unique $u \in H^2(a, b)$ such that

$$-(pu')' + qu' + ru = f \quad \text{almost everywhere in } (a, b)$$
$$u'(a) = u'(b) = 0.$$

Exercise 5.7 Let $-\infty < a < b < \infty$ and $f \in H^2(a, b)$.
(a) Show that $f'' = 0$ if and only if there exist constants $c_0, c_1 \in \mathbb{R}$ such that
$f(x) = c_0 + c_1 x$ almost everywhere in (a, b).
(b) Show that $f(x) = f(a) + f'(a)(x - a) + \int_a^x (x - y) f''(y) \, dy$ almost everywhere.

Exercise 5.8 (Locally integrable weak derivatives) Let $-\infty \le a < b \le \infty$. In this
exercise, we will consider the vector space

$$L_{1,\text{loc}}(a, b) := \left\{ f : (a, b) \to \mathbb{R} \text{ measurable:} \right.$$
$$\left. \int_c^d |f(x)| \, dx < \infty \text{ if } a < c < d < b \right\}$$

of *locally integrable* functions on (a, b).
(a) Let $f \in L_{1,\text{loc}}(a, b)$ be such that $\int_a^b fv \, dx = 0$ for all $v \in C_c^1(a, b)$. Show that
then $f(x) = 0$ almost everywhere. *Suggestion:* Use Lemma 5.1.
(b) We say that $f \in L_{1,\text{loc}}(a, b)$ is *weakly differentiable* if there exists a function
$f' \in L_{1,\text{loc}}(a, b)$ such that

$$-\int_a^b f(x) v'(x) \, dx = \int_a^b f'(x) v(x) \, dx$$

for all $v \in C_c^1(a, b)$. Show that given f, there can be at most one such weak
derivative f'.
(c) Show that

$$W_{1,\text{loc}}^1(a, b) := \{ f \in L_{1,\text{loc}}(a, b) : f \text{ is weakly differentiable} \}$$

is a vector space, and that $f \mapsto f' : W_{1,\text{loc}}^1(a, b) \to L_{1,\text{loc}}(a, b)$ is linear.

(d) Let $f \in W^1_{1,\text{loc}}(a, b)$ be such that $f' = 0$ almost everywhere. Show that there exists $c \in \mathbb{R}$ such that $f(x) = c$ almost everywhere. *Suggestion:* Modify the proof of Lemma 5.7.

(e) Let $f \in W^1_{1,\text{loc}}(a, b)$ and $x_0 \in (a, b)$. Show that there exists $c \in \mathbb{R}$ such that $f(x) = c + \int_{x_0}^{x} f'(y)\, dy$ almost everywhere. *Suggestion:* Proceed analogously to the proof of Theorem 5.9.

(f) Let $f \in W^1_{1,\text{loc}}(a, b)$ and $c_1 \in \mathbb{R}$, such that $f'(x) = c_1$ almost everywhere. Show that there exists a constant $c_0 \in \mathbb{R}$ such that $f(x) = c_1 x + c_0$ almost everywhere.

Exercise 5.9 (Harmonic functions) Let $-\infty \leq a < b \leq \infty$, and suppose that $f \in L_2(a, b)$ satisfies

$$\int_a^b f(x) v''(x)\, dx = 0 \quad \text{for all } v \in C_c^2(a, b).$$

Show that $f(x) = c_0 + c_1 x$ almost everywhere, for some constants $c_0, c_1 \in \mathbb{R}$.
Note: Here $C_c^k := \{v : (a, b) \to \mathbb{R} : k \text{ times continuously differentiable}, \exists a < c < d < b, \text{ such that } v(x) = 0, \text{ if } x \notin [c, d]\}, k \in \mathbb{N} \cup \{\infty\}$. Look for inspiration from the proof of Lemma 5.7 and from Exercise 5.7.

Exercise 5.10 Let $\lambda \in \mathbb{R}$, $u_0 \in \mathbb{R}$ and $f \in L_2(0, T)$, where $0 < T < \infty$, and set $u(t) = e^{-\lambda t} u_0 + \int_0^t e^{-\lambda(t-s)} f(s)\, ds$. Show that u is the unique solution of the problem

$$u \in H^1(0, T), \quad u'(t) = -\lambda u(t) + f(t), \quad u(0) = u_0. \tag{5.23}$$

Exercise 5.11 (a) Prove by induction that the inclusion

$$H^{k+1}(a, b) \subset C^k([a, b])$$

holds for all $k = 0, 1, 2, \ldots$, where $C^0([a, b]) := C([a, b])$.

(b) Show that $C^k([a, b])$ is a Banach space with respect to the norm

$$\|f\|_{C^k} = \sum_{m=0}^{k} \|f^{(m)}\|_{C([a,b])},$$

where $f^{(0)} = f$.

(c) Show using the closed graph theorem (Theorem A.2) that $H^{k+1}(a, b) \hookrightarrow C^k([a, b])$, that is, show that this imbedding is continuous.

Exercise 5.12 (Absolute convergence of Fourier series)

(a) Let $f = u + iv$, where $u, v \in H^1(0, 2\pi)$ and $f(0) = f(2\pi)$. Show that the Fourier series of f is absolutely convergent (to f).
Suggestion: If c_k is the kth Fourier coefficient of f (corresponding to (3.17)) and c'_k is the kth Fourier coefficient of $f' := u' + iv'$, then $c'_k = ikc_k$.

(b) Deduce *Dirichlet's theorem* from (a): if $f \in C_{2\pi}$ is *piecewise continuously differentiable* (that is, there exist $0 = t_0 < t_1 < \cdots < t_n = 2\pi$ and $g_i \in C^1([t_{i-1}, t_i])$ such that $f = g_i$ on $(t_{i-1}, t_i), i = 1, \ldots, n$), then the Fourier series of f converges uniformly to f.
   Suggestion: Use Theorem 5.14.

Exercise 5.13 (Weak solution of the wave equation) Consider

$$u_{tt} = u_{xx}, \quad t, x \in \mathbb{R} \tag{5.24}$$

on $\mathbb{R}^2$, where $u = u(t, x)$.
(a) Suppose that $u \in C^2(\mathbb{R}^2)$ is a solution of (5.24). Show that then

$$\int_{\mathbb{R}} \int_{\mathbb{R}} u(t, x)(\varphi_{tt}(t, x) - \varphi_{xx}(t, x)) \, dx \, dt = 0 \tag{5.25}$$

for all $\varphi \in C_c^2(\mathbb{R}^2)$.
(b) Let $f : \mathbb{R} \to \mathbb{R}$ be continuous and define $u : \mathbb{R}^2 \to \mathbb{R}^2$ by $u(t, x) = \frac{1}{2}(f(x + t) + f(x - t))$, so that u is also continuous. Show that u is a *weak solution* of (5.24), that is, (5.25) holds for all $\varphi \in C_c^2(\mathbb{R}^2)$.

Exercise 5.14 Define $H^1(\mathbb{R})$ analogously to how it is defined bounded intervals. Show that $H^1(\mathbb{R}) \subset C_0(\mathbb{R}) := \{u \in C(\mathbb{R}) : \lim_{|x| \to \infty} u(x) = 0\}$.

Chapter 6
Hilbert space methods for elliptic equations

We started by introducing Sobolev spaces in dimension one only, or more precisely for functions defined on an open interval in $\mathbb{R}$, in order to try and convey the essential points of the mathematical theory in as elementary a way as possible. In one space dimension, H^1-functions are automatically continuous, and they can be characterized by an indefinite integral (Theorem 5.9). This is no longer true in higher dimensions, and new arguments are necessary in order to deduce properties of weakly differentiable functions and Sobolev spaces.

In this chapter we will introduce Sobolev spaces in arbitrary space dimensions. We limit ourselves to the Hilbert space theory, which is to say Sobolev spaces which are built on $L_2(\Omega)$ (and not $L_p(\Omega)$ for $p \neq 2$). With the aid of the Lax–Milgram theorem, we will then be in a position to study numerous elliptic problems. One can also characterize Sobolev spaces on the whole space $\Omega = \mathbb{R}^d$ in terms of Fourier transforms; this can for example be used to prove interior regularity of solutions.

In this chapter we will restrict ourselves to Dirichlet boundary conditions. In doing so, the statements which we make here require no deep analysis of the boundary and are therefore comparatively simple to prove. A more detailed study of the boundary will be undertaken in the next chapter.

Chapter overview

© The Author(s), under exclusive license to Springer Nature Switzerland AG 2023
W. Arendt, K. Urban, *Partial Differential Equations*, Graduate Texts in
Mathematics 294, https://doi.org/10.1007/978-3-031-13379-4_6

6.1 Mollifiers

In this section we will introduce an important technique which will permit us to approximate integrable functions by smooth functions. It is based on convolution with smooth functions.

Let Ω be an open set in $\mathbb{R}^d$. We say that the function $u : \Omega \to \mathbb{R}$ has *compact support* if there exists a compact set $K \subset \Omega$ such that $u(x) = 0$ for all $x \in \Omega \setminus K$. If u is also continuous, then as before (cf. (2.1)) we call the set

$$\operatorname{supp} u := \overline{\{x \in \Omega : u(x) \neq 0\}}$$

the *support* of u. We denote by $C_c(\Omega)$ the space of all continuous real-valued functions on Ω with compact support, cf. (3.105). For $k \in \mathbb{N}$ we also set

$$C_c^k(\Omega) := C^k(\Omega) \cap C_c(\Omega), \quad C_c^0(\Omega) := C_c(\Omega). \tag{6.1}$$

The space

$$\mathcal{D}(\Omega) := C_c^\infty(\Omega) := C^\infty(\Omega) \cap C_c(\Omega) \tag{6.2}$$

plays a special role; its elements are called *test functions* on Ω. The following special test function $\varrho \in \mathcal{D}(\mathbb{R}^d)$ will be of particular importance:

$$\varrho(x) := \begin{cases} c \exp\left(\frac{1}{|x|^2 - 1}\right), & \text{if } |x| < 1, \\ 0, & \text{if } |x| \geq 1, \end{cases} \tag{6.3}$$

where $c > 0$ is chosen in such a way that

$$\int_{\mathbb{R}^d} \varrho(x)\,dx = 1. \tag{6.4}$$

With the usual notation

$$B(x,r) := \{y \in \mathbb{R}^d : |x - y| < r\}, \qquad \bar{B}(x,r) := \{y \in \mathbb{R}^d : |x - y| \leq r\}$$

for $x \in \mathbb{R}^d$ and $r > 0$, we then have:

Lemma 6.1 *The function ϱ belongs to $\mathcal{D}(\mathbb{R}^d)$ and* $\operatorname{supp}\varrho = \bar{B}(0,1)$.

Proof Set

$$g(r) := \begin{cases} c \exp\left(\frac{1}{r-1}\right), & \text{for } r < 1, \\ 0, & \text{for } r \geq 1. \end{cases}$$

Then $g \in C^\infty(\mathbb{R})$. The function $b(x) := |x|^2 = x_1^2 + \cdots + x_d^2$ is in $C^\infty(\mathbb{R}^d)$, and so $\varrho = f \circ b \in C^\infty(\mathbb{R}^d)$. It is immediate from the definition that $\operatorname{supp}\varrho = \bar{B}(0,1)$. $\square$

We can define a whole sequence of test functions based on ϱ.

Definition 6.2 For $n \in \mathbb{N}$ we set

$$\varrho_n(x) := n^d \varrho(nx), \quad x \in \mathbb{R}^d. \tag{6.5}$$

We call $(\varrho_n)_{n\in\mathbb{N}}$ a sequence of *mollifiers* (one sometimes sees the name *Friedrichs mollifier*). We will always denote the functions defined in (6.5) and (6.3) by ϱ_n and ϱ, respectively. $\qquad \triangle$

Figure 6.1 shows a few examples of the functions ϱ_n in dimensions $d = 1, 2$. Obviously, $\varrho_1 = \varrho$,

$$0 \le \varrho_n \in \mathcal{D}(\mathbb{R}^d), \qquad \mathrm{supp}\varrho_n \subset \bar{B}\left(0, \frac{1}{n}\right), \tag{6.6}$$

and

$$\int_{\mathbb{R}^d} \varrho_n(x)\, dx = 1. \tag{6.7}$$

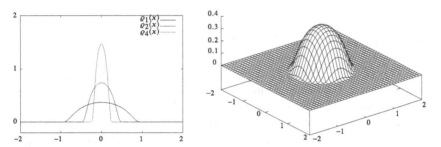

Fig. 6.1 Mollifiers in one and two dimensions.

We will use these functions to smoothen non-smooth functions, as the name "mollifier" already obliquely indicates. This is done using convolutions, a process which we now describe in detail. For $f, g \in L_2(\mathbb{R}^d)$ we define the *convolution* $f * g$ of f and g analogously to the case $d = 1$ in Chapter 3, by

$$(f * g)(x) := \int_{\mathbb{R}^d} f(x - y)g(y)\, dy \tag{6.8}$$

for all $x \in \mathbb{R}^d$. One should observe that $f(x - \cdot) \in L_2(\mathbb{R}^d)$ for almost every $x \in \mathbb{R}^d$, so that $(f * g)(x) = (f(x - \cdot), g)_{L_2(\mathbb{R}^d)}$ and

$$|(f * g)(x)| \le \|f\|_{L_2} \|g\|_{L_2} \tag{6.9}$$

for all $x \in \mathbb{R}^d$. In what follows, we will denote the sum of two sets $K_1, K_2 \subset \mathbb{R}^d$ by $K_1 + K_2 := \{x + y : x \in K_1,\ y \in K_2\}$. Then the following statement holds.

Lemma 6.3 *Let* $f, g \in L_2(\mathbb{R}^d)$ *be such that* $f(x) = 0$ *whenever* $x \notin K_1$, *and* $g(x) = 0$ *whenever* $x \notin K_2$. *Then* $(f * g)(x) = 0$ *whenever* $x \notin (K_1 + K_2)$. *If* f *and* g *are continuous, then this may be written in the form*

$$\text{supp}\,(f * g) \subset \text{supp}\, f + \text{supp}\, g. \tag{6.10}$$

Proof If $f(x - y)g(y) \neq 0$, then $x - y \in K_1$ and $y \in K_2$, that is, $x \in K_1 + K_2$. So if $x \notin K_1 + K_2$, then $f(x - y)g(y) = 0$ for all $y \in \mathbb{R}^d$ and thus $(f * g)(x) = 0$. $\qquad\square$

One can also make sense of formula (6.10) even if f and g are not continuous; see Exercise 6.1. We next define the space

$$C_0(\mathbb{R}^d) = \left\{ f \in C(\mathbb{R}^d) : \lim_{|x| \to \infty} f(x) = 0 \right\}. \tag{6.11}$$

This is a Banach space with respect to the norm

$$\|f\|_\infty := \sup_{x \in \mathbb{R}^d} |f(x)|.$$

Theorem 6.4 *Let* $f, g \in L_2(\mathbb{R}^d)$. *Then* $f * g \in C_0(\mathbb{R}^d)$ *and* $f * g = g * f$.

Proof Choose step functions f_n and g_n such that $f_n \to f$ and $g_n \to g$ in $L_2(\mathbb{R}^d)$. It follows by the Dominated Convergence Theorem that $f_n * g_n : \mathbb{R}^d \to \mathbb{R}$ is continuous. Indeed, if Q_1 and Q_2 are hyperrectangles in $\mathbb{R}^d$ (see Section A.3), then $\mathbb{1}_{Q_1} * \mathbb{1}_{Q_2}(x) = \int_{Q_2} \mathbb{1}_{Q_1}(x - y)\, dy$ is continuous in x, since, if $\lim_{n \to \infty} x_n = x$, then $\mathbb{1}_{Q_1}(x_n - y)$ converges to $\mathbb{1}_Q(x - y)$ for all $y \in \mathbb{R}^d \setminus \partial Q_1$. Since the boundary ∂Q_1 of Q_1 has measure zero, the Dominated Convergence Theorem now implies that $\lim_{n \to \infty} \mathbb{1}_{Q_1} * \mathbb{1}_{Q_2}(x_n) = \mathbb{1}_{Q_1} * \mathbb{1}_{Q_2}(x)$. This proves the continuity of $f_n * g_n$.

Since f_n and g_n vanish outside the ball $B(0, r)$, $f_n * g_n$ vanishes outside $B(0, 2r)$, and so $f_n * g_n \in C_c(\mathbb{R}^d) \subset C_0(\mathbb{R}^d)$ (cf. Definition 6.1), and

$$|(f_n * g_n)(x) - (f * g)(x)| \leq |(f_n * (g_n - g))(x)| + |((f_n - f) * g)(x)|$$
$$\leq \|f_n\|_{L_2}\|g_n - g\|_{L_2} + \|f_n - f\|_{L_2}\|g\|_{L_2}$$
$$\to 0, \quad n \to \infty$$

uniformly in $x \in \mathbb{R}^d$. It follows that $f * g \in C_0(\mathbb{R}^d)$. Fubini's theorem now implies that $f * g = g * f$. $\qquad\square$

Convolutions regularize functions in the following sense.

Theorem 6.5 *Let* $f \in L_2(\mathbb{R}^d)$ *and* $\varphi \in C_c^1(\mathbb{R}^d)$. *Then* $\varphi * f \in C^1(\mathbb{R}^d)$ *and*

$$\frac{\partial}{\partial x_j}(\varphi * f) = \frac{\partial \varphi}{\partial x_j} * f, \quad j = 1, \dots, d. \tag{6.12}$$

In particular, if $\varphi \in \mathcal{D}(\mathbb{R}^d)$, then $\varphi * f \in C^\infty(\mathbb{R}^d)$.

For the proof we will use the following estimate together with the chain rule and the Fundamental Theorem of Calculus. Let $x, v \in \mathbb{R}^d$, then

$$\varphi(x + v) - \varphi(x) = \int_0^1 \frac{\partial}{\partial t} \varphi(x + tv) \, dt = \int_0^1 \nabla \varphi(x + tv) \cdot v \, dt,$$

where $\nabla \varphi(y) = (D_1 \varphi(y), \dots, D_d \varphi(y)) \in \mathbb{R}^d$ is the *gradient* of φ at the point $y \in \mathbb{R}^d$ and $x \cdot y := x^T y$ denotes the scalar product of $x, y \in \mathbb{R}^d$. We thus have

$$\nabla \varphi(x + tv) \cdot v = \sum_{j=1}^d \frac{\partial \varphi}{\partial x_j}(x + tv)v_j,$$

and the estimate

$$|\varphi(x + v) - \varphi(x)| \leq \|\nabla \varphi\|_\infty \cdot |v| \tag{6.13}$$

follows by the Cauchy–Schwarz inequality, where the norm of the gradient is given by

$$\|\nabla \varphi\|_\infty = \sup_{x \in \mathbb{R}^d} |\nabla \varphi(x)|, \quad |\nabla \varphi(x)| = \left(\sum_{j=1}^d (D_j \varphi)(x)^2 \right)^{1/2}.$$

We now come to the proof of the differentiation rule (6.12).

Proof (of Theorem 6.5) Let $e_j = (0, \dots, 0, 1, 0, \dots, 0)^T$ be the jth canonical basis vector of $\mathbb{R}^d$. Then $\lim_{h \to 0} \frac{1}{h}(\varphi(x + he_j - y) - \varphi(x - y)) = \frac{\partial \varphi}{\partial x_j}(x - y)$. By (6.13),

$$\left| \frac{1}{h}(\varphi(x + he_j - y) - \varphi(x - y)) \right| \leq \|\nabla \varphi\|_\infty.$$

Moreover,

$$\frac{(\varphi * f)(x + he_j) - (\varphi * f)(x)}{h} = \int_{\mathbb{R}^d} \frac{\varphi(x + he_j - y) - \varphi(x - y)}{h} f(y) \, dy.$$

Passing to the limit as $h \to 0$, it follows from the Dominated Convergence Theorem that

$$\frac{\partial}{\partial x_j}(\varphi * f)(x) = \left(\frac{\partial \varphi}{\partial x_j} * f \right)(x),$$

as asserted.   □

The following theorem shows how a function can be approximated by smooth functions with the help of mollifiers.

Theorem 6.6 (Smoothing) *Let $f \in L_2(\mathbb{R}^d)$. Then $\varrho_n * f \in C^\infty(\mathbb{R}^d) \cap L_2(\mathbb{R}^d)$ and $\lim_{n \to \infty} \varrho_n * f = f$ in $L_2(\mathbb{R}^d)$.*

Proof The proof proceeds in five steps.

1. It follows from Theorem 6.5 and an induction argument that $\varrho_n * f \in C^\infty(\mathbb{R}^d)$ for all $n \in \mathbb{N}$.

2. We will show that

$$\int_{\mathbb{R}^d} |(\varrho_n * f)(x)|^2 \, dx \le \int_{\mathbb{R}^d} |f(x)|^2 \, dx. \tag{6.14}$$

Indeed, by Hölder's inequality and (6.7),

$$|(\varrho_n * f)(x)| \le \int_{\mathbb{R}^d} \varrho_n(x-y)^{1/2} \varrho_n(x-y)^{1/2} |f(y)| \, dy$$

$$\le \left(\int_{\mathbb{R}^d} \varrho_n(x-y) \, dy \right)^{1/2} \left(\int_{\mathbb{R}^d} \varrho_n(x-y) f(y)^2 \, dy \right)^{1/2}$$

$$= \left(\int_{\mathbb{R}^d} \varrho_n(x-y) f(y)^2 \, dy \right)^{1/2}.$$

Hence, by Fubini's theorem,

$$\int_{\mathbb{R}^d} (\varrho_n * f)(x))^2 \, dx \le \int_{\mathbb{R}^d} \int_{\mathbb{R}^d} \varrho_n(x-y) f(y)^2 \, dy \, dx$$

$$= \int_{\mathbb{R}^d} \int_{\mathbb{R}^d} \varrho_n(x-y) \, dx \, f(y)^2 \, dy = \int_{\mathbb{R}^d} f(y)^2 \, dy,$$

where we have used (6.7). This establishes (6.14).

3. Let $f = \mathbb{1}_Q$ be the characteristic function of a d-dimensional rectangle Q; then $(\varrho_n * \mathbb{1}_Q)(x) = \int_{|y| < 1/n} \mathbb{1}_Q(x-y) \varrho_n(y) \, dy$ converges to $\mathbb{1}_Q(x)$ as $n \to \infty$ for all $x \in \mathbb{R}^d \setminus \partial Q$. Since ∂Q has measure zero, it follows from the Dominated Convergence Theorem that $\lim_{n \to \infty} \varrho_n * \mathbb{1}_Q = \mathbb{1}_Q$.

4. We see from 3. that $\varrho_n * f \to f$ in $L_2(\mathbb{R}^d)$ whenever f is a *step function*, that is, a linear combination of characteristic functions of rectangles.

5. Finally, let $f \in L_2(\mathbb{R}^d)$ and $\varepsilon > 0$. By definition of the Lebesgue integral of f, there is a step function g such that $\|f - g\|_{L_2} \le \varepsilon$ (see Theorem A.11). Hence, by (6.14),

$$\|f * \varrho_n - f\|_{L_2} \le \|f * \varrho_n - g * \varrho_n\|_{L_2} + \|g * \varrho_n - g\|_{L_2} + \|g - f\|_{L_2}$$

$$\le \|(f - g) * \varrho_n\|_{L_2} + \|g * \varrho_n - g\|_{L_2} + \varepsilon$$

$$\le \|f - g\|_{L_2} + \|g * \varrho_n - g\|_{L_2} + \varepsilon$$

$$\le 2\varepsilon + \|g * \varrho_n - g\|_{L_2}.$$

By 3., we have $g * \varrho_n \to g$ in $L_2(\mathbb{R}^d)$ as $n \to \infty$, whence

$$\limsup_{n \to \infty} \|f * \varrho_n - f\|_{L_2} \le 2\varepsilon.$$

Since $\varepsilon > 0$ was arbitrary, the claim follows. $\qquad\square$

Convolution with ϱ_n permits us to define a large number of test functions, with just a single function (namely ϱ from (6.3)) as our starting point. The following test functions will be particularly useful.

Lemma 6.7 *Let* $\Omega \subset \mathbb{R}^d$ *be open and* $K \subset \Omega$ *compact. Then there exists a function* $\varphi \in \mathcal{D}(\Omega)$ *such that* $0 \le \varphi \le 1$ *and* $\varphi(x) = 1$ *for all* $x \in K$.

Proof For $n \in \mathbb{N}$ such that $\frac{2}{n} < \operatorname{dist}(K, \partial\Omega)$, the set

$$K_n := K + \bar{B}(0, 1/n) = \{y \in \mathbb{R}^d : \exists x \in K \text{ such that } |x - y| \le 1/n\}$$

is compact. By Theorem 6.6, $\varphi := \mathbb{1}_{K_n} * \varrho_n \in C^\infty(\mathbb{R}^d)$. Applying Lemma 6.3, we have $\operatorname{supp}\varphi \subset K_n + \bar{B}(0, 1/n) \subset \Omega$. Now if $x \in K$, then $x - y \in K_n$ for all $y \in B(0, 1/n)$, and so

$$\varphi(x) = \int_{|y|<1/n} \mathbb{1}_{K_n}(x - y)\varrho_n(y)\, dy = \int_{|y|<1/n} \varrho_n(y)\, dy = 1,$$

which proves the claim. $\qquad\square$

We will often use the following notation. We will write

$$U \Subset \Omega \qquad\qquad (6.15)$$

if U is open and bounded, and $\bar{U} \subset \Omega$. This means in particular that $\bar{U}$ is compact; it also means that U has positive distance to $\partial\Omega$, that is,

$$\operatorname{dist}(U, \partial\Omega) := \inf\{|x - y| : x \in U, y \in \partial\Omega\} > 0.$$

In addition to the mollified functions $\varrho_n * f$ we also wish to implement a cut-off procedure in order to be able to localize our arguments; this is of particular value when f does not have compact support. To this end, we define

$$\Omega_n := \left\{x \in \Omega : \operatorname{dist}(x, \partial\Omega) > \frac{1}{n}\right\} \cap B(0, n). \qquad (6.16)$$

Then $\Omega_n \Subset \Omega_{n+1} \Subset \Omega$ and $\bigcup_{n \in \mathbb{N}} \Omega_n = \Omega$. By Lemma 6.7 we can find functions $\eta_n \in \mathcal{D}(\mathbb{R}^d)$, $n \in \mathbb{N}$, such that

$$0 \le \eta_n(x) \le 1, \quad x \in \mathbb{R}^d, \qquad (6.17)$$

$$\operatorname{supp}\eta_n \subset \Omega_{n+1}, \qquad (6.18)$$

$$\eta_n(x) = 1, \qquad x \in \Omega_n. \qquad (6.19)$$

Given a function $f \in L_2(\Omega)$, we define the extension of f by zero by

$$\tilde{f}(x) := \begin{cases} f(x) & \text{if } x \in \Omega, \\ 0 & \text{if } x \in \mathbb{R}^d \setminus \Omega. \end{cases} \qquad (6.20)$$

Then $\eta_n(\varrho_n * \tilde{f}) \in \mathcal{D}(\Omega)$ for all $n \in \mathbb{N}$ and we arrive at the following result.

Theorem 6.8 (Smoothing and cut-off procedure) *For $f \in L_2(\Omega)$, we have*

$$\lim_{n \to \infty} \eta_n(\varrho_n * \tilde{f}) = f \text{ in } L_2(\Omega).$$

Proof We have

$$\|\eta_n(\varrho_n * \tilde{f}) - \tilde{f}\|_{L_2(\mathbb{R}^d)} \leq \|\eta_n(\varrho_n * \tilde{f} - \tilde{f})\|_{L_2(\mathbb{R}^d)} + \|\eta_n \tilde{f} - \tilde{f}\|_{L_2(\mathbb{R}^d)}$$

$$\leq \|\varrho_n * \tilde{f} - \tilde{f}\|_{L_2(\mathbb{R}^d)} + \|\eta_n \tilde{f} - \tilde{f}\|_{L_2(\mathbb{R}^d)}$$

by (6.17). Since $\lim_{n \to \infty} \eta_n(x) = 1$ for all $x \in \Omega_n$, it follows from the Dominated Convergence Theorem that $\lim_{n \to \infty} \eta_n \tilde{f} = \tilde{f}$ in $L_2(\mathbb{R}^d)$. Now Theorem 6.6 implies the claim. $\square$

This theorem has two direct consequences.

Corollary 6.9 *The space $\mathcal{D}(\Omega)$ of test functions is dense in $L_2(\Omega)$.*

Corollary 6.10 *Suppose that $f \in L_2(\Omega)$ satisfies $\int_\Omega f(x) \varphi(x) \, dx = 0$ for all $\varphi \in \mathcal{D}(\Omega)$. Then $f = 0$ almost everywhere.*

To finish, we will show that if we mollify a uniformly continuous function, then we obtain uniform convergence to the function as $n \to \infty$.

Theorem 6.11 *Let $u : \mathbb{R}^d \to \mathbb{R}$ be uniformly continuous. Then $\varrho_n * u \in C^\infty(\mathbb{R}^d)$ and*

$$\lim_{n \to \infty} (\varrho_n * u)(x) = u(x)$$

uniformly in $x \in \mathbb{R}^d$.

Here the convolution $\varrho_n * u$ is defined by (6.8), as before.

Proof Fix $x \in \mathbb{R}^d$ and choose $\eta \in \mathcal{D}(\mathbb{R}^d)$ such that $\eta = 1$ on $\bar{B}(x, 2)$. Then $u * \varrho_n = (\eta u) * \varrho_n$ holds in a neighborhood of x. Since $\eta u \in L_2(\mathbb{R}^d)$, it follows from Theorem 6.5 that $u * \varrho_n \in C^\infty(\mathbb{R}^d)$.

Now let $\varepsilon > 0$. Then there exists an $n_0 \in \mathbb{N}$ such that $|u(x - y) - u(x)| \leq \varepsilon$ whenever $|y| \leq 1/n_0$, and so, for all $n \geq n_0$,

$$|(\varrho_n * u)(x) - u(x)| = \left| \int_{|y|<1/n} \Big(u(x-y) - u(x) \Big) \varrho_n(y) \, dy \right|$$

$$\leq \varepsilon \int_{|y|\leq 1/n} \varrho_n(y) \, dy = \varepsilon,$$

which proves the claim. $\qquad\square$

6.2 Sobolev spaces on $\Omega \subseteq \mathbb{R}^d$

After the preparatory work undertaken in Section 6.1 we can now introduce Sobolev spaces on general domains in $\mathbb{R}^d$. So let Ω be an open set in $\mathbb{R}^d$; then as in Chapter 5 we shall define weak derivatives of functions on Ω via integration by parts. To start with, we will suppose that f is a function which is continuously differentiable in the classical sense, that is, $f \in C^1(\Omega)$ with partial derivatives $\frac{\partial f}{\partial x_j}$, $j = 1, \ldots, d$.

Lemma 6.12 (Integration by parts) *For $f \in C^1(\Omega)$ and $\varphi \in C_c^1(\Omega)$ we have*

$$-\int_\Omega f(x) \frac{\partial \varphi}{\partial x_j}(x) \, dx = \int_\Omega \frac{\partial f}{\partial x_j} \varphi(x) \, dx. \qquad (6.21)$$

Proof 1st case: $\Omega = \mathbb{R}^d$. In this case, we have

$$-\int_{\mathbb{R}^d} f(x) \frac{\partial \varphi}{\partial x_j}(x) \, dx = -\int_{\mathbb{R}} \cdots \int_{\mathbb{R}} f(x_1, \ldots, x_d) \frac{\partial \varphi}{\partial x_j}(x_1, \ldots, x_d) \, dx_1 \cdots dx_d$$

$$= \int_{\mathbb{R}} \cdots \int_{\mathbb{R}} \frac{\partial f}{\partial x_j}(x_1, \ldots, x_d) \varphi(x_1, \ldots, x_d) \, dx_1 \cdots dx_d$$

by integration by parts in the jth integral.

2nd case: Now suppose Ω is arbitrary. Choose $U \Subset \Omega$ such that $\operatorname{supp} \varphi \subset U$, and also choose $\eta \in \mathcal{D}(\Omega)$ such that $\eta = 1$ on $\operatorname{supp} \varphi$ and $\operatorname{supp} \eta \subset U$ (see Lemma 6.7). Then $\widetilde{f\eta} \in C^1(\mathbb{R}^d)$ (where, as in the previous section, $\widetilde{f\eta} = (f\eta)^\sim$ denotes the extension of the function $f\eta$ by zero in accordance with (6.20)). By the 1st case, with the relevant extensions,

$$-\int_\Omega f(x) \frac{\partial \varphi}{\partial x_j}(x) \, dx = -\int_{\mathbb{R}^d} \widetilde{f\eta}(x) \frac{\partial \varphi}{\partial x_j}(x) \, dx$$

$$= \int_{\mathbb{R}^d} \frac{\partial (\widetilde{f\eta})(x)}{\partial x_j} \varphi(x) \, dx = \int_\Omega \frac{\partial f}{\partial x_j}(x) \varphi(x) \, dx,$$

since $\eta = 1$ on $\operatorname{supp} \varphi$ and $\operatorname{supp} \varphi \subset U \Subset \Omega$. $\qquad\square$

We now take the identity (6.21) as the definition of the weak derivatives of a function in $L_2(\Omega)$: let $f \in L_2(\Omega)$ and $j \in \{1, \ldots, d\}$. If $g_j \in L_2(\Omega)$ satisfies

$$- \int_\Omega f \frac{\partial \varphi}{\partial x_j}(x)\, dx = \int_\Omega g_j \varphi\, dx \qquad (6.22)$$

for all $\varphi \in \mathcal{D}(\Omega)$, then we say that g_j is the *weak jth partial derivative* of f. Weak jth partial derivatives are unique, if they exist: if (6.22) holds both for g_j and some other function $\hat{g}_j \in L_2(\Omega)$, then $\int_\Omega g_j \varphi\, dx = \int_\Omega \hat{g}_j \varphi\, dx$ for all $\varphi \in \mathcal{D}(\Omega)$. Corollary 6.10 implies that $g_j = \hat{g}_j$ in $L_2(\Omega)$. We will denote the weak jth partial derivative of f, if it exists, by $D_j f$.

If $f \in L_2(\Omega)$ and $j \in \{1, \ldots, d\}$ then for brevity we will simply write $D_j f \in L_2(\Omega)$ to indicate that the weak jth partial derivative exists in the above sense. With this convention, the *Sobolev space* $H^1(\Omega)$ may be defined as follows:

$$H^1(\Omega) := \{u \in L_2(\Omega) : D_j u \in L_2(\Omega) \text{ for all } j = 1, \ldots, d\}.$$

If $u \in H^1(\Omega)$, then $D_j u$ is thus the unique function in $L_2(\Omega)$ for which

$$- \int_\Omega u \frac{\partial \varphi}{\partial x_j}\, dx = \int_\Omega D_j u\, \varphi\, dx$$

for all $\varphi \in \mathcal{D}(\Omega)$.

Remark 6.13 (Comparison of classical and weak derivatives) Take $u \in L_2(\Omega) \cap C^1(\Omega)$. Then, invoking Lemma 6.12 we see that u is in $H^1(\Omega)$ if and only if $\frac{\partial u}{\partial x_j} \in L_2(\Omega)$ for all $j = 1, \ldots, d$. In this case, the weak and classical partial derivatives agree, that is, $D_j u = \frac{\partial u}{\partial x_j}$, $j = 1, \ldots, d$ (where equality is in the sense of $L_2(\Omega)$-functions, that is, almost everywhere). Thus $D_j u$ is a generalization of the classical partial derivatives. For functions $u \in L_2(\Omega) \cap C^1(\Omega)$ we will use both symbols $\frac{\partial u}{\partial x_j}$ and $D_j u$, but $\frac{\partial}{\partial x_j}$ will be used exclusively for the classical partial derivative. △

Remark 6.14 (Test functions) We used the space $\mathcal{D}(\Omega)$ when defining weak derivatives and thus also $H^1(\Omega)$. This is particularly convenient when higher derivatives are needed, as they will be in some arguments. However, the "integration by parts formula" is automatically valid for the larger class $C_c^1(\Omega)$, that is,

$$- \int_\Omega u \frac{\partial \varphi}{\partial x_j}\, dx = \int_\Omega D_j u\, \varphi\, dx$$

for all $u \in H^1(\Omega)$, $\varphi \in C_c^1(\Omega)$, $j = 1, \ldots, d$, see Exercise 6.2. △

It follows immediately from the uniqueness of $D_j u$ that $H^1(\Omega)$ is a vector space and $D_j : H^1(\Omega) \to L_2(\Omega)$ is a linear mapping for all $j = 1, \ldots, d$. Moreover,

$$(u, v)_{H^1} := (u, v)_{L_2} + \sum_{j=1}^d (D_j u, D_j v)_{L_2}$$

$$= \int_\Omega u(x)v(x)\, dx + \int_\Omega \nabla u(x) \nabla v(x)\, dx$$

defines an inner product on $H^1(\Omega)$, whose corresponding norm is given by

$$\|u\|_{H^1(\Omega)}^2 := \int_\Omega u(x)^2 \, dx + \int_\Omega |\nabla u(x)|^2 \, dx.$$

Here we have taken the *gradient* ∇u of $u \in H^1(\Omega)$ to be the vector

$$\nabla u(x) := (D_1 u(x), \dots, D_d u(x)).$$

If $y = (y_1, \dots, y_d), z = (z_1, \dots, z_d) \in \mathbb{R}^d$, then we denote the natural inner product and the Euclidean norm on $\mathbb{R}^d$ by $y \cdot z = y^T z$ and $|y| := \sqrt{y \cdot y}$, respectively. For $u, v \in H^1(\Omega)$, we therefore write $\nabla u(x) \cdot \nabla v(x) = \sum_{j=1}^d D_j u(x) D_j v(x)$, $x \in \Omega$. Moreover, $(\nabla u \cdot \nabla v)(x) := \nabla u(x) \cdot \nabla v(x)$, $x \in \Omega$, defines a function $\nabla u \cdot \nabla v \in L_1(\Omega)$.

Theorem 6.15 *When equipped with the above inner product, $H^1(\Omega)$ is a Hilbert space.*

Proof The proof of Theorem 6.15 is completely analogous to the proof of Theorem 5.6 in the one-dimensional case. □

From the definition of the norm, we immediately obtain the following characterization of convergence in $H^1(\Omega)$: if $u_n, u \in H^1(\Omega)$, then $\lim_{n \to \infty} u_n = u$ in $H^1(\Omega)$ if and only if

$$\lim_{n \to \infty} u_n = u \text{ and } \lim_{n \to \infty} D_j u_n = D_j u \text{ in } L_2(\Omega) \text{ for all } j = 1, \dots, d.$$

We will make repeated use of this characterization. We next show how functions in $H^1(\Omega)$ can be approximated by C^∞-functions.

Theorem 6.16 (Approximation theorem) *Let $u \in H^1(\Omega)$. Then there exist functions $u_n \in \mathcal{D}(\mathbb{R}^d)$, $n \in \mathbb{N}$, such that*

$$\lim_{n \to \infty} u_n = u \quad \text{in } L_2(\Omega) \tag{6.23}$$

$$\text{and } \lim_{n \to \infty} \frac{\partial u_n}{\partial x_j} = D_j u \quad \text{in } L_2(U) \tag{6.24}$$

for all $U \Subset \Omega$ and all $j = 1, \dots, d$.

Proof We define $u_n := \eta_n(\varrho_n * \tilde{u})$ just as we did in the lead-up to Theorem 6.8, where $\tilde{u}$ denotes the extension of u by zero to $\mathbb{R}^d$ as in (6.20). Then $u_n \in \mathcal{D}(\mathbb{R}^d)$ and $\lim_{n \to \infty} u_n = u$ in $L_2(\Omega)$ by Theorem 6.8. Now fix $U \Subset \Omega$; we will show that

$$\frac{\partial u_n}{\partial x_j} = \varrho_n * \widetilde{D_j u} \tag{6.25}$$

holds on U, for all sufficiently large n. To this end, we choose an $n_0 \in \mathbb{N}$ such that $U \subset \Omega_{n_0}$ (cf. (6.16)) and fix $x \in U$. Then for $n \geq n_0$ we have $\operatorname{supp} \varrho_n(x - \cdot) \subset \Omega$ and

$(\varrho_n * \tilde{u})(x) = \int_\Omega u(y)\varrho_n(x - y)\,dy$. Hence, using Theorem 6.5 and the definition of $D_j u$, we obtain

$$\frac{\partial}{\partial x_j}(\varrho_n * \tilde{u})(x) = \int_\Omega u(y)\frac{\partial \varrho_n}{\partial x_j}(x - y)\,dy = -\int_\Omega u(y)\frac{\partial \varrho_n}{\partial y_j}(x - y)\,dy$$

$$= \int_\Omega D_j u(y)\varrho_n(x - y)\,dy = (\varrho_n * \widetilde{D_j u})(x).$$

It now follows from Theorem 6.6 that $\lim_{n\to\infty} \frac{\partial u_n}{\partial x_j} = D_j u$ in $L_2(U)$. $\square$

Remark 6.17 It is in general impossible to guarantee that convergence of the derivatives (6.24) holds in the whole space $L_2(\Omega)$ without imposing additional regularity conditions on the boundary of Ω. Indeed, there exist bounded open sets Ω for which $C(\overline{\Omega}) \cap H^1(\Omega)$ is not dense in $H^1(\Omega)$ (for example $\Omega = (0, 1) \cup (1, 2)$ in $\mathbb{R}$); see Exercise 6.8. However, the theorem of Meyers–Serrin (see [50] or, e.g., [28, Sec. 5.3.1]) states that $C^\infty(\Omega) \cap H^1(\Omega)$ is dense in $H^1(\Omega)$, for any open set $\Omega \subset \mathbb{R}^d$. $\triangle$

There is also a converse of the approximation theorem (Theorem 6.16); with this converse statement, we can completely describe the space $H^1(\Omega)$ in terms of its approximation properties.

Theorem 6.18 *Let $u, f_1, \ldots, f_d \in L_2(\Omega)$, and suppose that there exist functions $u_n \in C_c^1(\mathbb{R}^d)$ such that $u_n \to u$ and $\frac{\partial u_n}{\partial x_j} \to f_j$ as $n \to \infty$ in $L_2(U)$, for all $U \Subset \Omega$ and $j = 1, \ldots, d$. Then $u \in H^1(\Omega)$ and $D_j u = f_j$ for all $j = 1, \ldots, d$.*

Proof For $\varphi \in \mathcal{D}(\Omega)$, we have

$$-\int_\Omega u(x)\frac{\partial \varphi}{\partial x_j}(x)\,dx = \lim_{n\to\infty} -\int_\Omega u_n(x)\frac{\partial \varphi}{\partial x_j}(x)\,dx$$

$$= \lim_{n\to\infty} \int_\Omega \frac{\partial}{\partial x_j}u_n(x)\,\varphi(x)\,dx = \int_\Omega f_j(x)\varphi(x)\,dx$$

for all $j = 1, \ldots, d$. The claim now follows from the definition of $H^1(\Omega)$. $\square$

We will now collect some rules of differentiation and, in the process, test the suitability of our weak partial derivatives as defined above.

Theorem 6.19 *Let $\Omega \subset \mathbb{R}^d$ be open and connected, and let $u \in H^1(\Omega)$ be such that $D_j u(x) = 0$ almost everywhere in Ω, for all $j = 1, \ldots, d$. Then there exists a constant $c \in \mathbb{R}$ such that $u(x) = c$ almost everywhere.*

Proof The proof consists of two steps.

1. Let $\bar{B}(x, r) \subset \Omega$ for some $x \in \Omega$ and $r > 0$. We will show that there exists a constant $c \in \mathbb{R}$ such that $u(y) = c$ almost everywhere in $B(x, r)$. To this end, take $u_n := \eta_n(\varrho_n * \tilde{u})$ in accordance with Theorem 6.8 and choose n large enough that $\frac{1}{n} < \text{dist}(\bar{B}(x, r), \partial\Omega)$. Then by (6.25) we have $\frac{\partial}{\partial x_j}u_n = \frac{\partial}{\partial x_j}(\varrho_n * \tilde{u}) = \varrho_n * \widetilde{D_j u} = 0$

in $B(x, r)$, for all $j = 1, \ldots, d$. It follows that u_n is constant in $B(x, r)$, since by (6.13) $|u_n(y) - u_n(x)| \le \|\nabla u_n\|_\infty \cdot |x - y| = 0$ for all $y \in B(x, r)$. Since $u_n \to u$ in $L_2(\Omega)$ by Theorem 6.11, the claim follows.

2. By 1., there exists a constant c such that

$$U := \{x \in \Omega : u(y) = c \ \text{ for almost all } \ y \text{ in a neighborhood of } x\}$$

is not empty. The set U is clearly open; by 1., U is also relatively closed in Ω. Since Ω is connected, it follows that $U = \Omega$. $\qquad\qquad\square$

We next derive the product rule.

Theorem 6.20 (Product rule) *Let* $u, v \in H^1(\Omega)$ *be such that* $v, D_j v \in L_\infty(\Omega)$, $j = 1, \ldots, d$. *Then* $uv \in H^1(\Omega)$ *and*

$$D_j(uv) = (D_j u)v + u \, D_j v. \tag{6.26}$$

Proof By assumption, $uv \in L_2(\Omega)$ and $(D_j u)v + u \, D_j v \in L_2(\Omega)$. We need to show that

$$- \int_\Omega u(x) \, v(x) \, \frac{\partial}{\partial x_j} \varphi(x) \, dx = \int_\Omega (D_j u(x) \, v(x) + u(x) \, D_j v(x)) \varphi(x) \, dx \tag{6.27}$$

for all $\varphi \in \mathcal{D}(\Omega)$.

1st case: The function u has an extension in $\mathcal{D}(\mathbb{R}^d)$. Then $u\varphi \in \mathcal{D}(\Omega)$ for all $\varphi \in \mathcal{D}(\Omega)$ and thus, by definition of $D_j v$,

$$
\begin{aligned}
- \int_\Omega u(x) \, v(x) \, \frac{\partial}{\partial x_j} \varphi(x) \, dx &= - \int_\Omega v(x) \, \frac{\partial}{\partial x_j} (u(x) \, \varphi(x)) \, dx \\
&\quad + \int_\Omega v(x) \left(\frac{\partial}{\partial x_j} u(x) \right) \varphi(x) \, dx \\
&= \int_\Omega (D_j v(x) \, u(x) \, \varphi(x) + D_j u(x) \, v(x)) \varphi(x) \, dx.
\end{aligned}
$$

This establishes (6.27) in this case.

2nd case: Now let $u \in H^1(\Omega)$ be arbitrary. Then by Theorem 6.16 there exist functions $u_n \in \mathcal{D}(\mathbb{R}^d)$ such that $u_n \to u$ in $L_2(\Omega)$ and $D_j u_n \to D_j u$ in $L_2(U)$ for all $U \Subset \Omega$. The 1st case implies that $D_j(u_n \, v) = (D_j \, u_n)v + u_n \, D_j v$, and so $D_j(u_n \, v)$ converges to $(D_j u) \, v + u \, D_j v$ in $L_2(U)$ as $n \to \infty$, for all $U \Subset \Omega$. Since $u_n v \to uv$ in $L_2(\Omega)$ as $n \to \infty$, the claim follows from Theorem 6.18. $\qquad\square$

Theorem 6.21 (Chain rule) *Suppose that* $f \in C^1(\mathbb{R})$ *satisfies* $f(0) = 0$ *and* $|f'(r)| \le M$ *for all* $r \in \mathbb{R}$, *where* $M > 0$. *Then* $f \circ u \in H^1(\Omega)$ *and* $D_j(f \circ u) = (f' \circ u)D_j u$, $j = 1, \ldots, d$, *for all* $u \in H^1(\Omega)$.

Proof Fix $u \in H^1(\Omega)$. Firstly, by assumption $|f(r)| = \left| \int_0^r f'(s)\, ds \right| \le M|r|$ for all $r \in \mathbb{R}$, whence $|(f \circ u)(x)| \le M|u(x)|$ for all $x \in \Omega$ and so

$$\|f \circ u\|_{L_2}^2 = \int_\Omega |(f \circ u)(x)|^2\, dx \le M^2 \|u\|_{L_2}^2 < \infty.$$

Thus $f \circ u \in L_2(\Omega)$. Since f' is bounded, $f' \circ u \in L_\infty(\Omega)$ and $(f' \circ u)D_j u \in L_2(\Omega)$, where the latter assertion also uses that $u \in H^1(\Omega)$. Let $\varphi \in \mathcal{D}(\Omega)$; we need to show that

$$-\int_\Omega (f \circ u)(x)\frac{\partial \varphi}{\partial x_j}(x)\, dx = \int_\Omega (f' \circ u)(x)D_j u(x)\varphi(x)\, dx. \tag{6.28}$$

By the approximation theorem, Theorem 6.16, there exist functions $u_n \in \mathcal{D}(\mathbb{R}^d)$ such that $u_n \to u$ and $D_j u_n \to D_j u$ in $L_2(U)$ for all $U \Subset \Omega$. Choose $U \Subset \Omega$ such that $\operatorname{supp}\varphi \subset U$. By passing to a subsequence if necessary, we may assume that there exists $g \in L_2(\Omega)$ such that $|u_n| \le g$ and $|D_j u_n| \le g$ for all $n \in \mathbb{N}$. We may also assume that $u_n \to u$ and $D_j u_n \to D_j u$ as $n \to \infty$ almost everywhere in U (see the converse of the Dominated Convergence Theorem, Theorem A.10). By the classical chain rule, $\frac{\partial}{\partial x_j}(f \circ u_n) = (f' \circ u_n)\frac{\partial u_n}{\partial x_j}$. Hence, by the integration by parts formula (6.21),

$$-\int_\Omega (f \circ u_n)(x)\frac{\partial \varphi}{\partial x_j}(x)\, dx = \int_\Omega (f' \circ u_n)(x)\frac{\partial u_n}{\partial x_j}(x)\,\varphi(x)\, dx \tag{6.29}$$

for all $\varphi \in \mathcal{D}(\Omega)$. Since $f \circ u_n \to f \circ u$, $f' \circ u_n \to f' \circ u$ and $D_j u_n \to D_j u$ almost everywhere as $n \to \infty$, the Dominated Convergence Theorem allows us to pass to the limit in (6.29), and we obtain (6.28). $\square$

Next we wish to prove a substitution rule for Sobolev spaces.

Theorem 6.22 (Substitution rule) *Let $\Omega_1, \Omega_2 \subset \mathbb{R}^d$ be open and let $F : \Omega_2 \to \Omega_1$ be bijective and satisfy the conditions $F \in C^1(\Omega_2, \mathbb{R}^d)$, $F^{-1} \in C^1(\Omega_1, \mathbb{R}^d)$ with $\sup_{x \in \Omega_2}\|DF(x)\| < \infty$ and $\sup_{y \in \Omega_1}\|DF^{-1}(y)\| < \infty$. If $u \in H^1(\Omega_1)$, then $u \circ F \in H^1(\Omega_2)$ and*

$$D_j(u \circ F) = \sum_{i=1}^d (D_i u) \circ F \cdot \frac{\partial F_i}{\partial y_j}.$$

Here $F(y) = (F_1(y), \dots, F_d(y))^T$, $y \in \Omega_2$, and $(DF)(y) = \left(\frac{\partial F_i}{\partial y_j}(y) \right)_{i,j=1,\dots,d}$ is the Jacobian matrix of F at y.

Proof Let $u \in H^1(\Omega_1)$. Then by the approximation theorem, Theorem 6.16, there exist $u_n \in \mathcal{D}(\mathbb{R}^d)$ such that $u_n \to u$ in $L_2(\Omega_1)$ and $D_j u_n \to D_j u$ in $L_2(U)$ for any $U \Subset \Omega_1$. Hence we have the convergence

$$(D_i u_n \circ F) \cdot \frac{\partial F_i}{\partial y_j} \to (D_i u \circ F) \cdot \frac{\partial F_i}{\partial y_j}$$

in $L_2(U)$ as $n \to \infty$, for all $U \Subset \Omega_2$. Given $\varphi \in \mathcal{D}(\Omega_2)$, choose $\mathrm{supp}\varphi \subset U \Subset \Omega_2$, then

$$\int_V (u_n \circ F) \frac{\partial \varphi}{\partial y_j} \, dy = -\int_V \sum_{i=1}^{d} \left(\frac{\partial u_n}{\partial x_i} \circ F \right) \frac{\partial F_i}{\partial y_j} \cdot \varphi \, dy.$$

The result follows by passing to the limit as $n \to \infty$. □

A special case is when $F : \mathbb{R}^d \to \mathbb{R}^d$ is an affine mapping, given by $F(y) = By + b$ for an invertible $d \times d$ matrix B and $b \in \mathbb{R}^d$. For $\Omega_2 \subset \mathbb{R}^d$, in this case $F(\Omega_2) =: \Omega_1$ is open and the substitution rule can be applied. Of particular interest is the case where B is orthogonal: then F is *isometric*, that is, $\|F(y) - F(z)\| = \|y - z\|$ for all $y, z \in \mathbb{R}^d$ (and, conversely, every isometric mapping of $\mathbb{R}^d$ to itself is of the form $F(y) = By + b$ for an orthogonal matrix B).

Corollary 6.23 *Let B be an orthogonal $d \times d$ matrix, $b \in \mathbb{R}^d$, and $F(y) = By + b$ for $y \in \mathbb{R}^d$. If $\Omega_2 \subset \mathbb{R}^d$ is open and $\Omega_1 := F(\Omega_2)$, then the mapping $u \mapsto u \circ F$ defines a unitary operator from $H^1(\Omega_1)$ to $H^1(\Omega_2)$.*

We will now show the following connection between weak and classical derivatives.

Theorem 6.24 *Let $u \in H^1(\Omega)$ and suppose that $D_j u \in C(\Omega)$ for all $j = 1, \ldots, d$. Then $u \in C^1(\Omega)$.*

Proof Let $U \Subset \Omega$; we wish to show that $u \in C^1(U)$. Now we may assume that u has compact support in Ω, since otherwise we may replace u by ηu for a suitable cut-off function $\eta \in \mathcal{D}(\Omega)$ and apply the product rule of Theorem 6.20.

By Theorem 6.27 we can also assume that $\Omega = \mathbb{R}^d$, since otherwise we may replace u by its extension $\tilde{u}$. So we have that $u \in H^1(\mathbb{R}^d)$ and u has compact support. By assumption, $D_j u \in C_c(\mathbb{R}^d)$. Consider the function $u_n := \varrho_n * u \in \mathcal{D}(\mathbb{R}^d)$. Then $\lim_{n \to \infty} u_n = u$ in $L_2(\mathbb{R}^d)$ by Theorem 6.6 and $D_j u_n = \varrho_n * D_j u$ by Theorem 6.5, and so

$$|D_j u_n(x)| = \left| \int_{\mathbb{R}^d} \varrho_n(x - y) D_j u(y) \, dy \right| \le \|D_j u\|_\infty.$$

Thus $\| |\nabla u_n| \|_\infty \le c$ for all $n \in \mathbb{N}$ and some $c \ge 0$, since $D_j u \in C_c(\mathbb{R}^d)$. It follows from (6.13) that $|u_n(y) - u_n(x)| \le \| |\nabla u_n| \|_\infty \cdot |y - x| \le c |y - x|$ for all $n \in \mathbb{N}$ and all $x, y \in \mathbb{R}^d$. This establishes that the sequence $(u_n)_{n \in \mathbb{N}}$ is equicontinuous. Since $u_n \to u$ in $L_2(\mathbb{R}^d)$, we may assume that $u_n(x) \to u(x)$ almost everywhere (since otherwise we pass to a subsequence; see Theorem A.10). In particular, there exist $x_0 \in \mathbb{R}^d$ and $b \ge 0$ such that $|u_n(x_0)| \le b$ for all $n \in \mathbb{N}$. Then $|u_n(x)| \le b + c|x - x_0|$ for all $n \in \mathbb{N}$ and all $x \in \mathbb{R}^d$. This means that the sequence $(u_n)_{n \in \mathbb{N}}$ is bounded on every compact subset of $\mathbb{R}^d$. By the theorem of Arzelà–Ascoli (Theorem A.6), for each compact subset K of $\mathbb{R}^d$ there exists a subsequence of $(u_n)_{n \in \mathbb{N}}$ which converges uniformly in K to a continuous function $w \in C(K)$. Since $u_n(x) \to u(x)$ almost everywhere, the only possibility is that $u = w$ almost everywhere. Thus u is continuous (possibly after modification on a set of zero measure). For the jth canonical basis vector of $\mathbb{R}^d$, e_j, and for $x \in \mathbb{R}^d$, we have

$$u_n(x + t e_j) = u_n(x) + \int_0^t D_j u_n(x + s e_j)\, ds.$$

Since $D_j u_n = \varrho_n * D_j u$ converges uniformly to $D_j u$ by Theorem 6.11, passing to the limit as $n \to \infty$ we obtain

$$u(x + t e_j) = u(x) + \int_0^t D_j u(x + s e_j)\, ds.$$

Hence $\frac{\partial u}{\partial x_j}(x) = D_j u(x)$ and thus $u \in C^1(\mathbb{R}^d)$. $\qquad\qquad\square$

We finish this section by defining Sobolev spaces of higher order:

$$H^2(\Omega) := \{u \in H^1(\Omega) : D_j u \in H^1(\Omega),\ j = 1, \dots, d\}.$$

That is, if $u \in H^2(\Omega)$, then $D_i D_j \in L_2(\Omega)$ for all $i, j \in \{1, \dots, d\}$. It is easy to show that

$$D_i D_j u = D_j D_i u \qquad\qquad (6.30)$$

holds for all $i, j \in \{1, \dots, d\}$, since the same is true for all test functions by Schwarz's theorem (see Exercise 6.3). We then define inductively

$$H^{k+1}(\Omega) := \{u \in H^1(\Omega) : D_j u \in H^k(\Omega),\ j = 1, \dots, d\}. \qquad (6.31)$$

This defines $H^k(\Omega)$ for all $k \in \mathbb{N}$.

6.3 The space $H_0^1(\Omega)$

We now wish to study to Sobolev spaces with weak homogeneous boundary conditions on an open set $\Omega \subset \mathbb{R}^d$. If $d \geq 2$, then not all functions in $H^1(\Omega)$ are continuous (see Exercise 6.5). Since in general the boundary $\partial \Omega$ of Ω has measure zero and we identify functions in $H^1(\Omega)$ which agree almost everywhere, it is meaningless to write $u_{|\partial \Omega} = 0$. The solution is to define a weak version of the boundary condition "$u_{|\partial \Omega} = 0$". To this end we set $H_0^1(\Omega) := \overline{\mathcal{D}(\Omega)}^{H^1(\Omega)}$, that is,

$$H_0^1(\Omega) = \{u \in H^1(\Omega) : \exists\, \varphi_n \in \mathcal{D}(\Omega) \text{ such that } \lim_{n \to \infty} \varphi_n = u \text{ in } H^1(\Omega)\}.$$

Then $H_0^1(\Omega)$ is a closed subspace of $H^1(\Omega)$ and hence itself a Hilbert space. For $u \in H^1(\Omega)$ we interpret the condition $u \in H_0^1(\Omega)$ as a weak form of the statement that u vanishes on the boundary of Ω. We now wish to study the space $H_0^1(\Omega)$. If $\Omega = \mathbb{R}^d$, then Ω does not have a boundary, and indeed in this case $H_0^1(\mathbb{R}^d) = H^1(\mathbb{R}^d)$:

Theorem 6.25 *The space of test functions $\mathcal{D}(\mathbb{R}^d)$ is dense in $H^1(\mathbb{R}^d)$.*

Proof Let $u \in H^1(\mathbb{R}^d)$. Then $\varrho_n * u \in C^\infty(\mathbb{R}^d) \cap L_2(\mathbb{R}^d)$, and by Theorem 6.6 we have $\lim_{n \to \infty} \varrho_n * u = u$ in $L_2(\mathbb{R}^d)$. One also sees as in the proof of (6.25) that

$$\frac{\partial}{\partial x_j}(\varrho_n * u) = \varrho_n * D_j u.$$

Thus $\varrho_n * u \in H^1(\mathbb{R}^d) \cap C^\infty(\mathbb{R}^d)$ and $\lim_{n \to \infty} \varrho_n * u = u$ in $H^1(\mathbb{R}^d)$. Now let $\eta \in \mathcal{D}(\mathbb{R}^d)$ be such that $0 \le \eta \le 1$, $\eta(x) = 1$ for all $|x| \le 1$, and $\eta(x) = 0$ for all $|x| \ge 2$. Also let $\eta_k(x) := \eta(\frac{x}{k})$. Then $\eta_k \in \mathcal{D}(\mathbb{R}^d)$ satisfies $\eta_k(x) = 1$ for all $|x| \le k$ and $0 \le \eta_k \le 1$. It follows that $\lim_{k \to \infty} \eta_k(\varrho_n * u) = \varrho_n * u$ in $L_2(\mathbb{R}^d)$ by the Dominated Convergence Theorem. Moreover, as $k \to \infty$,

$$\frac{\partial}{\partial x_j}(\eta_k(\varrho_n * u))(x) = \left(\frac{\partial}{\partial x_j}\eta_k\right)(x)(\varrho_n * u)(x) + \eta_k(x)\frac{\partial}{\partial x_j}(\varrho_n * u)(x)$$

$$= \frac{1}{k}\left(\frac{\partial}{\partial x_j}\eta\right)\left(\frac{x}{k}\right)(\varrho_n * u)(x) + \eta_k(x)(\varrho_n * D_j u)(x)$$

converges to $\varrho_n * D_j u$ in $L_2(\mathbb{R}^d)$ since $\eta_k(x) = 1$ for $|x| \le k$ and the first term tends to zero. Hence $\lim_{k \to \infty} \eta_k(\varrho_n * u) = \varrho_n * u$ in $H^1(\mathbb{R}^d)$. Since $\eta_k(\varrho_n * u) \in \mathcal{D}(\mathbb{R}^d)$, we finally obtain $\varrho_n * u \in H_0^1(\mathbb{R}^d)$, and so the theorem is proved upon passing to the limit as $n \to \infty$. $\qquad\square$

One can easily extend the proof of Theorem 6.25 to the second derivative when $u \in H^2(\mathbb{R}^d)$. The following corollary of the proof will be useful later.

Corollary 6.26 *Let $u \in H^2(\mathbb{R}^d)$. Then there exist $u_n \in \mathcal{D}(\mathbb{R}^d)$ such that* $\lim_{n \to \infty} u_n = u$, $\lim_{n \to \infty} \frac{\partial}{\partial x_j} u_n = D_j u$, *and* $\lim_{n \to \infty} \frac{\partial}{\partial x_i}\frac{\partial}{\partial x_j} u_n = D_i D_j u$ *in $L_2(\mathbb{R}^d)$ for all $i, j \in \{1, \ldots, d\}$.*

If $f : \Omega \to \mathbb{R}^d$ is defined on an open set $\Omega \subset \mathbb{R}^d$, then analogously to (6.20) we may define the extension

$$\tilde{f}(x) := \begin{cases} f(x) & \text{if } x \in \Omega, \\ 0 & \text{in } \mathbb{R}^d \setminus \Omega. \end{cases} \tag{6.32}$$

If f is differentiable in Ω, then $\tilde{f}$ will generally not be differentiable on the boundary of Ω; nevertheless, for H_0^1-functions we have the following theorem.

Theorem 6.27 *Let $u \in H_0^1(\Omega)$. Then $\tilde{u} \in H^1(\mathbb{R}^d)$ and $D_j\tilde{u} = \widetilde{D_j u}$ for all $j = 1, \ldots, d$.*

Proof Let $u \in H_0^1(\Omega)$. Then there exists a sequence of functions $u_n \in \mathcal{D}(\Omega)$ such that $u_n \to u$ and $\frac{\partial u_n}{\partial x_j} \to D_j u$ in $L_2(\Omega)$ as $n \to \infty$ for all $j = 1, \ldots, d$. By (6.21) we have, for all $\varphi \in \mathcal{D}(\mathbb{R}^d)$,

$$-\int_{\mathbb{R}^d} \tilde{u}(x) \frac{\partial \varphi}{\partial x_j}(x) \, dx = \lim_{n \to \infty} -\int_{\Omega} u_n(x) \frac{\partial \varphi}{\partial x_j}(x) \, dx = \lim_{n \to \infty} \int_{\Omega} \frac{\partial u_n}{\partial x_j}(x) \, \varphi(x) \, dx$$

$$= \int_{\Omega} D_j u(x) \, \varphi(x) \, dx = \int_{\mathbb{R}^d} \widetilde{D_j u}(x) \, \varphi(x) \, dx.$$

This proves the claim. $\qquad\qquad\qquad\qquad\qquad\qquad\qquad\qquad\qquad\qquad\qquad\qquad\qquad$ □

Corollary 6.28 *Let* $u \in H_0^1(\Omega)$ *be such that* $D_j u = 0$ *for all* $j = 1, \ldots, d$. *Then* $u = 0$.

Proof By Theorem 6.27 we have $D_j \tilde{u} = 0$ for all $j = 1, \ldots, d$. It follows from Theorem 6.19 that $\tilde{u}$ is constant. Since $\tilde{u} \in L_2(\mathbb{R}^d)$, the only possibility is that $\tilde{u} = 0$.
□

Let $\Omega \subset \mathbb{R}^d$ be open. We now wish to compare the classical condition $u_{|\partial\Omega} = 0$ with the weak one, that is, $u \in H_0^1(\Omega)$. We will begin by showing that any function $u \in H^1(\Omega)$ which vanishes identically in a neighborhood of $\partial\Omega$ is in $H_0^1(\Omega)$. To this end, we define

$$H_c^1(\Omega) := \{u \in H^1(\Omega) : \exists K \subset \Omega \text{ compact such that } u = 0 \text{ in } \Omega \setminus K\}.$$

Theorem 6.29 *We have* $H_c^1(\Omega) \subset H_0^1(\Omega)$.

Proof Let $u \in H_c^1(\Omega)$, that is, $u(x) = 0$ for all $x \in \Omega \setminus K$, where $K \subset \Omega$ is compact. Let $n \in \mathbb{N}$ be so large that $\frac{1}{n} < \mathrm{dist}(K, \partial\Omega)$. Then by Lemma 6.3 $(\varrho_n * \tilde{u})(x) = 0$ whenever $x \notin K + \bar{B}(0, 1/n)$. Since $K + \bar{B}(0, 1/n) \subset \Omega$, we have $\varrho_n * \tilde{u} \in \mathcal{D}(\Omega)$. Moreover, for all $x \in \Omega$,

$$\frac{\partial}{\partial x_j}(\varrho_n * \tilde{u})(x) = \int_{\mathbb{R}^d} \frac{\partial}{\partial x_j} \varrho_n(x - y) \, \tilde{u}(y) \, dy = -\int_{\mathbb{R}^d} \frac{\partial}{\partial y_j} \varrho_n(x - y) \, \tilde{u}(y) \, dy$$

$$= -\int_{\Omega} \frac{\partial}{\partial y_j} \varrho_n(x - y) \, u(y) \, dy = \int_{\Omega} \varrho_n(x - y) D_j u(y) \, dy$$

$$= \int_{\mathbb{R}^d} \varrho_n(x - y) \widetilde{D_j u}(y) \, dy = (\varrho_n * \widetilde{D_j u})(x).$$

It then follows from Theorem 6.6 that $\frac{\partial}{\partial x_j}(\varrho_n * \tilde{u}) \to D_j u$ in $L_2(\Omega)$ and $\varrho_n * \tilde{u} \to u$ in $L_2(\Omega)$ as $n \to \infty$. We have thus approximated u by the test functions $\varrho_n * \tilde{u} \in \mathcal{D}(\Omega)$ in $H^1(\Omega)$. Thus $u \in H_0^1(\Omega)$. $\qquad\qquad\qquad\qquad\qquad\qquad\qquad\qquad$ □

For a bounded open set Ω we set

$$C_0(\Omega) := \{f \in C(\overline{\Omega}) : f_{|\partial\Omega} = 0\}. \tag{6.33}$$

The following theorem shows that H^1-functions with zero boundary conditions in the classical sense are in $H_0^1(\Omega)$.

Theorem 6.30 *If $\Omega \subset \mathbb{R}^d$ is open and bounded, then $H^1(\Omega) \cap C_0(\Omega) \subset H_0^1(\Omega)$.*

We will delay the proof until the next section since we will require lattice operations which will only be introduced at a later point.

Remark 6.31 The converse statement to Theorem 6.30, $H_0^1(\Omega) \cap C(\overline{\Omega}) \subset C_0(\Omega)$, is false in general; here is an example. Let $\Omega := B \setminus \{0\}$, where $B := B(0, 1) = \{x \in \mathbb{R}^3 : |x| < 1\}$ is the ball in $\mathbb{R}^3$ of unit radius. Then $\overline{\Omega} = \bar{B}(0, 1)$. Let $u \in \mathcal{D}(B)$; then $u \in C(\overline{\Omega})$, but $u \in C_0(\Omega)$ only if $u(0) = 0$. But even if $u(0) \neq 0$, we still have $u \in H_0^1(\Omega)$. To see this, first choose a function $f \in C^\infty(\mathbb{R})$ such that $f(r) = 1$ for $r \geq 1$ and $f(r) = 0$ for $r \leq \frac{1}{2}$.

Set $u_n(x) := f(n|x|)u(x)$. Then $u_n(x) = 0$ for all $|x| \leq \frac{1}{2n}$ and thus $u_n \in \mathcal{D}(\Omega)$. Since $u_n(x) = u(x)$ for $|x| \geq \frac{1}{n}$, the function u_n converges to u in $L_2(\Omega)$ by the Dominated Convergence Theorem. Moreover, by the product rule,

$$D_j u_n(x) = f(n|x|)D_j u(x) + n f'(n|x|)\frac{x_j}{|x|}u(x).$$

The first term on the right-hand side converges to $D_j u$ in $L_2(\Omega)$ as $n \to \infty$; the second tends to 0 in $L_2(\Omega)$ as $n \to \infty$ since

$$n^2 \int_\Omega f'(n|x|)^2 \, dx = n^2 \int_{|x|<1/n} f'(|nx|)^2 \, dx = \int_{|x|<1} f'(|y|)^2 \frac{dy}{n}.$$

We thus have $\lim_{n\to\infty} u_n = u$ in $H^1(\Omega)$ and so $u \in H_0^1(\Omega)$ since $H_0^1(\Omega)$ is closed in $H^1(\Omega)$.

However, the implication is correct if Ω does not have any such "holes", that is, if Ω is equal to the interior of $\overline{\Omega}$; see [14].   △

Finally, we wish to prove a very useful and important inequality for the L^2-norm of H_0^1-functions. We say that Ω is *contained in a strip*, or *bounded in one direction*, if there exist $j_0 \in \{1, \dots, d\}$ and $M > 0$ such that $|x_{j_0}| \leq M$ for all $x \in \Omega$. Obviously, every bounded domain has this property.

Theorem 6.32 (Poincaré inequality) *Let $\Omega \subset \mathbb{R}^d$ be open and contained in a strip. Then with $M > 0$ as above we have*

$$\|u\|_{L_2} \leq 2M \, |u|_{H^1} \tag{6.34}$$

for all $u \in H_0^1(\Omega)$, where $|u|_{H^1} := \left(\int_\Omega |\nabla u(x)|^2 \, dx \right)^{1/2} = \|\nabla u\|_{L_2}$.

Corollary 6.33 *If Ω is contained in a strip, then $|\cdot|_{H^1}$ defines an equivalent norm on $H_0^1(\Omega)$.*

Proof (of Theorem 6.32) Without loss of generality suppose that $j_0 = 1$. Let $h \in C^1([-M, M])$ be a function such that $h(-M) = 0$. Then by the Cauchy–Schwarz inequality

$$\int_{-M}^{M} h(x)^2 \, dx = \int_{-M}^{+M} \left| \int_{-M}^{x} h'(y) \, dy \right|^2 dx \leq \int_{-M}^{M} \int_{-M}^{x} h'(y)^2 \, dy \cdot (x + M) \, dx$$

$$\leq (2M)^2 \int_{-M}^{M} h'(y)^2 \, dy.$$

Hence the inequality

$$\int_{\Omega} (u(x))^2 \, dx \leq \int_{\mathbb{R}} \cdots \int_{\mathbb{R}} \int_{-M}^{+M} (2M)^2 \left(\frac{\partial \tilde{u}}{\partial x_1}(x_1, \ldots, x_d) \right)^2 dx_1 \cdots dx_d$$

$$\leq (2M)^2 \int_{\Omega} |\nabla u|^2 \, dx$$

holds for all $u \in \mathcal{D}(\Omega)$. Since $\mathcal{D}(\Omega)$ is dense in $H_0^1(\Omega)$, we conclude that (6.34) holds for all $u \in H_0^1(\Omega)$.   □

Remark 6.34 The statement of the Poincaré inequality also remains valid for bounded Ω in the somewhat more general case that u only vanishes on a non-trivial part $\Gamma_D \subset \partial\Omega$ of the boundary in an appropriate sense (here D stands for Dirichlet boundary conditions). We denote the corresponding space by $H_D^1(\Omega)$. See Exercise 7.11 for a precise statement and hints for the proof. This space plays an important role in the study of problems with mixed boundary conditions, and in particular in numerics (cf. [16, Ch. II]).   △

6.4 Lattice operations on $H^1(\Omega)$

Given a function $f : \Omega \to \mathbb{R}$ we set

$$f^+(x) := \begin{cases} f(x), & \text{if } f(x) > 0, \\ 0, & \text{if } f(x) \leq 0, \end{cases}$$

as well as $f^- := (-f)^+$ and $|f| := f^+ + f^-$. It is immediate from the definitions that $f = f^+ - f^-$. Moreover, if $f \in L_2(\Omega)$, then also $|f|, f^+, f^- \in L_2(\Omega)$ and $\|f\|_{L_2} = \| |f| \|_{L_2}$. One calls the mappings $f \mapsto |f|, f^+, f^-$ *lattice operations* on $L_2(\Omega)$. We also define the *sign function* of f by

$$(\text{sign } f)(x) := \begin{cases} 1, & \text{if } f(x) > 0, \\ 0, & \text{if } f(x) = 0, \\ -1, & \text{if } f(x) < 0, \end{cases}$$

as well as the characteristic function

$$\mathbb{1}_{\{f>0\}}(x) := \begin{cases} 1, & \text{if } f(x) > 0, \\ 0, & \text{if } f(x) \leq 0, \end{cases}$$

and $\mathbb{1}_{\{f<0\}} := \mathbb{1}_{\{-f>0\}}$. We then have the following relations for the weak derivatives of H^1-functions.

Theorem 6.35 (Derivatives under lattice operations) *If $u \in H^1(\Omega)$, then also $|u|, u^+, u^- \in H^1(\Omega)$ and*

$$D_j u^+ = \mathbb{1}_{\{u>0\}} D_j u \tag{6.35}$$
$$D_j u^- = -\mathbb{1}_{\{u<0\}} D_j u \tag{6.36}$$
$$D_j |u| = (\text{sign } u) D_j u \tag{6.37}$$

for all $j = 1, \ldots, d$.

Proof Let $u \in H^1(\Omega)$. We will show that $u^+ \in H^1(\Omega)$ and (6.35) holds; the other claims then follow immediately from this case, since $u^- = (-u)^+$ and $|u| = u^+ + u^-$. So set

$$f_n(r) := \begin{cases} (r^2 + n^{-2})^{1/2} - n^{-1} & \text{if } r > 0, \\ 0 & \text{if } r \leq 0; \end{cases}$$

then $f_n \in C^1(\mathbb{R})$, $f_n(0) = 0$, and

$$f_n'(r) = \begin{cases} r(r^2 + n^{-2})^{-1/2} & \text{if } r > 0, \\ 0 & \text{if } r \leq 0. \end{cases}$$

In particular, $0 \leq f_n'(r) \leq 1$ for all $r \in \mathbb{R}$; moreover, $0 \leq f_n(r) \leq r^+$ for all $n \in \mathbb{N}$ and $r \in \mathbb{R}$, and $\lim_{n \to \infty} f_n(r) = r^+$. Now take $\varphi \in \mathcal{D}(\Omega)$. By the chain rule,

$$-\int_\Omega (f_n \circ u(x)) \frac{\partial \varphi}{\partial x_j}(x) \, dx = \int_\Omega (f_n' \circ u)(x) (D_j u)(x) \, \varphi(x) \, dx. \tag{6.38}$$

By the monotonicity of f_n, we have $0 \leq f_n \circ u \leq u^+$. Since $f_n \circ u \to u^+$ almost everywhere, the Dominated Convergence Theorem implies that the left-hand side of (6.38) converges to $-\int_\Omega u^+(x) \frac{\partial \varphi}{\partial x_j}(x) \, dx$. Since $|(f_n' \circ u)(x)| \leq 1$ and $\lim_{n \to \infty} (f_n' \circ u)(x) = \mathbb{1}_{\{u>0\}}(x)$ for all $x \in \Omega$, the right-hand side of (6.38) converges to

$$\int_\Omega (f_n' \circ u)(x) (D_j u)(x) \, \varphi(x) \, dx \xrightarrow{n \to \infty} \int_\Omega \mathbb{1}_{\{u>0\}}(x) D_j u(x) \varphi(x) \, dx.$$

Hence, passing to the limit on both sides of (6.38), we obtain

$$-\int_\Omega u^+(x) \frac{\partial \varphi}{\partial x_j}(x) \, dx = \int_\Omega \mathbb{1}_{\{u>0\}}(x) (D_j u)(x) \varphi(x) \, dx,$$

and we have proved the statement of the theorem for u^+.   □

The following corollary is quite remarkable.

Corollary 6.36 (Stampacchia's lemma) *Suppose that* $u \in H^1(\Omega)$, $c \in \mathbb{R}$ *and* $j \in \{1, \ldots, d\}$. *Then* $(D_j u)(x) = 0$ *almost everywhere in the set* $\{x \in \Omega : u(x) = c\}$.

Proof The proof proceeds in two steps.

1. If Ω is bounded, then we may assume that $c = 0$. Indeed, otherwise we replace u by $u - c$. By (6.35) and (6.36), on $B := \{x \in \Omega : u(x) = 0\}$ we have $D_j u^+(x) = D_j u^-(x) = 0$ for all $x \in B$. The claim now follows since $D_j u = D_j u^+ - D_j u^-$.

2. For Ω arbitrary we have $\Omega = \bigcup_{n \in \mathbb{N}} \Omega_n$, where $\Omega_n := \Omega \cap B(0, n)$. For each $n \in \mathbb{N}$, by 1., the set $M_n := \{x \in \Omega_n : u(x) = c, D_j u(x) \neq 0\}$ has measure zero, since Ω_n is bounded. Hence $\bigcup_{n \in \mathbb{N}} M_n = \{x \in \Omega : u(x) = c, D_j u(x) \neq 0\}$ likewise has measure zero, and we have proved the claim.   □

We will now show that the lattice operations are continuous.

Theorem 6.37 (Continuity of lattice operations) *The mappings* $u \mapsto u^+, u^-, |u|$: $H^1(\Omega) \to H^1(\Omega)$ *are continuous.*

Proof Let $u_n \to u$ be a convergent sequence in $H^1(\Omega)$. We will show that $|u_n| \to |u|$ in $H^1(\Omega)$; since $u^+ = \frac{1}{2}(u + |u|)$ and $u^- = (-u)^+$, it will then follow directly that $u_n^+ \to u^+$ and $u_n^- \to u^-$ in $H^1(\Omega)$ as $n \to \infty$. Now since $||u_n| - |u|| \leq |u_n - u|$, we certainly have $|u_n| \to |u|$ in $L_2(\Omega)$, and it remains to show that $D_j |u_n| \to D_j |u|$ in $L_2(\Omega)$ for all $j \in \{1, \ldots, d\}$. By Lemma 4.40, it suffices to prove this for a subsequence.

By the converse of the Dominated Convergence Theorem (Theorem A.10), we may therefore assume that $u_n(x) \to u(x), D_j u_n(x) \to D_j u(x)$ for all $x \in \Omega \setminus N$, where N is a set of measure zero. We may also assume that there exists a function $g \in L_2(\Omega)$ such that $|D_j u_n| \leq g$ and $|u_n| \leq g$ for all $n \in \mathbb{N}$.

Set $B := \{x \in \Omega : u(x) = 0\}$, then by Stampacchia's lemma (Corollary 6.36), $D_j u(x) = 0$ almost everywhere in B, and so

$$\lim_{n \to \infty} D_j |u_n(x)| = \lim_{n \to \infty} \text{sign}\,(u_n(x)) D_j u_n(x) = 0 = D_j |u(x)|$$

almost everywhere in Ω. If $x \notin B$, that is, $u(x) \neq 0$, then $\lim_{n \to \infty} \text{sign}\, u_n(x) = \text{sign}\, u(x)$ whenever $x \notin N$. We thus have

$$D_j |u_n| = (\text{sign}(u_n)) D_j u_n \to (\text{sign}\, u) D_j u = D_j |u|$$

as $n \to \infty$ almost everywhere in Ω. It now follows from the Dominated Convergence Theorem that $\lim_{n \to \infty} D_j |u_n| = D_j |u|$ in $L_2(\Omega)$, and we have shown that $|u_n| \to |u|$ in $H^1(\Omega)$.   □

Remark 6.38 Here is a second, abstract proof of Theorem 6.37. Let $u_n \to u$ in $H^1(\Omega)$. By Lemma 4.40 it suffices to show that $|u_n| \to |u|$ in $H^1(\Omega)$ up to a subsequence. Now $\||u_n|\|_{H^1} = \| |u_n| \|_{H^1}$, and so $(|u_n|)_{n \in \mathbb{N}}$ is bounded in $H^1(\Omega)$. By Theorem 4.35 we may thus assume that $|u_n|$ converges weakly in $H^1(\Omega)$ to some $w \in H^1(\Omega)$ (after passing to a subsequence if necessary). Then $|u_n|$ also converges weakly to w in $L_2(\Omega)$.

But since $|u_n| \to |u|$ in $L_2(\Omega)$ (as established at the beginning of the above proof), it follows that $w = |u|$. We have thus shown that $|u_n| \rightharpoonup |u|$ in $H^1(\Omega)$. Since $\| |u_n| \|_{H^1} = \|u_n\|_{H^1} \to \|u\|_{H^1} = \| |u| \|_{H^1}$, it follows from Theorem 4.33 that indeed $|u_n| \to |u|$ in $H^1(\Omega)$. △

Theorem 6.37 states that $H^1(\Omega)$ is a sublattice of $L_2(\Omega)$ and that the lattice operations are continuous. In particular, for $u, v \in H^1(\Omega)$, the functions

$$u \vee v := \max\{u, v\} = \frac{u + v + |u - v|}{2}, \tag{6.39a}$$

$$u \wedge v := \min\{u, v\} = \frac{u + v - |u - v|}{2} \tag{6.39b}$$

are also in $H^1(\Omega)$. We will now show that $H_0^1(\Omega)$ is a *lattice ideal* in $H^1(\Omega)$.

Theorem 6.39 *(a) If $u \in H_0^1(\Omega)$, then $|u|, u^+, u^- \in H_0^1(\Omega)$.*
(b) If $v \in H^1(\Omega)$ and $u \in H_0^1(\Omega)$, with $0 \le v \le u$, then $v \in H_0^1(\Omega)$.

Proof (a) Let $u_n \in \mathcal{D}(\Omega)$ converge to u in $H^1(\Omega)$. By Theorem 6.37, also

$$\lim_{n \to \infty} |u_n| = |u|, \quad \lim_{n \to \infty} u_n^+ = u^+, \quad \lim_{n \to \infty} u_n^- = u^- \text{ in } H^1(\Omega).$$

Since $|u_n|, u_n^+, u_n^- \in H_c^1(\Omega) \subset H_0^1(\Omega)$ (see Theorem 6.29), the claim follows since $H_0^1(\Omega)$ is closed.

(b) Let $\varphi_n \in \mathcal{D}(\Omega)$ converge to u in $H^1(\Omega)$, and set

$$v_n := (\varphi_n \wedge v) \vee 0 = \max\{\min\{\varphi_n, v\}, 0\}.$$

Then $v_n \in H_c^1(\Omega)$ since $\varphi_n \in \mathcal{D}(\Omega)$, $v \in H_0^1(\Omega)$ and $v \ge 0$. Moreover,

$$\lim_{n \to \infty} v_n = \left(\left(\lim_{n \to \infty} \varphi_n \right) \wedge v \right) \vee 0 = (u \wedge v) \vee 0 = v$$

since $u \ge v \ge 0$. By Theorem 6.29 and the fact that $H_0^1(\Omega)$ is closed, we conclude that $v \in H_0^1(\Omega)$. □

The following statements will be useful for the further analysis of elliptic problems. We define the *positive cones* of the spaces $\mathcal{D}(\Omega)$ and $H_0^1(\Omega)$ by

$$\mathcal{D}(\Omega)_+ := \{\varphi \in \mathcal{D}(\Omega) : \varphi \ge 0\},$$

$$H_0^1(\Omega)_+ := \{u \in H_0^1(\Omega) : u \geq 0 \text{ almost everywhere}\},$$

respectively. Now by definition $\mathcal{D}(\Omega)$ is dense in $H_0^1(\Omega)$; the same is true of their positive cones.

Theorem 6.40 *The space $\mathcal{D}(\Omega)_+$ is dense in $H_0^1(\Omega)_+$.*

Proof Let $u \in H_0^1(\Omega)_+$ and choose $u_n \in \mathcal{D}(\Omega)$ such that $\lim_{n\to\infty} u_n = u$ in $H_0^1(\Omega)$. Then, by continuity of the lattice operations, $\lim_{n\to\infty} u_n^+ = u^+ = u$ in $H^1(\Omega)$ as well. Since $0 \leq u_n^+ \in H_c^1(\Omega)$, we have $\varrho_k * u_n^+ \geq 0$ for the mollifier ϱ_k from Definition 6.2. But since $\varrho_k * u_n^+ \in \mathcal{D}(\Omega)$ for k large enough and $\varrho_k * u_n^+ \to u_n^+$ in $H^1(\Omega)$ as $k \to \infty$ (note (6.25)), the theorem is proved. □

We finish by giving the proof of Theorem 6.30.

Proof (of Theorem 6.30) Let $u \in H^1(\Omega) \cap C_0(\Omega)$. It follows from Theorem 6.35 that $u^+, u^- \in H^1(\Omega) \cap C_0(\Omega)$. As such it is sufficient to consider positive functions, so let $u \geq 0$. Since Ω is bounded, we have $u - \frac{1}{n} \in H^1(\Omega)$ for all $n \in \mathbb{N}$, and also $u_n := (u - \frac{1}{n})^+ \in H^1(\Omega)$ (by Theorem 6.35). Since $u \in C_0(\Omega)$, the set $\{x \in \Omega : u(x) \geq \frac{1}{n}\}$ is compact, and thus $u_n \in H_c^1(\Omega) \subset H_0^1(\Omega)$ by Theorem 6.29. Since the lattice operations are continuous (Theorem 6.37), $\lim_{n\to\infty} u_n = u$ in $H^1(\Omega)$. Hence $u \in H_0^1(\Omega)$ since $H_0^1(\Omega)$ is closed. □

6.5 The Poisson equation with Dirichlet boundary conditions

Let Ω be an open set and $\lambda \geq 0$. Given $f \in L_2(\Omega)$, our goal is to analyze the Poisson equation with reaction term

$$-\Delta u + \lambda u = f \text{ in } \Omega, \tag{6.40a}$$

$$u_{|\partial\Omega} = 0. \tag{6.40b}$$

We start by considering the equation (6.40a) by itself. Let us first suppose that u is a classical solution of (6.40a), that is, that $u \in C^2(\Omega)$ and (6.40a) holds for $\Delta u = \sum_{j=1}^d \frac{\partial^2 u}{\partial x_j}$. If we multiply this equation by a test function $\varphi \in \mathcal{D}(\Omega)$ and integrate by parts twice (see (6.21)), then we obtain

$$\lambda \int_\Omega u(x)\,\varphi(x)\,dx - \int_\Omega u(x)\Delta\varphi(x)\,dx = \int_\Omega f(x)\,\varphi(x)\,dx. \tag{6.41}$$

This identity is also meaningful if u is merely in $L_2(\Omega)$. In this case, we shall call u a weak solution of (6.40a).

Definition 6.41 Let $f \in L_2(\Omega)$. A function $u \in L_2(\Omega)$ is called a *weak solution* of (6.40a) (without considering the boundary condition (6.40b)) if

$$\lambda \int_\Omega u(x)\,\varphi(x)\,dx - \int_\Omega u(x)\,\Delta\varphi(x)\,dx = \int_\Omega f(x)\,\varphi(x)\,dx$$

for all $\varphi \in \mathcal{D}(\Omega)$. We also say that (6.40a) *holds weakly*. △

If u is actually in $H^1(\Omega)$, then we can bring its weak derivatives into play and obtain the following description of weak solutions.

Lemma 6.42 *The function* $u \in H^1(\Omega)$ *is a weak solution of* (6.40a) *if and only if*

$$\lambda \int_\Omega u(x)\,\varphi(x)\,dx + \int_\Omega \nabla u(x)\,\nabla\varphi(x)\,dx = \int_\Omega f(x)\,\varphi(x)\,dx \qquad (6.42)$$

for all $\varphi \in H_0^1(\Omega)$.

Proof Let $\varphi \in \mathcal{D}(\Omega)$, then also $\frac{\partial\varphi}{\partial x_j} \in \mathcal{D}(\Omega)$ for all $j = 1,\dots,d$ and hence, by definition of weak derivatives,

$$\int_\Omega \nabla u(x)\nabla\varphi(x)\,dx = \sum_{j=1}^d \int_\Omega D_j u(x)\frac{\partial\varphi}{\partial x_j}(x)\,dx = -\sum_{j=1}^d \int_\Omega u(x)\frac{\partial^2\varphi}{\partial x_j^2}(x)\,dx$$

$$= -\int_\Omega u(x)\,\Delta\varphi(x)\,dx.$$

Hence, if $\varphi \in \mathcal{D}(\Omega)$, then (6.42) and (6.41) are equivalent. If (6.42) holds for all $\varphi \in \mathcal{D}(\Omega)$, then the identity also holds for $\varphi \in H_0^1(\Omega)$ by density. □

We will henceforth interpret the boundary condition (6.40b) as requiring that $u \in H_0^1(\Omega)$. Then Lemma 6.42 suggests that we should consider the bilinear form

$$a(u,v) := \lambda \int_\Omega u(x)\,v(x)\,dx + \int_\Omega \nabla u(x)\,\nabla v(x)\,dx$$

on $H_0^1(\Omega) \times H_0^1(\Omega)$. This form is continuous since by the Cauchy–Schwarz inequality

$$|a(u,v)| \le \lambda\|u\|_{L_2}\|v\|_{L_2} + \left(\int_\Omega |\nabla u(x)|^2\,dx\right)^{\frac{1}{2}} \left(\int_\Omega |\nabla v(x)|^2\,dx\right)^{\frac{1}{2}}$$

$$\le (\lambda + 1)\|u\|_{H^1}\|v\|_{H^1}.$$

Moreover, if $\lambda > 0$, then $a(u) \ge \min\{\lambda, 1\}\|u\|_{H^1}^2$, and the form is coercive, recall (4.5). If Ω is contained in a strip, then the Poincaré inequality (6.34) holds, and so

$$\int_\Omega |\nabla u(x)|^2\,dx = \frac{1}{2}\int_\Omega |\nabla u(x)|^2\,dx + \frac{1}{2}\int_\Omega |\nabla u(x)|^2\,dx$$

$$\ge \frac{1}{8M^2}\int_\Omega u(x)^2\,dx + \frac{1}{2}\int_\Omega |\nabla u(x)|^2\,dx \ge \alpha\|u\|_{H^1}^2,$$

where $\alpha = \min\left\{\frac{1}{8M^2}, \frac{1}{2}\right\}$. Hence, in this case the form is also coercive for $\lambda = 0$.

So suppose that $\lambda > 0$, or that Ω is contained in a strip. For $f \in L_2(\Omega)$, the mapping $\varphi \mapsto \int_\Omega f(x)\,\varphi(x)\,dx$ is a continuous linear form on $H_0^1(\Omega)$ and so by the Lax–Milgram theorem there exists a unique $u \in H_0^1(\Omega)$ which satisfies the equation $a(u, \varphi) = \int_\Omega f(x)\,\varphi(x)\,dx$ for all $\varphi \in H_0^1(\Omega)$. Hence (6.41) is satisfied and we have proved the following theorem.

Theorem 6.43 *Let $\lambda \geq 0$ and suppose that at least one of the following two conditions is satisfied: (a) $\lambda > 0$; (b) Ω is contained in a strip. Then for each $f \in L_2(\Omega)$ there exists a unique weak solution $u \in H_0^1(\Omega)$ of (6.40).*

We have thus proven the well-posedness of the Poisson problem in a weak sense. The question of regularity is, however, rather more difficult. On the one hand there is the question of interior regularity, that is, for example, under which assumptions we actually have $u \in C^2(\Omega)$, so that the equation (6.40a) is then satisfied in the classical sense; and then the question of whether $u \in C(\overline{\Omega})$ and $u_{|\partial\Omega} = 0$, that is, whether the boundary condition is satisfied in the classical sense. It turns out that Fourier transforms will help us with the first problem; we will thus investigate these transforms in the next section with a particular view to their differentiability properties.

The second question is equivalent to the well-posedness of the Dirichlet problem, and we refer to Section 6.9. Here, we wish to obtain a weak version of the maximum principle. We recall that a function $u \in C^2(\Omega)$ is called subharmonic if $-\Delta u \leq 0$ (cf. Definition 3.24). This is equivalent to

$$-\int_\Omega u(x)\,\Delta\varphi(x)\,dx \leq 0 \text{ for all } \varphi \in \mathcal{D}(\Omega)_+. \tag{6.43}$$

But this condition also makes sense if $u \in L_2(\Omega)$. We therefore more generally call a function $u \in L_2(\Omega)$ *subharmonic* if it satisfies (6.43). If $u \in H^1(\Omega)$ and $c \in \mathbb{R}$, then we say that

$$u \leq c \text{ weakly on } \partial\Omega, \tag{6.44}$$

if $(u - c)^+ \in H_0^1(\Omega)$. This is a generalization of the classical version. Indeed, if Ω is bounded and $u \in H^1(\Omega) \cap C(\overline{\Omega})$ with $u(z) \leq c$ for all $z \in \partial\Omega$, then $(u - c)^+$ is in $H^1(\Omega) \cap C_0(\Omega)$ and so also in $H_0^1(\Omega)$ by Theorem 6.30. The following theorem generalizes the classical maximum principle, Theorem 3.25.

Theorem 6.44 (Weak maximum principle) *Let $u \in H^1(\Omega)$ be subharmonic and $c \in \mathbb{R}$. If $u \leq c$ weakly on $\partial\Omega$, then $u \leq c$ almost everywhere in Ω.*

Proof Since $u \in H^1(\Omega)$, it follows from (6.43) that $\int_\Omega \nabla u \nabla \varphi \, dx \leq 0$ for all $\varphi \in \mathcal{D}(\Omega)_+$. Since $\mathcal{D}(\Omega)_+$ is dense in $H_0^1(\Omega)_+$ (Theorem 6.40), this inequality continues to hold for all $\varphi \in H_0^1(\Omega)_+$. In particular, we may choose $\varphi = (u - c)^+$. It then follows from (6.35) that $D_j\varphi = \mathbb{1}_{\{u>c\}}D_j u$, and

$$0 \geq \int_\Omega \nabla u(x) \nabla (u - c)^+(x)\, dx = \int_\Omega \nabla (u - c)(x) \nabla (u - c)^+(x)\, dx$$

$$= \int_\Omega |\nabla (u - c)^+(x)|^2\, dx.$$

Thus $\nabla (u - c)^+ = 0$ almost everywhere. By Corollary 6.28, $(u - c)^+ = 0$ almost everywhere as well. □

6.6 Sobolev spaces and Fourier transforms

We have already seen that the Fourier transform converts differentiation into multiplication by the variable. Since it is also an isomorphism of $L_2(\mathbb{R}^d)$ to $L_2(\mathbb{R}^d)$, it should not be surprising that we can easily describe the image of our Sobolev spaces under the Fourier transform. In addition to this characterization we will also prove imbedding and regularity theorems: how often must a function be weakly differentiable in order to be continuous or even C^1? Such results are also known as *imbedding theorems*. Although they hold for $\mathbb{R}^d$, they can be extended to open sets $\Omega \subset \mathbb{R}^d$ via localization and can thus be used for the Poisson problem.

In this section the underlying field will always be $\mathbb{C}$. We define complex-valued L_p-spaces for $1 \leq p < \infty$ by

$$L_p(\mathbb{R}^d, \mathbb{C}) := \left\{ f : \mathbb{R}^d \to \mathbb{C} \text{ is measurable and } \int_{\mathbb{R}^d} |f(x)|^p\, dx < \infty \right\}.$$

Here, as usual, we identify two functions in $L_p(\mathbb{R}^d, \mathbb{C})$ if they agree almost everywhere. For $f \in L_1(\mathbb{R}^d, \mathbb{C})$ we define the *Fourier transform* $\hat{f} : \mathbb{R}^d \to \mathbb{C}$ by

$$\hat{f}(x) := \frac{1}{(2\pi)^{d/2}} \int_{\mathbb{R}^d} e^{-ix \cdot y} f(y)\, dy,$$

where as before, $x \cdot y = x^T y$ denotes the inner product in $\mathbb{R}^d$ of vectors $x = (x_1, \ldots, x_d)$, $y = (y_1, \ldots, y_d) \in \mathbb{R}^d$. For brevity, we set $x^2 = x \cdot x = \sum_{j=1}^d x_j^2$; then $|x| := \sqrt{x^2}$. If $f \in L_1(\mathbb{R}^d, \mathbb{C})$, then

$$\hat{f} \in C_0(\mathbb{R}^d, \mathbb{C}) := \left\{ g : \mathbb{R}^d \to \mathbb{C} \text{ is continuous and } \lim_{|x| \to \infty} g(x) = 0 \right\}, \qquad (6.45)$$

see Exercise 6.14. This property is often referred to as the Riemann–Lebesgue lemma. Thus the Fourier transform defines a linear operator from $L_1(\mathbb{R}^d, \mathbb{C})$ to $C_0(\mathbb{R}^d, \mathbb{C})$, which is injective but not surjective. The Fourier transform on the space $L_2(\mathbb{R}^d, \mathbb{C})$ is rather better behaved, as we shall now see. For the proof of the following theorem we refer, e.g., to [54, Thm. 9.13].

Theorem 6.45 (Plancherel's theorem) *There exists a uniquely determined unitary operator* $\mathcal{F} : L_2(\mathbb{R}^d, \mathbb{C}) \to L_2(\mathbb{R}^d, \mathbb{C})$ *such that* $\mathcal{F}f = \hat{f}$ *for all* $f \in L_1 \cap L_2$. *Moreover,* $(\mathcal{F}^{-1}f)(x) = (\mathcal{F}f)(-x)$ *almost everywhere, for all* $f \in L_2(\mathbb{R}^d, \mathbb{C})$.

We also call the mapping $\mathcal{F}$ the *Fourier transform*. In particular, we have

$$(\mathcal{F}f, \mathcal{F}g)_{L_2} = (f, g)_{L_2} \text{ for all } f, g \in L_2 \quad \text{(Parseval's identity)} \tag{6.46}$$

$$f(x) = (2\pi)^{-d/2} \int_{\mathbb{R}^d} e^{ix \cdot y} \mathcal{F}f(y)\, dy \tag{6.47}$$

whenever $f \in L_2(\mathbb{R}^d, \mathbb{C})$ and $\mathcal{F}f \in L_1(\mathbb{R}^d, \mathbb{C})$. It follows from Corollary 6.9 that the test functions $\mathcal{D}(\mathbb{R}^d, \mathbb{C}) := \mathcal{D}(\mathbb{R}^d) + i\mathcal{D}(\mathbb{R}^d)$ are dense in $L_2(\mathbb{R}^d, \mathbb{C})$. We now define the complex Sobolev space

$$H^1(\mathbb{R}^d, \mathbb{C}) := \Big\{ u \in L_2(\mathbb{R}^d, \mathbb{C}) : \text{ there exists } D_1 u, \dots, D_d u \in L_2(\mathbb{R}^d, \mathbb{C})$$

$$\text{such that } - \int_{\mathbb{R}^d} u(x) \frac{\partial \varphi}{\partial x_j}(x)\, dx = \int_{\mathbb{R}^d} D_j u(x) \varphi(x)\, dx$$

$$\text{for all } \varphi \in \mathcal{D}(\mathbb{R}^d) \Big\}.$$

It is obvious that $H^1(\mathbb{R}^d, \mathbb{C}) = \{v + iw : v, w \in H^1(\mathbb{R}^d)\}$ and $D_j(v + iw) = D_j v + i D_j w$ for all $v, w \in H^1(\mathbb{R}^d)$. It thus follows from Theorem 6.25 that $\mathcal{D}(\mathbb{R}^d, \mathbb{C})$ is also dense in $H^1(\mathbb{R}^d, \mathbb{C})$. Consequently, for $u \in H^1(\mathbb{R}^d, \mathbb{C})$ we have

$$- \int_{\mathbb{R}^d} u(x)\, D_j \varphi(x)\, dx = \int_{\mathbb{R}^d} D_j u(x)\, \varphi(x)\, dx \text{ for all } \varphi \in H^1(\mathbb{R}^d, \mathbb{C}). \tag{6.48}$$

We now define the weighted space

$$\hat{H}^1(\mathbb{R}^d) := L_2(\mathbb{R}^d, (1 + x^2) dx, \mathbb{C})$$

$$:= \Big\{ f : \mathbb{R}^d \to \mathbb{C} \text{ measurable: } \int_{\mathbb{R}^d} |f(x)|^2 (1 + x^2)\, dx < \infty \Big\}.$$

This is a Hilbert space when equipped with the inner product

$$(f, g)_{\hat{H}^1(\mathbb{R}^d)} := \int_{\mathbb{R}^d} f(x)\, \overline{g(x)}\, (1 + x^2)\, dx.$$

Theorem 6.46 *We have* $\mathcal{F}(H^1(\mathbb{R}^d, \mathbb{C})) = \hat{H}^1(\mathbb{R}^d)$, *and* $\mathcal{F}|_{H^1(\mathbb{R}^d, \mathbb{C})}$ *is a unitary operator. Moreover, for* $f \in H^1(\mathbb{R}^d, \mathbb{C})$, *we have*

$$\mathcal{F}(D_j f)(x) = ix\, \mathcal{F}f(x) \text{ almost everywhere.} \tag{6.49}$$

Proof Let $f \in \mathcal{D}(\mathbb{R}^d, \mathbb{C})$, then

$$(\mathcal{F}D_j f)(x) = (2\pi)^{-d/2} \int_{\mathbb{R}^d} e^{-ix\cdot y} \frac{\partial f}{\partial y_j}(y)\, dy = (ix_j)(\mathcal{F}f)(x),$$

as one sees upon integrating by parts. Since $\mathcal{D}(\mathbb{R}^d, \mathbb{C})$ is dense in $H^1(\mathbb{R}^d, \mathbb{C})$, this identity continues to hold for all $f \in H^1(\mathbb{R}^d, \mathbb{C})$. In particular, for $f, g \in H^1(\mathbb{R}^d, \mathbb{C})$, by (6.46) we have

$$\begin{aligned}
(f, g)_{H^1} &= (f, g)_{L_2} + \sum_{j=1}^{d} (D_j f, D_j g)_{L_2} \\
&= (\mathcal{F}f, \mathcal{F}g)_{L_2} + \sum_{j=1}^{d} (\mathcal{F}(D_j f), \mathcal{F}(D_j g))_{L_2} \\
&= \int_{\mathbb{R}^d} \left\{ \mathcal{F}f(x)\overline{\mathcal{F}g(x)} + \sum_{j=1}^{d} ix_j\, \mathcal{F}f(x)\, \overline{ix_j\, \mathcal{F}g(x)} \right\} dx \\
&= \int_{\mathbb{R}^d} \mathcal{F}f(x)\overline{\mathcal{F}g}(x)(1 + x^2)\, dx = (\mathcal{F}f, \mathcal{F}g)_{\hat{H}^1(\mathbb{R}^d)}.
\end{aligned}$$

In particular, $\mathcal{F} : H^1(\mathbb{R}^d, \mathbb{C}) \rightarrow \hat{H}^1(\mathbb{R}^d)$ is isometric. We will now show that $\mathcal{F}$ is surjective. Let $g \in \hat{H}^1(\mathbb{R}^d)$, then the function $x \mapsto ix_j g$ is in $L_2(\mathbb{R}^d, \mathbb{C})$ for all $j = 1, \ldots, d$. Hence, by Theorem 6.45, there exist functions $f, f_1, \ldots, f_d \in L_2(\mathbb{R}^d, \mathbb{C})$ such that $\mathcal{F}f = g$ and $(\mathcal{F}f_j)(x) = ix_j g(x)$ for all $j = 1, \ldots, d$. Now let $\varphi \in \mathcal{D}(\mathbb{R}^d)$ (which we may take real-valued), then by (6.46),

$$\begin{aligned}
-\int_{\mathbb{R}^d} f(x)D_j\varphi(x)\, dx &= -\int_{\mathbb{R}^d} f(x)\overline{D_j\varphi}(x)\, dx = -\int \mathcal{F}f(x)\overline{\mathcal{F}(D_j\varphi)}(x)\, dx \\
&= -\int_{\mathbb{R}^d} g(x)(-ix_j)\overline{\mathcal{F}\varphi}(x)\, dx \\
&= \int_{\mathbb{R}^d} \mathcal{F}f_j(x)\overline{\mathcal{F}\varphi}(x)\, dx = \int_{\mathbb{R}^d} f_j(x)\,\varphi(x)\, dx
\end{aligned}$$

since φ was chosen real-valued. Hence $f \in H^1(\mathbb{R}^d, \mathbb{C})$ and $D_j f = f_j$ for all $j = 1, \ldots, d$. $\qquad\square$

The above theorem explains the notation $\hat{H}^1(\mathbb{R}^d)$, since this space is the image of $H^1(\mathbb{R}^d, \mathbb{C})$ under the Fourier transform. We recall that the higher order Sobolev spaces $H^k(\mathbb{R}^d, \mathbb{C})$, $k \in \mathbb{N}$ are defined inductively by

$$H^{k+1}(\mathbb{R}^d, \mathbb{C}) := \left\{ u \in H^1(\mathbb{R}^d, \mathbb{C}) : D_j u \in H^k(\mathbb{R}^d, \mathbb{C}), j = 1, \ldots, d \right\}.$$

We denote by $\hat{H}^k(\mathbb{R}^d)$ the weighted space

$$\hat{H}^k(\mathbb{R}^d) := \left\{ f : \mathbb{R}^d \rightarrow \mathbb{C} \text{ measurable} : \int_{\mathbb{R}^d} |f(x)|^2(1 + x^2)^k\, dx < \infty \right\}.$$

These spaces are nested, that is,

$$\hat{H}^{k+1}(\mathbb{R}^d) \subset \hat{H}^k(\mathbb{R}^d) \subset L_2(\mathbb{R}^d, \mathbb{C}), \quad k \in \mathbb{N}. \tag{6.50}$$

Theorem 6.47 *We have* $\mathcal{F}(H^k(\mathbb{R}^d, \mathbb{C})) = \hat{H}^k(\mathbb{R}^d)$ *for all* $k \in \mathbb{N}$.

Proof The statement is true for $k = 1$ (see Theorem 6.46). Now suppose it is true for $k \geq 1$; we will show that it is also true for $k + 1$.

1. Let $u \in H^{k+1}(\mathbb{R}^d, \mathbb{C})$, then $u \in H^1(\mathbb{R}^d, \mathbb{C})$ and $D_j u \in H^k(\mathbb{R}^d, \mathbb{C})$. We have $\mathcal{F}(D_j u) = g_j$, where $g_j(x) = i x_j \mathcal{F} u(x)$. Now by the induction hypothesis, $g_j \in \hat{H}^k(\mathbb{R}^d)$, that is, for all $1 \leq j \leq d$ we have

$$\int_{\mathbb{R}^d} x_j^2 |\mathcal{F} u(x)|^2 (1 + x^2)^k \, dx < \infty$$

as well as $\mathcal{F} u \in \hat{H}^k(\mathbb{R}^d)$, whence $\int_{\mathbb{R}^d} |\mathcal{F} u(x)|^2 (1 + x^2)^k \, dx < \infty$. Summing over j and the expression for $\mathcal{F} u$ yields $\int_{\mathbb{R}^d} (x^2 + 1) |\mathcal{F} u(x)|^2 (1 + x^2)^k \, dx < \infty$, that is, $\mathcal{F} u \in \hat{H}^{k+1}(\mathbb{R}^d)$.

2. Conversely, let $g \in \hat{H}^{k+1}(\mathbb{R}^d)$, then the functions g_j defined by $g_j(x) := i x_j g(x)$ are in $\hat{H}^k(\mathbb{R}^d)$. Since also $g \in \hat{H}^k(\mathbb{R}^d)$, by the induction hypothesis there exist functions $u, u_j \in H^k(\mathbb{R}^d, \mathbb{C})$ such that $\mathcal{F} u = g$ and $\mathcal{F} u_j = g_j$ for all $j = 1, \ldots, d$. Since $\mathcal{F}(D_j u) = g_j$, it follows that $u_j = D_j u$. Thus $D_j u = u_j \in H^k(\mathbb{R}^d, \mathbb{C})$ and so by definition $u \in H^{k+1}(\mathbb{R}^d, \mathbb{C})$. □

We will use Theorem 6.47 to prove that $H^k(\mathbb{R}^d) \subset C_0(\mathbb{R}^d)$ for $k > d/2$. The assumption $k > d/2$ is optimal; it shows that in higher dimensions one needs more Sobolev regularity in order to guarantee continuity: the gap between H^k and C^0 grows with the dimension. Alternatively, one could say that H^k contains wilder and wilder functions in higher dimensions. For the proof of the claimed imbedding we first need the following lemma.

Lemma 6.48 *For* $k > d/2$ *we have* $\hat{H}^k(\mathbb{R}^d) \subset L_1(\mathbb{R}^d, \mathbb{C})$.

Proof Let $f \in \hat{H}^k(\mathbb{R}^d)$, then by Hölder's inequality

$$\int_{\mathbb{R}^d} |f(x)| \, dx = \int_{\mathbb{R}^d} |f(x)| (1 + x^2)^{k/2} (1 + x^2)^{-k/2} \, dx$$

$$\leq \left(\int_{\mathbb{R}^d} |f(x)|^2 (1 + x^2)^k \, dx \right)^{\frac{1}{2}} \left(\int_{\mathbb{R}^d} (1 + x^2)^{-k} \, dx \right)^{\frac{1}{2}}.$$

The first term is finite by assumption, so it remains to show for the second one that $\int_{\mathbb{R}^d} (1 + x^2)^{-k} \, dx < \infty$. But we have, using Theorem A.8,

$$\int_{|x| \geq 1} (1 + x^2)^{-k} \, dx \leq \int_{|x| \geq 1} (x^2)^{-k} \, dx = \sigma(S^{d-1}) \int_1^\infty r^{-2k} r^{d-1} \, dr.$$

Since $d - 2k < 0$, this last integral is clearly finite.   □

Now we come to the announced imbedding.

Theorem 6.49 (Sobolev imbedding theorem) *For $k > d/2$ we have $H^k(\mathbb{R}^d, \mathbb{C}) \subset C_0(\mathbb{R}^d, \mathbb{C})$.*

Proof Let $f \in H^k(\mathbb{R}^d, \mathbb{C})$, then $\mathcal{F}f \in L_1(\mathbb{R}^d, \mathbb{C})$ by Lemma 6.48. It thus follows from (6.45) that $f \in C_0(\mathbb{R}^d, \mathbb{C})$, since $f(x) = \mathcal{F}^{-1}(\mathcal{F}f)(x) = \mathcal{F}(\mathcal{F}f)(-x)$ by Plancherel's theorem (Theorem 6.45).   □

We thus have the following regularity result.

Corollary 6.50 *For $k \in \mathbb{N}, k > \frac{d}{2}$ and $m \in \mathbb{N}_0$ we have $H^{k+m}(\mathbb{R}^d, \mathbb{C}) \subset C^m(\mathbb{R}^d, \mathbb{C})$.*

Proof The statement holds for $m = 0$ by Theorem 6.49, where as usual we have defined $C^0(\mathbb{R}^d, \mathbb{C}) := C(\mathbb{R}^d, \mathbb{C})$. Now suppose the statement is true for some $m \in \mathbb{N}_0$ and take $u \in H^{k+m+1}(\mathbb{R}^d, \mathbb{C})$. By the induction hypothesis we have $D_j u \in H^{k+m}(\mathbb{R}^d, \mathbb{C}) \subset C^m(\mathbb{R}^d, \mathbb{C})$. It follows from Theorem 6.24 that $u \in C^1(\mathbb{R}^d, \mathbb{C})$; since $D_j u \in C^m(\mathbb{R}^d, \mathbb{C})$, we obtain $u \in C^{m+1}(\mathbb{R}^d, \mathbb{C})$.   □

We now wish to consider the Poisson equation in $\mathbb{R}^d$. For $u \in H^2(\mathbb{R}^d, \mathbb{C})$ we have the relations $\mathcal{F}(D_j u)(x) = ix_j \mathcal{F}u(x)$, $\mathcal{F}(D_j^2 u)(x) = -x_j^2 \mathcal{F}u(x)$ and $\mathcal{F}(\Delta u)(x) = -x^2 \mathcal{F}u(x)$, and hence $\mathcal{F}(u - \Delta u)(x) = (1 + x^2)\mathcal{F}u(x)$. Since $u \in H^2(\mathbb{R}^d, \mathbb{C})$, we also have $\mathcal{F}u \in \hat{H}^2(\mathbb{R}^d)$ and thus $(1 + x^2)\mathcal{F}u \in L_2(\mathbb{R}^d, \mathbb{C})$. Conversely, if $f \in L_2(\mathbb{R}^d, \mathbb{C})$, then $x \mapsto (1 + x^2)^{-1}\mathcal{F}f$ defines a function in $\hat{H}^2(\mathbb{R}^d)$. Thus there exists a unique $u \in H^2(\mathbb{R}^d, \mathbb{C})$ such that $\mathcal{F}u = (1+x^2)^{-1}\mathcal{F}f$. Finally, since $\mathcal{F}(u - \Delta u) = \mathcal{F}u + x^2\mathcal{F}u = (1 + x^2)\mathcal{F}u = \mathcal{F}f$, we conclude that $u - \Delta u = f$. This means we have shown the following result:

Theorem 6.51 *For all $f \in L_2(\mathbb{R}^d, \mathbb{C})$ there exists a unique $u \in H^2(\mathbb{R}^d, \mathbb{C})$ such that $u - \Delta u = f$.*

If $f \in L_2(\mathbb{R}^d)$ is a real-valued function, then $u \in H^2(\mathbb{R}^d)$ is also real-valued. In this case, u is then clearly also a weak solution of $u - \Delta u = f$, that is, u coincides with the unique weak solution which we found in Theorem 6.43 using the Lax–Milgram theorem. But here we have gained additional regularity, since u is actually in $H^2(\mathbb{R}^d)$. This means that the weak derivatives $D_i D_j$ are all in $L^2(\mathbb{R}^d)$ and $\Delta u = \sum_{j=1}^d D_j^2 u$. In this case we speak of a *strong solution*. The term "classical solution" is reserved for functions which are actually in $C^2(\mathbb{R}^d)$. We next wish to strengthen the statement of uniqueness in Theorem 6.51.

Lemma 6.52 *If $v, f \in L_2(\mathbb{R}^d, \mathbb{C})$ satisfy $v - \Delta v = f$ weakly, then $v \in H^2(\mathbb{R}^d, \mathbb{C})$, that is, v is the unique solution from Theorem 6.51.*

Proof Let $u \in H^2(\mathbb{R}^d, \mathbb{C})$ be the solution from Theorem 6.51 and set $w := u - v$.
Then

$$\int_{\mathbb{R}^d} w(x) \overline{(\varphi(x) - \Delta\varphi(x))} \, dx = 0 \text{ for all } \varphi \in \mathcal{D}(\mathbb{R}^d, \mathbb{C}). \quad (6.51)$$

Since the test functions are dense in $H^2(\mathbb{R}^d, \mathbb{C})$ (see Corollary 6.26), (6.51) holds
for all $\varphi \in H^2(\mathbb{R}^d, \mathbb{C})$. Choose a function $\varphi \in H^2(\mathbb{R}^d, \mathbb{C})$ such that $\varphi - \Delta\varphi = w$ (the
existence of such a function is guaranteed by Theorem 6.51). Substituting this into
(6.51), we obtain $\int_{\mathbb{R}^d} |w(x)|^2 \, dx = 0$ and hence $w = 0$. $\qquad\square$

With this preparation we can now prove the following result on regularity.

Theorem 6.53 (Maximal regularity) *Let* $f, u \in L_2(\mathbb{R}^d, \mathbb{C})$ *satisfy* $\Delta u = f$ *weakly.*
Then $u \in H^2(\mathbb{R}^d, \mathbb{C})$. *If* $f \in H^k(\mathbb{R}^d, \mathbb{C})$, *then* $u \in H^{k+2}(\mathbb{R}^d, \mathbb{C})$.

Proof Since $u - \Delta u = u - f \in L_2(\mathbb{R}^d)$, it follows from Lemma 6.52 that $u \in H^2(\mathbb{R}^d, \mathbb{C})$.

We will prove the second claim by induction on k. For $k = 0$, it corresponds exactly
to the first part if we set $H^0(\mathbb{R}^d, \mathbb{C}) = L_2(\mathbb{R}^d, \mathbb{C})$. So suppose that the statement is
true for some $k \in \mathbb{N}$ and let $f \in H^{k+1}(\mathbb{R}^d, \mathbb{C}) \subset H^k(\mathbb{R}^d, \mathbb{C})$. Then by assumption
$u \in H^{k+2}(\mathbb{R}^d, \mathbb{C})$, so that $u - \Delta u = u - f = g \in H^{k+1}(\mathbb{R}^d, \mathbb{C})$. It follows that
$(1 + x^2)\mathcal{F}u = \mathcal{F}(u - \Delta u) = \mathcal{F}g \in \hat{H}^{k+1}(\mathbb{R}^d)$, whence $\mathcal{F}u \in \hat{H}^{k+3}(\mathbb{R}^d)$, and so finally
$u \in H^{k+3}(\mathbb{R}^d, \mathbb{C})$. $\qquad\square$

Theorem 6.53 is also sometimes called a *shift theorem*; the reason is that the degree
of regularity (smoothness) of the solution is given by the degree of regularity of the
right-hand side plus the order of the differential operator. The index of regularity is
"shifted" by the order of the operator. We will return to this point in Chapter 9. Next,
however, we wish to use Fourier transforms to show that the imbedding of $H_0^1(\Omega)$ in
$L_2(\Omega)$ is compact for bounded open $\Omega \subset \mathbb{R}^d$.

Theorem 6.54 (Rellich imbedding theorem) *Let* $\Omega \subset \mathbb{R}^d$ *be bounded and open.*
Then the imbedding of $H_0^1(\Omega)$ *in* $L_2(\Omega)$ *is compact.*

Proof Let $u_n \in H_0^1(\Omega)$ with $\|u_n\|_{H^1} \le c$ for all $n \in \mathbb{N}$. We need to show that $(u_n)_{n\in\mathbb{N}}$
has a subsequence which converges in $L_2(\Omega)$. To this end we consider the extension
$f_n := \tilde{u}_n \in H^1(\mathbb{R}^d)$. Then also $\|f_n\|_{H^1} \le c$; see Theorem 6.27. We may assume that
$f_n \rightharpoonup f$ in $H^1(\mathbb{R}^d)$ (otherwise we pass to a subsequence; see Theorem 4.35). We
also assume that $f = 0$ since otherwise we may replace f_n by $f_n - f$. We wish to
show that $\lim_{n\to\infty} \int_{\mathbb{R}^d} |f_n(x)|^2 \, dx = 0$. By Plancherel's theorem, Theorem 6.45, this
is equivalent to

$$\int_{\mathbb{R}^d} |\mathcal{F} f_n(x)|^2 \, dx \to 0, \ n \to \infty.$$

We will consider the expressions

$$I_n(R) := \int_{|x|<R} |\mathcal{F} f_n(x)|^2 \, dx, \qquad J_n(R) := \int_{|x|\geq R} |\mathcal{F} f_n(x)|^2 \, dx,$$

for $R > 0$ to be determined.

1. Note that for fixed $x \in \mathbb{R}^d$,

$$F(g) := \frac{1}{(2\pi)^{d/2}} \int_\Omega e^{-ix\cdot y} g(y) \, dy, \quad g \in H_0^1(\Omega)$$

defines a continuous linear form $F \in H_0^1(\Omega)'$. Thus

$$\mathcal{F} f_n(x) = \frac{1}{(2\pi)^{d/2}} \int_\Omega e^{-ix\cdot y} f_n(y) \, dy = \frac{1}{(2\pi)^{d/2}} \int_{\mathbb{R}^d} \mathbb{1}_\Omega(y) e^{-ix\cdot y} f_n(y) \, dy$$

converges to 0 as $n \to \infty$ for all $x \in \mathbb{R}^d$, since $f_n \rightharpoonup 0$ in $H^1(\mathbb{R}^d)$. Since $|\mathcal{F} f_n(x)| \leq |\Omega|^{1/2}(2\pi)^{-d/2}\|f_n\|_{L_2} \leq c_1$ for all $n \in \mathbb{N}$ and the constant function $f(x) \equiv c_1$ is in $L_2(B(0,R))$, it now follows from the Dominated Convergence Theorem that $\lim_{n\to\infty} I_n(R) = 0$ for all $R > 0$.

2. Let $\varepsilon > 0$ and choose $R > 0$ such that $c^2(1 + R^2)^{-1} \leq \varepsilon$. Then, for all $n \in \mathbb{N}$,

$$\begin{aligned}
J_n(R) &= \int_{|x|\geq R} |\mathcal{F} f_n(x)|^2 (1 + x^2)^{-1}(1 + x^2) \, dx \\
&\leq \frac{1}{1 + R^2} \int_{|x|\geq R} |\mathcal{F} f_n(x)|^2 (1 + x^2) \, dx \\
&\leq \frac{1}{1 + R^2} \|\mathcal{F} f_n\|_{\hat{H}^1(\mathbb{R}^d)}^2 = \frac{1}{1 + R^2} \|f_n\|_{H^1(\mathbb{R}^d)}^2 \leq \frac{1}{1 + R^2} c^2 \leq \varepsilon.
\end{aligned}$$

If we use 1., then we obtain

$$\limsup_{n\to\infty} \int_{\mathbb{R}^d} |\mathcal{F} f_n(x)|^2 \, dx \leq \limsup_{n\to\infty} (I_n(R) + J_n(R)) \leq \varepsilon.$$

Since $\varepsilon > 0$ was arbitrary, it follows that $\lim_{n\to\infty} \int_{\mathbb{R}^d} |\mathcal{F} f_n(x)|^2 \, dx = 0$. $\qquad\square$

The proof shows that the theorem remains true if instead of assuming that Ω is bounded we only assume that it has finite Lebesgue measure.

6.7 Local regularity

Using Fourier transforms we were able to show in the previous section that

$$H^{k+m}(\mathbb{R}^d) \subset C^m(\mathbb{R}^d) \quad \text{if} \quad k > d/2.$$

We now wish to prove a local version of this imbedding theorem. Suppose that Ω is a nonempty open set in $\mathbb{R}^d$. By *local* properties we mean statements which hold in the interior of Ω, or to be more precise, in every set $U \Subset \Omega$. In what follows, the underlying field will always be $\mathbb{K} = \mathbb{R}$. We define

$$L_{2,\text{loc}}(\Omega) := \left\{ f : \Omega \to \mathbb{R} \text{ measurable} : \int_U |f(x)|^2 \, dx < \infty \right.$$

$$\left. \text{whenever } U \Subset \Omega \right\},$$

$$H_{\text{loc}}^1(\Omega) := \left\{ u \in L_{2,\text{loc}}(\Omega) : \text{ there exists } D_j u \in L_{2,\text{loc}}(\Omega), j = 1, \ldots, d, \right.$$

$$\text{such that } \int_\Omega u(x) \frac{\partial \varphi}{\partial x_j}(x) \, dx = - \int_\Omega D_j u(x) \, \varphi(x) \, dx$$

$$\left. \text{for all } \varphi \in \mathcal{D}(\Omega) \right\}.$$

Thus for each $u \in H_{\text{loc}}^1(\Omega)$ there exists a weak partial derivative $D_j u \in L_{2,\text{loc}}(\Omega)$, which is defined via the integration by parts formula

$$- \int_\Omega u(x) \frac{\partial \varphi}{\partial x_j}(x) \, dx = \int_\Omega D_j u(x) \, \varphi(x) \, dx, \quad \varphi \in \mathcal{D}(\Omega). \tag{6.52}$$

Corollary 6.10 shows that $D_j u$ is unique. It follows from Lemma 6.12 that

$$C^1(\Omega) \subset H_{\text{loc}}^1(\Omega) \text{ and } D_j u = \frac{\partial \varphi}{\partial x_j}, \quad j = 1, \ldots, d. \tag{6.53}$$

We will frequently make use of the following observation: if $U \Subset \Omega$, then $u_{|U} \in H^1(U)$ for all $u \in H_{\text{loc}}^1(\Omega)$. The following characterization is essential in order to be able to use results for $\mathbb{R}^d$ here. Given a function $f : \Omega \to \mathbb{R}$, as usual we denote by $\tilde{f} : \mathbb{R}^d \to \mathbb{R}$ the extension of f by 0 from Ω to the whole of $\mathbb{R}^d$ (as in (6.20)).

Lemma 6.55 *We have $H_{\text{loc}}^1(\Omega) = \{u \in L_{2,\text{loc}}(\Omega) : \widetilde{\eta u} \in H^1(\mathbb{R}^d) \quad \forall \eta \in \mathcal{D}(\Omega)\}$. Moreover, for all $u \in H_{\text{loc}}^1(\Omega)$ and all $\eta \in \mathcal{D}(\Omega)$,*

$$D_j(\widetilde{\eta u}) = [(D_j \eta) u + \eta D_j u]^{\tilde{}}. \tag{6.54}$$

Proof Let $u \in H_{\text{loc}}^1(\Omega)$ and $\eta \in \mathcal{D}(\Omega)$. Then $\widetilde{\eta u}, [(D_j \eta) u + \eta D_j u]^{\tilde{}} \in L_2(\mathbb{R}^d)$, and for $\varphi \in \mathcal{D}(\mathbb{R}^d)$, if we apply the product rule to $\eta\varphi \in \mathcal{D}(\Omega)$, then we obtain

$$- \int_{\mathbb{R}^d} \widetilde{\eta u}(x) \frac{\partial \varphi}{\partial x_j}(x) \, dx = - \int_\Omega u(x) \frac{\partial(\eta\varphi)}{\partial x_j}(x) \, dx + \int_\Omega u(x) \frac{\partial \eta}{\partial x_j}(x) \, \varphi(x) \, dx$$

$$= \int_\Omega (D_j u)(x) \eta(x) \, \varphi(x) \, dx + \int_\Omega u(x) \, (D_j \eta)(x) \, \varphi(x) \, dx$$

$$= \int_{\mathbb{R}^d} [(D_j u) \eta + u D_j \eta]^{\tilde{}}(x) \varphi(x) \, dx.$$

This proves (6.54). Conversely, let $u \in L_{2,\text{loc}}(\Omega)$ be such that $\widetilde{\eta u} \in H^1(\mathbb{R}^d)$ for all $\eta \in \mathcal{D}(\Omega)$. Choose sets $U_k \Subset U_{k+1} \Subset \Omega$ such that $\bigcup_{k \in \mathbb{N}} U_k = \Omega$, and choose functions $\eta_k \in \mathcal{D}(\Omega)$ such that $\eta_k = 1$ on U_k. Then for all $\varphi \in \mathcal{D}(U_k)$ and $j = 1, \ldots, d$ we have

$$\int_{\mathbb{R}^d} \varphi(x) \, D_j(\eta_k u)^\sim(x) \, dx = -\int_{\mathbb{R}^d} (D_j \varphi)(x) \, \widetilde{\eta_k u}(x) \, dx$$

$$= -\int_{\mathbb{R}^d} (D_j \varphi)(x) \, \widetilde{\eta_m u}(x) \, dx = \int_{\mathbb{R}^d} \varphi(x) \, D_j(\widetilde{\eta_m u})(x) \, dx,$$

whence by Corollary 6.10

$$D_j(\eta_k u)^\sim = D_j(\eta_m u)^\sim \text{ on } U_k \text{ for all } m \geq k. \tag{6.55}$$

We can thus set $u_j := D_j(\eta_k u)^\sim$ on U_k as a well-defined function in $L_{2,\text{loc}}(\Omega)$. Now given $\varphi \in \mathcal{D}(\Omega)$, then there exists some $k \in \mathbb{N}$ such that $\text{supp}\varphi \subset U_k$, and so

$$-\int_\Omega u(x) \frac{\partial \varphi}{\partial x_j}(x) \, dx = -\int_{\mathbb{R}^d} \widetilde{u\eta_k}(x) \frac{\partial \varphi}{\partial x_j}(x) \, dx = \int_{\mathbb{R}^d} D_j(\widetilde{u\eta_k})(x) \, \varphi(x) \, dx$$

$$= \int_\Omega u_j(x) \, \varphi(x) \, dx.$$

Hence $u \in H^1_{\text{loc}}(\Omega)$ and $D_j u = u_j$. □

We now define the local Sobolev space of second order by

$$H^2_{\text{loc}}(\Omega) := \{u \in H^1_{\text{loc}} : D_j u \in H^1_{\text{loc}}(\Omega), \ j = 1, \ldots, d\}.$$

Thus for $u \in H^2_{\text{loc}}(\Omega)$ we have $D_i D_j u \in L_{2,\text{loc}}(\Omega)$ for all $i, j = 1, \ldots, d$. Since we know that $D_i D_j \varphi = D_j D_i \varphi$ for all test functions φ, we can deduce from the definition of weak derivatives that also

$$D_i D_j u = D_j D_i u. \tag{6.56}$$

More generally, we define higher order spaces recursively:

$$H^{k+1}_{\text{loc}}(\Omega) := \{u \in H^1_{\text{loc}}(\Omega) : D_j u \in H^k_{\text{loc}}(\Omega), \ j = 1, \ldots, d\}.$$

It follows from (6.53) that $C^k(\Omega) \subset H^k_{\text{loc}}(\Omega)$ for all $k \in \mathbb{N}$. We can also obtain a higher-order version of Lemma 6.55.

Lemma 6.56 *We have for all $k \in \mathbb{N}$ that*

$$H^k_{\text{loc}}(\Omega) = \{u \in L_{2,\text{loc}}(\Omega) : \widetilde{\eta u} \in H^k(\mathbb{R}^d) \text{ for all } \eta \in \mathcal{D}(\Omega)\}.$$

Proof For $k = 1$ this is exactly the statement of Lemma 6.55. So let us assume that the assertion is true for $k \geq 1$, and let $u \in H^{k+1}_{\text{loc}}(\Omega)$ and $\eta \in \mathcal{D}(\Omega)$. Then $\widetilde{\eta u} \in H^1(\mathbb{R}^d)$, and by (6.54) and the induction hypothesis we have $D_j(\widetilde{\eta u}) = [(D_j\eta)u + \eta D_j u]^\sim \in H^k(\mathbb{R}^d)$ since $D_j u \in H^k_{\text{loc}}(\Omega)$. Hence $\widetilde{\eta u} \in H^{k+1}(\mathbb{R}^d)$.

Conversely, let $u \in L_{2,\text{loc}}(\Omega)$ be such that $\widetilde{\eta u} \in H^{k+1}(\mathbb{R}^d)$ for all $\eta \in D(\Omega)$. Then $u \in H^1_{\text{loc}}(\Omega)$ and

$$\widetilde{\eta \, D_j u} = D_j(\widetilde{\eta u}) - \widetilde{(D_j\eta)\, u} \in H^k(\mathbb{R}^d).$$

By the induction hypothesis, it follows that $D_j u \in H^k_{\text{loc}}(\Omega)$, and so $u \in H^{k+1}_{\text{loc}}(\Omega)$ by definition.   □

We can now give the desired local imbedding theorem.

Theorem 6.57 *Let $\Omega \subset \mathbb{R}^d$ be open, and suppose $m \in \mathbb{N}_0$ and $k \in \mathbb{N}$ such that $k > d/2$. Then $H^{k+m}_{\text{loc}}(\Omega) \subset C^m(\Omega)$.*

Proof Let $u \in H^{k+m}_{\text{loc}}(\Omega)$ and $U \Subset \Omega$. Choose $\eta \in D(\Omega)$ such that $\eta = 1$ on U. Then by Corollary 6.50 $\widetilde{u\eta} \in H^{k+m}(\mathbb{R}^d) \subset C^m(\mathbb{R}^d)$ and thus $u \in C^m(U)$.   □

We now wish to study the regularity of solutions of the Poisson equation

$$\Delta u = f. \tag{6.57}$$

We will again obtain a shift theorem.

Theorem 6.58 (Local maximal regularity) *Given $f \in L_{2,\text{loc}}(\Omega)$, let $u \in L_{2,\text{loc}}(\Omega)$ be a weak solution of (6.57). Then $u \in H^2_{\text{loc}}(\Omega)$. Moreover, if $f \in H^k_{\text{loc}}(\Omega)$, then $u \in H^{k+2}_{\text{loc}}(\Omega)$. In particular, if $f \in C^\infty(\Omega)$, then $u \in C^\infty(\Omega)$ as well.*

We will postpone the proof for a moment to examine the statement of the theorem more closely. We recall that $u \in L_2(\Omega)$ is a weak solution of (6.57) if

$$\int_\Omega u(x)\, \Delta\varphi(x)\, dx = \int_\Omega f(x)\, \varphi(x)\, dx \text{ for all } \varphi \in D(\Omega). \tag{6.58}$$

Of course, here solutions are not going to be unique as we have not stipulated any boundary conditions. Theorem 6.58 asserts that a weak solution u is automatically in $H^2_{\text{loc}}(\Omega)$. As such, u has weak partial derivatives $D_i u$ and $D_i D_j u$, $i, j, = 1, \ldots, d$, which are all in $L_{2,\text{loc}}(\Omega)$; and Δu is given by

$$\Delta u = \sum_{j=1}^{d} D_j^2 u.$$

Thus Theorem 6.58 may be viewed as a statement about *interior* maximal regularity: all derivatives up to the order of the differential equation exist automatically as elements of the space $L_{2,\text{loc}}(\Omega)$, the function space in which we wish to solve the equation. We also say that u is a *strong solution* of (6.57) if $u \in H^2_{\text{loc}}(\Omega)$. Every classical solution $u \in C^2(\Omega)$ is also a strong solution.

If $\varphi, \eta \in C^2(\Omega)$, then the product rule for the Laplacian reads

$$\Delta(\eta\varphi) = (\Delta\eta)\varphi + 2\nabla\eta\,\nabla\varphi + \eta\,\Delta\varphi, \tag{6.59}$$

as one can easily verify. We will use this formula repeatedly in the following proof.

Proof (of Theorem 6.58) Let $u, f \in L_{2,\mathrm{loc}}(\Omega)$ satisfy $-\Delta u = f$ weakly. The proof is divided into three steps:

1. We first show that $u \in H^1_{\mathrm{loc}}(\Omega)$. To this end, take $\eta \in \mathcal{D}(\Omega)$; we have to show that $\widetilde{\eta u} \in H^1(\mathbb{R}^d)$. Now for $\varphi \in \mathcal{D}(\mathbb{R}^d)$ we have

$$
\begin{aligned}
F(\varphi) &:= \int_{\mathbb{R}^d} \widetilde{\eta u}(x)(\varphi - \Delta\varphi)(x)\,dx \\
&= \int_\Omega \{\eta u\varphi - u\Delta(\varphi\eta) + u(\Delta\eta)\varphi + 2u\nabla\varphi\nabla\eta\}dx \\
&= \int_\Omega (\eta u - f\eta + (\Delta\eta)u)(x)\,\varphi(x)\,dx + 2\int_\Omega (\nabla\varphi(x)\nabla\eta(x))\,u(x)\,dx.
\end{aligned}
$$

Hence the mapping $F : \mathcal{D}(\mathbb{R}^d) \to \mathbb{R}$ is linear and there exists a constant $c \geq 0$ such that $|F(\varphi)| \leq c\|\varphi\|_{H^1(\mathbb{R}^d)}$ for all $\varphi \in \mathcal{D}(\mathbb{R}^d)$. This means that F has a continuous linear extension which maps $H^1(\mathbb{R}^d)$ to $\mathbb{R}$, and so, by the theorem of Riesz–Fréchet, there exists a unique $v \in H^1(\mathbb{R}^d)$ such that

$$\int_{\mathbb{R}^d} v(x)\,\varphi(x)\,dx + \int_{\mathbb{R}^d} \nabla v(x)\nabla\varphi(x)\,dx = \int_{\mathbb{R}^d} \widetilde{\eta u}(x)(\varphi - \Delta\varphi)(x)\,dx$$

for all $\varphi \in \mathcal{D}(\mathbb{R}^d)$. Since $v \in H^1(\mathbb{R}^d)$, it follows that $\int_{\mathbb{R}^d} v(\varphi - \Delta\varphi)\,dx = \int_{\mathbb{R}^d} \widetilde{\eta u}(\varphi - \Delta\varphi)\,dx$ for all $\varphi \in \mathcal{D}(\mathbb{R}^d)$. Since $\mathcal{D}(\mathbb{R}^d)$ is dense in $H^2(\mathbb{R}^d)$ by Corollary 6.26, it follows that $\int_{\mathbb{R}^d} (v - \widetilde{\eta u})(\varphi - \Delta\varphi)\,dx = 0$ for all $\varphi \in H^2(\mathbb{R}^d)$. Now, by Theorem 6.51, there exists a $\varphi \in H^2(\mathbb{R}^d)$ such that $\varphi - \Delta\varphi = (v - \widetilde{\eta u})$. Thus $\int_{\mathbb{R}^d} (v - \widetilde{\eta u})^2\,dx = 0$, which finally allows us to conclude that $\widetilde{\eta u} = v$ is in $H^1(\mathbb{R}^d)$.

2. We next show that given $\eta \in \mathcal{D}(\Omega)$ we have $\widetilde{\eta u} \in H^2(\mathbb{R}^d)$. It is not hard to show using (6.59) that $\Delta(\widetilde{\eta u}) = [(\Delta\eta)u + 2\nabla\eta\,\nabla u + \eta f]\tilde{\ } =: g(\eta)$ weakly (see Exercise 6.15). Since by 1. we know that $D_j u \in L_{2,\mathrm{loc}}(\Omega)$, the function $\nabla\eta\,\nabla u = \sum_{j=1}^d D_j\eta D_j u$ is in $L_2(\Omega)$ and so $g(\eta) \in L_2(\mathbb{R}^d)$. It thus follows from Theorem 6.53 that $\widetilde{\eta u} \in H^2(\mathbb{R}^d)$.

3. We now prove Theorem 6.58 by induction. For $k = 0$ the statement is proved in 2. (where $H^0_{\mathrm{loc}}(\Omega) = L_{2,\mathrm{loc}}(\Omega)$). Now assume that the statement is true for $k \geq 0$ and let $f \in H^{k+1}_{\mathrm{loc}}(\Omega)$; we wish to show that $u \in H^{k+3}_{\mathrm{loc}}(\Omega)$. Let $\eta \in \mathcal{D}(\Omega)$, then by the induction hypothesis we have $u \in H^{k+2}_{\mathrm{loc}}(\Omega)$ and so $g(\eta) \in H^{k+1}(\mathbb{R}^d)$. It follows by Theorem 6.53 that $\widetilde{\eta u} \in H^{k+3}(\mathbb{R}^d)$.

This completes the proof of Theorem 6.58. $\qquad\square$

More generally, one can show that even if we only have $u \in L_{1,\mathrm{loc}}(\Omega)$ (in place of $u \in L_{2,\mathrm{loc}}(\Omega)$ as in Theorem 6.58) but $f \in C^\infty(\Omega)$, then $-\Delta u = f$ already implies that $u \in C^\infty(\Omega)$. The following theorem, which goes back to Sobolev, shows that the maximal regularity property of Theorem 6.58 does not hold for classical derivatives. This is one reason why the spaces $C^k(\Omega)$, $k = 0, 1, 2, \ldots$ are unsuited both to the numerics of partial differential equations and to the study of nonlinear problems. The negative statement of the following theorem illustrates clearly how advantageous Sobolev spaces are.

Theorem 6.59 *Let $\Omega \subset \mathbb{R}^d$ be an arbitrary nonempty open set in dimension $d \geq 2$. Then there exist $u, f \in C_c(\Omega)$ such that $\Delta u = f$ weakly, but $u \notin C^2(\Omega)$.*

Proof 1. We first give an example in $\mathbb{R}^2$. Let $B := \{(x, y) \in \mathbb{R}^2 : x^2 + y^2 < \frac{1}{4}\}$ and $u(x, y) := (x^2 - y^2) \log |\log r|$, $(x, y) \in B$, where $r = (x^2 + y^2)^{1/2}$. Then $u \in C^2(B \setminus \{0\})$. Since $|\log s| \leq \frac{1}{s}$ for $s > 0$ small and $|\log s| \leq s$ for $s > 0$ large, we have the estimate $\log |\log r| \leq |\log r| \leq \frac{1}{r}$ for $r > 0$ small and so $\lim_{r \to 0} u(x, y) = 0$. We will also denote the continuous extension of u to B by u. Now on $B \setminus \{0\}$ we have

$$u_x = 2x \log |\log r| + (x^3 - y^2 x) \frac{1}{r^2 \log r}$$

$$u_{xx} = 2 \log |\log r| + (5x^2 - y^2) \frac{1}{r^2 \log r} - (x^4 - x^2 y^2) \frac{2 \log r + 1}{r^4 (\log r)^2}.$$

Hence the function u_{xx} is unbounded in $B \setminus \{0\}$ since on the diagonal $\{(x, x) : |x| < \frac{1}{4}\}$ we have

$$\lim_{x \to 0} u_{xx}(x, x) = \lim_{x \to 0} \left\{ 2 \log |\log r| + \frac{2}{\log r} \right\} = \infty.$$

Since $u(x, y) = -u(y, x)$, we also have $u_{yy}(x, y) = -u_{xx}(y, x)$. This means that the singular term $2 \log |\log r|$ cancels in the expression for $\Delta u = u_{xx} + u_{yy}$ and

$$\Delta u = (x^2 - y^2) \left(\frac{4}{r^2 \log r} - \frac{1}{r^2 (\log r)^2} \right).$$

(cf. Exercise 6.24). It follows that $\lim_{r \to 0} \Delta u(x, y) = 0$.

Now let g denote the continuous extension of Δu to 0; we claim that $\Delta u = g$ weakly in B. To see this, we first observe that u_x and u_y are bounded in $B \setminus \{0\}$ (in fact $u \in C^1(B)$). We set $B_\varepsilon := \{(x, y) \in B : x^2 + y^2 > \varepsilon^2\}$ for given $0 < \varepsilon < \frac{1}{4}$ and let $\varphi \in \mathcal{D}(B)$. Then by Green's second identity (Corollary 7.7(c))

$$\int_{B_\varepsilon} u \, \Delta \varphi \, dx = \int_{B_\varepsilon} g \varphi \, dx + \int_{\partial B_\varepsilon} \left(u \frac{\partial \varphi}{\partial \nu} - \frac{\partial u}{\partial \nu} \varphi \right) d\sigma.$$

Letting $\varepsilon \to 0$, we obtain

$$\int_B u \, \Delta \varphi \, dx = \int_B g \varphi \, dx.$$

This proves that $\Delta u = g$ weakly. Finally, we take $\eta \in \mathcal{D}(B)$ such that $\eta(x, y) = 1$ in a neighborhood of 0; then $\eta u \in C_c(B)$ and $f := \eta g + 2\nabla \eta \nabla u + (\Delta \eta)u$ is in $C(B)$. It is now easy to see that $\Delta(\eta u) = f$ weakly (Exercise 6.15). This proves the theorem for the set B.

2. By adding more coordinates and translating accordingly one easily obtains a general example as claimed in the statement of the theorem. □

Remark 6.60 The function u from Theorem 6.59 is however in $H^2(\Omega) \cap H_0^1(\Omega)$. This follows from Theorem 6.58 since u has compact support. △

6.8 Inhomogeneous Dirichlet boundary conditions

Let Ω be a bounded open set in $\mathbb{R}^d$ with boundary $\partial\Omega$. Given $f \in L_2(\Omega)$ and $g \in C(\partial\Omega)$ we seek a solution of the Poisson problem

$$-\Delta u = f \text{ in } \Omega, \tag{6.60a}$$

$$u_{|\partial\Omega} = g \text{ on } \partial\Omega. \tag{6.60b}$$

Here, both the equation (6.60a) and the boundary condition (6.60b) are inhomogeneous. If $u \in L_2(\Omega)$ is a weak solution of (6.60a), then u is automatically in $H_{\text{loc}}^2(\Omega)$ by Theorem 6.58 and

$$\Delta u = \sum_{j=1}^{d} D_j^2 u \in L_{2,\text{loc}}(\Omega).$$

When considering the boundary condition (6.60b), we might first think of a classical interpretation: if $u \in C(\overline{\Omega})$, then we may demand that $u(z) = g(z)$ for all $z \in \partial\Omega$. However, it is not always possible to find solutions of (6.60a) which are continuous up to the boundary. This leads us to define the condition (6.60b) in a weak sense. For this, we need another assumption on the function g.

Definition 6.61 Let $g \in C(\partial\Omega)$.
(a) An H^1-*extension* of g is a function $G \in C(\overline{\Omega}) \cap H^1(\Omega)$ such that $G_{|\partial\Omega} = g$.
(b) An H^1-*solution* u of (6.60) is a function $u \in H^1(\Omega) \cap H_{\text{loc}}^2(\Omega)$ such that (6.60a) holds *and* g has an H^1-extension G such that

$$u - G \in H_0^1(\Omega) \tag{6.61}$$

is satisfied. △

We interpret the condition (b) as a weak form of the inhomogeneous Dirichlet boundary condition (6.60b). We wish to show that this definition is independent of the choice of the H^1-extension G, that is, that if (6.61) holds for some H^1-extension

G of g, then it holds for all of them. To this end, observe that if $G_1 \in C(\overline{\Omega}) \cap H^1(\Omega)$ is another function which satisfies $G_{1|\partial\Omega} = g$, then $G_1 - G \in C_0(\Omega) \cap H^1(\Omega) \subset H^1_0(\Omega)$ by Theorem 6.30. Since $u - G = (u - G_1) + (G_1 - G)$, it follows that $u - G \in H^1_0(\Omega)$ if and only if $u - G_1 \in H^1_0(\Omega)$.

We will see in the next section that not every continuous function $g \in C(\partial\Omega)$ has an H^1-extension. However, we only require a very mild additional regularity assumption on g in order to guarantee the existence of an H^1-extension. If it has one, then we have the following theorem on existence and uniqueness of solutions.

Theorem 6.62 *Let $f \in L_2(\Omega)$ and suppose that $g \in C(\partial\Omega)$ has an H^1-extension. Then (6.60) admits a unique H^1-solution.*

Proof The idea of the proof is as follows. If $u \in H^1(\Omega)$ is an H^1-solution, then $v := u - G \in H^1_0(\Omega)$ satisfies $-\Delta v = f + \Delta G$ weakly, that is,

$$\int_\Omega \nabla v(x) \nabla \varphi(x) \, dx = \int_\Omega f(x) \varphi(x) \, dx - \int_\Omega \nabla G(x) \nabla \varphi(x) \, dx$$

for all $\varphi \in \mathcal{D}(\Omega)$. We will encounter this principle again when studying numerical methods; it is known as "reduction to homogeneous boundary conditions". It leads us to the following proof, which relies on the Lax–Milgram theorem.

1. Existence: let G be an $H^1(\Omega)$-extension of g, then

$$F(\varphi) := \int_\Omega f(x) \varphi(x) \, dx - \int_\Omega \nabla G(x) \nabla \varphi(x) \, dx, \quad \varphi \in H^1_0(\Omega),$$

defines a continuous linear form on $H^1_0(\Omega)$. As above,

$$a(v, \varphi) := \int_\Omega \nabla v(x) \nabla \varphi(x) \, dx, \quad v, \varphi \in H^1_0(\Omega)$$

defines a continuous and coercive bilinear form on $H^1_0(\Omega)$ (see Corollary 6.33). Hence by the Lax–Milgram theorem there exists a unique $v \in H^1_0(\Omega)$ such that $a(v, \varphi) = F(\varphi)$ for all $\varphi \in H^1_0(\Omega)$. That is, for all $\varphi \in H^1_0(\Omega)$ we have

$$\int_\Omega \nabla v(x) \nabla \varphi(x) \, dx = \int_\Omega f(x) \varphi(x) \, dx - \int_\Omega \nabla G(x) \nabla \varphi(x) \, dx.$$

Now set $u := G + v$, then $\int_\Omega \nabla u(x) \nabla \varphi(x) \, dx = \int_\Omega f(x) \varphi(x) \, dx$ for all $\varphi \in \mathcal{D}(\Omega)$, that is, $-\Delta u = f$ weakly. It follows from Theorem 6.58 that $u \in H^2_{\text{loc}}(\Omega)$.

2. Uniqueness: let u_1 and u_2 be two solutions, and let G_1 and G_2 be two H^1-extensions of g such that $u_1 - G_1, u_2 - G_2 \in H^1_0(\Omega)$. Then $u = u_1 - u_2 = (u_1 - G_1) - (u_2 - G_2) + (G_1 - G_2) \in H^1_0(\Omega)$ and $\Delta u = 0$, whence $u = 0$ by Theorem 6.43. $\square$

We next show that, for solutions of (6.60a) which are continuous up to the boundary, the classical boundary condition implies the weak one provided only that

g has an H^1-extension. This is remarkable because u could oscillate strongly on the boundary and thus it is not clear *a priori* whether its derivatives are in $L_2(\Omega)$.

Theorem 6.63 *Let* $u \in C(\overline{\Omega}) \cap H^2_{\mathrm{loc}}(\Omega)$ *satisfy* (6.60a) *and suppose that* $u(z) = g(z)$ *for all* $z \in \partial\Omega$. *If* g *has an* H^1-*extension* G, *then* $u \in H^1(\Omega)$ *and* $u - G \in H^1_0(\Omega)$.

Proof Let $G \in H^1(\Omega) \cap C(\overline{\Omega})$ be such that $G_{|\partial\Omega} = g$. Then $v := u - G$ is in $C_0(\Omega) \cap H^1_{\mathrm{loc}}(\Omega)$. Now set $v_n := (v - \frac{1}{n})^+$, $n \in \mathbb{N}$; then $v_n \in C_c(\Omega) \cap H^1_{\mathrm{loc}}(\Omega) \subset H^1_c(\Omega) \subset H^1_0(\Omega)$ by Theorem 6.29. Moreover, by Theorem 6.35 we have $D_j v_n = \mathbb{1}_{\{v > \frac{1}{n}\}} D_j v$, $j = 1, \ldots, d$. Since $-\Delta u = f$, we have

$$\int_\Omega \nabla v(x) \, \nabla \varphi(x) \, dx = \int_\Omega f(x)\, \varphi(x)\, dx - \int_\Omega \nabla G(x)\, \nabla \varphi(x)\, dx. \qquad (6.62)$$

for all $\varphi \in \mathcal{D}(\Omega)$. Fix $n \in \mathbb{N}$ and choose $U \Subset \Omega$ such that $v_n(x) = 0$ for all $x \notin U$. Then $v_n \in C_0(U) \cap H^1(U) \subset H^1_0(U)$, meaning that v_n can be approximated in $H^1(U)$ by test functions in $\mathcal{D}(U)$. Since $v_{|U} \in H^1(U)$, (6.62) also holds with $\varphi = v_n$. Since $\nabla v \, \nabla v_n = \sum_{j=1}^d D_j v \, D_j v \, \chi_{\{v > \frac{1}{n}\}} = |\nabla v_n|^2$, it follows that

$$\int_\Omega |\nabla v_n(x)|^2 \, dx = \int_\Omega \nabla v(x) \nabla v_n(x) \, dx$$

$$= \int_\Omega f(x)\, v_n(x)\, dx - \int_\Omega \nabla G(x) \nabla v_n(x)\, dx$$

$$\leq \|f\|_{L_2} \|v_n\|_{L_2} + c_1 \left(\int_\Omega |\nabla v_n(x)|^2 \, dx \right)^{\frac{1}{2}}$$

$$\leq \|f\|_{L_2} \|v_n\|_{L_2} + \frac{c_1^2}{2} + \frac{1}{2} \int_\Omega |\nabla v_n(x)|^2 \, dx$$

by Young's inequality, where $c_1 := \left(\int_\Omega |\nabla G(x)|^2 \, dx \right)^{\frac{1}{2}} = |G|_{H^1(\Omega)}$. Since $\|v_n\|_{L_2} \leq \|v\|_{L_2}$, it follows that $(v_n)_{n \in \mathbb{N}}$ is bounded in $H^1_0(\Omega)$. Hence, appealing to Theorem 4.35, passing to a subsequence if necessary, we may assume that $v_n \rightharpoonup w$ for some $w \in H^1_0(\Omega)$. But since by definition $v_n \to v^+$ in $L_2(\Omega)$, it follows that $v^+ = w \in H^1_0(\Omega)$. By replacing u by $-u$, we see that $v^- = (-v)^+ \in H^1_0(\Omega)$ as well. We conclude that $v \in H^1_0(\Omega)$ and thus $u = v + G \in H^1(\Omega)$, which completes the proof. $\qquad \square$

The essential point in Theorem 6.63 is that we have complete freedom when choosing H^1-extensions of g. Nevertheless, the solution itself is always in $H^1(\Omega)$ as long as such an extension exists.

6.9 The Dirichlet problem

In this section we will study the Dirichlet problem, that is, the special case where $f = 0$ in (6.60). Our goals are threefold:

(a) We will show that weak solutions can be characterized by minimal energy (that is, minimal H^1-seminorm $| \cdot |_{H^1}$).
(b) We will give an example to show that even for very simple regular domains, there exist classical solutions with infinite energy.
(c) We will investigate more closely which conditions on the domain guarantee that the Dirichlet problem always admits a classical solution, for all boundary values.

Let $\Omega \subset \mathbb{R}^d$ be bounded and open and denote by $\partial\Omega$ its boundary. Given $g \in C(\partial\Omega)$, the Dirichlet problem consists in finding a solution of

$$\Delta u = 0 \text{ in } \Omega, \qquad u_{|\partial\Omega} = g, \tag{6.63}$$

Let us briefly recall the interior regularity theorem, Theorem 6.58, which we obtained for solutions of the Laplace equation: whenever $u \in L_{2,\text{loc}}(\Omega)$ and $\Delta u = 0$ holds weakly in Ω, the function u is in fact in $C^\infty(\Omega)$. Thus we may automatically suppose that all solutions are C^∞-functions. We still need to investigate in which sense the boundary condition $u_{|\partial\Omega} = g$ is satisfied. If g has an H^1-extension, that is, if there exists a function $G \in C(\overline{\Omega}) \cap H^1(\Omega)$ which coincides with g on $\partial\Omega$, then (6.63) admits exactly one H^1-solution by Theorem 6.62, namely, a function $u \in H^1(\Omega) \cap C^\infty(\Omega)$ such that $\Delta u = 0$ and $u - G \in H_0^1(\Omega)$. This H^1-solution may be characterized by the following minimality property.

Theorem 6.64 (Dirichlet principle) *Let $g \in C(\partial\Omega)$ have an H^1-extension G and let u be the H^1-solution of (6.63). Then*

$$\int_\Omega |\nabla u(x)|^2 \, dx < \int_\Omega |\nabla w(x)|^2 \, dx$$

for all $w \in H^1(\Omega)$ such that $w - G \in H_0^1(\Omega)$ and $w \neq u$. That is, the solution u is exactly the function with minimal energy

$$|u|_{H^1(\Omega)}^2 = \int_\Omega |\nabla u(x)|^2 \, dx$$

among all functions w in $H^1(\Omega)$ which satisfy the boundary condition $w = g$ in the weak sense.

Proof Let $w \in H^1(\Omega)$ be such that $w - G \in H_0^1(\Omega)$ and $w \neq u$, then $\varphi := w - u = w - G + G - u \in H_0^1(\Omega)$. Since $\varphi \neq 0$, we have $\int_\Omega |\nabla\varphi|^2 \, dx > 0$. Since $\Delta u = 0$ and $\varphi \in H_0^1(\Omega)$, it follows that $\int_\Omega \nabla u(x)\nabla\varphi(x) \, dx = 0$, and so

$$\int_\Omega |\nabla w(x)|^2 \, dx = \int_\Omega |\nabla(u + \varphi)(x)|^2 \, dx$$

$$= \int_\Omega |\nabla u(x)|^2 \, dx + 2 \int_\Omega \nabla u(x) \nabla \varphi(x) \, dx + \int_\Omega |\nabla \varphi(x)|^2 \, dx$$

$$= \int_\Omega |\nabla u(x)|^2 \, dx + \int_\Omega |\nabla \varphi(x)|^2 \, dx > \int_\Omega |\nabla u(x)|^2 \, dx,$$

which is exactly the claim.   □

By definition, H^1-solutions of the Dirichlet problem have finite energy. We now wish to consider classical solutions: given $g \in C(\partial\Omega)$, a *classical solution* of (6.63) is a function $u \in C^2(\Omega) \cap C(\overline{\Omega})$ such that $\Delta u = 0$ in Ω and $u_{|\partial\Omega} = g$.

Theorem 6.65 *Let u be a classical solution. Then the following three assertions are equivalent:*
(i) The function u has finite energy, that is, $|u|_{H^1(\Omega)} < \infty$.
(ii) u is an H^1-solution.
(iii) g has an H^1-extension.

Proof (i) $\Rightarrow$ (iii): Since $u \in C(\overline{\Omega})$, $G = u$ is an H^1-extension of u.
(iii) $\Rightarrow$ (ii): This is exactly Theorem 6.63.
(ii) $\Rightarrow$ (i) is clear since $u \in H^1(\Omega)$.   □

We will now illustrate with an example that there exist classical solutions which do not have finite energy and thus are not H^1-solutions. To this end we consider the unit disk $\mathbb{D}$ in $\mathbb{R}^2$, that is, $\mathbb{D} = \{x \in \mathbb{R}^2 : |x| < 1\}$ with the unit circle $\partial\mathbb{D} = \{x \in \mathbb{R}^2 : |x| = 1\}$ as its boundary. Given $g \in C(\partial D)$, the Dirichlet problem (6.63) admits the classical solution $u \in C^\infty(\mathbb{D}) \cap C(\overline{\mathbb{D}})$ given by

$$u(r\cos\theta, r\sin\theta) = c_0 + \sum_{k=1}^\infty r^k \left(a_k \cos(k\theta) + b_k \sin(k\theta) \right) \tag{6.64}$$

for $0 \le r < 1$ and $\theta \in \mathbb{R}$, where

$$a_k = \frac{1}{\pi} \int_0^{2\pi} g(\cos\theta, \sin\theta) \cos(k\theta) d\theta, \quad b_k = \frac{1}{\pi} \int_0^{2\pi} g(\cos\theta, \sin\theta) \sin(k\theta) d\theta,$$

$$c_0 = \frac{1}{2\pi} \int_0^{2\pi} g(\cos\theta, \sin\theta) d\theta;$$

see Theorem 3.30. We can now express the energy of u in terms of the Fourier coefficients a_k and b_k as follows.

Lemma 6.66 *Let u be defined as in (6.64). Then*

$$\int_\mathbb{D} |\nabla u(x)|^2 \, dx = \pi \sum_{k=1}^\infty k(a_k^2 + b_k^2). \tag{6.65}$$

Proof We recall that for continuous functions $f : \mathbb{D} \to [0, \infty)$ we have

$$\int_{\mathbb{D}} f(x)\, dx = \int_0^1 \int_0^{2\pi} f(r \cos\theta, r \sin\theta) d\theta\, r\, dr, \tag{6.66}$$

see Theorem A.12. Set $v(r, \theta) := u(r \cos\theta, r \sin\theta)$. By differentiating v in r and θ and solving for u_x and u_y, we can show that

$$|\nabla u|^2 (r \cos\theta, r \sin\theta) = \left(v_r^2 + \frac{v_\theta^2}{r^2} \right)(r, \theta), \tag{6.67}$$

cf. Exercise 6.16. Plugging this into (6.66), we obtain

$$\int_{\mathbb{D}} |\nabla u(x)|^2\, dx = \int_0^1 r \int_0^{2\pi} \left(v_r^2 + \frac{v_\theta^2}{r^2} \right) d\theta\, dr. \tag{6.68}$$

Now consider the function $v_k(r, \theta) := r^k (a_k \cos(k\theta) + b_k \sin(k\theta))$. Its partial derivatives v_{kr} und $v_{k\theta}$ satisfy $v_{kr}^2 + \frac{v_{k\theta}^2}{r^2} = k^2 r^{2k-2}(a_k^2 + b_k^2)$, and so

$$\int_{\mathbb{D}} |\nabla u(x)|^2\, dx = \sum_{k=1}^\infty k^2 (a_k^2 + b_k^2) \int_0^1 r \int_0^{2\pi} r^{2k-2} d\theta\, dr = \pi \sum_{k=1}^\infty k(a_k^2 + b_k^2),$$

which proves the claim. $\square$

We can now describe the classical example of Hadamard from 1906.

Example 6.67 (Hadamard (1906)) Set

$$g(\cos\theta, \sin\theta) := \sum_{n=1}^\infty 2^{-n} \cos(2^{2n}\theta).$$

Then $g \in C(\partial\mathbb{D})$ since the series converges normally. Let u be the classical solution of (6.63) with boundary value g (cf. Theorem 3.30). We claim that $\int_{\mathbb{D}} |\nabla u|^2\, dx = \infty$, which means in particular that $u \notin H^1(\mathbb{D})$. In fact, no function $G \in H^1(\mathbb{D}) \cap C(\overline{\mathbb{D}})$ exists which also satisfies $G_{|\partial\mathbb{D}} = g$. $\triangle$

Proof We have $b_k = 0$ for all $k \in \mathbb{N}$, while $a_k = 2^{-n}$ for $k = 2^{2n}$ and $a_k = 0$ for all $k \notin \{2^{2n} : n \in \mathbb{N}\}$. By (6.65) we thus have

$$\int_{\mathbb{D}} |\nabla u(x)|^2\, dx = \pi \sum_{n=1}^\infty 2^{2n} 2^{-2n} = \infty.$$

The last assertion of the example follows from Theorem 6.65. $\square$

We now know that even for a domain as regular as the unit disk the space

$$W(\partial\Omega) := \{g \in C(\partial\Omega) : \exists\, G \in C(\overline{\Omega}) \cap H^1(\Omega) \text{ such that } G_{|\partial\Omega} = g\}$$

is a proper subspace of $C(\partial\Omega)$. It is however dense; in fact, we have:

Lemma 6.68 *The space $\mathcal{D}(\partial\Omega) := \{G_{|\partial\Omega} : G \in \mathcal{D}(\mathbb{R}^d)\}$ is dense in $C(\partial\Omega)$.*

We will give two proofs of Lemma 6.68:

Proof (1st proof of Lemma 6.68) The space $\mathcal{D}(\partial\Omega)$ is a subalgebra of $C(\partial\Omega)$ which in particular contains the constant functions. Let $y, \bar{y} \in \partial\Omega$ be any two distinct points on the boundary, $y \neq \bar{y}$. Then there exists an index $j \in \{1, \ldots, d\}$ for which $y_j \neq \bar{y}_j$. Let $\eta \in \mathcal{D}(\mathbb{R}^d)$ be such that $\eta \equiv 1$ on $\overline{\Omega}$ (see Lemma 6.7) and set $G(x) := \eta \cdot x_j, x \in \mathbb{R}^d$; then $G \in \mathcal{D}(\mathbb{R}^d)$ and $G(y) \neq G(\bar{y})$. This shows that $\mathcal{D}(\partial\Omega)$ separates points in $\partial\Omega$ and so the claim follows from the Stone–Weierstrass theorem (see Theorem A.5 on page 426). □

Proof (2nd proof of Lemma 6.68) Let $g \in C(\partial\Omega)$, then by the Tietze extension theorem there exists a $G \in C_c(\mathbb{R}^d)$ such that $G_{|\partial\Omega} = g$. Let $G_n = \varrho_n * G$, then $G_n \in \mathcal{D}(\mathbb{R}^d)$, and $\lim_{n\to\infty} G_n = G$ holds uniformly in $\mathbb{R}^d$, see Theorem 6.11. □

If $g \in W(\partial\Omega)$, then the H^1-solution of (6.63) is bounded and continuous on Ω.

Lemma 6.69 *Given $g \in W(\partial\Omega)$, let u be the H^1-solution of (6.63). Then*

$$u(x) \leq \max_{z \in \partial\Omega} g(z) \text{ for all } x \in \Omega. \tag{6.69}$$

Bear in mind that $u \in C^\infty(\Omega)$.

Proof Set $c := \max_{z \in \partial\Omega} g(z)$ and suppose that $G \in C(\overline{\Omega}) \cap H^1(\Omega)$ satisfies $G_{|\partial\Omega} = g$. Let $v \in H_0^1(\Omega)$ be the unique solution of

$$\int_\Omega \nabla v(x)\, \nabla\varphi(x)\, dx = \int_\Omega \nabla G(x)\, \nabla\varphi(x)\, dx, \quad \varphi \in H_0^1(\Omega).$$

Then $u := G + v$ is the H^1-solution of (6.63); see the proof of Theorem 6.62. Since $g \leq c$, we have $(G - c)^+ \in C_0(\Omega) \cap H^1(\Omega) \subset H_0^1(\Omega)$; moreover, since $(u - c)^+ \leq (G - c)^+ + v^+$, it follows from Theorem 6.39(b) that $(u - c)^+ \in H_0^1(\Omega)$. It now follows from the weak maximum principle (Theorem 6.44) that $u(x) \leq c$ for all $x \in \Omega$, that is, (6.69) holds. □

If we apply (6.69) to $-g$ in place of g, then we obtain that

$$\min_{z \in \Omega} g(z) \leq u(x) \leq \max_{z \in \Omega} g(z) \tag{6.70}$$

for all $x \in \Omega$, where u is again the H^1-solution of (6.63). Let us now consider the mapping $T : W(\partial\Omega) \to C^b(\Omega)$ which assigns to each $g \in W(\partial\Omega)$ the H^1-solution of (6.63) which has boundary value g; here we have written

$$C^b(\Omega) := \{v : \Omega \to \mathbb{R} : v \text{ is bounded and continuous}\},$$

which is a Banach space with respect to the supremum norm

$$\|v\|_\infty := \sup_{x \in \Omega} |v(x)|.$$

It follows from the uniqueness of H^1-solutions that T is linear.

Furthermore, (6.70) implies that the mapping $T : W(\partial\Omega) \to C^b(\Omega)$ is contractive, that is,

$$\|Tg\|_\infty \le \|g\|_{C(\partial\Omega)}, \, g \in W(\partial\Omega). \tag{6.71}$$

Since $W(\partial\Omega)$ is dense in $C(\partial\Omega)$, by Theorem A.2 the mapping T has a unique contractive extension $\tilde{T} : C(\partial\Omega) \to C^b(\Omega)$ to the whole of $C(\partial\Omega)$. Here we are using the expression "contraction" in the non-strict sense, that is, for operators of norm ≤ 1. Such operators are also sometimes called "non-expansive".

Definition 6.70 Given $g \in C(\partial\Omega)$, the function $u_g := \tilde{T}g$ is called the *Perron solution* of (6.63). △

The Perron solution is harmonic and satisfies the maximum principle.

Theorem 6.71 *Given $g \in C(\partial\Omega)$, let u_g be the Perron solution of (6.63). Then*
(a) $u_g \in C^\infty(\Omega)$ and $\Delta u_g = 0$ in Ω;
(b) $\min_{z \in \partial\Omega} g(z) \le u_g(x) \le \max_{z \in \partial\Omega} g(z)$ for all $x \in \Omega$.

Proof (a) Let $g_n \in W(\partial\Omega)$ be such that $g_n \to g$ in $C(\partial\Omega)$. Then it follows from the definitions that $u_g(x) = \lim_{n\to\infty} u_{g_n}(x)$ uniformly in Ω. Since $\Delta u_{g_n} = 0$, we have that

$$\int_\Omega u_g(x) \Delta\varphi(x)\,dx = \lim_{n\to\infty} \int_\Omega u_{g_n}(x) \Delta\varphi(x)\,dx = 0$$

for all $\varphi \in \mathcal{D}(\Omega)$. This establishes that $\Delta u_g = 0$ weakly in Ω. It now follows from Theorem 6.58 that $u_g \in C^\infty(\Omega)$.

(b) The claim follows by passing to the limit in (6.70). □

Remark 6.72 One can show that if u is a classical solution of (6.63), then $u_g = u$. Thus the Perron solution is a generalization of the classical solution. We refer to the additional comments at the end of the chapter for more on this point. △

We now introduce an analyticity condition for points $z \in \partial\Omega$ on the boundary of our domain.

Definition 6.73 A *barrier function*, or *barrier* for short, at the point $z \in \partial\Omega$ is a function $b \in C(\overline{\Omega \cap B})$, where $B = B(z, r)$ is a ball of centre z and radius $r > 0$, which satisfies the following three conditions:
(a) b is *superharmonic* in $\Omega \cap B$, that is,

$$\int_\Omega b(x) \Delta\varphi(x)\,dx \le 0 \text{ for all } 0 \le \varphi \in \mathcal{D}(\Omega \cap B);$$

(b) $b(x) > 0$ for all $x \in \overline{\Omega \cap B} \setminus \{z\}$; and

(c) $b(z) = 0$.

The function b is called an H^1-*barrier* if additionally $b \in H^1(B \cap \Omega)$. $\triangle$

The following theorem then holds.

Theorem 6.74 *Given $z \in \partial\Omega$, suppose that there exists an H^1-barrier at z. Then for every $g \in C(\partial\Omega)$ we have*

$$\lim_{\Omega \ni x \to z} u_g(x) = g(z), \tag{6.72}$$

where u_g is the Perron solution of (6.63).

Proof We first consider a special case.

1st case: there exists $G \in \mathcal{D}(\mathbb{R}^d)$ such that $G_{|\partial\Omega} = g$. Let $v \in H_0^1(\Omega)$ satisfy $-\Delta v = \Delta G$; then $u_g = G + v$ by definition of the H^1-solution of (6.63). Now fix $\varepsilon > 0$; we will show that

$$\limsup_{\Omega \ni x \to z} u_g(x) \le g(z) + \varepsilon. \tag{6.73}$$

Let $B = B(z, r)$ be a ball which is so small that, in addition to the existence of an H^1-barrier $b \in H^1(\Omega \cap B) \cap C(\overline{\Omega \cap B})$, the inequality

$$G(x) - g(z) - \varepsilon \le 0 \text{ for all } x \in \overline{B} \tag{6.74}$$

holds. This is possible since G is continuous at z and $G(z) = g(z)$.

Now set $w(x) := u_g(x) - g(z) - \varepsilon - b(x)$, $x \in \Omega \cap B$. By property (b) of barriers, by multiplying b by a positive constant if necessary, we may assume that

$$b(x) \ge \|u_g\|_{L_\infty(\Omega)} - g(z) \text{ for all } x \in \partial B \cap \overline{\Omega}. \tag{6.75}$$

Then $w^+ \in H_0^1(\Omega \cap B)$, as we will show presently. Since $w \in H^1(\Omega \cap B)$ and $-\Delta w = \Delta b \le 0$, it then follows from the weak maximum principle (Theorem 6.44) that $w \le 0$, whence $u_g \le g(z) + \varepsilon + b$. Since $b(z) = 0$, we finally obtain that $\limsup_{\Omega \ni x \to z} u_g(x) \le g(z) + \varepsilon$.

It remains to show that $w^+ \in H_0^1(\Omega \cap B)$. Now we have $w \in H^1(\Omega \cap B)$ and $w = v + G - G(z) - \varepsilon - b$. Since $G - G(z) - \varepsilon \le 0$ and $b \ge 0$ in $\Omega \cap B$, it follows that $w^+ \le v^+$. Moreover, by (6.75), the inequality $w(x) = u_g(x) - G(z) - \varepsilon - b(x) \le -\varepsilon$ holds for all $x \in \partial B \cap \overline{\Omega}$. Since $v \in H_0^1(\Omega)$, there exist functions $\varphi_n \in \mathcal{D}(\Omega)$ such that $\varphi_n \to v$ in $H^1(\Omega)$. Then the sequence $\varphi_n^+ \wedge w^+$ converges to $v^+ \wedge w^+ = w^+$ in $H^1(\Omega \cap B)$ (see Theorem 6.37). Now $\varphi_n^+ \wedge w^+ \in C_c(\Omega \cap B) \cap H^1(\Omega \cap B)$ since $\varphi_n^+ \in C_c(\Omega)$ and $w^+ = 0$ in a neighborhood of $\partial B \cap \overline{\Omega}$. Thus $\varphi_n^+ \wedge w^+ \in H_0^1(\Omega \cap B)$, and so finally $w^+ \in H_0^1(\Omega \cap B)$. This completes the proof of (6.73).

2nd case: take $g \in C(\partial\Omega)$ arbitrary and fix $\varepsilon > 0$. Then there exists a function $h \in \mathcal{D}(\partial\Omega)$ such that $\|h - g\|_{C(\partial\Omega)} \le \varepsilon/2$. Hence, by Theorem 6.71(b),

$$|u_g(x) - g(z)| \leq |u_g(x) - u_h(x)| + |u_h(x) - h(z)| + |h(z) - g(z)|$$
$$\leq 2\|g - h\|_{C(\partial\Omega)} + |u_h(x) - h(z)| \leq \varepsilon + |u_h(x) - h(z)|$$

for all $x \in \Omega$. It thus follows from the 1st case that $\limsup_{\Omega \ni x \to z} |u_g(x) - g(z)| \leq \varepsilon$. Since $\varepsilon > 0$ was arbitrary, the proof is complete. $\qquad\square$

It turns out that the barrier criterion is also a necessary condition for existence, which allows us to formulate the following major theorem.

Theorem 6.75 *The following assertions are equivalent.*
(i) For each $g \in C(\partial\Omega)$ there exists a classical solution $u \in C^2(\Omega) \cap C(\overline{\Omega})$ of (6.63).
(ii) At every point $z \in \partial\Omega$ there exists an H^1-barrier.
(iii) At every point $z \in \partial\Omega$ there exists a barrier.

Proof (i) $\Rightarrow$ (ii): Given $z \in \partial\Omega$, define $g(x) = |z - x|^2$; then $g \in C^\infty(\mathbb{R}^d)$. Let u be the solution (6.63). By applying the strong maximum principle [28, 2.2 Theorem 4] on every connected component of Ω, we see that $u(x) > 0$ for all $x \in \overline{\Omega} \setminus \{z\}$. Thus u is an H^1-barrier at z.
(ii) $\Rightarrow$ (iii): This follows immediately from the definitions.
We refer to [25] for the implication (iii) $\Rightarrow$ (i). $\qquad\square$

We call the domain Ω *Dirichlet regular* if for every $g \in C(\partial\Omega)$ there exists a classical solution of the Dirichlet problem (6.63). By Theorem 6.75, Ω is thus Dirichlet regular if and only if there exists an H^1-barrier at every point $z \in \partial\Omega$.

Example 6.76 In dimension $d = 1$, every bounded open subset Ω of $\mathbb{R}$ is Dirichlet regular. $\qquad\triangle$

Proof Let $z \in \partial\Omega$ and set $b(x) := |x - z|$. Then $b \in C^\infty(\mathbb{R} \setminus \{z\})$ and $\Delta b = 0$ in $\mathbb{R} \setminus \{z\}$. Hence b is an H^1-barrier at z. $\qquad\square$

In the plane, the *segment condition* is sufficient for Dirichlet regularity.

Definition 6.77 The set $\Omega \subset \mathbb{R}^d$ satisfies the *segment condition* if for every $z \in \partial\Omega$ there exists an $x_0 \in \mathbb{R}^d \setminus \{z\}$ such that $\lambda x_0 + (1 - \lambda)z \notin \Omega$ for all $\lambda \in [0, 1]$, cf. Figure 6.2 (page 229). $\qquad\triangle$

Theorem 6.78 *If $\Omega \subset \mathbb{R}^2$ satisfies the segment condition, then Ω is Dirichlet regular.*

Proof Let $z \in \partial\Omega$. We may assume without loss of generality that $z = 0$ and $\Omega \cap B(0, r_0) \subset \{re^{i\theta} : 0 < r < r_0, -\pi < \theta < \pi\}$ (otherwise we translate and rotate Ω). Then

$$b(r, \theta) := \frac{-\log r}{(\log r)^2 + \theta^2}, \quad -\pi < \theta < \pi, 0 < r < r_0$$

defines a barrier. $\qquad\square$

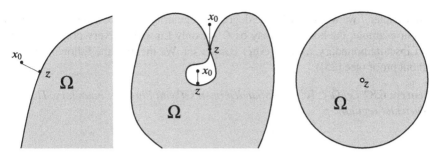

Fig. 6.2 The segment condition is satisfied for the domains on the left and in the center, but not for the domain on the right (the punctured disk).

Remark 6.79 (a) In the plane there is a very general and practical sufficient condition for Dirichlet regularity: every bounded, open, simply connected set is Dirichlet regular (see [23]).

(b) The punctured disk $\Omega = \{x \in \mathbb{R}^2 : 0 < |x| < 1\}$ (see Figure 6.2) is not Dirichlet regular; see Exercise 3.19.   △

We now consider higher dimensions, and start with a rather crude criterion.

Theorem 6.80 *Let $\Omega \subset \mathbb{R}^d$ be bounded and open, where $d \geq 2$, and suppose that Ω satisfies the* exterior sphere condition, *that is, for every $z \in \partial\Omega$ there exists an $x_0 \in \mathbb{R}^d \setminus \{z\}$ such that*

$$\overline{\Omega} \cap B(x_0, |x_0 - z|) = \{z\}, \tag{6.76}$$

cf. Figure 6.3. Then Ω is Dirichlet regular.

Proof Let $d \geq 3$ and $z \in \partial\Omega$. Choose an $x_0 \in \mathbb{R}^d \setminus \{z\}$ such that (6.75) holds. Then

$$b(x) := \left\{ \frac{1}{|z - x_0|^{d-2}} - \frac{1}{|x - x_0|^{d-2}} \right\}$$

defines a barrier. For $d = 2$ see [25, 4.18, p. 274].   □

Corollary 6.81 *Every bounded, open convex set in $\mathbb{R}^d$ is Dirichlet regular.*

Proof Let $z \in \partial\Omega$. The Hahn–Banach theorem (see, e.g., [54, Thm. 12.12]) implies the existence of a $c \in \mathbb{R}^d$ such that $x \cdot c < c \cdot z$ for all $x \in \Omega$. We suppose without loss of generality that $z = 0$ and then choose $x_0 = c$ and $r = |x_0|$. We claim that then $\overline{\Omega} \cap B(x_0, |x_0 - z|) = \{z\} = \{0\}$. Indeed, suppose that $|x - c| \leq |x_0 - z| = r$ and $x \neq 0$; then $x \cdot c > 0$ and so $x \notin \overline{\Omega}$. Now observe that $|x|^2 - 2x \cdot c + r^2 = (x - c) \cdot (x - c) \leq r^2$.   □

In Chapter 7 we will describe the degree of regularity of the boundary of an open set. For example, the boundary may be C^1 or only Lipschitz. Every polygon in $\mathbb{R}^2$ has Lipschitz boundary, as does every convex set. We mention the following result without proof (see [25]).

Theorem 6.82 *Let $\Omega \subset \mathbb{R}^d$ be a bounded open set with Lipschitz boundary. Then Ω is Dirichlet regular.*

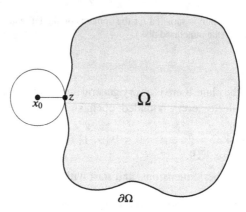

Fig. 6.3 Depiction of the exterior sphere condition for $d = 2$.

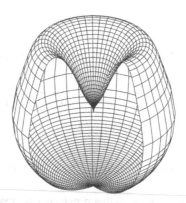

Fig. 6.4 The Lebesgue cusp set.

We mention that in $\mathbb{R}^3$ the continuity of the boundary does *not* automatically imply Dirichlet regularity. By rotating a cusp, one can generate a domain in $\mathbb{R}^3$ which is not Dirichlet regular if the cusp is pointy enough and points into the domain (see Figure 6.4). Such an example was given by Lebesgue in 1912. We refer to [10], where the details can be found and Fig. 6.4 is produced by Maple®. This shows in particular that the regularity of the given boundary function g does not imply the existence of a classical solution: in this example, it is possible to choose the function $g \in \partial\Omega$ in such a way that at every point $z \in \partial\Omega$ it has a harmonic extension to a neighborhood of z in $\mathbb{R}^3$, but nevertheless no solution of the Dirichlet problem exists. The Perron solution u_g oscillates in the vicinity of the cusp point z_0 and does not have a continuous extension to z_0. This example also shows that in $\mathbb{R}^3$ the segment condition is not sufficient for the Dirichlet regularity of the domain.

6.10 Elliptic equations with Dirichlet boundary conditions

In this section, in place of the Laplacian we wish to consider more general elliptic operators with variable coefficients. Let $\Omega \subset \mathbb{R}^d$ be a bounded open set and suppose that $a_{ij}, b_j, c \in L_\infty(\Omega)$, where the a_{ij} satisfy, for all $\xi \in \mathbb{R}^d$ and $x \in \Omega$,

$$\sum_{i,j=1}^d a_{ij}(x)\xi_i\xi_j \geq \alpha|\xi|^2, \tag{6.77}$$

for some fixed $\alpha > 0$. We refer to (6.77) as a *uniform ellipticity condition*.

Remark 6.83 In the symmetric case, that is, when $a_{ij} = a_{ji}$, the condition (6.77) is equivalent to the condition that the smallest eigenvalue of the coefficient-matrix $(a_{ij}(x))_{i,j=1,\ldots,d}$ is at least as large as α. We wish, however, to allow also non-symmetric matrices as well. △

Our goal is to study boundary value problems of the form

$$-\sum_{i,j=1}^d D_i(a_{ij}D_j u) + \sum_{j=1}^d b_j D_j u + cu = f, \tag{6.78a}$$

$$u_{|\partial\Omega} = 0, \tag{6.78b}$$

where $f \in L_2(\Omega)$ is given. If $a_{ij}(x) = 1$ for $i = j$ and $a_{ij} = 0$ for $i \neq j$, $b_j = c = 0$, then (6.78) reduces to the Poisson equation $-\Delta u = f$ which we considered above.

The first thing to strike us about (6.78) is that under our very general assumptions, the expression $D_i(a_{ij}D_j u)$ may not be well defined, even for twice continuously differentiable functions u. We will, however, once again work with weak solutions. We first suppose that $a_{ij} \in C^1(\Omega)$. In this case, if $u \in C_c^2(\Omega)$ is a solution of (6.78a), then by integration by parts we obtain

$$\int_\Omega \left\{ \sum_{i,j=1}^d a_{ij}(x)D_j u(x)D_i \varphi(x) + \sum_{j=1}^d b_j(x)D_j u(x)\varphi(x) + c(x)u(x)\varphi(x) \right\} dx$$

$$= \int_\Omega f(x)\varphi(x)\,dx \tag{6.79}$$

for all $\varphi \in \mathcal{D}(\Omega)$. Now observe that this expression is well defined as soon as $u, \varphi \in H^1(\Omega)$.

Definition 6.84 A *weak solution* of (6.78) is a function $u \in H_0^1(\Omega)$ for which (6.79) holds for all $\varphi \in \mathcal{D}(\Omega)$.   △

This definition leads us to the bilinear form

$$a(u, v) := \int_\Omega \left\{ \sum_{i,j=1}^d a_{ij}(x)D_ju(x)D_iv(x) + \sum_{j=1}^d b_j(x)D_ju(x)v(x) \right.$$
$$\left. + c(x)u(x)v(x) \right\} dx,$$

$u, v \in H_0^1(\Omega)$. Clearly, the form $a : H_0^1(\Omega) \times H_0^1(\Omega) \to \mathbb{R}$ is bilinear and continuous. It is also coercive provided that the coefficients satisfy appropriate conditions. In this case we obtain the existence and uniqueness of solutions of (6.78).

Theorem 6.85 *Suppose, in addition to* (6.77), *that one of the following two conditions holds:*
(a) $\sum_{j=1}^d b_j(x)^2 \le 2\alpha\, c(x),\quad x \in \Omega,\quad$ *or*
(b) $b_j \in C^1(\Omega)$ *and* $\sum_{j=1}^d (D_jb_j)(x) \le 2c(x),\quad x \in \Omega.$
Then for every $f \in L_2(\Omega)$ *there exists a unique solution of* (6.78).

Proof Let $f \in L_2(\Omega)$. By definition, a function $u \in H_0^1(\Omega)$ is a weak solution of (6.78) if and only if

$$a(u, \varphi) = \int_\Omega f(x)\varphi(x)\,dx \quad \text{for all } \varphi \in \mathcal{D}(\Omega). \tag{6.80}$$

Since $\mathcal{D}(\Omega)$ is dense in $H_0^1(\Omega)$, this is equivalent to (6.80) holding for all $\varphi \in H_0^1(\Omega)$. The mapping $\varphi \mapsto \int_\Omega f(x)\varphi(x)\,dx$ defines a continuous linear form on $H_0^1(\Omega)$, and so by the Lax–Milgram theorem there exists a unique weak solution, provided that $a(\cdot, \cdot)$ is coercive. We will now show that this is indeed the case if (a) or (b) is satisfied.

For (a), we apply Young's inequality (5.19) to obtain

$$ub_jD_ju \le \frac{\alpha}{2}(D_ju)^2 + \frac{1}{2\alpha}b_j^2u^2.$$

Hence, by (6.77),

$$a(u) \ge \alpha \int_\Omega |\nabla u(x)|^2\,dx - \frac{\alpha}{2}\int_\Omega |\nabla u(x)|^2\,dx - \frac{1}{2\alpha}\int_\Omega \sum_{j=1}^d b_j(x)^2u(x)^2\,dx$$

$$+ \int_\Omega c(x)u(x)^2\,dx \geq \frac{\alpha}{2} \int_\Omega |\nabla u(x)|^2\,dx.$$

Since the expression

$$\left(\frac{\alpha}{2} \int_\Omega |\nabla u(x)|^2\,dx \right)^{\frac{1}{2}} = \sqrt{\frac{\alpha}{2}} \, |u|_{H^1}(\Omega)$$

defines an equivalent norm on H_0^1 by the Poincaré inequality (Theorem 6.32), this proves the coercivity of $a(\cdot,\cdot)$.

For (b) we integrate by parts and obtain, for $u \in \mathcal{D}(\Omega)$,

$$\int_\Omega b_j(x)D_j u(x)\,u(x)\,dx = \int_\Omega b_j \frac{1}{2} D_j u^2\,dx = -\frac{1}{2} \int_\Omega D_j b_j(x)u(x)^2\,dx$$

for all $j = 1, \ldots, d$, whence

$$\int_\Omega \sum_{j=1}^d b_j(x)D_j u(x)u(x)\,dx + \int_\Omega c(x)u(x)^2\,dx$$

$$= -\frac{1}{2} \int_\Omega \sum_{j=1}^d D_j b_j(x)u(x)^2\,dx + \int_\Omega c(x)u(x)^2\,dx \geq 0.$$

This shows that $a(u) \geq \alpha \int_\Omega |\nabla u(x)|^2\,dx$ for all $u \in \mathcal{D}(\Omega)$ and hence for all $u \in H_0^1(\Omega)$. $\qquad\square$

6.11 H^2-regularity

Let Ω be a bounded open set in $\mathbb{R}^d$. We have seen that solutions of the Poisson equation always belong to $H_{\text{loc}}^2(\Omega)$. Here we wish to pursue the question of what conditions actually guarantee that the solutions are in $H^2(\Omega)$. This revolves around the H^2-regularity at the boundary. This question is also important for error estimates for numerical methods (cf. Chapter 9) and depends in general on the regularity of the boundary itself. A set Ω is said to be *convex* if whenever $x, y \in \Omega$, we also have $\lambda x + (1 - \lambda)y \in \Omega$ for all $\lambda \in (0, 1)$. We recall that for every $f \in L_2(\Omega)$ there exists a unique solution

$$u \in H_0^1(\Omega) \cap H_{\text{loc}}^2(\Omega), \tag{6.81a}$$

$$- \Delta u = f, \tag{6.81b}$$

of the Poisson equation (Theorems 6.43 and 6.58).

Theorem 6.86 *If Ω is convex, then for every $f \in L_2(\Omega)$ the solution u of* (6.81) *is in $H^2(\Omega)$. Moreover, there exists a constant $c_2 > 0$ such that $\|u\|_{H^2(\Omega)} \leq c_2 \|f\|_{L_2(\Omega)}$.*

We refer to [33] for the proof of the first assertion. The existence of the constant c_2 is then an immediate consequence of the Closed Graph Theorem.

Remark 6.87 The conclusion of Theorem 6.86 continues to hold when the open set Ω has C^2-boundary (see Definition 7.1) without necessarily being convex; see [18, Theorem 9.25] or [28, §6.3, Theorem 4]. △

We next wish to give an example to show that the solution of (6.81) need not be in $H^2(\Omega)$ without any further assumptions on the boundary. We will consider the following domain in $\mathbb{R}^2 = \mathbb{C}$: for $\pi < \alpha < 2\pi$ and $r_0 > 0$ we set

$$\Omega_{\alpha, r_0} := \{re^{i\theta} : 0 \leq r < r_0, \ 0 < \theta < \alpha\},$$

cf. Figure 6.5. We start by considering the Dirichlet problem. Let $h(z) := z^\beta$, where

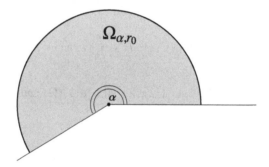

Fig. 6.5 The set Ω_{α, r_0}.

$\beta = \frac{\pi}{\alpha}$. Then h is holomorphic in Ω_{α, r_0} and continuous in $\overline{\Omega}_{\alpha, r_0}$. Hence the imaginary part $v = \operatorname{Im} h$ of h is harmonic. We have $v(r\cos\theta, r\sin\theta) = r^\beta \sin(\beta\theta)$ and $v_{|\partial\Omega_{\alpha, r_0}} = g$, where $g(r_0 e^{i\theta}) = r_0^\beta \sin(\beta\theta)$ for $0 < \theta < \alpha$, and $g(re^{i\alpha}) = g(r) = 0$ for $0 \leq r \leq r_0$. As with every holomorphic function, we have $h'(z) = (\operatorname{Im} h)_y(z) + i(\operatorname{Im} h)_x(z)$, and so $v_y + iv_x = \beta z^{\beta-1}$ for our function v. If we differentiate this function again, the same argument yields $\beta(\beta - 1)z^{\beta-2} = v_{xy} + iv_{xx}$. Since

$$\int_\Omega |z^{\beta-2}|^2 \, dz = \int_0^{r_0} r\alpha r^{(\beta-2)2} \, dr = \alpha \int_0^{r_0} r^{2\beta-3} \, dr = \infty,$$

it follows that $v \notin H^2(\Omega_{\alpha, r_0})$. However, it follows from Theorem 6.63 that $v \in H^1(\Omega_{\alpha, r_0})$ (which one can also check directly using (6.67)). We next modify this example to exhibit a solution of the Poisson equation which is not in $H^2(\Omega)$.

Let $\eta \in \mathcal{D}(\mathbb{R}^2)$ satisfy $\eta(z) = 0$ for $|z| \leq \frac{r_0}{3}$ and $\eta(z) = 1$ for $z \in \Omega_{\alpha,r_0}$ such that $|z| \geq \frac{r_0}{2}$. Then for $G(z) := \eta(z) \, \mathrm{Im} \, h(z)$ we have $G \in C^\infty(\overline{\Omega_{\alpha,r_0}})$ and $G_{|\partial\Omega} = g$. Now let $u \in H_0^1(\Omega_{\alpha,r_0})$ be the unique function for which $-\Delta u = \Delta G =: f$ in Ω_{α,r_0}. Then we know from Theorem 6.63 that $v = G + u$. Hence $u \notin H^2(\Omega_{\alpha,r_0})$, since $v \notin H^2(\Omega_{\alpha,r_0})$. We have thus shown the following result.

Example 6.88 There exists a function $f \in C^\infty(\overline{\Omega_{\alpha,r_0}})$ such that the solution u of (6.81) (with $\Omega = \Omega_{\alpha,r_0}$) satisfies $u \in C(\overline{\Omega_{\alpha,r_0}})$, but $u \notin H^2(\Omega_{\alpha,r_0})$. △

One can extend this example to every bounded open set with a non-convex (that is, inward-pointing) corner.

Definition 6.89 Let $\Omega \subset \mathbb{R}^2$ be bounded and open and $z \in \partial\Omega$. We say that Ω *has a non-convex corner at* z, if in a suitably chosen Cartesian coordinate system there exist an angle $\pi < \alpha < 2\pi$ and some $\varepsilon > 0$ such that

$$z + re^{i\theta} \in \Omega \text{ for } 0 < r \leq \varepsilon, \ \theta \in (0, \alpha),$$
$$z + re^{i\theta} \notin \Omega \text{ for } 0 \leq r < \varepsilon, \ \alpha \leq \theta \leq 2\pi.$$

Compare Figure 6.6. △

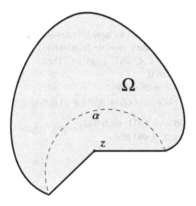

Fig. 6.6 A domain Ω with a non-convex corner at z.

Theorem 6.90 *Let* $\Omega \subset \mathbb{R}^2$ *be a bounded open set with a non-convex corner. Then there exists a function* $u \in H_0^1(\Omega) \cap H_{\mathrm{loc}}^2(\Omega)$ *such that* $-\Delta u =: f \in L_2(\Omega)$, *but* $u \notin H^2(\Omega)$.

Proof Let $U := \{re^{i\theta} + z : 0 < r < \varepsilon, \ 0 < \theta < \alpha\}$. By Example 6.88 there exists a function $w \in H_0^1(U) \cap H_{\mathrm{loc}}^2(U)$ such that $-\Delta w = g \in L_2(U)$, but $w \notin H^2(U_0)$, where $U_0 := U \cap B\left(z, \frac{\varepsilon}{2}\right)$. Extend w by 0 on $\Omega \setminus U$; then $w \in H_0^1(\Omega)$ by Theorem 6.27. Now let $\eta \in \mathcal{D}(\mathbb{R}^d)$ satisfy $\mathrm{supp}\,\eta \subset B(z, \varepsilon)$ and $\eta \equiv 1$ on $B\left(z, \frac{\varepsilon}{2}\right)$, and set $u := w\eta$. Then $u \notin H^2(\Omega)$, but $u \in H_0^1(\Omega) \cap H_{\mathrm{loc}}^2(\Omega)$ and $-\Delta u = -\eta g - 2\nabla w \nabla \eta - w\Delta\eta \in L_2(\Omega)$. □

In Chapter 9 we will return to the example of the L-shaped domain $\Omega = (-1, 1)^2 \setminus [0, 1)^2$. Clearly, Ω has a non-convex corner at the origin, meaning that the Poisson problem is not H^2-regular, that is, the conclusion of Theorem 6.86 does not hold.

6.12* Comments on Chapter 6

The Dirichlet problem is one of the oldest and most frequently studied problems in analysis. The name was coined by Riemann in honor of his teacher Johann P.G.L. Dirichlet (1805–1859), who lectured at the University of Berlin from 1831 to 1855 and succeeded Gauss (Gauß) at the University of Göttingen in 1855. In his lectures on potential theory, he spoke about the Dirichlet principle (also later named after him by Riemann). This led to a sustained mathematical dispute after Weierstrass gave an example in 1869 which showed that the minimum of energy integrals need not be attained. It was Arzelà in 1896 and Hilbert in 1900 who, completely independently of each other, gave the first rigorous solution of the Dirichlet problem using the variational principle. But the question of whether every function $u \in C(\overline{\Omega})$ such that $\Delta u = 0$ in Ω has finite energy $\int_\Omega |\nabla u(x)|^2 \, dx$, remained open. A first counterexample was obtained by Friedrich Prym in 1871; the example given by Hadamard in 1906 became famous, and it is this example which we have reproduced here.

Mollifiers were introduced by K.O. Friedrichs in 1944 to big effect in the theory of partial different equations. In [31, Vol. 1], Peter Lax, of Lax–Milgram fame, recounts the story of how they got their name: it is an allusion, among other things, to the eponymous protagonist of Daniel Defoe's 1722 novel *Moll Flanders*, who managed to mollify her last husband after having led a rather irregular life.

The definition of Perron solutions which we have given is adapted to our Sobolev space approach. Oskar Perron (1880–1975) originally gave another (equivalent) definition in 1905: let $\Omega \subset \mathbb{R}^d$ be bounded and open and suppose $g \in C(\partial\Omega)$ is given. Then a function $v \in C(\overline{\Omega})$ is called a *subsolution* of the Dirichlet problem if
(a) $\limsup_{\Omega \ni x \to z} v(x) \le g(z)$, $z \in \partial\Omega$ and
(b) $-\Delta v \le 0$ weakly, that is, $\int_\Omega v(x)\Delta\varphi(x)\, dx \le 0$ for all $0 \le \varphi \in \mathcal{D}(\Omega)$.

A *supersolution* is a function $w \in C(\overline{\Omega})$ such that
(a) $\limsup_{\Omega \ni x \to z} w(x) \ge g(z)$, $z \in \partial\Omega$ and
(b) $-\Delta w \ge 0$ weakly.
Then

$$\overline{u}(x) := \inf\{w(x) : w \text{ is a supersolution}\} \quad \text{and} \quad \underline{u}(x) := \sup\{v(x) : v \text{ is a subsolution}\}$$

exist for all $x \in \Omega$. Furthermore, $\overline{u} = \underline{u}, \overline{u} \in C^\infty(\Omega)$ is bounded, and $\Delta\overline{u} = 0$. Finally, for all $x \in \Omega$ we have

$$\min_{z \in \partial\Omega} g(z) \le \overline{u}(x) \le \max_{z \in \partial\Omega} g(z).$$

If (6.63) has a classical solution u, then $u = \overline{u}$. For all these statements we refer to [25]. In the literature, the function $\overline{u}$ is called the *Perron solution* of the Dirichlet problem. One can show that this notion of solution coincides with the one given in Definition 6.70 (see [10]). In particular, $u = u_g$ if u is a classical solution of (6.63).

Oskar Perron was a remarkable mathematician. He served as a professor in the German cities of Tübingen, Heidelberg and Munich, and contributed to many different areas of mathematics. Not only is his solution of the Dirichlet problem (as just described) famous, so too is his work on positive matrices (the Perron–Frobenius theory). Perron was in addition one of the German scientists who were brave enough to take clear anti-Nazi positions during the Nazi dictatorship.

The most important operator in this book is the Laplacian. It bears the name of Pierre-Simon Laplace (1749–1827), who studied the orbits of the planets; and in his famous work *Traité de Mécanique Céleste*, which he published in 1799, the Laplace equation plays a decisive role. An important question which Laplace addressed is the stability of the planetary system, which he solved using Newtonian mechanics (he thought). Isaac Newton (1643–1727) himself did not believe in its stability; rather, he thought that from time to time divine intervention was necessary in order to "regulate" things.

We quote [29] on the subject: "When citizen Laplace showed General Napoleon the first edition of his *Exposition du Système du monde*, the general said to him, 'Newton spoke of God in his book. I have looked through yours and did not find this word once.' To which Laplace answered, 'Citizen and First Consul, I never needed this hypothesis.'" Laplace had already met Napoleon much earlier: in 1784 he examined the 16-year-old Napoleon for the entrance exam of the latter to the École Militaire in Paris. How different might the history of Europe and even the world have been had Laplace not passed him? Napoleon, for his part, showed admiration for mathematics and in 1799 even appointed Laplace minister of the interior, although unlike Fourier he enjoyed little success in such a political role: Napoleon dismissed him after only six weeks, later uttering the famous remark that Laplace had brought the spirit of the "infinitesimally small" into the administration. Laplace would eventually be named a marquis, in 1817.

Laplace is known for his formula for computing the determinant of a matrix in terms of its minors, which he discovered as a young man, as well as for his seminal work on elementary probability theory. His magnum opus on celestial mechanics was highly successful – but does not make easy reading. His colleagues, and even Laplace himself, often had difficulties following its correct but often sketchy arguments. He had the habit of replacing arguments by the phrase "Il est aisé à voir" – it is easy to see. This stylistic device has become commonplace in mathematics, often to the detriment of the reader.

6.13 Exercises

Exercise 6.1 (Support of a measurable function) Let $\Omega \subset \mathbb{R}^d$ be open and $f : \Omega \to \mathbb{R}$ measurable. Set $O_f := \{x \in \Omega : \exists \varepsilon > 0 \text{ such that } f(x) = 0 \text{ almost everywhere in } B(x, \varepsilon)\}$. The set supp $f := \Omega \setminus O_f$ is called the *support* of f.
(a) Show that O_f is the largest open set in Ω on which f vanishes almost everywhere.
(b) Let $f, g \in L_2(\mathbb{R}^d)$. Show that then supp $f * g \subset \overline{\text{supp } f + \text{supp } g}$.

Exercise 6.2 Let $\Omega \subset \mathbb{R}^d$ be open and $j \in \{1, \ldots, d\}$. Show that

$$-\int_\Omega u \, D_j v \, dx = \int_\Omega D_j u \, v \, dx$$

for all $u \in H^1(\Omega)$ and $v \in H_0^1(\Omega)$.

Exercise 6.3 Let $u \in H^2(\Omega)$, where $\Omega \subset \mathbb{R}^d$ is open. Show that $D_i D_j u = D_j D_i u$ for all $i, j = 1, \ldots, d$.

Exercise 6.4 Let $\Omega = \mathbb{D} := \{x \in \mathbb{R}^d : |x| < 1\}$, $d \geq 3$, and let $u \in C^1(\Omega \setminus \{0\})$ be such that $\int_{0<|x|<1} |u(x)|^2 \, dx < \infty$ und $\int_{0<|x|<1} |\nabla u(x)|^2 \, dx < \infty$. Show that $u \in H^1(\Omega)$ and $(D_j u)(x) = \frac{\partial u}{\partial x_j}(x)$ for all $x \neq 0$.

Suggestion: Let $\varphi \in \mathcal{D}(\Omega)$. Show that there exists $\varphi_n \in \mathcal{D}(\Omega \setminus \{0\})$ such that $\varphi_n \to \varphi$ in $H^1(\Omega)$. To this end, choose $\varphi_n(x) := f(n|x|)\varphi(x)$ as in Remark 6.31.

Exercise 6.5 (a) Let $\Omega = \mathbb{D} = \{x \in \mathbb{R}^d : |x| < 1\}$ and $u(x) = |x|^{2\alpha}$. Determine for which $\alpha \in \mathbb{R}$ the function u is in $H^1(\Omega)$.
(b) Let $d \geq 3$, and let $\Omega \subset \mathbb{R}^d$ be open and nonempty. Show that $H^1(\Omega)$ contains a function which is not continuous.

Exercise 6.6 (a) Let $\Omega = \mathbb{D} := \{x \in \mathbb{R}^2 : |x| < 1\}$ and suppose that $\eta \in \mathcal{D}(\Omega)$ satisfies $\eta(x) = 1$ for $|x| \leq \frac{1}{2}$. Let $u(x) = \left(\log \frac{1}{|x|}\right)^{1/4} \cdot \eta(x)$. Show that $u \in H^1(\Omega)$.
(b) Let $\Omega \subset \mathbb{R}^2$ be open and nonempty. Show that $H^1(\Omega)$ contains a function which is not continuous.

Exercise 6.7 Show that $\mathcal{D}(\mathbb{R}^d)$ is dense in $H^2(\mathbb{R}^d)$.
Suggestion: Adapt the proof of Theorem 6.25.

Exercise 6.8 Let $\Omega = (0, 1) \cup (1, 2)$. Show that $C(\overline{\Omega}) \cap H^1(\Omega)$ is not dense in $H^1(\Omega)$.
Suggestion: Consider the function $\mathbb{1}_{(0,1)} \in H^1(\Omega)$.

Exercise 6.9 (Variant of the product rule) Let $\Omega \subset \mathbb{R}^d$ be open. Show that if u and v are in $H^1(\Omega) \cap L_\infty(\Omega)$, then the product uv is in $H^1(\Omega)$ and $D_j(uv) = D_j u \, v + u \, D_j v$.

Exercise 6.10 (Variant of the chain rule) Let $f \in C^1(\mathbb{R})$ such that $\sup_{x \in \mathbb{R}} |f'(x)| < \infty$ and let $\Omega \subset \mathbb{R}^d$ be bounded. Show that $\varphi(u) := f \circ u$ defines a mapping $\varphi : H^1(\Omega) \to H^1(\Omega)$, and that $D_j \varphi(u) = (f' \circ u) D_j u$.

Exercise 6.11 Let $\Omega \subset \mathbb{R}^d$ be open and u and v in $H^1(\Omega)$.
(a) Show that $w := u \wedge v$, defined by $w(x) := \min\{u(x), v(x)\}$, is also in $H^1(\Omega)$, and determine its weak derivative.
(b) Show that $(-u) \wedge (-v) = -(u \vee v)$, and conclude then that $(u \vee v)(x) = \max\{u(x), v(x)\}$ also defines a function $u \vee v \in H^1(\Omega)$.

Exercise 6.12 Let $\Omega \subset \mathbb{R}^d$ be open (but not necessarily bounded). Show:
(a) If $u \in H^1(\Omega)$, then also $u \wedge \mathbb{1}_\Omega \in H^1(\Omega)$. Determine the weak derivative of $u \wedge \mathbb{1}_\Omega$.
(b) $L_\infty(\Omega) \cap H^1(\Omega)$ is dense in $H^1(\Omega)$.

Exercise 6.13 Consider the setup of Theorem 6.43 with $\lambda > 0$, and suppose that $f(x) \leq 1$ almost everywhere. Show that then also $\lambda u(x) \leq 1$ almost everywhere.

Exercise 6.14 (Riemann–Lebesgue lemma) Let $d \in \mathbb{N}$ and as usual let $C_0(\mathbb{R}^d) := \{u \in C(\mathbb{R}^d) : \lim_{|x| \to \infty} u(x) = 0\}$. Show:
(a) If $u \in L_1(\mathbb{R}^d)$, then $\mathcal{F}u \in C(\mathbb{R}^d)$.
(b) The mapping $u \mapsto \mathcal{F}u$ is linear and continuous from $L_1(\mathbb{R}^d)$ into $L_\infty(\mathbb{R}^d)$.

(c) Let $u \in \mathcal{D}(\mathbb{R}^d)$. Then there exists $c \in \mathbb{R}$ such that $|\mathcal{F}u(x)| \leq \frac{c}{1+|x|^2}$ for all $x \in \mathbb{R}^d$. In particular, $\mathcal{F}u \in C_0(\mathbb{R}^d)$.
   Suggestion: Make use of the rules for calculating Fourier transforms.
(d) For every $u \in L_1(\mathbb{R}^d)$ we have $\mathcal{F}u \in C_0(\mathbb{R}^d)$.
   Suggestion: Recall that $C_0(\mathbb{R}^d)$ is closed in $L_\infty(\Omega)$. One may also use without proof that $\mathcal{D}(\mathbb{R}^d)$ is dense in $L_1(\mathbb{R}^d)$; this theorem can be shown similarly to the case $L_2(\mathbb{R}^d)$, cf. Corollary 6.9.

Exercise 6.15 Let $\Omega \subset \mathbb{R}^d$ be open, and let $u \in H^1_{loc}(\Omega)$ as well as $f \in L_{2,loc}(\Omega)$ be such that $\Delta u = f$ weakly. Show that for $\eta \in \mathcal{D}(\Omega)$ the following identity holds
$\Delta(\widetilde{\eta u}) = [(\Delta \eta)u + 2\nabla \eta \, \nabla u + f\eta]^\sim$.
   Suggestion: Use (6.59).

Exercise 6.16 Prove the identity (6.67).

Exercise 6.17 Show directly, without using H^1-barriers, that every bounded open subset Ω of $\mathbb{R}$ is Dirichlet regular.
   Suggestion: Use that Ω is a countable union of open intervals.

Exercise 6.18 Let $\Omega \subset \mathbb{R}^d$ be bounded, open and convex. Given $g \in C(\partial\Omega)$, let $u \in C(\overline{\Omega}) \cap C^2(\Omega)$ be a classical solution of the Dirichlet problem (6.63). Suppose that there exists $G \in H^2(\Omega) \cap C(\overline{\Omega})$ such that $G_{|\partial\Omega} = g$. Show that then $u \in H^2(\Omega)$.

Exercise 6.19 Prove Corollary 6.23.

Exercise 6.20 (Invariance of the Poisson equation under isometries) Let B be an orthogonal $d \times d$ matrix, $b \in \mathbb{R}^d$, and $F(y) = By + b$ for all $y \in \mathbb{R}^d$. Let $\Omega_2 \subset \mathbb{R}^d$ be an open set and $\Omega_1 := F(\Omega_2)$. Suppose that $f \in L_2(\Omega_1)$ and $u \in H^1_0(\Omega_1) \cap H^2_{loc}(\Omega_1)$ satisfy $-\Delta u = f$ weakly. Show that $u \circ F \in H^1_0(\Omega_2) \cap H^2_{loc}(\Omega_2)$ and $-\Delta(u \circ F) = f \circ F$.
   Suggestion: Use that the function $u \in H^1_0(\Omega)$ is the unique solution of the problem $\int_\Omega \nabla u \, \nabla v \, dx = \int_\Omega fv \, dx$, $v \in H^1_0(\Omega)$.

Exercise 6.21 (More general linear elliptic differential operators) Let $\Omega \subset \mathbb{R}^d$ be bounded and open, and assume that a_{ij}, b_j, c_i and e are in $L_\infty(\Omega)$. For the formal differential operator $Lu := -\sum_{j=1}^d D_j \left(\sum_{i=1}^d a_{ij} D_i u + b_j u \right) + \sum_{i=1}^d c_i D_i u + eu$ and a given function $f \in L_2(\Omega)$, one understands a *weak solution* of the problem

$$(D) \qquad Lu = f \quad \text{in } \Omega, \qquad u = 0 \quad \text{on } \partial\Omega,$$

to be a function $u \in H_0^1(\Omega)$ such that

$$a_L(u, v) := \sum_{i,j=1}^d \int_\Omega a_{ij} D_i u D_j v + \sum_{j=1}^d \int_\Omega b_j u D_j v + \sum_{i=1}^d \int_\Omega c_i D_i u v + \int_\Omega e u v$$

$$= \int_\Omega f v$$

for all test functions $v \in \mathcal{D}(\Omega)$. Finally, suppose that there exists $\alpha > 0$ such that $\sum_{i,j=1}^d a_{ij} \xi_i \xi_j \geq \alpha |\xi|^2$ for all $\xi \in \mathbb{R}^d$. Show:

(a) The form a_L is bilinear and continuous on $H_0^1(\Omega) \times H_0^1(\Omega)$..

(b) If there exists $\delta < 2\alpha$ such that $\sum_{j=1}^d (b_j + c_j)^2 \leq 2\delta e$ almost everywhere, then (D) has a unique weak solution.
    Suggestion: For $x, y \in \mathbb{R}$ and $\varepsilon > 0$ we have $xy \leq \frac{\varepsilon}{2} x^2 + \frac{1}{2\varepsilon} y^2$ (Young's inequality, Lemma 5.22).

(c) If b_i and c_i are continuously differentiable and if $\sum_{i=1}^d (D_i b_i + D_i c_i) \leq 2e$ holds almost everywhere, then (D) has a unique weak solution.

(d) If all functions are sufficiently regular and u is a weak solution of the problem, then the expression Lu given above is well defined. Moreover, in this case a regular function u which is also in $H_0^1(\Omega)$ is a weak solution if and only if $Lu = f$ holds in the classical sense. Try to find regularity assumptions which are as weak as possible.

Exercise 6.22 (Composition of H^1-functions) Consider an open set $\Omega \subset \mathbb{R}^2$ and a straight line $G \subset \mathbb{R}^2$. Let $u \in C(\overline{\Omega})$ be differentiable in $\Omega \setminus G$, with bounded partial derivatives. Show that then $u \in H^1(\Omega)$.
Suggestion: After rotating and translating, one may assume that $G = \mathbb{R} \times \{0\}$.

Exercise 6.23 (Sobolev imbedding)

(a) Let $\Omega \subset \mathbb{R}^d$ be open and assume that $1 < p_1, p_2, r < \infty$ are such that $\frac{1}{p_1} + \frac{1}{p_2} = \frac{1}{r}$. Further let $f_1 \in L_{p_1}(\Omega_1)$ and $f_2 \in L_{p_2}(\Omega_2)$. Show that $f_1 \cdot f_2 \in L_r(\Omega)$.
    Suggestion: Use Hölder's inequality: if $1 < p < \infty$, $\frac{1}{p} + \frac{1}{p'} = 1$, $g \in L_p(\Omega)$, and $h \in L_{p'}(\Omega)$, then $g \cdot h \in L_1(\Omega)$.

(b) Let $d = 2$ and $g(x) = (1 + |x|)^{-1}$. Show that $g \in L_q(\mathbb{R}^2)$ for all $2 < q < \infty$.

(c) Show that $\hat{H}_1(\mathbb{R}^2) \subset L_p(\mathbb{R}^2)$ for all $1 < p \leq 2$, where $\hat{H}_1(\mathbb{R}^2) = \mathcal{F} H^1(\mathbb{R}^2)$ (see Theorem 6.46).

(d) Show that $H^1(\mathbb{R}^2) \subset L_q(\mathbb{R}^2)$ for all $q \in [2, \infty)$.
    Suggestion: Use that $\mathcal{F}^{-1}(L_2(\mathbb{R}^d) \cap L_p(\mathbb{R}^d)) \subset L_{p'}(\mathbb{R}^d)$ for $1 < p < 2$ (Hausdorff–Young theorem [54, Thm. 12.12]).

Exercise 6.24 Check the calculations in the proof of Theorem 6.59.

Chapter 7
Neumann and Robin boundary conditions

In this chapter we first study elliptic partial differential equations with Neumann boundary conditions $\frac{\partial u}{\partial \nu} = 0$. These require somewhat different techniques from the ones we saw when treating Dirichlet boundary conditions. We start by defining C^1-domains and proving Gauss's theorem. This is followed by analytic properties of the Sobolev space $H^1(\Omega)$ such as extension properties and the trace theorem, which we will draw on when studying boundary value problems with Neumann and Robin conditions.

Chapter overview

© The Author(s), under exclusive license to Springer Nature Switzerland AG 2023
W. Arendt, K. Urban, *Partial Differential Equations*, Graduate Texts in
Mathematics 294, https://doi.org/10.1007/978-3-031-13379-4_7

7.1 Gauss's theorem

The aim of this section is to generalize the Fundamental Theorem of Calculus to functions of more than one variable. We will also take the opportunity to define regularity properties of the boundary of a domain. Such properties play a decisive role when it comes to the regularity of (weak) solutions of the Poisson problem, among other partial differential equations.

Given $x, y \in \mathbb{R}^d$, as before we denote their inner product by $x \cdot y = x^T y$ and by $|x| = \sqrt{x \cdot x}$ the Euclidean norm in $\mathbb{R}^d$. Throughout this section, $\Omega \subset \mathbb{R}^d$ will be a bounded open set with boundary $\partial\Omega$.

Definition 7.1 Let $U \subset \mathbb{R}^d$ be open.
(a) We say that $U \cap \partial\Omega$ is a *normal C^1-graph* (for Ω) if there exist a function $g \in C^1(\mathbb{R}^{d-1})$ and constants $r > 0$ and $h > 0$ such that

$$U = \{(y, g(y) + s) : y \in \mathbb{R}^{d-1}, |y| < r, s \in \mathbb{R}, |s| < h\},$$

and such that for all $x = (y, g(y) + s) \in U$ we have (cf. Figure 7.1)

$$x \in \Omega \text{ if and only if} \qquad\qquad s > 0,$$
$$x \in \partial\Omega \text{ if and only if} \qquad\qquad s = 0,$$
$$x \notin \overline{\Omega} \text{ if and only if} \qquad\qquad s < 0.$$

(b) The set $U \cap \partial\Omega$ is called a *C^1-graph* (for Ω) if there exist an orthogonal $d \times d$ matrix B and a vector $b \in \mathbb{R}^d$ such that $\phi(U) \cap \partial(\phi(\Omega))$ is a normal C^1-graph for $\phi(\Omega)$, where $\phi(x) := B(x) + b$, $x \in \mathbb{R}^d$. Thus $U \cap \partial\Omega$ is a C^1-graph for Ω if $U \cap \partial\Omega$ is a normal C^1-graph for Ω with respect to a different Cartesian coordinate system.
(c) We say that Ω has a *boundary of class C^1*, or just *C^1-boundary* for short, if for every $z \in \partial\Omega$ there exists an open neighborhood U of z such that $U \cap \partial\Omega$ is a C^1-graph (for Ω). △

Remark 7.2 (C^k-boundary and Lipschitz boundary)
(a) Given $k \in \mathbb{N}$, we define a *C^k-graph* by requiring in part (a) of the above definition that $g \in C^k(\mathbb{R}^{d-1})$.
(b) We obtain a *Lipschitz graph* if we require that $g : \mathbb{R}^{d-1} \to \mathbb{R}$ be *Lipschitz continuous* (that is, that there exist $L \geq 0$ such that $|g(x) - g(y)| \leq L|x - y|$ for all $x, y \in \mathbb{R}^{d-1}$).
(c) As before, we speak of a *continuous graph* if g is continuous.
(d) We say that Ω has *C^k-boundary* (or *Lipschitz boundary, continuous boundary*, respectively) if for every $z \in \partial\Omega$ there exists an open neighborhood $U \subset \mathbb{R}^d$ of z such that $U \cap \partial\Omega$ is a *C^k-graph* (or a Lipschitz or continuous graph, respectively) for Ω.
(e) A *C^k-domain* is a bounded, connected open set with *C^k-boundary*. We define *continuous domains* and *Lipschitz domains* analogously.

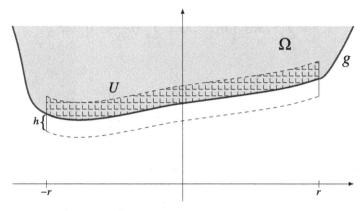

Fig. 7.1 A normal C^1-graph in $\mathbb{R}^2$.

(f) We say that Ω has C^∞-*boundary* if the boundary is C^k for all $k \in \mathbb{N}$.

(g) Finally, we call Ω a C^∞-*domain* if it is a C^k-domain for all $k \in \mathbb{N}$.   △

We have thus defined a hierarchy of classes of regularity. Note that since every function $g \in C^1(\mathbb{R}^{d-1})$ is Lipschitz continuous on every compact set in $\mathbb{R}^{d-1}$, a C^1-boundary is automatically also Lipschitz. Every hyperrectangle in $\mathbb{R}^d$ has Lipschitz but not C^1 boundary, as does every polygon in $\mathbb{R}^2$.

In what follows we suppose $\Omega \subset \mathbb{R}^d$ to be a bounded open set with C^1-boundary. We wish to formulate Gauss's theorem for such sets. To this end, we first need outer normal vectors. These can be defined easily via local graphs which parametrize the boundary (see (7.2)). However, in order to circumvent the need to prove the independence of the definition from the choice of parametrization, we prefer to give an intrinsic definition using tangent spaces.

Theorem 7.3 (Tangent space) *Let* $z \in \partial\Omega$. *A vector* $v \in \mathbb{R}^d$ *is called a* tangent vector *to* $\partial\Omega$ *at* z *if there exist* $\varepsilon > 0$ *and* $\psi \in C^1((-\varepsilon, \varepsilon), \mathbb{R}^d)$ *such that* $\psi(t) \in \partial\Omega$ *for all* $t \in (-\varepsilon, \varepsilon)$, $\psi(0) = z$ *and* $\psi'(0) = v$. *The set*

$$T_z := \{v \in \mathbb{R}^d : v \text{ is a tangent vector to } \partial\Omega \text{ at } z\}$$

is a $(d-1)$-*dimensional subspace of* $\mathbb{R}^d$. *It is called the* tangent space *to* $\partial\Omega$ *at* z.

Proof Let $U \cap \partial\Omega$ be a C^1-graph for Ω, where U is an open neighborhood of z. We may assume that the graph is normal and use the notation from Definition 7.1(a). We set $z' := (z_1, \ldots, z_{d-1})$, that is, $z = (z', z_d)$. Then $|z'| < r$ and $z_d = g(z')$. Let e_j be the jth canonical basis vector in $\mathbb{R}^{d-1}$ and $u_j := (e_j, \frac{\partial g}{\partial x_j}(z'))$, $j = 1, \ldots, d-1$. Then the vectors $u_1, \ldots, u_{d-1}$ are linearly independent. We will show that $T_z = \operatorname{span}\{u_1, \ldots, u_{d-1}\}$. To this end, let $u \in \operatorname{span}\{u_1, \ldots, u_{d-1}\}$; then there exist $\lambda_1, \ldots, \lambda_{d-1} \in \mathbb{R}$ such that

$$u = \sum_{j=1}^{d-1} \lambda_j\, u_j = \left(\lambda_1, \ldots, \lambda_{d-1}, \sum_{j=1}^{d-1} \lambda_j \frac{\partial g}{\partial x_j}(z')\right) \in \mathbb{R}^d.$$

Now set, for $t \in \mathbb{R}$,

$$\psi(t) := \left(z' + t \sum_{j=1}^{d-1} \lambda_j\, e_j, \quad g\!\left(z' + t \sum_{j=1}^{d-1} \lambda_j\, e_j\right)\right).$$

Then, by Definition 7.1(a) $\psi(t) \in \partial\Omega$ for all $|t| < \varepsilon$, if $\varepsilon > 0$ is chosen sufficiently small (since $s = 0$); and we have $\psi(0) = z$, as well as

$$\psi'(0) = \left(\lambda_1, \ldots, \lambda_{d-1}, \sum_{j=1}^{d-1} \lambda_j \frac{\partial g}{\partial x_j}(z')\right) = u.$$

This shows that $u \in T_z$.

Conversely, let $u \in T_z$. Then there exists a $\psi \in C^1((-\varepsilon, \varepsilon), \mathbb{R}^d)$ such that $\psi(t) \in \partial\Omega$ for all $|t| < \varepsilon$; and $\psi(0) = z$, $\psi'(0) = u$. Then $\psi_d(t) = g(\psi_1(t), \ldots, \psi_{d-1}(t))$ for $|t| < \varepsilon$, where $g \in C^1(\mathbb{R}^{d-1})$ is the function from Definition 7.1(a). By the chain rule,

$$\psi_d'(0) = \sum_{j=1}^{d-1} \psi_j'(0) \frac{\partial g}{\partial x_j}(z'). \tag{7.1}$$

Now set $\lambda_j := \psi_j'(0)$ for all $j = 1, \ldots, d-1$; then

$$u = \psi'(0) = \left(\lambda_1, \ldots, \lambda_{d-1}, \sum_{j=1}^{d-1} \lambda_j \frac{\partial g}{\partial x_j}(z')\right) \in \mathrm{span}\{u_1, \ldots, u_{d-1}\}.$$

We have thus shown that $T_z = \mathrm{span}\{u_1, \ldots, u_{d-1}\}$. □

We can now define the outer normal vector intrinsically, that is, without using parametrizations of the boundary. Since T_z is a $(d-1)$-dimensional subspace of $\mathbb{R}^d$, the orthogonal space $T_z^{\perp} := \{w \in \mathbb{R}^d : w \cdot v = 0 \quad \forall v \in T_z\}$ is one-dimensional.

Theorem 7.4 (Outer unit normal vector) *Let $\Omega \subset \mathbb{R}^d$ be a bounded open set with C^1-boundary and let $z \in \partial\Omega$. Then there exists a unique vector $v(z) \in \mathbb{R}^d$ such that*
(a) $|v(z)| = 1$;
(b) $v(z) \in T_z^{\perp}$;
(c) $\exists \varepsilon > 0$ such that $z + tv(z) \notin \overline{\Omega}$ if $0 < t < \varepsilon$ and $z + tv(z) \in \Omega$ if $-\varepsilon < t < 0$.
The mapping $v : \partial\Omega \to \mathbb{R}^d$, $z \mapsto v(z)$, is continuous.

The vector $v(z)$ is called the *outer unit normal vector* to Ω at z. We often abbreviate this to *outer unit normal* or even just *outer normal*.

Proof We keep the notation from the previous proof. Let

$$w := (\nabla g(z'), -1) = \left(\frac{\partial g}{\partial x_1}(z'), \ldots, \frac{\partial g}{\partial x_{d-1}}(z'), -1 \right).$$

1. We will show that $T_z = w^\perp$. To this end, let $b \in T_z$; then there exists a function $\psi \in C^1((-\varepsilon, \varepsilon), \mathbb{R}^d)$ such that $\psi(t) \in \partial\Omega$ for all $|t| < \varepsilon$, $b = \psi'(0)$ and $z = \psi(0)$. Hence, by (7.1), $\psi_d'(0) = \sum_{j=1}^{d-1} \psi_j'(0) \frac{\partial g}{\partial x_j}(z')$, that is, $b \cdot w = 0$. We have thus shown that $T_z \subset w^\perp$ holds. But since $\dim T_z = d - 1 = \dim w^\perp$, the equality $T_z = w^\perp$ follows.

2. Now we will show that $z + tw \notin \Omega$ if $0 \leq t < \varepsilon$ and $z + tw \in \Omega$ if $-\varepsilon < t < 0$, for suitable $\varepsilon > 0$. We set $w' = (w_1, \ldots, w_{d-1})$ and apply Taylor's theorem to g to obtain $g(z'+tw') = g(z')+\nabla g(z')\cdot tw'+o(t) = z_d+t|\nabla g(z')|^2+o(t)$, where $o(t)$ represents an error term with the property that $\lim_{t\to 0} \frac{o(t)}{t} = 0$. Thus $(z+tw)_d = z_d - t < g(z'+tw')$ four $0 < t < \varepsilon$ and $z_d - t > g(z' + tw')$ for $-\varepsilon < t < 0$, if $\varepsilon > 0$ is chosen small enough. This establishes the claim.

3. If we choose $v(z) := \frac{w}{|w|}$, then $v(z)$ satisfies the conditions of the theorem. Since $\dim T_z^\perp = 1$, $v(z)$ is unique (since its sign is determined by (c)). This concludes the proof. $\qquad\square$

We observe that

$$v(z) = \frac{(\nabla g(z'), -1)}{\sqrt{|\nabla g(z')|^2 + 1}} \tag{7.2}$$

for $z \in \partial\Omega \cap U$ if U is a normal C^1-graph for Ω (with the notation from Definition 7.1(a), and where $z = (z', z_d)$, $z' = (z_1, \ldots, z_{d-1})$).

We can now formulate Gauss's theorem, also sometimes known the Gauss–Ostrogradsky theorem, among other names. (Also note that we will derive corollaries of this theorem which are also often called Gauss's theorem in the literature.) We will denote by $C^1(\overline{\Omega})$ the set of all functions $u \in C^1(\Omega) \cap C(\overline{\Omega})$, such that for every $j \in \{1, \ldots, d\}$ the function $D_j u$ admits a continuous extension to $\overline{\Omega}$, where $D_j u = \frac{\partial u}{\partial x_j}$ denotes the partial derivative of u in the jth coordinate direction. We will also use the notation $D_j u$ for this continuous extension to $\overline{\Omega}$.

Theorem 7.5 (Gauss's theorem) *There exists a unique Borel measure σ on $\partial\Omega$ such that*

$$\int_\Omega (D_j u)(x)\, dx = \int_{\partial\Omega} u(z) v_j(z)\, d\sigma(z) \quad \text{for all } j = 1, \ldots, d \tag{7.3}$$

and for all $u \in C^1(\overline{\Omega})$. We call σ the surface measure *on $\partial\Omega$. Here $v \in C(\partial\Omega)$ is the outer unit normal vector with components $v(z) = (v_1(z), \ldots, v_d(z))$, $z \in \partial\Omega$.*

Gauss's theorem may be regarded as the Fundamental Theorem of Calculus for functions of two or more variables. To make the analogy clear we consider the special case of a bounded interval $\Omega = (a, b)$; then $\partial\Omega = \{a, b\}$. If we define $v(a) = -1, v(b) = 1$ and $\sigma(a) = \sigma(b) = 1$, then by the one-dimensional fundamental theorem applied to the function $u \in C^1[a, b]$, we have $\int_a^b u'(x)\, dx = u(b) - u(a) = \int_{\{a,b\}} v(z)u(z)\, d\sigma(z)$. We will prove Gauss's theorem in the next section. In practice, for the applications, it is sufficient to know the statement of the above Theorem 7.5 and its consequences; the actual construction of the measure is usually irrelevant.

We now turn to various corollaries of Gauss's theorem. Often the theorem is formulated for vector fields, that is, for functions $u \in C^1(\overline{\Omega}, \mathbb{R}^d)$, more precisely $u = (u_1, \ldots, u_d)$ with $u_j \in C^1(\overline{\Omega})$, $j = 1, \ldots, d$. Then we can define the *divergence* div u of u by

$$\operatorname{div} u(x) := \sum_{j=1}^d D_j u_j(x), \quad x \in \overline{\Omega},$$

so that div $u \in C(\overline{\Omega})$. We obtain the following equivalent formulation of Gauss's theorem, which is often known as the divergence theorem (as well as the Gauss–Ostrogradsky theorem, etc.). It is popular among physicists because it has a number of direct physical interpretations (see Section 1.6.1).

Corollary 7.6 (Divergence theorem) *Let $u \in C^1(\overline{\Omega}, \mathbb{R}^d)$, then*

$$\int_\Omega \operatorname{div} u(x)\, dx = \int_{\partial\Omega} u(z) \cdot v(z)\, d\sigma(z).$$

Proof By Theorem 7.5,

$$\int_\Omega \operatorname{div} u(x)\, dx = \sum_{j=1}^d \int_\Omega (D_j u_j)(x)\, dx = \sum_{j=1}^d \int_{\partial\Omega} u_j(z) v_j(z)\, d\sigma(z)$$

$$= \int_{\partial\Omega} u(z) \cdot v(z)\, d\sigma(z),$$

where $u(z) \cdot v(z)$ denotes the inner product in $\mathbb{R}^d$.   □

In what follows we will sometimes omit the variables of integration if they are clear from the context.

Corollary 7.7 (Integration by parts) *If $u, v \in C^1(\overline{\Omega})$, then*

$$\int_\Omega D_j u \cdot v\, dx = -\int_\Omega u\, D_j v\, dx + \int_{\partial\Omega} u v\, v_j\, d\sigma, \quad j = 1, \ldots, d.$$

Proof By Theorem 7.5 and the product rule,

$$
\int_{\partial\Omega} u(x)v(x)v_j\, d\sigma = \int_\Omega D_j(uv)(x)\, dx
$$

$$
= \int_\Omega (D_j u)(x)v(x)\, dx + \int_\Omega u(x)D_j v(x)\, dx,
$$

which proves the claim. $\square$

We define $C^2(\overline{\Omega}) := \{u \in C^1(\overline{\Omega}) : D_j u \in C^1(\overline{\Omega}),\ j = 1,\ldots,d\}$. Thus, if $u \in C^2(\overline{\Omega})$ then its derivatives $D_i D_j u$ are in $C(\overline{\Omega})$ for all $i, j = 1,\ldots,d$. For $u \in C^1(\overline{\Omega})$ the function $\frac{\partial u}{\partial v} : \partial\Omega \to \mathbb{R}$ given by

$$
\frac{\partial u}{\partial v}(z) := \nabla u(z) \cdot v(z) = \sum_{j=1}^d D_j u(z)v_j(z) \tag{7.4}
$$

is called the *(outer) normal derivative* of u at z; for $u \in C^1(\overline{\Omega})$ we always have $\frac{\partial u}{\partial v} \in C(\partial\Omega)$.

Corollary 7.8 (Green's identities) *If $u \in C^2(\overline{\Omega})$, then*

(a) $\displaystyle \int_\Omega (\Delta u)(x)\, dx = \int_{\partial\Omega} \frac{\partial u}{\partial v}(z)\, d\sigma(z),$

(b) $\displaystyle \int_\Omega (\Delta u)(x)v(x)\, dx + \int_\Omega \nabla u(x)\nabla v(x)\, dx = \int_{\partial\Omega} \frac{\partial u}{\partial v}(z)v(z)\, d\sigma(z),\ v \in C^1(\overline{\Omega}),$

(c) $\displaystyle \int_\Omega (v\Delta u - u\Delta v)\, dx = \int_{\partial\Omega} \left(\frac{\partial u}{\partial v}v - u\frac{\partial v}{\partial v} \right) d\sigma(z),\ v \in C^2(\overline{\Omega}).$

Part (b) is generally known as *Green's first identity*, (c) is *Green's second identity*.

Proof We start with (b). By Theorem 7.5 we have

$$
\int_\Omega D_j u\, D_j v\, dx = -\int_\Omega (D_j^2 u)v\, dx + \int_\Omega D_j(D_j u\, v)\, dx
$$

$$
= -\int_\Omega (D_j^2 u)v\, dx + \int_{\partial\Omega} D_j u(z)v(z)v_j(z)\, d\sigma(z).
$$

Summing over $j = 1,\ldots,d$ yields (b). For $v \equiv 1$, (b) reduces to (a). If in (b) one swaps the functions u and v and subtracts the resulting identity from (b), then one obtains (c). $\square$

7.2 Proof of Gauss's theorem

In order to prove Gauss's theorem, it will be useful first to characterize measures as positive linear forms.

Definition 7.9 Let $K \subset \mathbb{R}^d$ be a compact set and $C(K)$ the vector space of continuous real-valued functions on K. A *positive linear form* on $C(K)$ is a linear mapping $\varphi : C(K) \to \mathbb{R}$ such that $\varphi(f) \geq 0$ for all $f \geq 0$. Here we write $f \geq 0$ if $f \in C(K)$ satisfies $f(x) \geq 0$ for all $x \in K$.   △

Every positive linear form is continuous with respect to the supremum norm on $C(K)$ (see Exercise 7.1). If μ is a Borel measure on K, then

$$\varphi(f) := \int_K f(x) \, d\mu(x), \quad f \in C(K) \tag{7.5}$$

defines a positive linear form φ on $C(K)$. The following theorem due to Riesz establishes the converse: every positive linear form admits such a representation.

Theorem 7.10 (Riesz representation theorem) *Let $K \subset \mathbb{R}^d$ be compact and let $\varphi : C(K) \to \mathbb{R}$ be a positive linear form on $C(K)$. Then there exists a unique Borel measure μ on K such that (7.5) holds.*

Proof For the proof we refer to [54, Theorems 2.14 and 2.18].   □

We first show that the surface measure on Ω is uniquely determined by the conditions in Gauss's theorem. Throughout this section, Ω will be a bounded open set with C^1-boundary and $\nu = (\nu_1, \ldots, \nu_d)^T \in C(\partial\Omega, \mathbb{R}^d)$ its outer unit normal.

Lemma 7.11 (Uniqueness of surface measures) *Let $\varphi : C(\partial\Omega) \to \mathbb{R}$ be a continuous linear form which satisfies*

$$\varphi(\nu_j u_{|\partial\Omega}) = 0 \quad \text{for all } u \in C^1(\overline{\Omega}), \ j = 1, \ldots, d. \tag{7.6}$$

Then $\varphi = 0$.

Proof The set $\mathcal{A} := \{u_{|\partial\Omega} : u \in C^1(\mathbb{R}^d)\}$ is a subalgebra of $C(\partial\Omega)$ which contains all constant functions and which separates points in $\partial\Omega$. Hence, by the Stone–Weierstrass theorem, Theorem A.5, $\mathcal{A}$ is dense in $C(\partial\Omega)$. It follows that $\varphi(g\nu_j) = 0$ for all $g \in C(\partial\Omega)$, $j = 1, \ldots, d$. By replacing g by $g\nu_j \in C(\partial\Omega)$, we see that $\varphi(g\nu_j^2) = 0$ for all $g \in C(\partial\Omega)$. Since $\sum_{j=1}^d \nu_j(z)^2 = 1$ for all $z \in \partial\Omega$, it follows that $\varphi(g) = 0$ for all $g \in C(\partial\Omega)$.   □

The uniqueness of the surface measure as described in Theorem 7.5, now comes about as follows: let σ_1, σ_2 be two Borel measures on $\partial\Omega$ which both satisfy (7.3). Then $\varphi(f) := \int_{\partial\Omega} f(z) \, d\sigma_1(z) - \int_{\partial\Omega} f(z) \, d\sigma_2(z)$ defines a continuous linear form on $C(\partial\Omega)$ for which (7.6) holds; thus $\varphi = 0$ by Lemma 7.11. Uniqueness in the Riesz representation theorem now implies that $\sigma_1 = \sigma_2$.

For the proof of existence we use what is known as a partition of unity. This is a construction which in many cases permits one to derive global properties from local ones. Here, it will allow us to construct a positive linear form on $C(\partial\Omega)$ starting with ones which are only defined for functions with support in a small set.

Theorem 7.12 (Partition of unity) *Let $K \subset \mathbb{R}^d$ be a compact set and let $U_m \subset \mathbb{R}^d$, $m = 0, 1, \ldots, M$ be open sets such that $K \subset \bigcup_{m=0}^M U_m$. Then there exist $\eta_m \in \mathcal{D}(\mathbb{R}^d)$ such that*

(a) $0 \leq \eta_m(x) \leq 1$, $x \in \mathbb{R}^d$, $m = 0, \ldots, M$;
(b) supp $\eta_m(x) \subset U_m$, $m = 0, \ldots, M$;
(c) $\sum_{m=0}^M \eta_m(x) = 1$ for all $x \in K$.

We call the collection of functions $\eta_0, \ldots, \eta_M$ a *partition of unity* of K subordinate to the open cover $U_0, \ldots, U_M$.

Proof For each $x \in K$ there exists at least one m such that $x \in U_m$. Hence we can find an $r > 0$ such that $\bar{B}(x, r) \subset U_m$. Since K is compact, it is covered a finite collection of these balls, say $B_1, \ldots, B_k$. Let $K_m := \bigcup_{\bar{B}_\ell \subset U_m} \bar{B}_\ell$. Then K_m is compact and $K_m \subset U_m$, $m = 0, \ldots, M$; moreover, $K \subset \bigcup_{m=0}^M K_m$. Now choose, for each $m \in \{0, 1, \ldots, M\}$, a function $\psi_m \in \mathcal{D}(\mathbb{R}^d)$ such that $0 \leq \psi_m \leq 1$, supp $\psi_m \subset U_m$ and $\psi_m(x) = 1$ for all $x \in K_m$ (see Lemma 6.7), and set

$$\eta_0 := \psi_0,$$
$$\eta_1 := (1 - \psi_0)\psi_1,$$
$$\eta_2 := (1 - \psi_0)(1 - \psi_1)\psi_2,$$
$$\vdots$$
$$\eta_M := (1 - \psi_0)(1 - \psi_1)\cdots(1 - \psi_{M-1})\psi_M.$$

Then we have $\sum_{m=0}^n \eta_m + (1 - \psi_0)(1 - \psi_1)\cdots(1 - \psi_n) = 1$ for all $n \in \{0, 1, \ldots, M\}$, as is easy to prove by induction. In particular,

$$\sum_{m=0}^M \eta_m + (1 - \psi_0)(1 - \psi_1)\cdots(1 - \psi_M) = 1,$$

whence $0 \leq \sum_{m=0}^M \eta_m = 1$. Since for each $x \in K$ there exists an $m \in \{0, 1, \ldots, M\}$ such that $\psi_m(x) = 1$, we have $\sum_{m=0}^M \eta_m(x) = 1$ for all $x \in K$. $\square$

We now construct our surface measure. Let $U \subset \mathbb{R}^d$ be open with $U \cap \partial\Omega \neq \emptyset$. We set

$$C_c(U \cap \partial\Omega) := \{u \in C(\partial\Omega) : \exists K \subset U \text{ compact}, u(z) = 0 \text{ for } z \in \partial\Omega \setminus K\};$$

then $C_c(U \cap \partial\Omega)$ is a vector space. By

$$C_c(U \cap \partial\Omega)'_+ := \{\varphi : C_c(U \cap \partial\Omega) \to \mathbb{R} \text{ linear},$$
$$\text{such that } \varphi(f) \geq 0 \text{ for } f \geq 0\}$$

we denote the set of *positive linear forms* on $C_c(U \cap \partial\Omega)$. Now let $U \cap \partial\Omega$ be a normal C^1-graph for Ω. Using the notation from Definition 7.1(a), we set

$$\varphi(f) := \int_{|y|<r} f(y, g(y))\sqrt{1 + |\nabla g(y)|^2}\, dy \qquad (7.7)$$

for all $f \in C_c(U \cap \partial\Omega)$. Here we integrate over the $(d-1)$-dimensional ball of radius r. Observe that $(y, g(y)) \in \partial\Omega \cap U$ for all $y \in \mathbb{R}^{d-1}$ with $|y| < r$, and so $\varphi \in C_c(U \cap \partial\Omega)'_+$. With this notation we have:

Lemma 7.13 *Suppose $u \in C^1(\overline{\Omega})$ with supp $u \subset U$. Then*

$$\int_\Omega D_j u(x)\, dx = \varphi(u_{|\partial\Omega} \nu_j), \quad j = 1, \ldots, d. \qquad (7.8)$$

Observe that $u_{|\partial\Omega} \in C_c(U \cap \partial\Omega)$ since supp $u \subset U$.

Proof Let $B'_r := \{y \in \mathbb{R}^{d-1} : |y| < r\}$ and $h > 0$ be as in Definition 7.1. The mapping $\phi : B'_r \times (0, h) \to U \cap \Omega$ defined by $\phi(y, s) = (y, g(y)+s)$ is a diffeomorphism since $g \in C^1(\mathbb{R}^{d-1})$, and we have

$$\frac{\partial\phi_i}{\partial y_j}(y, s) = \frac{\partial}{\partial y_j} y_i = \delta_{i,j}, \quad 1 \le i \le d-1, \quad 1 \le j \le d-1,$$

$$\frac{\partial\phi_d}{\partial y_j}(y, s) = \frac{\partial}{\partial y_j}(g(y) + s) = (D_j g)(y), \quad 1 \le j \le d-1,$$

$$\frac{\partial\phi_i}{\partial s}(y, s) = 0, \quad 1 \le i \le d-1, \quad \frac{\partial\phi_d}{\partial s}(y, s) = 1.$$

Hence the Jacobian matrix of ϕ has the form

$$D\phi(y, s) = \begin{bmatrix} 1 & & 0 & 0 \\ & \ddots & & \vdots \\ & & \ddots & \vdots \\ 0 & & 1 & 0 \\ \hline D_1 g(y) & \cdots \cdots & D_{d-1}g(y) & 1 \end{bmatrix} \in \mathbb{R}^{d \times d},$$

with $\det D\phi(y, s) = 1$. Hence, if $f : \mathbb{R}^d \to \mathbb{R}$ is any continuous function, then

$$\int_{B'_r} \int_0^h f(y, g(y) + s)\, ds\, dy = \int_{|y|<r} \int_0^h f(y, g(y) + s)\, ds\, dy$$

$$= \int_{U \cap \Omega} f(x)\, dx. \qquad (7.9)$$

Since supp $u \subset U$, it follows that for any $1 \le j \le d-1$

$$\int_\Omega D_j u(x)\, dx = \int_{\Omega \cap U} D_j u(x)\, dx = \int_{|y|<r} \int_0^h (D_j u)(y, g(y) + s)\, ds\, dy.$$

By the chain rule,

$$\frac{\partial}{\partial y_j} u(y, g(y) + s) = (D_j u)(y, g(y) + s) + (D_d u)(y, g(y) + s)(D_j g)(y),$$

and so

$$\int_\Omega D_j u(x)\, dx = \int_{|y|<r} \int_0^h \left\{ \frac{\partial}{\partial y_j} u(y, g(y) + s) \right.$$

$$\left. - (D_d u)(y, g(y) + s)(D_j g)(y) \right\} ds\, dy.$$

For the first term we use the Fundamental Theorem of Calculus in one variable (namely y_j) and obtain

$$\int_{|y|<r} \int_0^h \frac{\partial}{\partial y_j} u(y, g(y) + s)\, ds\, dy = \int_0^h \int_{|y|<r} \frac{\partial}{\partial y_j} u(y, g(y) + s)\, dy\, ds = 0,$$

since supp $u \subset U = \{(y, g(y) + s) : |y| < r, |s| < h\}$. For the second term we again use the fundamental theorem in one variable (this time s) to obtain

$$-\int_{|y|<r} \int_0^h (D_d u)(y, g(y) + s)(D_j g)(y)\, ds\, dy = \int_{|y|<r} u(y, g(y))(D_j g)(y)\, dy$$

and thus $\int_\Omega D_j u(x)\, dx = \int_{|y|<r} u(y, g(y))(D_j g)(y)\, dy$. By (7.2), for all $1 \le j \le d-1$ we have

$$v_j(y, g(y)) = \frac{D_j g(y)}{\sqrt{1 + |\nabla g(y)|^2}}$$

and hence (cf. (7.7))

$$\int_\Omega D_j u(x)\, dx = \int_{|y|<r} u(y, g(y)) v_j(y, g(y)) \sqrt{1 + |\nabla g(y)|^2}\, dy = \varphi(v_j u_{|\partial\Omega}).$$

When $j = d$ we have $\frac{\partial}{\partial s} u(y, g(y) + s) = (D_d u)(y, g(y) + s)$. Thus, using (7.9) and similar arguments to the first case we obtain

$$\int_\Omega D_d u(x)\, dx = \int_{|y|<r} \int_0^h (D_d u)(y, g(y) + s)\, ds\, dy$$

$$= \int_{|y|<r} \int_0^h \frac{\partial}{\partial s} u(y, g(y) + s)\, ds\, dy = -\int_{|y|<r} u(y, g(y))\, dy$$

$$= \int_{|y|<r} u(y, g(y)) v_d(y, g(y)) \sqrt{1 + |\nabla g(y)|^2}\, dy = \varphi(u_{|\partial\Omega} v_d)$$

since by (7.2) $v_d(y, g(y)) = \frac{-1}{\sqrt{1 + |\nabla g(y)|^2}}$. $\qquad\qquad\square$

The statement remains true if instead of normal C^1-graphs we consider just C^1-graphs. For the sake of completeness we will give the argument.

Let $U \subset \mathbb{R}^d$ be open and let $U \cap \partial\Omega$ be a C^1-graph for Ω, that is, we suppose that there exist an orthogonal matrix B and a vector $b \in \mathbb{R}^d$ such that $\partial\widetilde{\Omega} \cap \widetilde{U}$ is a normal C^1-graph, where $\phi(x) = Bx + b$ and $\widetilde{\Omega} := \phi(\Omega)$, $\widetilde{U} := \phi(U)$. We may assume that $\det B = 1$ (otherwise we exchange two variables). The outer unit normal to $\widetilde{\Omega}$ at $\phi(z)$ is given by $\tilde{\nu}(\phi(z)) = B\nu(z)$, for all $z \in \partial\Omega$. Let $\tilde{\varphi}$ be the positive linear form on $C_c(\widetilde{U} \cap \partial\widetilde{\Omega})$ which we constructed in Lemma 7.13. Now define $\varphi \in C_c(U \cap \partial\Omega)'_+$ by

$$\varphi(f) := \tilde{\varphi}(f \circ \phi^{-1}), \quad f \in C_c(U \cap \partial\Omega). \tag{7.10}$$

Then we have the following analog of Lemma 7.13.

Lemma 7.14 Suppose $u \in C^1(\overline{\Omega})$ with supp $u \subset U$. Then

$$\int_\Omega D_j u(x)\, dx = \varphi(u_{|\partial\Omega} \nu_j), \quad j = 1, \ldots, d. \tag{7.11}$$

Proof Let $B = (b_{kj})_{k,j=1,\ldots,d}$, $\tilde{u} = u \circ \phi^{-1}$. Then $\tilde{u} \in C^1(\overline{\widetilde{\Omega}})$, and by Lemma 7.13,

$$\int_\Omega D_j u(x)\, dx = \int_\Omega \frac{\partial}{\partial x_j}(\tilde{u} \circ \phi)(x)\, dx = \int_\Omega \sum_{k=1}^d (D_k \tilde{u})(\phi(x)) b_{kj}\, dx$$

$$= \int_{\widetilde{\Omega}} \sum_{k=1}^d D_k \tilde{u}(y) b_{kj}\, dy = \sum_{k=1}^d b_{kj} \tilde{\varphi}(\tilde{u}_{|\partial\widetilde{\Omega}} \tilde{\nu}_k)$$

$$= \varphi((u \circ \phi^{-1})_{|\partial\widetilde{\Omega}}(B^{-1}\tilde{\nu})_j) = \varphi(u_{|\Gamma} \nu_j),$$

which proves the claim. $\qquad\square$

We can now use partitions of unity to combine our positive linear forms in order to obtain one on $C(\partial\Omega)$ which yields the surface measure on $\partial\Omega$ via the Riesz representation theorem.

Proof (of Theorem 7.5) For each point $z \in \partial\Omega$ we can find an open neighborhood U of z such that $U \cap \partial\Omega$ is a C^1-graph. Since $\partial\Omega$ is compact, we can thus choose open sets $U_1, \ldots, U_M$ such that $\partial\Omega \subset \bigcup_{m=1}^M U_m$ and such that $U_m \cap \partial\Omega$ is a C^1-graph for Ω. Now choose an open set $U_0 \subset\subset \Omega$ such that $\overline{\Omega} \subset \bigcup_{m=0}^M U_m$. Let $(\eta_m)_{m=0,\ldots,M}$ be a partition of unity subordinate to this covering. Then $\eta_m \in \mathcal{D}(\mathbb{R}^d)$, $0 \le \eta_m \le 1$, supp$\eta_m \subset U_m$ for $m = 0, \ldots, M$ and $\sum_{j=0}^M \eta_j(x) = 1$ for all $x \in \overline{\Omega}$. By Lemma 7.14, for all $j = 1, \ldots, d$, $m = 1, \ldots, M$ and $u \in C^1(\overline{\Omega})$ with supp $u \subset U_m$ there exists a function $\varphi_m \in C_c(U_m \cap \partial\Omega)'_+$ such that

$$\int_\Omega D_j u(x)\, dx = \varphi_m(\nu_j u_{|\partial\Omega \cap U_m}). \tag{7.12}$$

We now define a positive linear form φ on $C(\partial\Omega)$ by

$$\varphi(f) := \sum_{m=1}^{M} \varphi_m(\eta_m f), \quad f \in C(\partial\Omega). \tag{7.13}$$

By the Riesz representation theorem there exists a Borel measure σ on $\partial\Omega$ such that $\varphi(f) = \int_{\partial\Omega} f(z)\,d\sigma(z)$ for all $f \in C(\partial\Omega)$. Let $u \in C^1(\overline{\Omega})$; then $u\eta_0 \in C_c^1(\Omega)$. By Lemma 6.12, it follows that $\int_\Omega D_j(u\eta_0)(x)\,dx = 0$ for all $j \in \{1,\ldots,d\}$ (see Exercise 6.2). Since $u\eta_m \in C_c(U_m \cap \partial\Omega)$ for all $m \in \{1,\ldots,M\}$, it follows by Lemma 7.14 that $\int_\Omega D_j(\eta_m u)(x)\,dx = \varphi_m(\eta_m u_{|\partial\Omega} v_j)$, $j = 1,\ldots,d$. Hence

$$\int_\Omega D_j u(x)\,dx = \int_\Omega D_j \left(\sum_{m=0}^{M} \eta_m u \right)(x)\,dx = \sum_{m=0}^{M} \int_\Omega D_j(\eta_m u)(x)\,dx$$

$$= \sum_{m=1}^{M} \int_\Omega D_j(\eta_m u)(x)\,dx = \sum_{m=1}^{M} \varphi_m(\eta_m u_{|\partial\Omega} v_j)$$

$$= \varphi(u_{|\partial\Omega} v_j) = \int_{\partial\Omega} u_{|\partial\Omega}(z) v_j(z)\,d\sigma,$$

and our Borel measure satisfies the conditions of Gauss's theorem. We have already proved uniqueness; in particular, the above construction is independent both of the choice of graphs and the partition of unity. □

This completes the proof of Gauss's theorem. The proof also gives us the following local description of the surface measure σ on $\partial\Omega$.

Theorem 7.15 *Let $U \cap \partial\Omega$ be a normal C^1-graph for Ω. Then, keeping the notation from Definition 7.1,*

$$\int_{\partial\Omega} f(z)\,d\sigma(z) = \int_{|y|<r} f(y, g(y))\sqrt{1 + |\nabla g(y)|^2}\,dy \tag{7.14}$$

for every continuous function $f : \partial\Omega \cap U \to \mathbb{R}$ such that $\mathrm{supp}\, f \subset U$.

Corollary 7.16 *Let $z \in \partial\Omega$ and $r > 0$. Then $\sigma(B(z,r) \cap \partial\Omega) > 0$.*

We will always take the surface measure on $\partial\Omega$. The space $L_2(\partial\Omega)$ is then the space of Borel-measurable functions $b : \partial\Omega \to \mathbb{R}$ such that $\int_{\partial\Omega} |b(z)|^2\,d\sigma(z) < \infty$. Here we identify functions whenever they agree σ-almost everywhere. We next note the following density properties.

Theorem 7.17 *The space $F := \{\varphi_{|\partial\Omega} : \varphi \in \mathcal{D}(\mathbb{R}^d)\}$ is dense in $L_2(\partial\Omega)$.*

Proof By the Stone–Weierstrass theorem (Theorem A.5), F is dense in $C(\partial\Omega)$ with respect to the supremum norm $\|\cdot\|_\infty$. It also follows from general measure theory that $C(\partial\Omega)$ is dense in $L_2(\partial\Omega)$ (see for example [54, Theorem 3.14]). □

7.3 The extension property

We saw earlier that a function $u \in H_0^1(\Omega)$, when extended by zero, becomes a function in $H^1(\mathbb{R}^d)$. However, doing this for functions in $H^1(\Omega)$ does not generally yield weakly differentiable functions. But there is another possibility for extending $H^1(\Omega)$-functions, as long as the boundary satisfies certain regularity conditions. We start by giving the desired property a name.

Definition 7.18 An open set $\Omega \subset \mathbb{R}^d$ is said to have the *extension property* if for every $u \in H^1(\Omega)$ there exists a $w \in H^1(\mathbb{R}^d)$ such that $w_{|\Omega} = u$.   △

A simple criterion for such sets Ω is as follows.

Theorem 7.19 *Let $\Omega \subset \mathbb{R}^d$ be a bounded open set with Lipschitz boundary. Then Ω has the extension property.*

In particular Ω has the extension property if the boundary is C^1. We refer to [2, A8.12], [28, Sec 5.5] or [18] for a complete proof and instead sketch the idea of the proof here.

Proof (Idea) We shall assume for simplicity that Ω has C^1-boundary. As in Section 7.2, by using a partition of unity we can reduce the problem to the exercise of constructing a local extension. Now let $g \in C^1(\mathbb{R}^{d-1})$, $h > 0$, $r > 0$, $U := \{(y, g(y) + s) : y \in \mathbb{R}^{d-1}, |y| < r, |s| < h\}$, and $U_+ := \{(y, g(y) + s) \in U, s > 0\}$, where $\Omega \cap U = U_+$. Also let $u \in H^1(\Omega)$ satisfy $u(x) = 0$ for all $x \in \Omega \setminus K$, where $K \subset U$ is compact. We define $\tilde{u} : U \to \mathbb{R}$ by

$$\tilde{u}(y, g(y) + s) := \begin{cases} u(y, g(y) + s), & s > 0, \\ u(y, g(y) - s), & s \le 0, \end{cases}$$

and extend $\tilde{u}$ to $\mathbb{R}^d$ by 0. Then it can be proved that $\tilde{u} \in H^1(\mathbb{R}^d)$, see Exercise 7.10. For the case of Lipschitz boundary we refer to [2, A8.12, page 275].   □

If Ω has the extension property, then we can also extend in a linear and continuous fashion.

Theorem 7.20 *Let Ω be a bounded open set in $\mathbb{R}^d$ which has the extension property, and let $U \subset \mathbb{R}^d$ be open, such that $\overline{\Omega} \subset U$. Then there exists a continuous linear operator $E : H^1(\Omega) \to H_0^1(U)$ such that $(Eu)_{|\Omega} = u$ for all $u \in H^1(\Omega)$.*

We call E an *extension operator*.

Proof Let $T : H^1(\mathbb{R}^d) \to H^1(\Omega)$ be given by $Tv = v_{|\Omega}$; then T is linear, continuous and by assumption surjective, and $H^1(\mathbb{R}^d) = \text{Ker } T \oplus (\text{Ker } T)^\perp$, where the orthogonal complement is taken in the Hilbert space $H^1(\mathbb{R}^d)$. It follows that $T_{|(\text{Ker } T)^\perp} : (\text{Ker } T)^\perp \to H^1(\Omega)$ is an isomorphism, whose inverse $S : H^1(\Omega) \to (\text{Ker } T)^\perp \subset H^1(\mathbb{R}^d)$ is an extension operator for $U = \mathbb{R}^d$. Observe that S is continuous by Theorem A.4. Now let U be an arbitrary open set such that $\overline{\Omega} \subset U$ and choose a test function $\psi \in \mathcal{D}(\mathbb{R}^d)$ such that supp $\psi \subset U$ and $\psi(x) = 1$ for all $x \in \overline{\Omega}$ (see Lemma 6.7). We now define $Eu := \psi Su$; then $Eu \in H^1_c(U) \subset H^1_0(U)$, see Theorem 6.29. Clearly, $(Eu)_{|\Omega} = u$ for all $u \in H^1(\Omega)$. $\qquad\square$

An important consequence of the extension property is the density of test functions on $\mathbb{R}^d$ in $H^1(\Omega)$. This is to be compared with the weaker density statement of Theorem 6.16, which is valid for every open set in $\mathbb{R}^d$.

Theorem 7.21 *Let Ω be a bounded open set in $\mathbb{R}^d$ with the extension property and let $U \subset \mathbb{R}^d$ be an open set such that $\overline{\Omega} \subset U$. Then $\{\varphi_{|\Omega} : \varphi \in \mathcal{D}(U)\}$ is dense in $H^1(\Omega)$.*

Proof Let $u \in H^1(\Omega)$; then $Eu \in H^1_0(U)$, where E is the extension operator from Theorem 7.20. Hence there exist $\varphi_n \in \mathcal{D}(U)$ such that $\varphi_n \to Eu$ in $H^1(U)$. Since the extension mapping is continuous, it follows that $\varphi_{n|\Omega} \to u = (Eu)_{|\Omega}$. $\qquad\square$

A further consequence is the compactness of the imbedding $H^1(\Omega) \hookrightarrow L_2(\Omega)$.

Theorem 7.22 *Let $\Omega \subset \mathbb{R}^d$ be a bounded open set with the extension property. Then the imbedding of $H^1(\Omega)$ in $L^2(\Omega)$ is compact, as is the imbedding $H^2(\Omega) \hookrightarrow H^1(\Omega)$.*

Proof 1. Let U be bounded and open with $\overline{\Omega} \subset U$. We know by the Rellich imbedding theorem that the imbedding $H^1_0(U) \hookrightarrow L_2(U)$ is compact (Theorem 6.54). Let $E : H^1(\Omega) \to H^1_0(U)$ be an extension operator and suppose $u_n \in H^1(\Omega)$ form a bounded sequence, say, $\|u_n\|_{H^1(\Omega)} \le c$. Then $\|Eu_n\|_{H^1_0(U)} \le c\|E\|$, and so there exists a subsequence $(u_{n_k})_{k \in \mathbb{N}}$ such that Eu_{n_k} converges in $L_2(U)$. Hence $u_{n_k} = (Eu_{n_k})_{|\Omega}$ converges in $L_2(\Omega)$.

2. To prove the second assertion we use Theorem 4.39. Let $u_n \rightharpoonup u$ in $H^2(\Omega)$; then $D_j u_n \rightharpoonup D_j u$ in $H^1(\Omega)$ for all $j \in \{1, \ldots, d\}$. Since $H^1(\Omega) \hookrightarrow L_2(\Omega)$ is compact, we deduce that $D_j u_n \to D_j u$ in $L_2(\Omega)$. Since $u_n \rightharpoonup u$ in $H^1(\Omega)$, we have that $u_n \to u$ in $L_2(\Omega)$ as well. Hence $u_n \to u$ in $H^1(\Omega)$. It now follows from Theorem 4.39 that the imbedding $H^2(\Omega) \hookrightarrow H^1(\Omega)$ is compact. $\qquad\square$

For Lipschitz domains a more general extension property holds, which also allows a similar conclusion for the higher derivatives. To show this requires a different method from the simple reflection technique we sketched in the context of Theorem 7.19, which no longer works here.

Theorem 7.23 (General extension property) *Let Ω be a Lipschitz domain. Then there exists a continuous linear operator $E : H^1(\Omega) \to H^1(\mathbb{R}^d)$ such that*
(a) $(Eu)_{|\Omega} = u$ for all $u \in H^1(\Omega)$;
(b) $EH^k(\Omega) \subset H^k(\mathbb{R}^d)$ for all $k \in \mathbb{N}$.

It follows directly from the closed graph theorem that the restriction E_k of E to $H^k(\Omega)$ is a continuous linear operator from $H^k(\Omega)$ to $H^k(\mathbb{R}^d)$. We refer to [56, Chapter 6] for the proof of Theorem 7.23.

This more general extension theorem allows us to transfer the imbedding theorems which hold for $\mathbb{R}^d$ to Ω.

Corollary 7.24 *Let $\Omega \subset \mathbb{R}^d$ be a Lipschitz domain, and suppose $k \in \mathbb{N}$ satisfies $k > \frac{d}{2}$. Then $H^{k+m}(\Omega) \subset C^m(\overline{\Omega})$ for all $m \in \mathbb{N}_0$, where $C^0(\overline{\Omega}) = C(\overline{\Omega})$.*

Proof We know from Corollary 6.50 that $H^{k+m}(\mathbb{R}^d) \subset C^m(\mathbb{R}^d)$. Thus if $u \in H^{k+m}(\Omega)$ then $Eu \in C^m(\mathbb{R}^d)$. The claim then follows since $u = (Eu)|_\Omega$.   □

We wish to formulate the case $d = 2$ explicity, as we will need it in this form in Chapter 9.

Corollary 7.25 *Let $\Omega \subset \mathbb{R}^2$ be a Lipschitz domain (for example a polygon). Then $H^2(\Omega) \subset C(\overline{\Omega})$.*

The following domain has continuous boundary but does *not* enjoy the extension property.

Example 7.26 Given $\alpha > 1$, let $\Omega = \{(x, y) \in \mathbb{R}^2 : 0 < x < 1, |y| < x^\alpha\}$ (see Figure 7.2). Then Ω does not have the extension property; we refer to Exercise 7.7 for the proof. In the limit case $\alpha = 1$, Ω becomes a triangle, which of course does have the extension property. For $\alpha > 1$, the Lipschitz condition is violated only in the vicinity of the point $(0, 0)$.   △

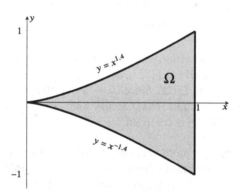

Fig. 7.2 Ω is a domain without the extension property (the figure corresponds to $\alpha = 1.4$).

However, both the compactness of the imbedding of $H^1(\Omega)$ in $L_2(\Omega)$ and the density statement of Theorem 7.21 continue to hold for the set Ω of Example 7.26. In fact, these two properties require only the mildest regularity conditions on the boundary which we have defined; for the proof of this statement we refer to [26, V. Theorem 4.17, page 267 and V. Theorem 4.7, page 248].

Theorem 7.27 *Let $\Omega \subset \mathbb{R}^d$ be a bounded open set with continuous boundary. Then the imbedding of $H^1(\Omega)$ in $L_2(\Omega)$ is compact; moreover, the space $\{\varphi_{|\Omega} : \varphi \in \mathcal{D}(\mathbb{R}^d)\}$ is dense in $H^1(\Omega)$.*

A consequence of the compactness of the imbedding is the following, second, Poincaré inequality. While the first (Theorem 6.32) holds for functions in $H_0^1(\Omega)$, in the second we consider functions in $H^1(\Omega)$ with mean value 0.

Theorem 7.28 (Second Poincaré inequality) *Let $\Omega \subset \mathbb{R}^d$ be a bounded domain with continuous boundary. Then there exists a constant $c > 0$ such that*

$$\int_\Omega |u(x)|^2 \, dx \leq c \int_\Omega |\nabla u(x)|^2 \, dx \tag{7.15}$$

for all $u \in H^1(\Omega)$ such that $\int_\Omega u(x) \, dx = 0$.

The subspace

$$H_m^1(\Omega) := \left\{ u \in H^1(\Omega) : \int_\Omega u(x) \, dx = 0 \right\} \tag{7.16}$$

of $H^1(\Omega)$ is closed and thus itself a Hilbert space. The second Poincaré inequality states that

$$|u|_{H^1(\Omega)} := \left(\int_\Omega |\nabla u(x)|^2 \, dx \right)^{\frac{1}{2}} \tag{7.17}$$

defines an equivalent norm on $H_m^1(\Omega)$, as long as Ω is a bounded domain with continuous boundary. On $H^1(\Omega)$ the quantity $| \cdot |_{H^1(\Omega)}$ is merely a semi-norm since $|1_\Omega|_{H^1(\Omega)} = 0$.

Proof (of Theorem 7.28) Suppose that the assertion is false. Then we can find functions $u_n \in H^1(\Omega)$ such that

$$\int_\Omega u_n(x) \, dx = 0 \quad \text{and} \quad \lim_{n \to \infty} \int_\Omega |\nabla u_n(x)|^2 \, dx = 0$$

but $\|u_n\|_{L_2(\Omega)} = 1$ for all $n \in \mathbb{N}$. Since the imbedding $H^1(\Omega) \hookrightarrow L_2(\Omega)$ is compact by Theorem 7.27, there exists a subsequence $(u_{n_k})_{k \in \mathbb{N}}$ such that $u := \lim_{k \to \infty} u_{n_k}$ exists as a function in $L_2(\Omega)$, and $\|u\|_{L_2(\Omega)} = 1$. Now let $\varphi \in \mathcal{D}(\Omega)$; then for all $j = 1, \ldots, d$

$$\int_\Omega u(x) D_j \varphi(x) \, dx = \lim_{k \to \infty} \int_\Omega u_{n_k}(x) D_j \varphi(x) \, dx$$

$$= \lim_{k \to \infty} - \int_\Omega D_j u_{n_k}(x) \varphi(x) \, dx = 0,$$

since $\lim_{n\to\infty} \int_\Omega |\nabla u_n(x)|^2\,dx = 0$. Hence $u \in H^1(\Omega)$ and $\nabla u = 0$. Since Ω is connected, it now follows from Theorem 6.19 that u is constant. But since

$$\int_\Omega u(x)\,dx = \lim_{k\to\infty} \int_\Omega u_{n_k}(x)\,dx = 0,$$

we can only have $u = 0$. This is a contradiction to $\|u\|_{L_2(\Omega)} = 1$.   □

We now give an example to show that the imbedding $H^1(\Omega) \hookrightarrow L_2(\Omega)$ is not always compact.

Example 7.29 Let $d = 1$ and $\Omega := (0,1) \setminus \{2^{-n} : n \in \mathbb{N}\}$. Then the imbedding $H^1(\Omega) \hookrightarrow L_2(\Omega)$ is not compact. To see this, we consider the sequence given by $u_n := c_n \mathbb{1}_{(2^{-(n+1)}, 2^{-n})}$, where $c_n := 2^{(n+1)/2}$, also $\|u_n\|_{L_2(\Omega)} = 1$. Then $u_n \in H^1(\Omega)$ and $u_n' = 0$, whence $\|u_n\|_{H^1(\Omega)} = 1$. But since u_n is orthogonal in $L_2(\Omega)$ to u_m whenever $n \neq m$, we always have $\|u_n - u_m\|_{L_2(\Omega)}^2 = 2$ for all $n \neq m$. As such, $(u_n)_{n\in\mathbb{N}}$ cannot admit a subsequence which converges in $L_2(\Omega)$.   △

It is equally possible to find a connected bounded open set $\Omega \subset \mathbb{R}^d$ for which the imbedding of $H^1(\Omega)$ into $L_2(\Omega)$ is not compact.

7.4 The Poisson equation with Neumann boundary conditions

In this section we will consider the Poisson equation with Neumann boundary conditions. We wish to start by introducing a "reaction term" into the equation. Let Ω be a bounded open set with C^1-boundary, $\lambda \in \mathbb{R}$ and $f \in L_2(\Omega)$. Then we seek a solution of the problem

$$\lambda u - \Delta u = f \text{ in } \Omega, \tag{7.18a}$$

$$\frac{\partial u}{\partial \nu} = 0 \text{ on } \partial\Omega. \tag{7.18b}$$

We call u a *classical solution* of (7.18) if $u \in C^2(\overline{\Omega})$ and (7.18) holds pointwise. Here the expression

$$\frac{\partial u}{\partial \nu}(z) = \nabla u(z) \cdot \nu(z) = \sum_{j=1}^{d} \frac{\partial u}{\partial x_j}(z)\nu_j(z)$$

is the normal derivative of u at $z \in \partial\Omega$. If a classical solution exists, then f is necessarily in $C(\overline{\Omega})$.

We wish to apply the Hilbert space techniques developed in Chapter 4, more precisely, the theorem of Riesz–Fréchet or, more generally, the Lax–Milgram theorem. To this end we need to introduce weak solutions of (7.18) which we can then interpret in terms of an appropriate bilinear form. We multiply (7.18a) by $\varphi \in C^1(\overline{\Omega})$, integrate and apply Green's first identity (Corollary 7.8(b)) to obtain

$$
\int_\Omega f(x)\varphi(x)\,dx = \lambda \int_\Omega u(x)\varphi(x)\,dx - \int_\Omega \Delta u(x)\varphi(x)\,dx
$$
$$
= \lambda \int_\Omega u(x)\varphi(x)\,dx + \int_\Omega \nabla u(x)\nabla\varphi(x)\,dx
$$
$$
- \int_{\partial\Omega} \frac{\partial u}{\partial \nu}(z)\varphi(z)\,d\sigma(z).
$$

If we now use the boundary condition (7.18b), then this reduces to

$$
\int_\Omega f(x)\varphi(x)\,dx = \lambda \int_\Omega u(x)\varphi(x)\,dx + \int_\Omega \nabla u(x)\nabla\varphi(x)\,dx \qquad (7.19)
$$

for all $\varphi \in C^1(\overline{\Omega})$. Since $C^1(\overline{\Omega})$ is dense in $H^1(\Omega)$ by Theorem 7.21, it follows that (7.19) holds for all $\varphi \in H^1(\Omega)$. But the equation (7.19) is meaningful for all $\varphi \in H^1(\Omega)$ provided only that $u \in H^1(\Omega)$. Moreover, the outer normal no longer appears, and so weak solutions can even be defined for arbitrary open sets.

Definition 7.30 Let $\Omega \subset \mathbb{R}^d$ be open, $\lambda \in \mathbb{R}$ and $f \in L_2(\Omega)$. A *weak solution* of (7.18) is a function $u \in H^1(\Omega)$ for which (7.19) holds for all $\varphi \in H^1(\Omega)$. $\triangle$

We have seen that in the case where Ω has C^1-boundary, every classical solution is also a weak solution. Conversely, if u is a weak solution, then u is already a classical solution provided that u is sufficiently regular. The following theorem makes precise these assertions.

Theorem 7.31 *Let Ω be a bounded open set with C^1-boundary and $f \in L_2(\Omega)$, $\lambda \in \mathbb{R}$.*
(a) If u is a classical solution of (7.18), then u is also a weak solution.
(b) If $u \in C^2(\overline{\Omega})$ is a weak solution of (7.18), then $f \in C(\overline{\Omega})$ and u is a classical solution.

Proof We only need to prove (b). Let $u \in C^2(\overline{\Omega})$ be a weak solution of (7.18). By applying Green's first identity, Corollary 7.8(b), to (7.19), we obtain

$$
\int_\Omega f(x)\varphi(x)\,dx = \lambda \int_\Omega u(x)\varphi(x)\,dx - \int_\Omega \Delta u(x)\varphi(x)\,dx
$$

for all $\varphi \in \mathcal{D}(\Omega)$. We thus have $\int_\Omega (f(x)-\lambda u(x)+\Delta u(x))\varphi(x)\,dx = 0$ for all $\varphi \in \mathcal{D}(\Omega)$. It now follows from Corollary 6.10 that $f - \lambda u + \Delta u = 0$ almost everywhere, that is, $f = \lambda u - \Delta u$ almost everywhere. Since $\lambda u - \Delta u \in C(\overline{\Omega})$, f has a unique continuous representative. If we identify f with this representative, then (7.18a) is satisfied pointwise. To show that the boundary condition (7.18b) also holds, we now consider

(7.19) for $\varphi \in C^1(\overline{\Omega})$ and exploit the fact that (as we have just shown) $\lambda u - \Delta u = f$. Thus, starting from the fact that u is a weak solution,

$$
\begin{aligned}
0 &= \int_\Omega (\lambda u(x) - \Delta u(x)) \varphi(x) \, dx = \int_\Omega f(x) \varphi(x) \, dx \\
&= \int_\Omega \lambda u(x) \varphi(x) \, dx + \int_\Omega \nabla u(x) \nabla \varphi(x) \, dx \\
&= \int_\Omega \lambda u(x) \varphi(x) \, dx - \int_\Omega \Delta u(x) \varphi(x) \, dx + \int_{\partial\Omega} \frac{\partial u}{\partial \nu}(z) \varphi(z) \, d\sigma(z)
\end{aligned}
$$

for all $\varphi \in C^1(\overline{\Omega})$, where we have again used Green's first identity, Corollary 7.8(b). Hence $\int_{\partial\Omega} \frac{\partial u}{\partial \nu}(z) \varphi(z) \, d\sigma(z) = 0$ for all $\varphi \in C^1(\overline{\Omega})$. Since the set $\{\varphi_{|\partial\Omega} : \varphi \in C^1(\overline{\Omega})\}$ is dense in $L_2(\partial\Omega, d\sigma)$ by Theorem 7.17, it follows that $\frac{\partial u}{\partial \nu} = 0$ in $L_2(\partial\Omega, d\sigma)$, that is, $\frac{\partial u}{\partial \nu}(z) = 0$ almost everywhere. But since $\frac{\partial u}{\partial \nu} \in C(\partial\Omega)$ and by Corollary 7.16 $\nu(U) > 0$ for every nonempty relatively open set U in $\partial\Omega$, we conclude that $\frac{\partial u}{\partial \nu}(z) = 0$ for all $z \in \partial\Omega$. □

Theorem 7.31 allows us to split the analysis of problem (7.18) into two parts:
(a) the proof of existence and uniqueness of weak solutions, and
(b) the study of the regularity of u.
We can solve part (a) quite easily via the theorem of Riesz–Fréchet; however, a detailed investigation of regularity properties of weak solutions would take us well outside the scope of this book, and so we restrict ourselves to a few remarks. For the existence and uniqueness of weak solutions we require no regularity assumptions whatsoever on Ω; the outer normal vector to Ω appears nowhere in Definition 7.30, and it is not even required to exist.

Theorem 7.32 *Let $\Omega \subset \mathbb{R}^d$ be an arbitrary open set, $\lambda > 0$ and $f \in L_2(\Omega)$. Then problem (7.18) admits exactly one weak solution.*

Proof We first define $F \in H^1(\Omega)'$ by $F(v) := \int_\Omega f(x) v(x) \, dx$, and observe that $a(u, v) := \lambda \int_\Omega u(x) v(x) \, dx + \int_\Omega \nabla u(x) \nabla v(x) \, dx$, $u, v \in H^1(\Omega)$, defines an equivalent inner product on $H^1(\Omega)$, that is, the norm induced by $a(\cdot, \cdot)$ is equivalent to $\| \cdot \|_{H^1}$. By the theorem of Riesz–Fréchet (Theorem 4.21) there exists exactly one $u \in H^1(\Omega)$ such that $a(u, v) = F(v)$ for all $v \in H^1(\Omega)$. But this means exactly that (7.19) is satisfied. This completes the proof. □

On the question of the regularity of the weak solution u, we know from Theorem 6.58 that $u \in H^2_{\mathrm{loc}}(\Omega)$. This means that the partial derivatives $D_i D_j$ are in $L_{2,\mathrm{loc}}(\Omega)$ and the equation $\lambda u - \Delta u = f$ is an identity between functions in $L_{2,\mathrm{loc}}(\Omega)$. We will not prove any further regularity properties; we do however mention that the weak solution u of (7.18) is in $C^\infty(\overline{\Omega})$ provided $f \in C^\infty(\overline{\Omega})$ and Ω has C^∞-boundary (see [18, Théorème IX.26] or [32, Theorem 6.30]).

We next consider the case $\lambda = 0$. If $\Omega \subset \mathbb{R}^d$ is bounded and open, then every constant function is a weak solution of (7.18) with $\lambda = 0$ and $f = 0$. As a consequence, weak solutions cannot be unique. But there is more: if, given $f \in L_2(\Omega)$, there exists a weak solution u of (7.18), then, by choosing $\varphi \equiv 1$ in (7.19), we deduce that

$$\int_\Omega f(x)\,dx = 0, \tag{7.20}$$

that is, (7.20) is a necessary condition for the existence of a weak solution of (7.18), if $\lambda = 0$. It turns out that this condition is also sufficient, as long as the second Poincaré inequality is available.

Theorem 7.33 *Let Ω be a bounded domain with continuous boundary and suppose $f \in L_2(\Omega)$ satisfies $\int_\Omega f(x)\,dx = 0$. If $\lambda = 0$, then (7.18) has a unique weak solution $u \in H^1(\Omega)$, and this solution satisfies $\int_\Omega u(x)\,dx = 0$.*

Proof The set $L_{2,0}(\Omega) := \{v \in L_2(\Omega) : \int_\Omega v(x)\,dx = 0\}$ forms a closed subspace of $L_2(\Omega)$ and is thus a Hilbert space. The space $H_m^1(\Omega) = H^1(\Omega) \cap L_{2,0}(\Omega)$ from (7.16) is likewise closed in $H^1(\Omega)$ and thus also a Hilbert space. We now show that $H_m^1(\Omega)$ is dense in $L_{2,0}(\Omega)$. Let $g \in L_{2,0}(\Omega)$ be such that $(w, g)_{L_2(\Omega)} = 0$ for all $w \in H_m^1(\Omega)$. By Corollary 4.18, it suffices to show that $g = 0$. Set $e_1 := |\Omega|^{-1/2} \mathbb{1}_\Omega$, where $|\Omega|$ is the Lebesgue measure of Ω. If $v \in H^1(\Omega)$, then $w := v - (v, e_1)_{L_2(\Omega)} e_1 \in H_m^1(\Omega)$, since

$$\int_\Omega w(x)\,dx = \int_\Omega v(x)\,dx - \left(|\Omega|^{-1/2} \int_\Omega v(x)\,dx\right)\left(|\Omega|^{-1/2} \int_\Omega \mathbb{1}_\Omega(x)\,dx\right) = 0.$$

Since $(g, e_1)_{L_2(\Omega)} = 0$ and $(g, w)_{L_2(\Omega)} = 0$, it follows that $(g, v)_{L_2(\Omega)} = 0$. Summarizing, we have shown that $(g, v)_{L_2(\Omega)} = 0$ for all $v \in H^1(\Omega)$; since $H^1(\Omega)$ is dense in $L_2(\Omega)$, it follows that $g = 0$. Hence $H_m^1(\Omega)$ is indeed dense in $L_{2,0}(\Omega)$.

By (7.19), $a(u, v) := \int_\Omega \nabla u(x) \nabla v(x)\,dx$ defines an equivalent inner product on $H_m^1(\Omega)$. Let $F \in H_m^1(\Omega)'$ be given by $F(x) := \int_\Omega f(x) w(x)\,dx$, $w \in H_m^1(\Omega)$. By the Lax–Milgram theorem there exists a unique $u \in H_m^1(\Omega)$ such that

$$a(u, w) = \int_\Omega f(x) w(x)\,dx, \quad w \in H_m^1(\Omega). \tag{7.21}$$

Equivalently, u is a weak solution of (7.18), as can be seen as follows: we have $\int_\Omega \nabla u(x) \nabla e_1(x)\,dx = 0 = \int_\Omega f(x) e_1(x)\,dx$. Since every $v \in H^1(\Omega)$ may be written in the form $v = v - (v, e_1)_{L_2(\Omega)} e_1 + (v, e_1)_{L_2(\Omega)} e_1$, where $v - (v, e_1)_{L_2(\Omega)} e_1 \in H_m^1(\Omega)$, it follows from (7.21) that

$$\int_\Omega f(x) v(x)\,dx = \int_\Omega f(x)(v(x) - (v, e_1)_{L_2(\Omega)} e_1(x))\,dx + (v, e_1)_{L_2(\Omega)}(f, e_1)_{L_2(\Omega)}$$

$$= a(u, v - (v, e_1)_{L_2(\Omega)} e_1) + (v, e_1)_{L_2(\Omega)} a(u, e_1) = a(u, v),$$

which means that u is a weak solution for $\lambda = 0$. $\qquad\square$

7.5 The trace theorem and Robin boundary conditions

Throughout this section we will take $\Omega \subset \mathbb{R}^d$ to be a bounded open set with C^1-boundary, and we will denote by σ surface measure on $\partial\Omega$; the space $L_2(\partial\Omega) = L_2(\partial\Omega, \sigma)$ is constructed with this measure.

Theorem 7.34 (Trace theorem) *There exists a unique continuous linear operator* $T : H^1(\Omega) \to L_2(\partial\Omega)$ *such that* $Tu = u_{|\partial\Omega}$ *for all* $u \in C(\overline{\Omega}) \cap H^1(\Omega)$.

This operator is called the *trace operator*. Now by Theorem 7.21, the space $C^1(\overline{\Omega})$ is dense in $H^1(\Omega)$. We will prove below that there exists a constant $c > 0$ such that

$$\|u_{|\partial\Omega}\|_{L_2(\partial\Omega)} \le c\|u\|_{H^1(\Omega)} \tag{7.22}$$

for all $u \in C^1(\overline{\Omega})$. It follows from Theorem A.2 that there exists a unique operator $T \in \mathcal{L}(H^1(\Omega), L_2(\partial\Omega))$ such that $Tu = u_{|\partial\Omega}$ for all $u \in C^1(\overline{\Omega})$. That this identity remains true for all $u \in C(\overline{\Omega}) \cap H^1(\Omega)$ requires an additional argument.

Proof (of Theorem 7.34) 1. Let $U \subset \mathbb{R}^d$ be open, such that $U \cap \partial\Omega$ is a C^1-graph for Ω. We will show that there exists a constant $c > 0$ for which (7.22) holds for all $u \in C^1(\overline{\Omega})$ such that $\operatorname{supp} u \subset U$. In order to prove this we may assume that U is a normal C^1 graph (otherwise we change coordinate system). We will use the notation from Definition 7.1(a). So let $u \in C^1(\overline{\Omega})$ satisfy $\operatorname{supp} u \subset U$ and set

$$c_1 := \sup_{|y| \le r} \sqrt{1 + |\nabla g(y)|^2};$$

then by (7.14),

$$
\begin{aligned}
\|u\|_{L_2(\partial\Omega)}^2 &= \int_{|y|<r} u(y, g(y))^2 \sqrt{1 + |\nabla g(y)|^2}\, dy \le c_1 \int_{|y|<r} u(y, g(y))^2\, dy \\
&= c_1 \int_{|y|<r} \int_0^h -\frac{\partial}{\partial s} u(y, g(y) + s)^2\, ds\, dy \\
&= c_1 \int_{|y|<r} \int_0^h -2u(y, g(y) + s) D_d u(y, g(y) + s)\, ds\, dy \\
&\le c_1 \int_{|y|<r} \int_0^h (u(y, g(y) + s)^2 + D_d u(y, g(y) + s)^2)\, ds\, dy \\
&= c_1 \int_\Omega (u(x)^2 + (D_d u(x))^2)\, dx \quad \le c_1 \|u\|_{H^1(\Omega)}^2,
\end{aligned}
$$

where for the last equality we use the relation (7.9) and for the inequality just before that we use Young's inequality $2\alpha\beta \le \alpha^2 + \beta^2$ (Lemma 5.22). This establishes (7.22) for all $u \in C^1(\overline{\Omega})$ with $\operatorname{supp} u \subset U$.

2. Since $\partial\Omega$ is compact, we can find open sets $U_1, \ldots, U_m \subset \mathbb{R}^d$ such that $\partial\Omega \cap U_k$ is a C^1-graph for Ω for each k, and $\bigcup_{k=1}^m U_k \supset \partial\Omega$. By 1., for every k there exists

a $c_k \geq 0$ such that (7.22) is satisfied for all $u \in C^1(\overline{\Omega})$ such that supp $u \subset U_k$, with $c = c_k$. Consider a partition of unity of $\partial\Omega$ subordinate to the cover $U_1, \ldots, U_k$, that is, let $\eta_k \in \mathcal{D}(\mathbb{R}^d)$ be such that $0 \leq \eta_k \leq 1$, supp$\eta_k \subset U_k$ and $\sum_{k=0}^m \eta_k(x) = 1$ for all $x \in \partial\Omega$. Now let $u \in C^1(\overline{\Omega})$; then $u_k := u \cdot \eta_k \in C^1(\overline{\Omega})$ satisfies supp $u_k \subset U_k$ and $\sum_{k=1}^m u_k = u$ on $\partial\Omega$. Hence

$$\|u\|_{L_2(\partial\Omega)} = \left\| \sum_{k=1}^m u_{k|\partial\Omega} \right\|_{L_2(\partial\Omega)} \leq \sum_{k=1}^m \|u_{k|\partial\Omega}\|_{L_2(\partial\Omega)}$$

$$\leq \sum_{k=1}^m c_k \|u_k\|_{H^1(\Omega)} \leq \left(\max_{k=1,\ldots,m} c_k \right) \sum_{k=1}^m \|u_k\|_{H^1(\Omega)}.$$

Since $D_j(\eta_k u) = (D_j\eta_k)u + \eta_k D_j u$, it follows that $\|\eta_k u\|_{H^1(\Omega)} \leq c\|u\|_{H^1(\Omega)}$ for some constant $c > 0$. We have thus shown that (7.22) holds for all $u \in C^1(\overline{\Omega})$, which in light of our earlier comments implies the existence of a unique $T \in \mathcal{L}(H^1(\Omega), L_2(\partial\Omega))$ such that $Tu = u_{|\partial\Omega}$ for all $C^1(\overline{\Omega})$.

If $u \in C(\overline{\Omega}) \cap H^1(\Omega)$, then we can find functions $u_n \in C^\infty(\overline{\Omega})$, $n \in \mathbb{N}$, which converge to u in both $H^1(\Omega)$ and $C(\overline{\Omega})$ (see [28, Proof of Theorem 3 in §5.3.3]). Thus $Tu = \lim_{n\to\infty} Tu_n = \lim_{n\to\infty} u_{n|\partial\Omega} = u_{|\partial\Omega}$, where the limits are in $L_2(\partial\Omega)$. This completes the proof.   □

We recall that $H_0^1(\Omega)$ is defined as the closure in $H^1(\Omega)$ of the space of test functions (see Section 6.3). It follows from Theorem 7.34 that $Tu = 0$ for all $u \in H_0^1(\Omega)$. But the converse is also true (see [28, Sec. 5.5]), meaning that we obtain a new characterization of $H_0^1(\Omega)$ (for bounded open sets Ω with C^1-boundary).

Theorem 7.35 We have $H_0^1(\Omega) = \{u \in H^1(\Omega) : Tu = 0\}$, where T is the trace operator of Theorem 7.34.

Using the trace operator we can now define a weak form of the normal derivative $\frac{\partial u}{\partial \nu}$ and thus also more general boundary conditions. We continue to assume that Ω is a bounded open set in $\mathbb{R}^d$ with C^1-boundary. Now if $u \in C^2(\overline{\Omega})$, then by Corollary 7.8,

$$\int_\Omega \Delta u(x)v(x)\, dx + \int_\Omega \nabla u(x)\nabla v(x)\, dx = \int_{\partial\Omega} \frac{\partial u}{\partial \nu}(z)v(z)\, d\sigma(z) \tag{7.23}$$

for all $v \in C^1(\overline{\Omega})$. This motivates the following definition.

Definition 7.36 Suppose that $u \in H^1(\Omega) \cap H_{\text{loc}}^2(\Omega)$, with $\Delta u \in L_2(\Omega)$. We say that $\frac{\partial u}{\partial \nu}$ exists as a function in $L_2(\partial\Omega)$ if there exists some $b \in L_2(\partial\Omega)$ for which

$$\int_\Omega \Delta u(x)v(x)\, dx + \int_\Omega \nabla u(x)\nabla v(x)\, dx = \int_{\partial\Omega} b(z)(Tv)(z)\, d\sigma(z) \tag{7.24}$$

for all $v \in H^1(\Omega)$. In this case $b \in L_2(\partial\Omega)$ is determined uniquely and we set $\frac{\partial u}{\partial \nu} := b$.   △

It is of course sufficient to check (7.24) for all $v \in C(\overline{\Omega}) \cap H^1(\Omega)$ or even just for $v = \varphi_{|\Omega}$ for all $\varphi \in \mathcal{D}(\mathbb{R}^d)$, since these functions are dense in $H^1(\Omega)$ by Theorem 7.21. The claim that there can be at most one $b \in L_2(\partial\Omega)$ which satisfies (7.24) follows from Theorem 7.17 and Corollary 4.18. We can now obtain existence and uniqueness of solutions of the Poisson problem with Robin boundary conditions.

Theorem 7.37 *Let Ω be a bounded domain in $\mathbb{R}^d$ with C^1-boundary and suppose that $b : \partial\Omega \to [0, \infty)$ is a bounded measurable function such that $b(z)$ is different from 0 on $\partial\Omega$ on a set of positive surface measure. Then for every $f \in L_2(\Omega)$ there exists a unique solution $u \in H^1(\Omega) \cap H^2_{\mathrm{loc}}(\Omega)$ of*

$$-\Delta u = f, \tag{7.25a}$$

$$\frac{\partial u}{\partial v} + bTu = 0, \tag{7.25b}$$

where $T : H^1(\Omega) \to L_2(\partial\Omega)$ denotes the trace operator.

Proof We define the bilinear form $a : H^1(\Omega) \times H^1(\Omega) \to \mathbb{R}$ by $a(u, v) := \int_\Omega \nabla u(x) \nabla v(x) \, dx + \int_{\partial\Omega} b(z) Tu(z) Tv(z) \, d\sigma(z)$. Then $a(\cdot, \cdot)$ is clearly continuous; we now show that it is also coercive. If this were not the case, then we could find $u_n \in H^1(\Omega)$ such that $\|u_n\|_{H^1(\Omega)} = 1$ but $\lim_{n\to\infty} a(u_n) = 0$. By Theorem 4.35 we may assume that u_n converges weakly to $u \in H^1(\Omega)$. Since the imbedding $H^1(\Omega) \hookrightarrow L_2(\Omega)$ is compact (Theorem 7.22), it follows from Theorem 4.39 that $u_n \to u$ in $L_2(\Omega)$. Hence $\|u\|_{L_2(\Omega)} = 1$. Now since $\lim_{n\to\infty} a(u_n) = 0$ and $b \geq 0$, it follows that

$$\lim_{n\to\infty} \int_\Omega |\nabla u_n(x)|^2 \, dx = 0, \tag{7.26}$$

whence

$$\int_\Omega u(x) D_j \varphi(x) \, dx = \lim_{n\to\infty} \int_\Omega u_n(x) D_j \varphi(x) \, dx$$

$$= \lim_{n\to\infty} -\int_\Omega (D_j u_n(x)) \varphi(x) \, dx = 0$$

holds for all $\varphi \in \mathcal{D}(\Omega)$. It now follows from Theorem 6.19 that there exists $c \in \mathbb{R}$ such that $u(x) = c$ almost everywhere in Ω. Moreover, by (7.26) we have $\lim_{n\to\infty} u_n = u$ in $H^1(\Omega)$. Hence $\lim_{n\to\infty} Tu_n = Tu = c\mathbb{1}_{\partial\Omega}$ in $L_2(\partial\Omega)$ as well. Using the definition of the bilinear form $a(\cdot, \cdot)$, we obtain

$$c^2 \int_{\partial\Omega} b(z) \, d\sigma(z) = \lim_{n\to\infty} \int_{\partial\Omega} b(z) |Tu_n(z)|^2 \, d\sigma(z) = \lim_{n\to\infty} a(u_n) = 0.$$

By our assumptions on b it finally follows that $c = 0$, a contradiction to $\|u\|_{L_2(\Omega)} = 1$. We have thus proved that $a(\cdot, \cdot)$ is coercive. As usual $F(v) := \int_\Omega f(x) v(x) \, dx$ defines a continuous linear form $F \in H^1(\Omega)'$, and so by the Lax–Milgram theorem there exists a unique $u \in H^1(\Omega)$ such that

$$a(u, v) = \int_\Omega f(x)v(x)\,dx \quad \text{for all } v \in H^1(\Omega). \tag{7.27}$$

But this is equivalent to (7.25). Indeed, if (7.27) holds, then $\int_\Omega f(x)\varphi(x)\,dx = a(u, \varphi) = \int_\Omega \nabla u(x)\nabla\varphi(x)\,dx$ for all $\varphi \in \mathcal{D}(\Omega)$, whence $-\Delta u = f$. Hence also $u \in H^2_{\text{loc}}(\Omega)$ by Theorem 6.58. If we substitute the relation $f = -\Delta u$ into (7.27), then we see that $-\int_\Omega \Delta u(x)v(x)\,dx = \int_\Omega f(x)v(x)\,dx = a(u, v) = \int_\Omega \nabla u(x)\nabla v(x)\,dx + \int_{\partial\Omega} b(Tu)(Tv)\,d\sigma(z)$ for all $v \in H^1(\Omega)$. This means that $\frac{\partial u}{\partial \nu} = -bTu$ by Definition 7.36.

The proof that every solution of (7.25) satisfies condition (7.27) is similar.

$\square$

We call (7.25b) *Robin boundary conditions* or *boundary conditions of the third kind*. When $b = 0$ we recover Neumann boundary conditions.

7.6* Comments on Chapter 7

Gauss's theorem was first discovered by Joseph-Louis Lagrange in 1792, and then rediscovered independently by Carl Friedrich Gauss in 1813, George Green in 1825 and Mikhail V. Ostrogradsky in 1831. This is why this theorem can be found under so many different names in the literature. It is also possible to define surface measure for Lipschitz domains and correspondingly generalize Gauss's theorem to such domains. We refer to [2].

Neumann boundary conditions are named after Carl G. Neumann (1832–1925), who was a professor in Germany and Switzerland, at the Universities of Halle, Basel, Tübingen and Leipzig. He worked on the Dirichlet principle and is also responsible for the analog of geometric series for matrices. It is for this reason that series of the form $\sum_{k=0}^\infty T^k$ are known as *Neumann series*; if T is an operator with norm $\|T\| < 1$, then its Neumann series converges to $(I - T)^{-1}$.

Robin boundary conditions are named after Victor G. Robin (1855–1897), who held lectures on mathematical physics at the Sorbonne and worked in the area of thermodynamics. They had also been studied by Isaac Newton (1643–1727).

The Sobolev space H^1 plays a central role in Chapters 6 and 7. It was the Dirichlet problem that led Beppo Levi to study spaces of this kind for the first time in 1906 (shortly after the invention of the Lebesgue integral). For this reason, this space (with various equivalent definitions) was initially called a space of type (BL). Weak solutions started appearing in the 1930s; they are fundamental in the famous (and still current) work published by J. Leray in 1934, in which the existence of a weak global solution of the Navier–Stokes equation is shown. (Proving uniqueness would imply the existence of a classical *global* solution, which is an open Millennium Problem.) Weak solutions of the wave equation were considered by S. L. Sobolev (1908–1989) in 1936. It was also Sobolev who undertook a *systematic* investigation of Sobolev spaces, which reached its pinnacle in his influential book published in 1950. The fundamental Sobolev imbedding theorem goes back to 1938. In this context we can formulate a special case as follows: the space $H^1_0(\Omega)$ is a subspace of $L_p(\Omega)$ for $p = 2d/(d - 2)$, if Ω is a bounded open set in $\mathbb{R}^d$ (see Exercise 7.7(a)).

Sobolev came from the famous mathematical school of St. Petersburg and joined the Steklov Institute of Mathematics in Moscow in 1934, becoming director of the Institute for the Theory of

Partial Differential Equations there in 1935. Sobolev received a large number of distinctions; aged only 31 he became a member of the Soviet Academy of Sciencies and was awarded the highest possible distinctions in the Soviet Union (the Stalin Prize and the title Hero of Socialist Labor).

The Göttingen school (to which for example K. Friedrichs belonged) contented itself in the 1930s and 1940s with spaces of classically differentiable functions, without making use of the fact that these can be completed to make a Hilbert space (of Lebesgue integrable functions). This Hilbert space was long known under the letter H, while the Sobolev space $H^1(\Omega)$ was denoted by W. It was only shown in 1964 thanks to an ingenious trick of Meyers and Serrin tbat these two spaces are in fact the same (cf. Remark 6.17). The short publication of Meyers–Serrin to this effect, [50], has an appropriately short title: $H = W$.

7.7 Exercises

Exercise 7.1 Let $K \subset \mathbb{R}^d$ be compact. Show that every positive linear form on $C(K)$ is continuous (cf. Definition 7.9).

Exercise 7.2 (a) Let $\Omega \subset \mathbb{R}^d$ be bounded and open with C^1-boundary, $d \geq 2$, and let $z \in \Omega$. Given $u \in C(\overline{\Omega}) \cap C^1(\Omega \setminus \{z\})$, suppose that $\frac{\partial u}{\partial x_j}$ is bounded on $\Omega \setminus \{z\}$ for all $j = 1, \ldots, d$. Show that then $u \in H^1(\Omega)$.
Suggestion: Multiply $\frac{\partial u}{\partial x_j}$ by a function $\varphi \in \mathcal{D}(\Omega)$ and integrate by parts on $\Omega \setminus \bar{B}(z, \varepsilon)$.
(b) Let $B := \{(x, y) \in \mathbb{R}^2 : x^2 + y^2 < \frac{1}{4}\}$, $u(x, y) = (x^2 - y^2) \log |\log r|$, $r = (x^2 + y^2)^{1/2}$. Show that $u \in H^1(B)$ (cf. Theorem 6.59).

Exercise 7.3 Let $\Omega_1, \Omega_2 \subset \mathbb{R}^d$ be open, set $\Omega := \Omega_1 \cup \Omega_2$, and let $u, f \in L_{2,\text{loc}}(\Omega)$. Suppose that $\Delta u = f$ holds weakly in Ω_1 and in Ω_2 (see Definition 6.41). Show that $\Delta u = f$ holds weakly in Ω.
Suggestion: Use a partition of unity.

Exercise 7.4 (Poincaré inequality) For the open set $\Omega \subset \mathbb{R}^d$, suppose that the imbedding $H_0^1(\Omega) \hookrightarrow L_2(\Omega)$ is compact.
(a) Show that there exists a constant $c > 0$ such that

$$\int_\Omega |u|^2 \, dx \leq c \int_\Omega |\nabla u|^2 \, dx \text{ for all } u \in H_0^1(\Omega).$$

Where is the compactness of the imbedding needed? *Suggestion:* Use the proof of Theorem 7.28 for orientation.
(b) Give an example of an unbounded open set $\Omega \subset \mathbb{R}^d$ for which the imbedding is compact.
Suggestion: Use the remark after Theorem 6.54.
(c) Given an example of an open set for which the Poincaré inequality does not hold.

Exercise 7.5 (Gauss's theorem for triangles) A *segment* in $\mathbb{R}^2$ is a set of the form $[x, y] := \{\lambda x + (1 - \lambda)y : 0 \le \lambda \le 1\}$, where $x, y \in \mathbb{R}^2$, $x \ne y$ are the endpoints of $[x, y]$. We define a measure σ on $[x, y]$ by $\int_{[x,y]} f \, d\sigma = |y - x| \int_0^1 f(\lambda x + (1 - \lambda)y) \, d\lambda$ for all $f \in C[x, y]$ (cf. Theorem 7.10). If T is a triangle with corners $\{t_1, t_2, t_3\}$, then $\partial T = [t_1, t_2] \cup [t_2, t_3] \cup [t_1, t_3]$. Given $f \in C(\partial T)$ we define $\int_{\partial T} f \, d\sigma = \int_{[t_1, t_2]} f \, d\sigma + \int_{[t_2, t_3]} f \, d\sigma + \int_{[t_1, t_3]} f \, d\sigma$.

(a) Let $u \in C^1(\overline{T})$. Show that $\int_T u_{x_j} \, dx = \int_{\partial T} u v_j \, d\sigma$. Here v_j is the jth component of the outer unit normal v to T, $j = 1, 2$.
  Suggestion: After rotating and translating if necessary, one may assume that $t_1 = (0, 0)$, $t_2 = (b, 0)$, $t_3 = (d, c)$ for some $0 < d < b$, $c > 0$.

(b) Show that $\int_T (D_j u)v \, dx = - \int_T u D_j v \, dx + \int_{\partial T} v_j u v \, d\sigma$ holds for all $u, v \in C^1(\overline{T})$ and $j = 1, 2$.

Exercise 7.6 Let $\Omega \subset \mathbb{R}^2$ be a polygon and $\{T_k : k = 1, \dots, n\}$ an admissible triangulation of Ω (see Section 9.2.7). Let $u \in C(\overline{\Omega})$ be such that $u_{|T_k} \in C^1(T_k)$ with bounded derivatives, for all $k = 1, \dots, n$.

(a) Let $\varphi \in \mathcal{D}(\Omega)$. Show using Exercise 7.5(b) that

$$- \int_\Omega u \frac{\partial \varphi}{\partial x_j} \, dx = \sum_{k=1}^n \int_{T_k} \frac{\partial u}{\partial x_j} \varphi \, dx, \quad j = 1, 2.$$

(b) Deduce from (a) that $u \in H^1(\Omega)$ and $D_j u = \frac{\partial u}{\partial x_j}$ on T_k for $j = 1, 2$ and $k = 1, \dots, n$.

Exercise 7.7 (Domain with continuous boundary but not the extension property)

(a) Show that if a bounded domain $\Omega \subset \mathbb{R}^2$ has the extension property, then $H^1(\Omega) \subset L_q(\Omega)$ for all $q \in [2, \infty)$.
  Suggestion: Use Exercise 6.13.

(b) Given $\alpha > 0$, let $\Omega_\alpha := \{(x, y) \in \mathbb{R}^2 : 0 < x < 1, |y| < x^\alpha\}$, and for $\beta > 0$ set $u(x, y) := x^{-\beta}$. Show that $u \in H^1(\Omega_\alpha)$ if $\beta < \frac{\alpha - 1}{2}$, but $u \notin L_q(\Omega)$ if q is sufficiently large. Deduce that Ω_α does not have the extension property.

(c) Let $\alpha > 3$. Show that $u \in H^2(\Omega_\alpha)$ if $\beta < \frac{\alpha - 3}{2}$ and deduce that $H^2(\Omega_\alpha) \not\subset C(\overline{\Omega_\alpha})$.

Exercise 7.8 Given a curve $\Gamma = \{(x, \gamma(x)) : x \in [a, b]\}$ such that $\gamma \in C^1([a, b])$, we define

$$\int_\Gamma f \, d\sigma := \int_a^b f(x, \gamma(x))\sqrt{1 + \gamma'(x)^2} \, dx$$

for every Borel-measurable function $f : \Gamma \to [0, \infty]$. Let $\Omega = \Omega_4$ be the domain from Exercise 7.7 with $\alpha = 4$ and let $\Gamma := \{(x, x^4) : 0 \le x \le 1\}$. Then Γ is a compact subset of $\partial \Omega$.

(a) Show that $u(x, y) = \frac{1}{x}$ defines a function $u \in H^1(\Omega)$.

(b) Show that $\int_\Gamma u^2 \, d\sigma = \infty$.

Exercise 7.9 (a) In dimension $d = 1$, let $\Omega := (0, \frac{1}{2}) \cup (\frac{1}{2}, 1)$. Show that Ω does not have the extension property.

(b) Show that the connected open set $\Omega := (0, 1)^2 \setminus \{(1/2, y) : 0 < y < 1/2\}$ in $\mathbb{R}^2$ does not have the extension property (cf. Figure 7.3).
   Suggestion: construct a function $u \in H^1(\Omega)$ such that $u = 0$ on $(0, 1/2) \times (0, 1/4)$ and $u = 1$ on $(1/2, 1) \times (0, 1/4)$. Apply Stampacchia's lemma, Lemma 6.36.

(c) For the set Ω from (b), is the imbedding of $H^1(\Omega)$ into $L_2(\Omega)$ compact?

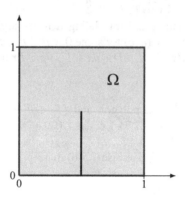

Fig. 7.3 The set Ω from Exercise 7.9 which does not have the extension property.

Exercise 7.10 (Proof that C^1-domains have the extension property) Let $\Omega \subset \mathbb{R}^d$ be a C^1-domain.

(a) Let $u \in C^1(\overline{\Omega})$ and let $\tilde{u}$ be the extension of u described in the sketch of the proof of Theorem 7.19. Show that $\tilde{u} \in H^1(\mathbb{R}^d)$ and $\|\tilde{u}\|_{H^1(\mathbb{R}^d)} \leq c\|u\|_{H^1(\Omega)}$, where the constant $c > 0$ is independent of u.

(b) Use a partition of unity to deduce from (a) that there exists a continuous linear mapping $E : C^1(\overline{\Omega}) \to H^1(\mathbb{R}^d)$ such that $(Eu)_{|\Omega} = u$, where $C^1(\overline{\Omega})$ is equipped the $H^1(\Omega)$-norm.

(c) Use the property that $C^1(\overline{\Omega})$ is dense in $H^1(\Omega)$ (see Theorem 7.27) to conclude that Ω has the extension property.

Remark: For a different, direct proof which does not use the density of $C^1(\overline{\Omega})$ in $H^1(\Omega)$, see [18, Theorem 9.7].

Exercise 7.11 (Poincaré inequality on $H_D^1(\Omega)$) Let $\Omega \subset \mathbb{R}^d$ be an open, connected, bounded set with C^1-boundary $\partial\Omega$. Let $\Gamma_D \subset \partial\Omega$ be a Borel set such that $\sigma(\Gamma_D) > 0$. Define $H_D^1(\Omega) := \{u \in H^1(\Omega) : (Tu)_{|\Gamma_D} = 0\}$, where $T : H^1(\Omega) \to L_2(\partial\Omega)$ is the trace operator.

(a) Show that $H_D^1(\Omega)$ is weakly closed in $H^1(\Omega)$, i.e., if $u_n \in H_D^1(\Omega)$ and $u_n \rightharpoonup u$ in $H^1(\Omega)$, then $u \in H_D^1(\Omega)$. Use Theorem 4.38.

(b) Show that there exists a constant $c > 0$ such that $\|u\|_{L_2(\Omega)} \leq c \|\nabla u\|_{L_2(\Omega)}$ for all $u \in H_D^1(\Omega)$. *Hint:* See the proof of Theorem 7.28.

Chapter 8
Spectral decomposition and evolution equations

The spectral decomposition of the Laplacian with suitable boundary conditions gives us precise information about the solvability of the equation

$$\lambda u - \Delta u = f.$$

But it also yields a method for proving the well-posedness of what are known as *evolution equations*. These are equations which, as the name suggests, describe the evolution of some quantity in time. We will treat the heat equation

$$u_t = \Delta u$$

as well as the wave equation

$$u_{tt} = \Delta u.$$

Here, $u = u(t, x)$ is a function which depends on a time variable $t \geq 0$ and a space variable $x \in \Omega$, where Ω is a domain in $\mathbb{R}^d$. In addition, the spectrum also gives us information about the asymptotic behavior of the solutions of such equations as $t \to \infty$.

Chapter overview

© The Author(s), under exclusive license to Springer Nature Switzerland AG 2023
W. Arendt, K. Urban, *Partial Differential Equations*, Graduate Texts in
Mathematics 294, https://doi.org/10.1007/978-3-031-13379-4_8

8.1 A vector-valued initial value problem

In this section we return to the situation covered by the spectral theorem (Theorem 4.46): we have an operator A which can be represented as a diagonal operator with respect to a certain orthonormal basis of the ambient Hilbert space; this operator defines an initial value problem whose solution can be determined easily as a series over the basis vectors. Our first major goal in this chapter is to apply these ideas in the case where A is the Laplacian with given boundary conditions. The inital value problem then becomes a partial differential equation, more precisely the heat equation. We start by collecting some simple properties of differentiable functions which take values in a Banach space.

Definition 8.1 Let E be a real Banach space and $I \subset \mathbb{R}$ an interval.
(a) A function $u : I \to E$ is said to be *differentiable* if

$$u'(t) = \lim_{I \ni s \to t} \frac{u(s) - u(t)}{s - t}$$

exists as an element of E, for every $t \in I$. The function is *continuously differentiable* if u' is additionally continuous. We denote by $C^1(I, E)$ the space of continuously differentiable functions and by $C(I, E)$ the space of continuous functions from I to E. Since the variable $t \in I$ often stands for time, we often write $\dot{u}(t)$ in place of $u'(t)$.
(b) We define recursively $C^{k+1}(I, E) := \{u \in C^1(I, E) : u' \in C^k(I, E)\}$ and set $u^{(k+1)}(t) := (u')^{(k)}(t)$, $k \in \mathbb{N}$. We also set $C^\infty(I, E) := \bigcap_{k \in \mathbb{N}} C^k(I, E)$ for $k = \infty$. △

We can define the *Riemann integral* of a continuous function $u : [a, b] \to E$ in the same way as for real-valued functions: By π we denote a couple consisting of a partition $a = t_0 < t_1 < \cdots < t_n = b$ and a set of intermediate points $s_i \in [t_{i-1}, t_i]$, $i = 1, \ldots, n$. Then $S(\pi, u) := \sum_{i=1}^n u(s_i)(t_i - t_{i-1})$ gives us a *Riemann sum* associated with π. Denote by $|\pi| := \max_{i=1,\ldots,n}(t_i - t_{i-1})$ the *mesh size* of such a partition with intermediate points π. Then it is possible to show, just as in the case $E = \mathbb{R}$, that there exists a unique $y \in E$ such that $y = \lim_{m \to \infty} S(\pi_m, u)$ for every sequence $(\pi_m)_{m \in \mathbb{N}}$ of partitions with intermediate points for which $\lim_{m \to \infty} |\pi_m| = 0$. This defines the integral $\int_a^b u(t) \, dt := y$. The usual basic properties of Riemann integrals continue to hold here; this includes the Fundamental Theorem of Calculus: for $u \in C^1(I, E)$ and $a, b \in I$ we have $\int_a^b u'(s) \, ds = u(b) - u(a)$. Conversely, if $w : I \to E$ is continuous and $a \in I$, $u_0 \in E$, then $u(t) := u_0 + \int_a^t w(s) \, ds$ defines a function $u \in C^1(I, E)$ for which $u' = w$. Rules for interchanging the order of differentiation and summation hold just as in the scalar case.

Lemma 8.2 *Let $u_n \in C^1(I, E)$, $n \in \mathbb{N}$, be such that the series $u(t) := \sum_{n=0}^\infty u_n(t)$ and $v(t) := \sum_{n=0}^\infty u_n'(t)$ both converge uniformly on I. Then $u \in C^1(I, E)$ and $u' = v$.*

We now wish to take the assumptions of the spectral theorem (Theorem 4.46). That is, we suppose that V and H are real Hilbert spaces, where H is infinite dimensional and V is compactly and densely imbedded in H, and we let $a : V \times V \to \mathbb{R}$ be a continuous H-elliptic bilinear form. We will also assume that $a(u) := a(u, u) \geq 0$ for all $u \in V$. Denote by A the operator on H associated with a; then the domain of A takes the form $D(A) = \{u \in V : \exists f \in H, \ a(u, v) = (f, v)_H \ \text{for all} \ v \in V\}$, and given $u \in D(A)$ the vector Au is the uniquely determined vector $f \in H$ for which $a(u, v) = (f, v)_H$ for all $v \in V$ (cf. Section 4.8, in particular (4.19) and (4.20) and the following lines). By Theorem 4.46 there exist an orthonormal basis $\{e_n : n \in \mathbb{N}\}$ of H and $0 \leq \lambda_1 \leq \cdots \leq \lambda_n \leq \lambda_{n+1} \leq \cdots$ with $\lim_{n \to \infty} \lambda_n = \infty$, such that

$$D(A) = \left\{ u \in H : \sum_{n=1}^{\infty} \lambda_n^2 |(u, e_n)_H|^2 < \infty \right\} \tag{8.1}$$

and

$$Au = \sum_{n=1}^{\infty} \lambda_n (u, e_n)_H \, e_n, \quad u \in D(A). \tag{8.2}$$

In particular, $e_n \in D(A)$ and $Ae_n = \lambda_n e_n$. It also follows from (8.2) that

$$(Au, e_n)_H = \lambda_n (u, e_n)_H \quad \text{for all} \ u \in D(A). \tag{8.3}$$

We will now consider the initial value problem defined by A: given $u_0 \in H$, the objective is to determine a function $u \in C^1((0, \infty), H) \cap C([0, \infty), H)$ such that $u(t) \in D(A)$ for all $t > 0$ and

$$\dot{u}(t) + Au(t) = 0, \ t > 0, \tag{8.4a}$$
$$u(0) = u_0. \tag{8.4b}$$

Such a function u will be called a *solution* of (8.4); this function should solve an ordinary differential equation with values in an infinite-dimensional Hilbert space. Once again, we are requiring minimal regularity of u: it should be differentiable in $t > 0$ and it should satisfy the differential equation (8.4a). In order for the initial condition (8.4b) to make sense, we also require that $u : [0, \infty) \to H$ be continuous. Since A is not defined on the whole of H, but rather only on its domain $D(A)$, the demand that $u(t) \in D(A)$ for all $t > 0$ is necessary. Having made these remarks, we now wish to prove the following theorem on existence and uniqueness of solutions.

Theorem 8.3 *For every $u_0 \in H$ the problem (8.4) admits a unique solution u. This solution is given by*

$$u(t) = \sum_{n=1}^{\infty} e^{-\lambda_n t} (u_0, e_n)_H \, e_n. \tag{8.5}$$

Moreover, $u \in C^\infty((0, \infty), H)$.

Proof 1. Uniqueness: Let u be a solution of (8.4). We consider the real-valued function $u_n(t) := (u(t), e_n)_H$, $n \in \mathbb{N}$. Then $u_n : [0, \infty) \to \mathbb{R}$ is continuous, differentiable on $(0, \infty)$, and satisfies

$$\dot{u}_n(t) = (\dot{u}(t), e_n)_H = -(Au(t), e_n)_H = -\lambda_n(u(t), e_n)_H = -\lambda_n u_n(t), \ t > 0,$$

where we have used (8.3). Thus u_n satisfies a linear differential equation on $(0, \infty)$ with initial value $u_n(0) = (u_0, e_n)_H$. It follows that $u_n(t) = u_n(0)e^{-\lambda_n t} = (u_0, e_n)_H \, e^{-\lambda_n t}$. By Theorem 4.9 on orthonormal expansions, it now follows that $u(t)$ is given by (8.5).

2. Existence: We define the function u by (8.5). By Theorem 4.13, the corresponding series converges in H for all $t \geq 0$, and moreover $u(0) = u_0$. We will show that $u \in C([0, \infty), H)$. To this end we observe that $T(t)u_0 := \sum_{n=1}^{\infty} e^{-\lambda_n t}(u_0, e_n)_H e_n$ defines a continuous linear operator $T(t) : H \to H$ which satisfies $\|T(t)\| \leq 1$. If $u_0 \in span\{e_n : n \in \mathbb{N}\}$, then obviously $\lim_{h \to 0} T(t + h)u_0 = T(t)u_0$. If $u_0 \in H$ is arbitrary, then one can use the following approximation argument: choose $u_1 \in span\{e_n : n \in \mathbb{N}\}$ such that $\|u_0 - u_1\|_H \leq \varepsilon$. Then

$$\begin{aligned}
\|T(t + h)u_0 - T(t)u_0\|_H &\leq \|T(t + h)u_0 - T(t + h)u_1\|_H \\
&\quad + \|T(t + h)u_1 - T(t)u_1\|_H + \|T(t)u_1 - T(t)u_0\|_H \\
&\leq \|T(t + h)\| \, \|u_0 - u_1\|_H + \|T(t + h)u_1 - T(t)u_1\|_H \\
&\quad + \|T(t)\| \, \|u_1 - u_0\|_H \\
&\leq \varepsilon + \|T(t + h)u_1 - T(t)u_1\|_H + \varepsilon.
\end{aligned}$$

Hence $\limsup_{h \to 0} \|T(t + h)u_0 - T(t)u_0\|_H \leq 2\varepsilon$. Since $\varepsilon > 0$ is arbitrary, it follows that $\lim_{h \to 0} \|T(t + h)u_0 - T(t)u_0\|_H = 0$. To show that $u \in C^\infty((0, \infty), H)$, note that for all $\varepsilon \in (0, 1)$ and $k \in \mathbb{N}$

$$\sup_{\varepsilon \leq t \leq \frac{1}{\varepsilon}} \sup_{n \in \mathbb{N}} \lambda_n^k e^{-\lambda_n t} < \infty. \tag{8.6}$$

Then for every $k \in \mathbb{N}$ the series

$$\sum_{n=1}^{\infty} (-1)^k \lambda_n^k e^{-\lambda_n t}(u_0, e_n)_H \, e_n \tag{8.7}$$

converges uniformly in the interval $[\varepsilon, \frac{1}{\varepsilon}]$, as long as $0 < \varepsilon < 1$. By Lemma 8.2, (8.7) is exactly the kth derivative of u. This shows that $u \in C^\infty([\varepsilon, \frac{1}{\varepsilon}], H)$ for all $0 < \varepsilon < 1$, which means that $u \in C^\infty((0, \infty), H)$. Finally, by (8.2), $\dot{u}(t) = \sum_{n=0}^{\infty} -\lambda_n e^{-\lambda_n t}(u_0, e_n)_H \, e_n = -Au(t)$ for all $t > 0$. $\qquad \square$

To prove uniqueness of the solution we only needed the assumption that $u \in C^1((0, \infty), H)$; however, the proof allows us to deduce that the solution is automatically C^∞. This is due to the special form (8.5) of the solution. We next wish to prove a stronger differentiability property, which we do by replacing H by a smaller space. This property will be used in the next section to prove regularity in the space variables. Natural smaller spaces to use are the domains of definition of powers of A. Their definition and properties are the subject of the next lemma, which follows directly from (8.1) and (8.2).

Lemma 8.4 *For $m = 1$ we set $A^1 := A$, and define the operators A^m, $m \in \mathbb{N}$, recursively by $D(A^{m+1}) := \{u \in D(A^m) : A^m u \in D(A)\}$, $A^{m+1}u := A(A^m u)$. Then we have*

$$D(A^m) = \left\{ u \in H : \sum_{n=1}^{\infty} \lambda_n^{2m}(u, e_n)_H^2 < \infty \right\}, \quad A^m u = \sum_{n=1}^{\infty} \lambda_n^m (u, e_n)_H\, e_n.$$

In particular, $D(A^m)$ is a Hilbert space with respect to the inner product $(u, v)_{A^m} := \sum_{n=1}^{\infty} \lambda_n^{2m}(u, e_n)_H (e_n, v)_H + (u, v)_H$. Finally, $D(A^{m+1}) \hookrightarrow D(A^m) \hookrightarrow H$ for all $m \in \mathbb{N}$.

We can now formulate a stronger regularity statement for the solutions of (8.4).

Theorem 8.5 *Given $u_0 \in H$, let u be the solution of (8.4). Then we have that $u \in C^{\infty}((0, \infty), D(A^m))$ for all $m \in \mathbb{N}$.*

Proof We have, for $k \in \mathbb{N}$,

$$u^{(k)}(t) = \sum_{n=1}^{\infty} e^{-\lambda_n t}(-\lambda_n)^k (u_0, e_n)_H\, e_n. \tag{8.8}$$

It follows that $\|u^{(k)}(t)\|_{D(A^m)}^2 = \sum_{n=1}^{\infty} (e^{-2\lambda_n t} \lambda_n^{2k} \lambda_n^{2m} + 1)|(u_0, e_n)_H|^2$. Since we still have $\sup_{n \in \mathbb{N}} e^{-\lambda_n \varepsilon}|\lambda_n|^p < \infty$ for all $p \in \mathbb{N}$ and $\varepsilon > 0$, the series (8.8) converges uniformly on $[\varepsilon, \frac{1}{\varepsilon}]$, $\varepsilon > 0$. The claim now follows from Lemma 8.2. □

We finish by giving some additional information about the solutions.

* Comment 8.6 (Semigroups) If as in (8.5) we define the operator $T(t)$ via the rule $T(t)x := \sum_{n=1}^{\infty} e^{-\lambda_n t}(x, e_n)_H e_n$ for $x \in H$, then we obtain an operator $T(t) \in \mathcal{L}(H)$. This operator satisfies (a) $T(t + s) = T(t)T(s)$, $t, s > 0$ and (b) $\lim_{t \downarrow 0} T(t)x = x$, $x \in H$. We call such a family $(T(t))_{t > 0}$ of linear operators on H a *continuous semigroup* (or C_0-*semigroup*) if (a) and (b) hold. Given such a semigroup, we call its *generator B* the operator given by

$$D(B) := \left\{ x \in H : \lim_{t \downarrow 0} \frac{T(t)x - x}{t} \text{ exists} \right\}, \quad Bx := \lim_{t \downarrow 0} \frac{T(t)x - x}{t}.$$

One can check that in our case $B = A$; see Exercise 8.8.

A further example of a semigroup is implicitly given in Theorem 3.43: the *Gauss semigroup* on $L_2(\mathbb{R}^d)$ defined by

$$(T(t)f)(x) := \frac{1}{(4\pi t)^{d/2}} \int_{\mathbb{R}^d} e^{-(x-y)^2/4t} f(y)\, dy,$$

$x \in \mathbb{R}^d, t > 0, f \in L_2(\mathbb{R}^d)$. Its generator is the operator B given by $D(B) := H^2(\mathbb{R}^d)$, $Bu := \Delta u$. We refer to [27, 41] and [6] for the theory of semigroups, which provides the right framework for studying evolution equations.

8.2 The heat equation: Dirichlet boundary conditions

We will now apply the spectral theorem from the previous section to a special operator, the *Dirichlet Laplacian*. Let $\Omega \subset \mathbb{R}^d$ be a bounded open set; in this section we will not assume any regularity of the boundary of Ω and consider the Hilbert spaces $H := L_2(\Omega)$ and $V := H_0^1(\Omega)$. Then $H_0^1(\Omega)$ is compactly and densely imbedded in $L_2(\Omega)$ (Theorem 6.54). Let $a : H_0^1(\Omega) \times H_0^1(\Omega) \to \mathbb{R}$ be given by $a(u, v) := \int_\Omega \nabla u \, \nabla v \, dx$, then a is continuous, symmetric and coercive (the latter property is a consequence of the Poincaré inequality, Theorem 6.32). Let A be the operator on $L_2(\Omega)$ associated with a; this operator can be described as follows.

Theorem 8.7 *We have* $D(A) = \{u \in H_0^1(\Omega) \cap H_{\text{loc}}^2(\Omega) : \Delta u \in L_2(\Omega)\}$ *and* $Au = -\Delta u$.

Proof Let $u \in D(A)$ with $Au = f$. Then by definition of A we have the identity $\int_\Omega \nabla u(x) \nabla \varphi(x) \, dx = \int_\Omega f(x)\varphi(x) \, dx$ for all $\varphi \in H_0^1(\Omega)$ and in particular for all $\varphi \in \mathcal{D}(\Omega)$. This means that $-\Delta u = f$ weakly. Now by Theorem 6.58 we have $u \in H_{\text{loc}}^2(\Omega)$. Conversely, if $u \in H_0^1(\Omega) \cap H_{\text{loc}}^2(\Omega)$ and $f := -\Delta u \in L_2(\Omega)$, then

$$\int_\Omega \nabla u(x) \nabla \varphi(x) \, dx = -\int_\Omega \Delta u(x)\varphi(x) \, dx = \int_\Omega f(x)\varphi(x) \, dx$$

for all $\varphi \in \mathcal{D}(\Omega)$. Since $\mathcal{D}(\Omega)$ is dense in $H_0^1(\Omega)$, it thus follows that $a(u, v) = \int_\Omega \nabla u \, \nabla v = \int_\Omega fv$ for all $v \in H_0^1(\Omega)$. Hence $u \in D(A)$ and $Au = f$. $\qquad\square$

We call the operator $\Delta^D := -A$ the *Laplacian with Dirichlet boundary conditions*, or for short the *Dirichlet Laplacian*. To summarize, we have

$$D(\Delta^D) := \{u \in H_0^1(\Omega) \cap H_{\text{loc}}^2(\Omega) : \Delta u \in L_2(\Omega)\},$$
$$\Delta^D u = \Delta u, \; u \in D(\Delta^D).$$

The spectral theorem (Theorem 4.46) yields the existence of an orthonormal basis of eigenvectors $\{e_n : n \geq 1\}$ of $L_2(\Omega)$ and associated eigenvalues $0 < \lambda_1^D \leq \lambda_2^D \leq \cdots \leq \lambda_n^D \leq \lambda_{n+1}^D \leq \cdots$ such that $\lim_{n\to\infty} \lambda_n^D = \infty$, $e_n \in D(\Delta^D)$ and $-\Delta^D e_n = \lambda_n^D e_n$. The set $\{\lambda_n^D : n \in \mathbb{N}\}$ represents the totality of the eigenvalues of the Dirichlet Laplacian (where some eigenvalues may be repeated); we often refer to them simply as the *Dirichlet eigenvalues*.

If $\lambda = \lambda_n^D$ for some $n \in \mathbb{N}$, then the Poisson equation with reaction term

$$\lambda v + \Delta v = f \tag{8.9}$$

with $v \in H_0^1(\Omega) \cap H_{\text{loc}}^2(\Omega)$ the unknown, no longer admits a unique solution: if v is a solution, then so too is $v + \alpha e_n$ for all $\alpha \in \mathbb{R}$. On the other hand, Corollary 4.48 shows that (8.9) has a unique solution whenever $\lambda \notin \{\lambda_n^D : n \in \mathbb{N}\}$.

By Theorem 4.50, the form a can be represented in diagonal form by

$$\int_\Omega |\nabla u|^2 \, dx = \sum_{n=1}^\infty \lambda_n^D |(u, e_n)_{L_2}|^2 \tag{8.10}$$

for all $u \in H_0^1(\Omega)$. Since on the other hand we have Parseval's identity (Theorem 4.9(b))

$$\int_\Omega |u|^2 \, dx = \sum_{n=1}^\infty |(u, e_n)_{L_2}|^2$$

for all $u \in L_2(\Omega)$, and $\lambda_n^D \leq \lambda_{n+1}^D$, we obtain

$$\int_\Omega |\nabla u|^2 \, dx \geq \lambda_1^D \int_\Omega |u|^2 \, dx \tag{8.11}$$

for all $u \in H_0^1(\Omega)$. This is exactly the Poincaré inequality (Theorem 6.32), but this time with the optimal constant. Indeed, if we choose $u = e_1$ in (8.11), then by (8.10) we obtain equality; the estimate (8.11) is thus sharp.

We now turn to our main objective, the analysis of the heat equation with Dirichlet boundary conditions. To this end we consider the following initial-boundary value problem. Let $u_0 \in L_2(\Omega)$ be a given initial value. The problem is to find a $u \in C^\infty((0, \infty) \times \Omega)$ such that

$$u(t, \cdot) \in H_0^1(\Omega), \qquad\qquad t > 0, \tag{8.12a}$$
$$u_t(t, x) = \Delta u(t, x), \qquad\qquad t > 0, \ x \in \Omega, \tag{8.12b}$$

$$\lim_{t \downarrow 0} \int_\Omega |u(t, x) - u_0|^2 \, dx = 0. \tag{8.12c}$$

Here the Laplacian is to be taken with respect to the space variable x, that is,

$$\Delta u(t, x) := \sum_{j=1}^d \frac{\partial^2 u(t, x)}{\partial x_j^2}, \qquad t > 0, \ x \in \Omega.$$

The condition (8.12b) states that u satisfies the heat equation on $(0, \infty) \times \Omega$, while the initial condition is specified in (8.12c). In fact, we have chosen an initial condition u_0 which is merely in $L_2(\Omega)$. As such, the convergence of $u(t, \cdot)$ to u_0 in quadratic mean as $t \downarrow 0$, as stipulated by (8.12c), is a natural condition. The requirement that $u(t, \cdot) \in H_0^1(\Omega)$, $t > 0$, means that u should satisfy the Dirichlet boundary condition in the usual weak sense. If we take an eigenfunction e_n as the initial condition, then

$$u(t, x) := e^{-\lambda_n^D t} e_n(x), \qquad t > 0, \ x \in \Omega$$

is a solution of (8.12a). Note that by Theorem 6.58 the function e_n possesses derivatives of all orders; and we see that this special solution has "separated variables".

In fact, our spectral decomposition method is nothing other than the method of *separation of variables* (space and time variables in our case).

Now if $u_0 \in L_2(\Omega)$ is an arbitrary initial condition, then we can create a special solution as a series of the above solutions. More precisely, by Theorem 4.9, we may write $u_0 = \sum_{n=1}^{\infty}(u_0, e_n)_{L_2(\Omega)} e_n$, where the series converges in $L_2(\Omega)$. The series

$$w(t) := \sum_{n=1}^{\infty} e^{-\lambda_n t}(u_0, e_n)_{L_2(\Omega)} e_n \tag{8.13}$$

likewise converges in $L_2(\Omega)$ for all $t > 0$. Now we know from Theorem 8.3 that w solves the abstract Cauchy problem (8.4), and in fact $w \in C^{\infty}((0, \infty), L_2(\Omega))$. We will now show using Theorem 8.5 that u defines a solution of (8.12). We stress that the function $w(t)$ defined by (8.13) is in $L_2(\Omega)$ for each $t > 0$.

Theorem 8.8 *We have* $w(t) \in C^{\infty}(\Omega)$ *for all* $t > 0$. *If we set*

$$u(t, x) := w(t)(x), \quad t > 0, \ x \in \Omega, \tag{8.14}$$

then $u \in C^{\infty}((0, \infty) \times \Omega)$ *and* u *satisfies* (8.12).

Proof 1. Let $A = -\Delta^D$ be the operator on $L_2(\Omega)$ associated with the form a; then $D(A) \subset H^2_{loc}(\Omega)$. We start by showing by induction on $m \in \mathbb{N}$ that $D(A^m) \subset H^{2m}_{loc}(\Omega)$. We know that the claim is true when $m = 1$. Now assume that it holds for some $m \in \mathbb{N}$ and let $v \in D(A^{m+1})$. Then by the induction hypothesis, $\Delta v = Av \in D(A^m) \subset H^{2m}_{loc}(\Omega)$. It follows by Theorem 6.58 that $v \in H^{2m+2}_{loc}(\Omega)$.

Now let $k \in \mathbb{N}$ be such that $k > d/4$, then by Theorem 6.58, we have the inclusions $D(A^{m+k}) \subset H^{2m+2k}_{loc}(\Omega) \subset C^m(\Omega)$ for all $m \in \mathbb{N}_0$. Now $D(A^{m+k})$ is a Hilbert space (Lemma 8.4) which is continuously imbedded in $L_2(\Omega)$. For $U \Subset \Omega$, the space $C^m(\overline{U})$ is a Banach space with respect to the norm

$$\|v\|_{C^m(\overline{\Omega})} := \sum_{|\alpha| \leq m} \|D^\alpha u\|_\infty,$$

where $D^\alpha u = \frac{\partial^{|\alpha|} u}{\partial x_1^{\alpha_1} \ldots \partial x_d^{\alpha_d}}$, $\alpha = (\alpha_1, \ldots, \alpha_d) \in \mathbb{N}_0^d$, $|\alpha| = \alpha_1 + \cdots + \alpha_d$ and $D^0 u = u$. In particular, $\|v\|_\infty \leq \|v\|_{C^m(\overline{\Omega})}$. By the closed graph theorem, it follows that the restriction mapping $v \mapsto v_{|\overline{U}} : D(A^{m+k}) \to C^m(\overline{U})$ is continuous. Indeed, if $v_n, v \in D(A^{m+k})$, $g \in C^m(\overline{U})$ satisfy $v_n \to v$ in $D(A^{m+k})$ and $v_{n|\overline{U}} \to g$ in $C^m(\overline{U})$ as $n \to \infty$, then v_n converges to v in $L_2(\Omega)$ and to g uniformly in $\overline{U}$. Hence $v = g$ almost everywhere in $\overline{U}$. But since v is continuous (or, more precisely, v has a continuous representative), we conclude that $v = g$, meaning that the graph of the restriction mapping is closed.

By Theorem 8.5, $w \in C^{\infty}((0, \infty), D(A^{m+k}))$; hence the mapping $t \mapsto w(t)|_{\overline{U}} : (0, \infty) \to C^m(\overline{U})$ has derivatives of all orders. In particular, $w(t) \in C(\Omega)$ for all

$t > 0$. If we now set $u(t, x) := w(t)(x), t > 0, x \in \Omega$, then it follows that u is infinitely differentiable in $t > 0$, and differentiable in $x \in \Omega$, and

$$u_t(t, x) = \dot{w}(t)(x) = \Delta w(t) = \sum_{j=1}^{d} \frac{\partial^2 u(t, x)}{\partial x_j^2} =: \Delta u(t, x).$$

Hence $u \in C^\infty((0, \infty) \times \Omega)$ and u solves the heat equation (8.12a).

2. We have $w(t) = \sum_{n=1}^{\infty} e^{-\lambda_n^D t}(u_0, e_n)_{L_2(\Omega)} e_n$. Since

$$\sum_{n=1}^{\infty} \lambda_n^D |e^{-\lambda_n^D t}(u_0, e_n)_{L_2(\Omega)}|^2 < \infty,$$

it follows from Theorem 4.50 that $w(t) \in H_0^1(\Omega)$ for all $t > 0$. Hence $u(t, \cdot) = w(t)$ satisfies the boundary condition in (8.12).

3. Finally, it follows from Parseval's identity (Theorem 4.9) that

$$\int_\Omega |u(t, x) - u_0(x)|^2 \, dx = \|w(t) - u_0\|_{L_2(\Omega)}^2 = \sum_{n=1}^{\infty} |(w(t) - u_0, e_n)_{L_2(\Omega)}|^2$$

$$= \sum_{n=1}^{\infty} (e^{-\lambda_n^D t} - 1)^2 |(u_0, e_n)_{L_2(\Omega)}|^2 \to 0, \quad \text{as } t \to 0.$$

The convergence here follows from the Dominated Convergence Theorem applied to the discrete space ℓ^2; alternatively, it can be shown as follows: given $\varepsilon > 0$ there exists an $N \in \mathbb{N}$ such that $\sum_{n=N+1}^{\infty} 4|(u_0, e_n)_{L_2(\Omega)}|^2 \leq \varepsilon/2$. There exists a $t_0 > 0$ such that $\sum_{n=1}^{N} (e^{-\lambda_n^D t} - 1)^2 |(u_0, e_n)_{L_2(\Omega)}|^2 \leq \varepsilon/2$ for all $0 \leq t \leq t_0$, whence

$$\sum_{n=1}^{\infty} (e^{-\lambda_n^D t} - 1)^2 |(u_0, e_n)_{L_2(\Omega)}|^2$$

$$= \sum_{n=1}^{N} (e^{-\lambda_n^D t} - 1)^2 |(u_0, e_n)_{L_2(\Omega)}|^2 + \sum_{n=N+1}^{\infty} (e^{-\lambda_n^D t} - 1)^2 |(u_0, e_n)_{L_2(\Omega)}|^2 \leq \varepsilon$$

for all $0 \leq t \leq t_0$. We have thus proved (8.12c). □

We next note a few further properties of the solution u defined by (8.13) and (8.14): we have

$$\int_\Omega |u(t, x)|^2 \, dx \leq e^{-2\lambda_1^D t} \int_\Omega |u_0(x)|^2 \, dx, \quad t > 0, \quad (8.15)$$

since

$$\int_\Omega |u(t, x)|^2 \, dx = \|w(t)\|_{L_2(\Omega)}^2 = \sum_{n=1}^{\infty} e^{-2\lambda_n^D t} |(u_0, e_n)_{L_2(\Omega)}|^2$$

$$\leq e^{-2\lambda_1^D t} \sum_{n=1}^{\infty} |(u_0, e_n)_{L_2(\Omega)}|^2 = e^{-2\lambda_1^D t} \|u_0\|_{L_2(\Omega)}^2,$$

where we have used Parseval's identity twice. The energy of u can be expressed as follows:

$$\int_{\Omega} |\nabla u(t, x)|^2 \, dx = \sum_{n=1}^{\infty} \lambda_n^D e^{-2\lambda_n^D t} |(u_0, e_n)_{L_2(\Omega)}|^2 < \infty; \qquad (8.16)$$

in particular, $\int_{\Omega} |\nabla u(t, x)|^2 \, dx$ is a monotonically decreasing function of $t > 0$. It is bounded in $t > 0$ if and only if the initial value u_0 is in $H_0^1(\Omega)$. By using the conditions (8.15) and (8.16), we can now show that our solution from (8.13) is unique.

Theorem 8.9 *Given $u_0 \in L_2(\Omega)$, let $u \in C^{1,2}((0, \infty) \times \Omega)$ be a solution of (8.12) such that*

$$\int_0^T \int_{\Omega} |u(t, x)|^2 \, dx \, dt < \infty, \quad T > 0, \qquad (8.17)$$

and

$$\int_{\varepsilon}^T \int_{\Omega} |\nabla u(t, x)|^2 \, dx \, dt < \infty, \quad 0 < \varepsilon < T < \infty. \qquad (8.18)$$

Then $u(t, x) = w(t)(x)$, $t > 0$, $x \in \Omega$, where w is given by (8.13).

Proof Fix $T > 0$, then for $\varphi \in C_c^1(0, T)$ and $v \in \mathcal{D}(\Omega)$ we have

$$-\int_0^T \varphi'(t) \int_{\Omega} u(t, x) v(x) \, dx \, dt = -\int_{\Omega} \int_0^T \varphi'(t) u(t, x) \, dt \, v(x) \, dx$$

$$= \int_{\Omega} \int_0^T \varphi(t) u_t(t, x) \, dt \, v(x) \, dx = \int_0^T \int_{\Omega} v(x) u_t(t, x) \, dx \, \varphi(t) \, dt$$

$$= \int_0^T \int_{\Omega} v(x) \Delta u(t, x) \, dx \, \varphi(t) \, dt = -\int_0^T \int_{\Omega} \nabla v(x) \nabla u(t, x) \, dx \, \varphi(t) \, dt,$$

where $\nabla u(t, x)$ denotes the gradient with respect to the space variables x. Now fix $n \in \mathbb{N}$ and set $u_n(t) := \int_{\Omega} u(t, x) e_n(x) \, dx$. Since we can approximate e_n by test functions v in the $H^1(\Omega)$-norm, it follows from the above identity that

$$-\int_0^T \varphi'(t) u_n(t) \, dt = -\int_0^T \int_{\Omega} \nabla e_n(x) \nabla u(t, x) \, dx \, \varphi(t) \, dt$$

$$= -\lambda_n \int_0^T u_n(t) \varphi(t) \, dt. \qquad (8.19)$$

By (8.17) we have $u_n \in L_2(0, T)$. Since (8.19) holds for all $\varphi \in C_c^1(0, T)$, we obtain that $u_n \in H^1(0, T)$ and $u_n' = -\lambda_n u_n$ (see Definition 5.2). It now follows from Corollary 5.11 that $u_n \in C^1([0, T])$. Hence the function u_n solves the given linear differential equation with initial condition $u_n(0) = (u_0, e_n)_{L_2(\Omega)}$, whence $u_n(t) = e^{-\lambda_n t}(u_0, e_n)_{L_2(\Omega)}$. Finally, we apply Theorem 4.9(a) on orthogonal expansions to conclude that $u(t, \cdot) = w(t)$ for all $t \geq 0$. $\qquad\square$

We recall the physical interpretation of the solution u of (8.12) which was derived in Section 1.6.2. Consider, say, the case $d = 3$; then Ω is a body in space and u_0 is an initial distribution of heat in Ω. The value $u(t, x)$ gives the temperature at the point x at time t; the Dirichlet boundary condition determines that the temperature at the boundary is always 0, at every point in time. We may imagine that the body is lying, say, in an ice bath. We see from equation (8.15) that the temperature converges to 0 exponentially as $t \to \infty$, which reflects what we might expect based on the model. An even more illuminating example is that of diffusion, such as of ink in water. If u_0 is the initial concentration, then $u(t, x)$ gives the concentration of the ink at time t. For a measurable set $U \subset \Omega$, the total amount of ink in the part U of the body Ω at time t is given by the integral $\int_U u(t, x)\, dx$.

* **Comment 8.10 (Classical Dirichlet boundary conditions)** In this section we have worked with arbitrary bounded open sets $\Omega \subset \mathbb{R}^d$, without imposing any conditions on the boundary. This means that the solution $u(t, x)$ need not always be continuous at the boundary. In fact, the following statements about the continuity of solutions at the boundary hold.

1. Continuity of the first eigenfunction at the boundary.
The set Ω is Dirichlet regular (see Section 6.8) if and only if $e_1 \in C_0(\Omega)$, that is, $\lim_{x \to z} e_1(x) = 0$ for all $z \in \partial\Omega$, [7].

2. Continuity of the solution from Theorem 8.8 at the boundary.
If Ω is Dirichlet regular, then the solution u from Theorem 8.8 is continuous on $(0, \infty) \times \overline{\Omega}$ if one sets $u(t, z) = 0$ pointwise for all $t > 0$ and $z \in \partial\Omega$.

3. Inhomogeneous heat boundary value problem.
Let Ω be Dirichlet regular and $T > 0$. Also let $g \in C(\partial^*\Omega_T)$, where we use $\partial^*\Omega_T = ([0, T] \times \partial\Omega) \cup (\{0\} \times \overline{\Omega})$ to denote the parabolic boundary of $(0, T) \times \Omega$. Then there exists a uniquely determined function $u \in C([0, T] \times \overline{\Omega})$ such that u is infinitely differentiable in $(0, T) \times \Omega$, $u_t = \Delta u$ holds, and $u_{|\partial^*\Omega_T} = g$ [6, Theorem 6.2.8]. Uniqueness follows directly from the parabolic maximum principle in Theorem 3.31.

* **Comment 8.11 (Positivity)** Let $\Omega \subset \mathbb{R}^d$ be a bounded domain; we also assume that Ω is connected, but make no assumptions on its boundary. Then the following statements hold.

1. Strict positivity.
If the initial value $u_0 \in L_2(\Omega)$ is a positive function (that is, $u_0(x) \geq$ almost everywhere) which is not 0 almost everywhere, then in fact $u(t, x) > 0$ for all $t > 0$ and all $x \in \Omega$.

2. Special role of the first eigenfunction.
Under these assumptions on Ω, we have $\lambda_1^D < \lambda_2^D$ (that is, the first eigenvalue is *simple*) and the first eigenfunction e_1 can be chosen in such a way that $e_1(x) > 0$ for all $x \in \Omega$.

For a systematic treatment of positivity in the context of the heat equation we refer to [5] and [11].

8.3 The heat equation: Robin boundary conditions

In this section we take $\Omega \subset \mathbb{R}^d$ to be a bounded domain with C^1-boundary (see Chapter 7). Our goal is to investigate the heat equation with Robin boundary conditions (see (8.26)). In the case of elliptic equations, we already treated these boundary conditions in Section 7.5. Here we also wish to consider the corresponding parabolic equation. We start by defining a realization of the Laplacian with Robin boundary conditions on $L_2(\Omega)$, to which we then apply the spectral theorem. For this, we specify a bounded Borel measurable function $b : \partial\Omega \to [0, \infty)$ and define the continuous bilinear form $a : H^1(\Omega) \times H^1(\Omega) \to \mathbb{R}$ by

$$a(u, v) := \int_\Omega \nabla u(x)\nabla v(x)\,dx + \int_{\partial\Omega} b\,Tu\,Tv\,d\sigma(z),$$

where $T : H^1(\Omega) \to L_2(\partial\Omega)$ denotes the trace operator from Theorem 7.34. Since

$$a(u) + \|u\|^2_{L_2(\Omega)} \geq \|u\|^2_{H^1(\Omega)}$$

for all $u \in H^1(\Omega)$, the form is $L_2(\Omega)$-elliptic (see (4.22) and (4.5)). Let A be the operator on $L_2(\Omega)$ associated with a. By definition, this means for $u, f \in L_2(\Omega)$ that $u \in D(A)$ and $Au = f$ if and only if $u \in H^1(\Omega)$ and

$$a(u, v) = \int_\Omega f(x)v(x)\,dx \tag{8.20}$$

for all $v \in H^1(\Omega)$. We now wish to describe the operator A. We will be using the definition of the weak normal derivative introduced in Definition 7.36.

Theorem 8.12 We have $D(A) = \{u \in H^1(\Omega) \cap H^2_{\mathrm{loc}}(\Omega) : \Delta u \in L_2(\Omega), \frac{\partial u}{\partial \nu} + b(Tu) = 0\}$, and $Au = -\Delta u$ for all $u \in D(A)$.

Proof Let $u \in D(A)$ and set $f := Au$. Then by (8.20),

$$\int_\Omega \nabla u(x)\nabla v(x)\,dx + \int_{\partial\Omega} b\,Tu\,Tv\,d\sigma = \int_\Omega f(x)v(x)\,dx \tag{8.21}$$

for all $v \in H^1(\Omega)$. Hence, in particular, $\int_\Omega \nabla u(x)\nabla v(x)\,dx = \int_\Omega f(x)v(x)\,dx$ for $v \in \mathcal{D}(\Omega)$. it follows from Theorem 6.58 that $u \in H^2_{\mathrm{loc}}(\Omega)$ and $-\Delta u = f$. If we now plug this expression for f into (8.21), then we obtain

$$\int_\Omega \nabla u(x)\nabla v(x)\,dx + \int_\Omega (\Delta u)(x)v(x)\,dx = -\int_{\partial\Omega} b\,Tu\,Tv\,d\sigma \tag{8.22}$$

for all $v \in H^1(\Omega)$. This means exactly that $\frac{\partial u}{\partial \nu} + b(Tu) = 0$ (see Definition 7.36). Conversely, suppose that $u \in H^1(\Omega) \cap H^2_{\mathrm{loc}}(\Omega)$ is such that $-f := \Delta u \in L_2(\Omega)$ and $\frac{\partial u}{\partial \nu} + b(Tu) = 0$. Then (8.22) holds for all $v \in H^1(\Omega)$ by Definition 7.36. Hence (8.20) is satisfied, that is, we have $u \in D(A)$ and $Au = f = -\Delta u$. $\qquad\square$

We shall call the operator $-A =: \Delta^b$ the *Laplacian with Robin boundary conditions*, or just *Robin Laplacian*. If $b = 0$ then we recover Neumann boundary conditions, and so we call the operator $\Delta^0 =: \Delta^N$ the *Neumann Laplacian*. Since the imbedding $H^1(\Omega) \hookrightarrow L_2(\Omega)$ is compact, we can apply the spectral theorem (Theorem 4.46) to obtain the existence of an orthonormal basis $\{e_n : n \in \mathbb{N}\}$ of $L_2(\Omega)$ and $\lambda_n^b \in \mathbb{R}$ such that $0 \leq \lambda_1^b \leq \cdots \leq \lambda_n^b \leq \lambda_{n+1}^b \leq \cdots$, $\lim_{n\to\infty} \lambda_n^b = \infty$ with $e_n \in D(\Delta^b)$ and

$$-\Delta^b e_n = \lambda_n^b e_n, \quad n \in \mathbb{N}. \tag{8.23}$$

If λ is a real number different from λ_n^b for all $n \in \mathbb{N}$, then by Corollary 4.48 the Poisson equation

$$\lambda v + \Delta v = f \quad \text{in } \Omega, \tag{8.24a}$$

$$\frac{\partial v}{\partial \nu} + bTv = 0 \quad \text{on } \partial\Omega, \tag{8.24b}$$

has, for every $f \in L_2(\Omega)$, a unique solution $v \in H^1(\Omega) \cap H^2_{\text{loc}}(\Omega)$.

We now turn to the parabolic equation. Let $u_0 \in L_2(\Omega)$ be a given initial value. If we set

$$w(t) := \sum_{n=1}^{\infty} e^{-\lambda_n^b t}(u_0, e_n)_{L_2(\Omega)} e_n, \tag{8.25}$$

then w is the unique solution of (8.4) for $A = -\Delta^b$. In particular, we have $w \in C^\infty((0, \infty), L_2(\Omega))$ and $w(t) \in H^1(\Omega) \cap H^2_{\text{loc}}(\Omega)$, $t > 0$, satisfies

$$\dot{w}(t) = \Delta w(t), \qquad t > 0, \tag{8.26a}$$

$$\frac{\partial w}{\partial \nu} + bTw(t) = 0, \quad t > 0, \tag{8.26b}$$

$$\lim_{t\downarrow 0} w(t) = u_0 \quad \text{in } L_2(\Omega). \tag{8.26c}$$

If we exploit the regularity properties of the Laplacian just as in the proof of Theorem 8.8, then we see that $w(t) \in C^\infty(\Omega)$, $t > 0$. We can thus define

$$u(t, x) := w(t)(x), \quad t > 0, x \in \Omega, \tag{8.27}$$

and obtain a function $u \in C^\infty((0, \infty) \times \Omega)$ which solves the heat equation

$$u_t(t, x) = \Delta u(t, x), \quad t > 0, \ x \in \Omega,$$

and which satisfies both the Robin boundary condition (8.26b) and the initial condition (8.26c).

We next wish to examine the asymptotic behavior of $w(t)$ as $t \to \infty$. We recall that Ω is connected. The following theorem asserts that every solution decays to 0 exponentially fast.

Theorem 8.13 *Assume that $b(z) \neq 0$ on a subset of $\partial\Omega$ of positive surface measure. Then the first eigenvalue $\lambda_1^b > 0$, and*

$$\|w(t)\|_{L_2(\Omega)}^2 \leq e^{-\lambda_1^b t} \|u_0\|_{L_2(\Omega)}^2, \quad t > 0. \tag{8.28}$$

Proof Suppose that $\lambda_1^b = 0$. Then

$$0 = a(e_1) = \int_\Omega |\nabla e_1|^2 \, dx + \int_{\partial\Omega} b(Te_1)^2 \, d\sigma.$$

Since $b \geq 0$, this implies that $\nabla e_1(z) = 0$ almost everywhere, and so, by Theorem 6.46, e_1 is a constant function, that is, $e_1(x) = c$ for all $x \in \Omega$. In particular, $0 = a(e_1) = c^2 \int_{\partial\Omega} b(z) \, d\sigma(z)$. If $c \neq 0$, then the only possibility is that $b(z) = 0$ σ-almost everywhere, which is a contradiction to the assumptions of the lemma. $\square$

In the case $b = 0$ we recover Neumann boundary conditions; in this case, we denote the eigenvalues by $\lambda_n^N := \lambda_n^0$. These are the eigenvalues of the Neumann Laplacian $-\Delta^N$, for short, the *Neumann eigenvalues*. The Neumann Laplacian Δ^N is not injective: its kernel consists exactly of the constant functions. Indeed, by definition, $u \in \ker \Delta^N$ if and only if $u \in H^1(\Omega)$ and $\int_\Omega \nabla u \nabla v \, dx = 0$ for all $v \in H^1(\Omega)$. By Theorem 6.19 this is equivalent to u being constant. Thus $\lambda_1^N = 0$ and $\lambda_2^N > 0$. The constant eigenfunction normalized to have $L^2(\Omega)$-norm 1 is $e_1 = |\Omega|^{-1/2} \mathbb{1}_\Omega$. We have established that in the case of Robin boundary conditions the solution decays exponentially to 0, regardless of how small the absorption b on the boundary may be. In the case of Neumann boundary conditions, that is, when the absorption b is equal to 0, we encounter completely different asymptotic behavior, namely convergence to a (nonzero) equilibrium.

Theorem 8.14 (Convergence to equilibrium) *In the case of Neumann boundary conditions, that is, when $b = 0$, we have*

$$\lim_{t \to \infty} w(t) = |\Omega|^{-1} \int_\Omega u_0 \, dx \cdot \mathbb{1}_\Omega \tag{8.29}$$

in $L_2(\Omega)$.

Proof The statement follows from the simplicity of the first eigenvalue. More precisely, since $\lambda_1^N = 0$ and $e_1 = |\Omega|^{-1/2} \mathbb{1}_\Omega$, we have

$$|\Omega|^{-1} \int_\Omega u_0 \, dx \mathbb{1}_\Omega = (u_0, e_1)_{L_2(\Omega)} e_1.$$

Then

$$w(t) - (u_0, e_1)_{L_2(\Omega)} e_1 = \sum_{n=2}^{\infty} e^{-\lambda_n^N t} (u_0, e_n)_{L_2(\Omega)} e_n.$$

By Parseval's identity, it follows that

$$\left\| w(t) - (u_0, e_1)_{L_2(\Omega)} e_1 \right\|_{L_2(\Omega)}^2 = \sum_{n=2}^{\infty} e^{-\lambda_n^N 2t} |(u_0, e_n)_{L_2(\Omega)}|^2$$

$$\leq e^{-\lambda_2^N 2t} \sum_{n=2}^{\infty} |(u_0, e_n)_{L_2(\Omega)}|^2$$

$$\leq e^{-\lambda_2^N 2t} \|u_0\|_{L_2(\Omega)}^2,$$

which was the claim. □

8.4 The wave equation

We start by analyzing the wave equation in an abstract form, just as we did with the heat equation in Section 8.1. We then interpret the results we obtain in the special case of the Laplacian with Dirichlet or Neumann boundary conditions. We suppose that two real Hilbert spaces V and H are given, such that V is compactly imbedded in H. We also take $a : V \times V \to \mathbb{R}$ to be a continuous, symmetric bilinear form which is H-elliptic (see (4.22)). We finally wish to assume that

$$a(u) := a(u, u) \geq 0, \quad u \in V. \tag{8.30}$$

Let A be the operator on H associated with $a(\cdot, \cdot)$. We shall study the following abstract second order initial value problem. Throughout this section we will assume that $0 < T < \infty$. Let $u_0 \in D(A)$ and $u_1 \in H$. The goal is to find a function $w \in C^2([0, T], H)$ such that $w(t) \in D(A)$ for $t \in [0, T]$ and

$$\ddot{w}(t) + Aw(t) = 0, \qquad t \in [0, T], \tag{8.31a}$$
$$w(0) = u_0, \ \dot{w}(0) = u_1. \tag{8.31b}$$

That is, u_0 is the given *initial value* and u_1 is the given *initial velocity*. By spectrally decomposing A, we can reduce problem (8.31) to a one-dimensional second order linear differential equation. Indeed, by the spectral theorem (Theorem 4.46) there exist an orthonormal basis $\{e_n : n \in \mathbb{N}\}$ of H and numbers $\lambda_n \in \mathbb{R}$ with $0 \leq \lambda_n \leq \lambda_{n+1}$, $\lim_{n \to \infty} \lambda_n = \infty$, such that A is given by

$$D(A) = \left\{ v \in H : \sum_{n=1}^{\infty} \lambda_n^2 |(v, e_n)_H|^2 < \infty \right\}, \quad Av = \sum_{n=1}^{\infty} \lambda_n (v, e_n)_H e_n; \tag{8.32}$$

in particular, $e_n \in D(A)$ and $Ae_n = \lambda_n e_n$ for all $n \in \mathbb{N}$. By (8.30), we have that $\lambda_n = (Ae_n, e_n)_H = a(e_n) \geq 0$, $n \in \mathbb{N}$. We now consider the following special initial values.

(a) Take $u_0 = e_n$ and $u_1 = 0$. Then $w(t) = \cos(\sqrt{\lambda_n}t)e_n$ is a solution of (8.31).

(b) Take $u_0 = 0$ and $u_1 = e_n$. If $\lambda_n > 0$, then $w(t) = \lambda_n^{-1/2} \sin(\sqrt{\lambda_n}t)e_n$ defines a solution of (8.31). If $\lambda_n = 0$, then $w(t) = te_n$ is a solution of (8.31).

We can now express the general solution of (8.31) as a series of solutions of these special types and prove the following theorem.

Theorem 8.15 *Let $u_0 \in D(A)$ and $u_1 \in V$. Then (8.31) admits a unique solution $w \in C^2([0, T], H)$ such that $w(t) \in D(A)$ for all $t \in [0, T]$.*

Proof Uniqueness: Let w be a solution of (8.31), and set $w_n(t) := (w(t), e_n)_H$ for $n \in \mathbb{N}$ and $t \in [0, T]$. Then $w_n : [0, T] \to \mathbb{R}$ is twice differentiable with derivative

$$\ddot{w}_n(t) = (\ddot{w}(t), e_n)_H = -(Aw(t), e_n)_H = -\sum_{k=1}^{\infty} \lambda_k (w(t), e_k)_H (e_k, e_n)_H$$

$$= -\lambda_n (w(t), e_n)_H = -\lambda_n w_n(t), \quad t \in (0, T).$$

Hence w_n satisfies the second order linear differential equation $\ddot{w}_n(t) = -\lambda_n w_n(t)$, $t \geq 0$, with initial conditions $w_n(0) = (u_0, e_n)_H$, $\dot{w}_n(0) = (u_1, e_n)_H$. It follows that

$$w_n(t) = \cos(\sqrt{\lambda_n}t)(u_0, e_n)_H + \frac{1}{\sqrt{\lambda_n}} \sin(\sqrt{\lambda_n}t)(u_1, e_n)_H, \text{if } \lambda_n > 0, \qquad (8.33)$$

and

$$w_n(t) = (u_0, e_n)_H + t(u_1, e_n)_H, \quad \text{if } \lambda_n = 0. \qquad (8.34)$$

Indeed, it is easy to check that w_n, thus defined, is a solution of the given differential equation. But it is the *unique* solution, as follows from simple properties of the theory of ordinary differential equations; alternatively, we may use the conservation of energy property obtained below in Theorem 8.16. Since $w(t) = \sum_{n=1}^{\infty} w_n(t)e_n$ is the unique expansion of $w(t)$ with respect to the orthonormal basis $\{e_n : n \in \mathbb{N}\}$, the uniqueness statement has been proved.

Existence: We now define $w_n(t)$ by (8.33). Since $\sum_{n=1}^{\infty} \lambda_n^2 |(u_0, e_n)_H|^2 < \infty$ and $\sum_{n=1}^{\infty} \lambda_n |(u_1, e_n)_H|^2 < \infty$ (see (4.26)), not only does $w(t) := \sum_{n=1}^{\infty} w_n(t)e_n$ converge uniformly in H, but so too do the series $\sum_{n=1}^{\infty} \dot{w}_n(t)e_n$ and $\sum_{n=1}^{\infty} \ddot{w}_n(t)e_n$.

Hence, by Lemma 8.2, $w \in C^2([0, T], H)$. Since $\sum_{n=1}^{\infty} \lambda_n^2 |w_n(t)|^2 < \infty$, we have $w(t) \in D(A)$ for all $t \in [0, T]$. Thus

$$\ddot{w}(t) = \sum_{n=1}^{\infty} \ddot{w}_n(t)e_n = \sum_{n=1}^{\infty} w_n(t)\lambda_n e_n = -Aw(t)$$

by definition of A. In addition,

$$w(0) = \sum_{n=1}^{\infty}(u_0, e_n)_H e_n = u_0 \quad \text{and} \quad \dot{w}(0) = \sum_{n=1}^{\infty}(u_1, e_n)_H e_n = u_1.$$

This shows that w solves (8.31), and we have proved existence of solutions. □

We also wish to note the following principle of conservation of energy.

Theorem 8.16 (Conservation of energy) *Let w be the solution of* (8.31). *Then,*

$$a(w(t)) + \|\dot{w}(t)\|_H^2 = a(u_0) + \|u_1\|_H^2 \quad \text{for all } t \in [0, T]. \tag{8.35}$$

One way to obtain (8.35) is via a direct calculation using the form of the solution which we obtained in the previous proof (Exercise 8.1). More interesting, however, is the following proof. Once we have (8.35), we can use it to give a second proof of uniqueness in Theorem 8.15 (see below). In order to prove Theorem 8.16 we will use the following chain rule.

Lemma 8.17 (Chain rule) *Let $b : V \times V \to \mathbb{R}$ be continuous, symmetric and bilinear and let $u \in C^1((a, b), V)$. Then*

$$\frac{d}{dt}b(u(t)) = 2\,b(\dot{u}(t), u(t)),$$

where we have again set $b(v) := b(v, v)$, $v \in V$.

Proof For arbitrary $t \in (a, b)$, we have

$$b(u(t+h)) - b(u(t)) = b(u(t+h) - u(t), u(t+h)) + b(u(t), u(t+h) - u(t)).$$

Division by $h \neq 0$ and then letting $h \to 0$ yields the claim. □

Proof (of Theorem 8.16) We apply the previous Lemma 8.17 to $a(w(t))$ and to $\|\dot{w}(t)\|_H^2 = (\dot{w}(t), \dot{w}(t))_H$ to obtain

$$\frac{d}{dt}\{a(w(t)) + \|\dot{w}(t)\|_H^2\} = 2a(w(t), \dot{w}(t)) + 2(\ddot{w}(t), \dot{w}(t))_H$$

$$= 2(Aw(t), \dot{w}(t))_H + 2(-Aw(t), \dot{w}(t))_H = 0.$$

Hence $a(w(t)) + \|\dot{w}(t)\|_H^2$ is constant in $t \in [0, T]$, and so by (8.31b),

$$a(w(t)) + \|\dot{w}(t)\|_H^2 = a(w(0)) + \|\dot{w}(0)\|_H^2 = a(u_0) + \|u_1\|_H^2,$$

which was the claim. □

Proof (2nd proof of uniqueness in Theorem 8.15) Let $\underline{w}$ and $\overline{w}$ be two solutions of (8.31). Then $w(t) := \underline{w}(t) - \overline{w}(t)$ defines a solution $w \in C^2([0,T], H)$ of (8.31a) with initial conditions $w(0) = \dot{w}(0) = 0$. By Theorem 8.16, $a(w(t)) + \|\dot{w}(t)\|_H^2 = 0$ for all $t \in [0,T]$. It follows that $\dot{w}(t) \equiv 0$, whence $w(t) = w(0) = 0$ for all $t \in [0,T]$, that is, $\overline{w} = \underline{w}$. $\qquad\square$

Let $u_0 \in D(A)$ and $u_1 \in V$. Then we know from Theorem 8.15 that the problem (8.31) has a unique solution $u \in C^2([0,T], H)$. For the numerical analysis in Section 9.5 we will need stronger regularity properties of w; to obtain these we need stronger assumptions on the initial values u_0 and u_1.

We recall the definition of $D(A)$ from (8.32) and will continue to use the orthonormal basis $\{e_n : n \in \mathbb{N}\}$ of H consisting of eigenvectors of A, i.e., $Ae_n = \lambda_n e_n$. Now the square of A is defined by

$$D(A^2) := \{v \in D(A) : Av \in D(A)\}, \qquad A^2 v := A(Av).$$

This leads to a hierarchy of spaces $D(A^2) \subset D(A) \subset V \subset H$.

Theorem 8.18 (Regularity) *Let* $u_0 \in D(A)$, $u_1 \in V$ *and assume that* $w \in C^2([0,T], H)$ *is the solution of* (8.31).
(a) If $u_0 \in D(A^2)$ *and* $u_1 \in D(A)$ *with* $Au_1 \in V$, *then* $w \in C^3([0,T], V)$.
(b) If $u_0 \in D(A^2)$ *with* $A^2 u_0 \in V$ *and* $u_1 \in D(A^2)$, *then* $w \in C^4([0,T], V)$.

The proof will use the following facts.

Lemma 8.19 *(a) Let* $v \in H$. *Then* $v \in V$ *if and only if* $\sum_{n=1}^{\infty} \lambda_n (v, e_n)_H^2 < \infty$.
(b) The quantity

$$\||v|\|_V^2 := \|v\|_H^2 + a(v,v) = \sum_{n=1}^{\infty} (1 + \lambda_n)(v, e_n)_H^2$$

defines an equivalent norm on V.
(c) Let $c_n \in C([0,T])$ *and* $\gamma_n \geq 0$ *such that* $\sum_{n=1}^{\infty} \gamma_n < \infty$. *If* $\lambda_n |c_n(t)|^2 \leq \gamma_n$ *for all* $t \in [0,T]$, $n \in \mathbb{N}$, *then* $w(t) := \sum_{n=1}^{\infty} c_n(t) e_n$ *converges in* V *uniformly with respect to* $t \in [0,T]$. *In particular,* $w \in C([0,T], V)$.
(d) Let $f \in C([0,T], V)$ *and* $f_n(t) := (f(t), e_n)_H$. *Then* $\sum_{n=1}^{\infty} f_n(t) e_n$ *converges in* V *to* $f(t)$ *uniformly with respect to* $t \in [0,T]$.

Proof Since $a(\cdot, \cdot)$ is H-elliptic there exist $\omega \geq 0$, $\alpha > 0$ such that $a(v,v) + \omega \|v\|_H^2 \geq \alpha \|v\|_V^2$ for all $v \in V$. This implies the existence of some $\varepsilon > 0$ such that

$$\||v|\|_V^2 = a(v,v) + \|v\|_H^2 \geq \varepsilon \|v\|_V^2 \qquad \text{for all } v \in V. \tag{8.36}$$

Indeed, otherwise we could find $(v_n)_{n \in \mathbb{N}} \subset V$, such that $a(v_n, v_n) + \|v_n\|_H^2 < \frac{1}{n} \|v_n\|_V^2$. We may assume that $\|v_n\|_V = 1$ (replacing v_n by $v_n / \|v_n\|_V$ otherwise). Since

$a(v_n, v_n) \geq 0$ it follows that $a(v_n, v_n) \to 0$ and $\|v_n\|_H^2 \to 0$ as $n \to \infty$. This contradicts the H-ellipticity, and (8.36) is proven. We have thus shown that $\|\| \cdot \|\|_V$ defines an equivalent norm on V. This norm is associated with the scalar product

$$[u, v]_V := (u, v)_H + a(u, v), \quad \text{i.e.,} \quad \|\|v\|\|_V^2 = [v, v]_V, \qquad u, v \in V.$$

Note that $[e_n, v]_V = (1 + \lambda_n)(e_n, v)_H$ for all $v \in V$. Let $\hat{e}_n := \frac{1}{\sqrt{1+\lambda_n}} e_n$, then, $(\hat{e}_n)_{n \in \mathbb{N}}$ is orthonormal in $(V, [\cdot, \cdot]_V)$; in fact, by Corollary 4.19, it is actually an orthonormal basis of $(V, [\cdot, \cdot]_V)$. Then by Theorem 4.9 and Theorem 4.13, for $\hat{c}_n \in \mathbb{R}$, the series $\sum_{n=1}^{\infty} \hat{c}_n \hat{e}_n$ converges in V if and only if $\sum_{n=1}^{\infty} \hat{c}_n^2 < \infty$. Thus, for $c_n \in \mathbb{R}$, the series $\sum_{n=1}^{\infty} c_n e_n = \sum_{n=1}^{\infty} (1+\lambda_n)^{1/2} c_n \hat{e}_n$ converges in V if and only if $\sum_{n=1}^{\infty} (1+\lambda_n) c_n^2 < \infty$, which is equivalent to $\sum_{n=1}^{\infty} \lambda_n c_n^2 < \infty$. Taking (4.27) in Theorem 4.50 into account completes the proof of (a) and (b).

In order to show (c), assume that $\lambda_n c_n(t)^2 \leq \gamma_n$ for all $n \in \mathbb{N}$ and $t \in [0, T]$. It follows from (b) that $w(t) \in V$ and

$$\left\| w(t) - \sum_{n=1}^{N} c_n(t) e_n \right\|_V^2 = \left\| \sum_{n=N+1}^{\infty} c_n(t) e_n \right\|_V^2 = \sum_{n=N+1}^{\infty} |c_n(t)|^2 (1 + \lambda_n)$$

$$\leq \sum_{n=N+1}^{\infty} \gamma_n \left(\frac{1}{\lambda_N} + 1 \right) \to 0 \quad \text{as } n \to \infty.$$

It remains to show (d). Since $\|\| f(\cdot) \|\|_V^2$ is a continuous function it follows from Dini's theorem[1] that the series

$$\sum_{n=1}^{\infty} (1 + \lambda_n) |f_n(t)|^2 = \|\| f(t) \|\|_V^2$$

converges uniformly on $[0, T]$. Thus, given $\varepsilon > 0$, there exists $N \in \mathbb{N}$ such that

$$\left\| f(t) - \sum_{n=1}^{N} f_n(t) e_n \right\|_V^2 = \sum_{n=N+1}^{\infty} (1 + \lambda_n) |f_n(t)|^2 \leq \varepsilon$$

for all $t \in [0, T]$.   □

Proof (of Theorem 8.18) We know from Theorem 8.15 and its proof that (8.31) has a unique solution $w \in C^2([0, T], H)$ given by $w(t) = \sum_{n=1}^{\infty} w_n(t) e_n$, where w_n, in turn, is given by (8.33) or (8.34). Observe that by (8.33) for $\lambda_n > 0$ (and trivially for $\lambda_n = 0$)

$$\lambda_n |w_n(t)|^2 \leq 2\lambda_n (u_0, e_n)_H^2 + 2 (u_1, e_n)_H^2 \qquad (8.37)$$

---

[1] If a monotone sequence of continuous functions converges pointwise on a compact set and if the limit function is also continuous, then the convergence is uniform.

Since $u_0 \in V$, we have $\sum_{n=1}^{\infty} \lambda_n (u_0, u_n)_H^2 < \infty$. Moreover, $\sum_{n=1}^{\infty} (u_1, e_n)_H^2 < \infty$. It follows from Lemma 8.19 (c) that the series $\sum_{n=1}^{\infty} w_n(t) e_n$ converges in V, uniformly in $[0, T]$. In particular, $w \in C([0, T], V)$.

Next we show that $w \in C^1([0, T], V)$. Observe that for $\lambda_n > 0$ by (8.33) and (8.34)

$$\dot{w}_n(t) = -\sqrt{\lambda_n}\, \sin(\sqrt{\lambda_n} t)\, (u_0, e_n)_H + \cos(\sqrt{\lambda_n} t)\, (u_1, e_n)_H.$$

Thus

$$\lambda_n\, |\dot{w}_n(t)|^2 \le 2\lambda_n^2\, (u_0, e_n)_H^2 + 2\lambda_n\, (u_1, e_n)_H^2. \tag{8.38}$$

Since $u_0 \in D(A)$, we have $\sum_{n=1}^{\infty} \lambda_n^2 (u_0, e_n)_H^2 < \infty$; and since $u_1 \in V$, we also have $\sum_{n=1}^{\infty} \lambda_n (u_1, e_n)_H^2 < \infty$. It follows from Lemma 8.19 (c) that $\sum_{n=1}^{\infty} \dot{w}_n(t) e_n$ converges in V uniformly in $[0, T]$. Lemma 8.2 implies that $w \in C^1([0, T], V)$ and $\dot{w}(t) = \sum_{n=1}^{\infty} \dot{w}_n(t) e_n$. Observe that so far we have only used that $u_0 \in D(A)$ and $u_1 \in V$.

In order to prove (a) we now assume that $u_0 \in D(A^2)$ and $u_1 \in D(A)$ with $Au_1 \in V$. Then $(A^2 u_0, e_n)_H = \lambda_n (Au_0, e_n)_H = \lambda_n^2 (u_0, e_n)_H$, whence

$$\sum_{n=1}^{\infty} \lambda_n^4 (u_0, e_n)_H^2 = \sum_{n=1}^{\infty} (A^2 u_0, e_n)_H^2 = \|A^2 u_0\|_H^2 < \infty \quad \text{and similarly}$$

$$\sum_{n=1}^{\infty} \lambda_n^3 (u_1, e_n)_H^2 = \sum_{n=1}^{\infty} \lambda_n (Au_1, e_n)_H^2 \le \|\!|Au_1|\!\|_V^2 < \infty.$$

Now $\ddot{w}_n = -\lambda_n w_n(t)$ and so, by (8.37),

$$\lambda_n\, |\ddot{w}_n(t)|^2 \le 2\lambda_n^3\, (u_0, e_n)_H^2 + 2\lambda_n^2\, (u_1, e_n)_H^2.$$

It follows from Lemma 8.19 (c) that $\sum_{n=1}^{\infty} \ddot{w}_n(t) e_n$ converges in V uniformly in $[0, T]$ and thus by Lemma 8.2 that $w \in C^2([0, T], V)$.

Finally, $w_n^{(3)}(t) = -\lambda_n \dot{w}_n(t)$. Thus by (8.38)

$$\lambda_n\, |w_n^{(3)}(t)|^2 \le 2\lambda_n^4\, (u_0, e_n)_H^2 + 2\lambda_n^3\, (u_1, e_n)_H^2.$$

Again appealing to Lemma 8.19 and Lemma 8.2 we deduce the fact that $w \in C^3([0, T], V)$. Thus assertion (a) is proved.

The proof of (b) is similar: if in fact $u_0 \in D(A^2)$, $A^2 u_0 \in V$ and $u_1 \in D(A^2)$, then

$$\sum_{n=1}^{\infty} \lambda_n^5 (u_0, e_n)_H^2 \le \|\!|A^2 u_0|\!\|_V^2 < \infty \quad \text{and}$$

$$\sum_{n=1}^{\infty} \lambda_n^4 (u_1, e_n)_H^2 \le \|\!|A^2 u_1|\!\|_V^2 < \infty.$$

Note that $w_n^{(4)}(t) = \lambda_n^2 w_n(t)$. Thus by (8.37)

$$\lambda_n |w_n^{(4)}(t)|^2 \leq 2\lambda_n^5 (u_0, e_n)_H^2 + 2\lambda_n^4 (u_1, e_n)_H^2.$$

Again Lemma 8.19 and Lemma 8.2 imply that $w \in C^4([0, T], V)$.   □

We next consider the inhomogeneous wave equation with initial values

$$\ddot{w}(t) + A w(t) = f(t), \tag{8.39a}$$
$$w(0) = u_0, \dot{w}(0) = u_1. \tag{8.39b}$$

We have the following result on existence and uniqueness.

Theorem 8.20 *Let $u_0 \in D(A)$, $u_1 \in V$ and $f \in C([0, T], V)$. Then there exists a unique function $w \in C^2([0, T], H)$ satisfying $w(t) \in D(A)$ for all $t \in [0, T]$ which solves (8.39).*

Proof Since the difference of two solution of (8.39) is a solution of the homogeneous problem (8.31), uniqueness follows from Theorem 8.15. Moreover, since the homogeneous problem has a solution, in order to prove existence for (8.39), we may and will assume that $u_0 = u_1 = 0$.

The function f_n given by $f_n(t) := (f(t), e_n)_H$ is continuous on $[0, T]$, and if w is a solution of (8.39), then $w_n(t) := (w(t), e_n)_H$ solves $\ddot{w}_n + \lambda_n w_n = f_n$, $w_n(0) = \dot{w}_n(0) = 0$. This means that necessarily

$$w_n(t) = \frac{1}{\sqrt{\lambda_n}} \int_0^t \sin(\sqrt{\lambda_n}(t - s)) f_n(s) \, ds, \qquad \text{if } \lambda_n > 0 \text{ and}$$

$$w_n(t) = \int_0^t (t - s) f_n(s) \, ds, \qquad \text{if } \lambda_n = 0,$$

as one verifies immediately. We now *define* w_n by these expressions and show that

$$w(t) := \sum_{n=1}^{\infty} (w_n(t), e_n)_H \, e_n \tag{8.40}$$

converges in H and defines a solution of (8.39). Since $\lim_{n \to \infty} \lambda_n = \infty$ there exists $N_0 \in \mathbb{N}$ such that $\lambda_n > 0$ for all $n \geq N_0$. By Hölder's inequality, we have, for $n \geq N_0$,

$$\lambda_n w_n(t)^2 \leq \int_0^T \sin(\sqrt{\lambda_n}(t - s))^2 \, ds \int_0^T f_n(s)^2 \, ds \leq T \int_0^T f_n(s)^2 \, ds =: \gamma_n.$$

Since

$$\sum_{n=N_0}^{\infty} \gamma_n \leq T \int_0^T \sum_{n=1}^{\infty} f_n(s)^2 \, ds = T \int_0^T \|f(s)\|_H^2 < \infty,$$

it follows from Lemma 8.19 (c) that (8.40) converges in V uniformly with respect to $t \in [0, T]$. Thus $w \in C([0, T], V)$. Moreover, for $n \geq N_0$,

$$\lambda_n^2 \, w_n(t)^2 \leq T \int_0^T \lambda_n \, f_n(s)^2 \, ds =: \gamma_n^1.$$

Since

$$\sum_{n=N_0}^\infty \gamma_n^1 = T \int_0^T \sum_{n=N_0}^\infty \lambda_n \, f_n(s)^2 \, ds \leq T \int_0^T \||f(s)|\|_V^2 \, ds < \infty,$$

it follows from the definition of A that $w(t) \in D(A)$ and that $\sum_{n=1}^\infty \lambda_n \, w_n(t) \, e_n$ converges to $Aw(t)$ in H uniformly with respect to $t \in [0, T]$. In particular, $Aw(\cdot) \in C([0, T], H)$. Next we show that $w \in C^1([0, T], V)$. Let $n \geq N_0$. Note that

$$\dot{w}_n(t) = \int_0^t \cos(\sqrt{\lambda_n}(t - s)) \, f_n(s) \, ds.$$

Thus by Hölder's inequality

$$\lambda_n \, \dot{w}_n(t)^2 \leq T \int_0^T \lambda_n \, f_n(s)^2 \, ds = \gamma_n^1.$$

It follows from Lemma 8.19 (c) that $\sum_{n=1}^\infty \dot{w}_n(t) \, e_n$ converges in V uniformly with respect to $t \in [0, T]$. Lemma 8.2 implies that $w \in C^1([0, T], V)$ and $\dot{w}(t) = \sum_{n=1}^\infty \dot{w}_n(t) \, e_n$.

In order to prove that $w \in C^2([0, T], H)$ we note that $\ddot{w}_n(t) = -\lambda_n \, w_n(t) + f_n(t)$. We have already seen that $\sum_{n=1}^\infty -\lambda_n \, w_n(t) \, e_n$ converges in H to $-Aw(t)$ uniformly with respect to $t \in [0, T]$. Since $f \in C([0, T], V)$ it follows from Lemma 8.19 (d) that $\sum_{n=1}^\infty f_n(t) \, e_n$ converges in V uniformly with respect to $t \in [0, T]$. Thus $w \in C^2([0, T], H)$ and $\ddot{w}(t) + Aw(t) = f(t)$ for all $t \in [0, T]$. $\square$

Next we establish the regularity results for the solutions of the inhomogeneous wave equation which will be needed in Section 9.5.

Theorem 8.21 *Let* $u_0 \in D(A)$, $u_1 \in V$, $f \in C([0, T], V)$ *and let* w *be the solution of problem* (8.39).
(a) *If* $u_0 \in D(A^2)$, $u_1 \in D(A)$, $Au_1 \in V$ *and* $f \in C^2([0, T], V)$ *with* $f(0) \in D(A)$,
     then $w \in C^3([0, T], V)$.
(b) *If* $u_0 \in D(A^2)$ *with* $A^2 u_0 \in V$, $u_1 \in D(A^2)$, $f \in C^3([0, T], V)$ *such that* $f(0) \in$
     $D(A)$, $Af(0) \in V$ *and* $\dot{f}(0) \in D(A)$, *then* $w \in C^4([0, T], V)$.

Proof In view of Theorem 8.18 we can and will assume that $u_0 = u_1 = 0$. Adopting the notation of Theorem 8.20, we already know from the proof of that theorem that $w \in C^1([0, T], V)$ and that $\dot{w}(t) = \sum_{n=1}^\infty \dot{w}_n(t) \, e_n$ converges in V uniformly with respect to $t \in [0, T]$.

1. Assume that $f \in C^1([0,T],V)$. We show that $w \in C^2([0,T],V)$ (and not merely in $C^2([0,T],H)$ as was the case in Theorem 8.20). Let $N_0 \in \mathbb{N}$ such that $\lambda_n > 0$ for all $n \geq N_0$. For $n \geq N_0$, we have

$$\dot{w}_n(t) = \int_0^t \cos(\sqrt{\lambda_n}(t-s)) f_n(s)\, ds = \int_0^t \cos(\sqrt{\lambda_n}s) f_n(t-s)\, ds.$$

Thus,

$$\ddot{w}_n(t) = \int_0^t \cos(\sqrt{\lambda_n}s)\, \dot{f}_n(t-s)\, ds + \cos(\sqrt{\lambda_n}t)\, f_n(0).$$

We shall apply Lemma 8.19 (c) and estimate these two terms:

$$\sum_{n=N_0}^{\infty} \lambda_n \left(\int_0^t \cos(\sqrt{\lambda_n}s)\, \dot{f}_n(t-s)\, ds \right)^2 \leq \sum_{n=N_0}^{\infty} \lambda_n T \left(\int_0^T \dot{f}_n^2\, ds \right)$$

$$= T \int_0^T \sum_{n=N_0}^{\infty} \lambda_n \dot{f}_n^2\, ds \leq T \int_0^T \|\dot{f}_n(s)\|_V^2\, ds < \infty$$

for the first term. Since $f(0) \in V$, we obtain that $\sum_{n=1}^{\infty} \lambda_n\, |f_n(0)|^2 < \infty$. The claim follows from Lemma 8.19 (c), which also shows that $\ddot{w}(t) = \sum_{n=1}^{\infty} \ddot{w}_n(t)\, e_n$ in V uniformly on $[0,T]$.

2. Now assume that $f \in C^2([0,T],V)$ and $f(0) \in D(A)$. We have

$$w_n^{(3)}(t) = \int_0^t \cos(\sqrt{\lambda_n}s)\, \ddot{f}_n(t-s)\, ds + \cos(\sqrt{\lambda_n}t)\dot{f}_n(0)$$
$$- \sqrt{\lambda_n}\, \sin(\sqrt{\lambda_n}t)\, f_n(0).$$

In order to apply Lemma 8.19 (c) we estimate the corresponding three terms:

(i) $\displaystyle \sum_{n=N_0}^{\infty} \lambda_n \left(\int_0^t \cos(\sqrt{\lambda_n}s)\, \ddot{f}_n(t-s)\, ds \right)^2 \leq \sum_{n=N_0}^{\infty} T \int_0^T \lambda_n\, \ddot{f}_n(s)^2\, ds$

$$\leq T \int_0^T \|\ddot{f}(s)\|_V^2\, ds < \infty;$$

(ii) $\displaystyle \sum_{n=N_0}^{\infty} \lambda_n |\dot{f}_n(0)|^2 < \infty$ since $\dot{f}_n(0) \in V$;

(iii) $\displaystyle \sum_{n=N_0}^{\infty} \lambda_n \left(\sqrt{\lambda_n}\, \sin(\sqrt{\lambda_n}t)\, f_n(0) \right)^2 \leq \sum_{n=N_0}^{\infty} \lambda_n^2\, f_n(0)^2 < \infty$

since $f(0) \in D(A)$.

It follows that $w \in C^3([0,T],V)$ and $w^{(3)}(t) = \sum_{n=1}^{\infty} w_n^{(3)}(t)\, e_n$ in V uniformly on $[0,T]$. This proves (a).

3. We now prove (b). We have

$$w_n^{(4)}(t) = \int_0^t \cos(\sqrt{\lambda_n}s)\, f_n^{(3)}(t-s)\, ds + \cos(\sqrt{\lambda_n}t)\, \ddot{f}_n(0)$$
$$- \sqrt{\lambda_n}\, \sin(\sqrt{\lambda_n}t)\dot{f}_n(0) - \lambda_n \cos(\sqrt{\lambda_n}t)f_n(0).$$

In order to apply Lemma 8.19 (c) we now have to estimate these four terms:

(i) $$\sum_{n=N_0}^{\infty} \lambda_n \left(\int_0^t \cos(\sqrt{\lambda_n}s)\, f_n^{(3)}(t-s)\, ds \right)^2 \le \sum_{n=N_0}^{\infty} T \int_0^T \lambda_n\, f_n^{(3)}(s)^2\, ds$$
$$\le T \int_0^T \||f^{(3)}(s)\||_V^2\, ds < \infty;$$

(ii) $$\sum_{n=N_0}^{\infty} \lambda_n \left(\cos(\sqrt{\lambda_n}t)\, \ddot{f}_n(0) \right)^2 \le \sum_{n=N_0}^{\infty} \lambda_n \ddot{f}_n(0)^2 < \infty \text{ since } \ddot{f}(0) \in V;$$

(iii) $$\sum_{n=N_0}^{\infty} \lambda_n \left(-\sqrt{\lambda_n}\, \sin(\sqrt{\lambda_n}t)\dot{f}_n(0) \right)^2 \le \sum_{n=N_0}^{\infty} \lambda_n^2 \dot{f}_n(0)^2 < \infty$$

since $\dot{f}(0) \in D(A)$;

(iv) $$\sum_{n=N_0}^{\infty} \lambda_n \left(\lambda_n \cos(\sqrt{\lambda_n}t)f_n(0) \right)^2 \le \sum_{n=1}^{\infty} \lambda_n^3\, f_n(0)^2 = \sum_{n=1}^{\infty} \lambda_n\, (Af(0), e_n(0))_H^2$$
$$\le \sum_{n=1}^{\infty} (1 + \lambda_n)\, (Af(0), e_n)_H^2 = \||Af(0)\||_V^2 < \infty$$

Again, Lemma 8.19 (c) shows that $w \in C^4([0,T], V)$. □

We now wish to apply Theorem 8.15 to the Laplacian; in doing so, we will limit ourselves to Dirichlet and Neumann boundary conditions. Let $\Omega \subset \mathbb{R}^d$ be a bounded domain; in the case of Neumann boundary conditions, we shall also assume that Ω is of class C^1. In both cases we will take $H = L_2(\Omega)$. In the case of Dirichlet boundary conditions, the following result is a direct corollary of Theorem 8.20.

Theorem 8.22 *Assume that Ω is convex. Let $u_0 \in H_0^1(\Omega) \cap H^2(\Omega)$ and let $u_1 \in H_0^1(\Omega)$ be given. If $f \in C([0,T], H_0^1(\Omega))$, then there exists a uniquely determined $w \in C^2([0,T], L_2(\Omega))$ for which $w(t) \in H_0^1(\Omega) \cap H^2(\Omega)$ for all $t \in [0,T]$, w satisfies $\ddot{w}(t) = \Delta w(t) + f(t)$ for all $t \in [0,T]$, and $w(0) = u_0$, $\dot{w}(0) = u_1$. Moreover, $w \in C([0,T], H^2(\Omega))$.*

Proof Let $A = \Delta^D$ it follows from Theorem 6.86 that $D(A) = H_0^1(\Omega) \cap H^2(\Omega)$. Thus the first assertion is a direct consequence of Theorem 8.20. It remains to prove the last statement. Recall that the form $a(\cdot, \cdot)$ is coercive (by the Poincaré inequality, Theorem 6.32). Thus the operator A is invertible. This implies that $D(A)$ is a Banach space with respect to the norm $\|v\|_A := \|Av\|_{L_2(\Omega)} = \|\Delta v\|_{L_2(\Omega)}$. Note that

$H_0^1(\Omega) \cap H^2(\Omega)$ is a closed subspace of $H^2(\Omega)$, and $\|v\|_A \leq \|v\|_{H^2(\Omega)}$. This means that the identity $I : H^2(\Omega) \cap H_0^1(\Omega) \to D(A)$ is continuous. By Theorem A.4 its inverse I^{-1} is also continuous, i.e., there exists a constant $\beta > 0$ such that $\|v\|_{H^2(\Omega)} \leq \beta \|v\|_A$. Thus the two norms $\| \cdot \|_A$ and $\| \cdot \|_{H^2(\Omega)}$ are equivalent on $H^2(\Omega) \cap H_0^1(\Omega)$. Since $w \in C^2([0,T], H)$ and $\ddot{w} = \Delta w + f$, it follows that $Aw(\cdot) = \Delta w(\cdot) \in C([0,T], H)$. This implies that $w \in C([0,T], D(A))$. Since the two norms are equivalent we finally obtain that $w \in C([0,T], H^2(\Omega))$. □

Theorem 8.21 yields the following regularity result, which will be useful for the numerical solution of the wave equation in Section 9.5. For further regularity results of the same kind, see also Exercise 8.11.

Theorem 8.23 *Assume that $\Omega \subset \mathbb{R}^d$ is bounded, open and convex.*
(a) Let $u_0 \in H^4(\Omega) \cap H_0^1(\Omega)$ such that $\Delta u_0 \in H_0^1(\Omega)$ and let $u_1 \in H^2(\Omega) \cap H_0^1(\Omega)$
    such that $\Delta u_1 \in H_0^1(\Omega)$. Finally, let $f \in C^2([0,T], H_0^1(\Omega))$ with $f(0) \in H^2(\Omega)$.
    Then $w \in C^3([0,T], H_0^1(\Omega))$, where w is the solution from Theorem 8.22.
(b) Let $u_0 \in H^4(\Omega) \cap H_0^1(\Omega)$ such that $\Delta u_0 \in H_0^1(\Omega)$, $\Delta^2 u_0 \in H_0^1(\Omega)$ and let
    $u_1 \in H^4(\Omega) \cap H_0^1(\Omega)$ such that $\Delta u_1 \in H_0^1(\Omega)$. Finally, let $f \in C^3([0,T], H_0^1(\Omega))$
    such that $f(0) \in H^2(\Omega)$, $\Delta f(0) \in H_0^1(\Omega)$ and $\dot{f}(0) \in H^2(\Omega)$. Then, we have that
    $w \in C^4([0,T], H_0^1(\Omega))$.

Proof Note that $D(A) = H^2(\Omega) \cap H_0^1(\Omega)$ by Theorem 6.86. This implies that $\{u \in H^4(\Omega) \cap H_0^1(\Omega) : \Delta u \in H_0^1(\Omega)\} \subset \{u \in H^2(\Omega) \cap H_0^1(\Omega) : \Delta u \in H^2(\Omega) \cap H_0^1(\Omega)\} = D(A^2)$. Now the claim follows from Theorem 8.22. □

We recall that the domain of the Laplacian with Neumann boundary conditions is defined as follows:

$$D(\Delta^N) = \left\{ u \in H^1(\Omega) \cap H_{\text{loc}}^2(\Omega) : \Delta u \in L_2(\Omega), \frac{\partial u}{\partial \nu} = 0 \right\};$$

see Theorem 8.12 and the subsequent definition. If we choose $A = \Delta^N$ in Theorem 8.15 we obtain:

Theorem 8.24 *Suppose that $\Omega \subset \mathbb{R}^d$ is a bounded, open set with C^1-boundary. Let $u_0 \in D(\Delta^N)$, $u_1 \in H^1(\Omega)$ and $f \in C([0,T], H^1(\Omega))$. Then there exists a uniquely determined function $w \in C^2([0,T], L_2(\Omega))$ for which $w(t) \in D(\Delta^N)$, $\ddot{w}(t) = \Delta w(t) + f(t)$ for all $t \in [0,T]$, and $w(0) = u_0$, $\dot{w}(0) = u_1$.*

The corresponding theorem for the case of Robin boundary conditions is formulated in Exercise 8.3. We next wish to compare the eigenvalues corresponding to Dirichlet and to Neumann boundary conditions, respectively. As before, we will denote by

$$0 < \lambda_1^D < \lambda_2^D \leq \lambda_3^D \leq \cdots \leq \lambda_n^D \leq \lambda_{n+1}^D \leq \cdots$$

the Dirichlet eigenvalues and by $\{e_n : n \in \mathbb{N}\}$ a corresponding orthonormal basis of $L_2(\Omega)$ consisting of eigenfunctions $e_n \in D(\Delta^D)$, that is, $-\Delta e_n = \lambda_n^D e_n$. Then for $n \in \mathbb{N}$, $w(t) = \cos(\sqrt{\lambda_n^D}t)e_n$, $t \geq 0$, is the solution of (8.30) with initial conditions $u_0 = e_n$, $u_1 = 0$. This is a *stationary solution* of the wave equation: its variation in time consists only of multiplication by the factor $\cos(\sqrt{\lambda_n^D}t)$. These solutions are often referred to as the *normal modes* of the domain Ω. If $d = 2$, then we may imagine Ω to be a membrane or a tambourine; the eigenvalues λ_n^D give the *natural frequencies* (or *resonant frequencies*) of the membrane.

Neumann boundary conditions describe the situation where the boundary is not fixed. The Neumann eigenvalues

$$0 = \lambda_1^N < \lambda_2^N \leq \lambda_3^N \leq \cdots \leq \lambda_n^N \leq \lambda_{n+1}^N \leq \cdots$$

represent the natural frequencies of a *free* membrane, when $d = 2$. The following theorem shows that such a free membrane has a deeper sound than the corresponding fixed membrane described by Dirichlet boundary conditions.

Theorem 8.25 *We have $\lambda_n^N \leq \lambda_n^D$ for all $n \in \mathbb{N}$.*

Proof Let $a(u) = \int_\Omega |\nabla u|^2 \, dx$ for $u \in H^1(\Omega)$. By Theorem 4.52,

$$\lambda_n^N = \max_{\substack{W \subset H^1(\Omega) \\ \operatorname{codim} W \leq n-1}} \min_{\substack{u \in W \\ \|u\|_{L_2}=1}} a(u), \quad \lambda_n^D = \max_{\substack{V \subset H_0^1(\Omega) \\ \operatorname{codim} V \leq n-1}} \min_{\substack{u \in V \\ \|u\|_{L_2}=1}} a(u).$$

Let $W \subset H^1(\Omega)$ be a subspace of $H^1(\Omega)$ with codim $W \leq n-1$. By Remark 4.53, this means exactly that $W \cap U \neq \{0\}$ for every subspace U of $H^1(\Omega)$ such that dim $U \geq n$. Now set $V := W \cap H_0^1(\Omega)$, then we claim that codim $V \leq n-1$ in $H_0^1(\Omega)$. Indeed, if U is a subspace of $H_0^1(\Omega)$ with dim $U \geq n$, then $U \cap W \neq \{0\}$ by choice of W. Hence $U \cap V = U \cap W \neq \{0\}$ as well, which implies that codim $V \leq n-1$ in $H_0^1(\Omega)$, as claimed. Now since $V \subset W$, for this W we have

$$\min_{\substack{u \in W \\ \|u\|_{L_2}=1}} a(u) \leq \min_{\substack{u \in V \\ \|u\|_{L_2}=1}} a(u) \leq \lambda_n^D.$$

Since W was an arbitrary subspace of $H^1(\Omega)$ for which codim$W \leq n-1$, it follows that $\lambda_n^N \leq \lambda_n^D$. □

---

* Comment 8.26 It is possible to say more, namely $\lambda_{n+1}^N < \lambda_n^D$ for all $n \in \mathbb{N}$; see [30].

The Dirichlet eigenvalues depend on the domain Ω; we may write $\lambda_n^D(\Omega)$ to emphasize this dependence: these give the resonant frequencies of the domain, which in principle can be heard. A famous problem is the question as to whether these eigenvalues actually *determine* the domain; more precisely, the question is whether the assumption that $\lambda_n^D(\Omega_1) = \lambda_n^D(\Omega_2)$ for all $n \in \mathbb{N}$ (where $\Omega_1, \Omega_2 \subset \mathbb{R}^d$ are two domains in $\mathbb{R}^d$), already implies that Ω_1 and Ω_2 are *congruent* (that is, whether there exist an orthogonal $d \times d$ matrix B and a vector $c \in \mathbb{R}^d$ such that $\Omega_2 = \{Bx + c : x \in \Omega_1\}$). In 1966, Marc Kac [39] formulated this question as follows: "can one hear the shape of a drum?" It was only in 1992 that a now-famous counterexample was given, by Gordon, Webb and Wolpert.

This takes the form of two polygons in $\mathbb{R}^2$ which are not congruent to each other but have the same Dirichlet and Neumann eigenvalues. We refer to [12] for a recent presentation of this result in the language of bilinear forms as used extensively throughout this book. However, the problem remains open for domains with C^∞-boundary in any dimension $d \geq 2$, as well as for convex domains in dimension $d = 2$ and 3.

8.5 Inhomogeneous parabolic equations

We next wish to study the inhomogeneous heat equation

$$u_t(t, x) = \Delta u(t, x) + f(t, x), \quad t > 0, \quad x \in \Omega \tag{8.41}$$

with suitable initial and boundary conditions. In contrast to the inhomogeneous wave equation, where the Riemann integral was used throughout, for the inhomogeneous parabolic equation (8.41) we will need the Lebesgue integral of functions defined on an interval with values in a Hilbert space H. We mention that this is a special case of the Bochner integral for Banach space-valued functions (see [6, Sec. 1.1]), but here we will not need this theory: we will follow a more direct approach, defining the integral with the help of an orthonormal basis of H. Once we have introduced H-valued Lebesgue integrals we are able to treat Sobolev spaces on an interval exactly as in Chapter 4; the problem itself will again be treated via spectral decomposition.

What we gain from this approach (in comparison with the theory of classical solutions in C^1) is remarkable: an existence and uniqueness theorem with maximal regularity, Theorem 8.32, which will be our main result for this section.

While in the case $f = 0$ all solutions of (8.41) are automatically C^∞ in space and time, the question of regularity becomes more subtle if $f \neq 0$. Yet this question is particularly important for the numerical treatment of (8.41), which we will discuss in Chapter 9; thus the maximal regularity assertion of Theorem 8.32, together with the estimates on solutions of (8.41), is of particular importance.

We first derive an abstract result which is formulated for coercive forms. For this, we need Hilbert space-valued Sobolev spaces on an interval.

Let H be a real separable Hilbert space and let $I \subset \mathbb{R}$ be a time interval. A function $f : I \to H$ is said to be (*weakly*) *measurable* if $(f(\cdot), v)_H$ is measurable for all $v \in H$. In this case, $\|f(\cdot)\|_H$ is also measurable as the supremum of a sequence of measurable functions (choose a sequence $\{v_n : n \in \mathbb{N}\}$ which is dense in the unit sphere of H and $f_n(t) := |(f(t), v_n)|_H$). We set

$$L_1(I, H) := \left\{ f : I \to H \text{ measurable: } \int_I \|f(t)\|_H \, dt < \infty \right\}.$$

Lemma 8.27 *For each $f \in L_1(I, H)$ there exists a unique $w \in H$ such that*

$$\int_I (f(t), v)_H \, dt = (w, v)_H, \quad v \in H.$$

We set $\int_I f(t)\,dt := w$. Lemma 8.27 is a simple consequence of the Riesz representation theorem. We next observe that the mapping

$$f \in L_1(I, H) \mapsto \int_I f(t)\,dt \in H$$

is a continuous linear operator. We will usually consider time intervals of the form $I = (0, T)$, where $0 < T \le \infty$. We will denote by $L_2((0, T), H)$ the space of measurable functions $u : (0, T) \to H$ for which $\int_0^T \|u(t)\|_H^2\,dt < \infty$. If one identifies functions which coincide almost everywhere (as we will always do), then $L_2((0, T), H)$ becomes a Hilbert space with respect to the inner product

$$(f, g)_{L_2((0,T),H)} := \int_0^T (f(t), g(t))_H\,dt.$$

We also have $L_2((0, T), H) \subset L_1((0, T), H)$ if $T < \infty$. For these and other, similar properties, we refer to the exercises. The space $L_2((0, T), H)$ can be described easily via the use of an orthonormal basis $\{e_n : n \in \mathbb{N}\}$ of H.

Lemma 8.28 (a) Let $u \in L_2((0, T), H)$ and set $u_n(t) := (u(t), e_n)_H$. Then

$$\sum_{n=1}^{\infty} \int_0^T |u_n(t)|^2\,dt < \infty. \tag{8.42}$$

(b) Conversely, let $u_n \in L_2(0, T)$ satisfy (8.42). Then there exists a uniquely determined $u \in L_2((0, T), H)$ for which $u_n = (u(\cdot), e_n)_H$ for all $n \in \mathbb{N}$.

Proof (a) We have

$$\sum_{n=1}^{\infty} \int_0^T |u_n(t)|^2\,dt = \int_0^T \sum_{n=1}^{\infty} |(u(t), e_n)_H|^2\,dt = \int_0^T \|u(t)\|_H^2\,dt < \infty.$$

(b) It follows from (8.42) and the monotone convergence theorem (Theorem A.8, applied to the partial sums) that $\sum_{n=1}^{\infty} |u_n(t)|^2 < \infty$ almost everywhere. By altering the functions on a set of measure zero if necessary, we may assume that $\sum_{n=1}^{\infty} |u_n(t)|^2 < \infty$ for all $t \in (0, T)$. Hence the series $u(t) := \sum_{n=1}^{\infty} u_n(t)e_n$ converges in H for all $t \in (0, T)$. Since $(u(t), v)_H = \sum_{n=1}^{\infty} u_n(t)(v, e_n)_H$ for all $v \in H$, we see that u is measurable. Finally, we have that $\int_0^T \|u(t)\|_H^2\,dt = \int_0^T \sum_{n=1}^{\infty} |u_n(t)|^2\,dt < \infty$. Hence $u \in L_2((0, T), H)$. □

Let $u \in L_2((0, T), H)$. We say that a function $w \in L_2((0, T), H)$ is a *weak derivative* of u if

$$-\int_0^T u(t)\dot\varphi(t)\,dt = \int_0^T w(t)\varphi(t)\,dt \quad \text{for all } \varphi \in C_c^1(0, T). \tag{8.43}$$

In this case w is unique, and we set $\dot u := w$. Here we use the notation $\dot u$ since we imagine the variable $t \in (0, T)$ to represent time. Obviously, the notation u'

could equally be used, as it was in Chapter 5. Note that we have used scalar test functions (that is, functions in $C_c^1(0, T) := C_c^1((0, T), \mathbb{R})$) for the definition of the weak derivative. We observe in passing that if (8.43) merely holds for all test functions $\varphi \in \mathcal{D}(0, T)$, then a regularization argument (e.g., using mollifiers) shows that it actually holds for all $\varphi \in C_c^1(0, T)$.

We now set $H^1((0, T), H) := \{u \in L_2((0, T), H) :$ the weak derivative $\dot{u}$ exists in $L_2((0, T), H)\}$. The following statement can be proved exactly as in the scalar case.

Theorem 8.29 (a) *The space $H^1((0, T), H)$ is a Hilbert space with respect to the inner product* $(u, v)_{H^1((0,T),H)} := \int_0^T \{(u(t), v(t))_H + (\dot{u}(t), \dot{v}(t))_H\} \, dt$.
(b) *Let $u \in H^1((0, T), H)$. Then there exists a unique $w \in C([0, T], H)$ such that $u(t) = w(t)$ almost everywhere. Moreover, $w(t) = w(0) + \int_0^t \dot{u}(t) \, dt$ for all $t \in [0, T)$.*

In what follows, in accordance with (b) we will always identify a function $u \in H^1((0, T), H)$ with its continuous representative w. With this identification, we may write

$$H^1((0, T), H) \subset C([0, T]), H), \tag{8.44}$$

and $H^1((0, T), H) \subset C([0, T], H)$ if $T < \infty$ (cf. Exercise 8.7). It is also possible to describe $H^1((0, T), H)$ by means of an orthonormal basis of H.

Lemma 8.30 *Let $\{e_n : n \in \mathbb{N}\}$ be an orthonormal basis of the Hilbert space H and let $u \in L_2((0, T), H)$ as well as $u_n := (u(\cdot), e_n)_H$. If $u_n \in H^1(0, T), n \in \mathbb{N}$, and $\sum_{n=1}^\infty \int_0^T \dot{u}_n(t)^2 \, dt < \infty$, then $u \in H^1((0, T), H)$ and*

$$\|\dot{u}\|_{L_2((0,T),H)}^2 = \sum_{n=1}^\infty \int_0^T \dot{u}_n(t)^2 \, dt.$$

This statement follows easily from Lemma 8.28. We next define

$$H^2((0, T), H) := \{u \in H^1((0, T), H) : \dot{u} \in H^1((0, T), H)\};$$

it follows from Theorem 8.29(b) that

$$H^2((0, T), H) \subset C^1([0, T], H). \tag{8.45}$$

With this background, we now consider a self-adjoint operator A on H, defined via a form in accordance with (4.19) and (4.20). More precisely, we take a further Hilbert space V which is compactly and densely imbedded in H and a continuous, coercive, symmetric bilinear form $a : V \times V \to \mathbb{R}$, and assume that A is the operator on H associated with a. By the spectral theorem (Theorem 4.46) there exist an orthonormal basis $\{e_n : n \in \mathbb{N}\}$ of H and numbers $0 < \lambda_1 \leq \lambda_2 \leq \cdots \leq \lambda_n \leq \lambda_{n+1}$ with $\lim_{n \to \infty} \lambda_n = \infty$, such that

$$D(A) = \left\{ u \in H : \sum_{n=1}^{\infty} \lambda_n^2 |(u, e_n)_H|^2 < \infty \right\}, \quad Au = \sum_{n=1}^{\infty} \lambda_n (u, e_n)_H e_n, \quad u \in D(A).$$

We also have

$$V = \left\{ u \in H : \sum_{n=1}^{\infty} \lambda_n |(u, e_n)_H|^2 < \infty \right\} \quad \text{und} \quad a(u, v) = \sum_{n=1}^{\infty} \lambda_n (u, e_n)_H (e_n, v)_H$$

for all $u, v \in V$. Since $\lambda_1 > 0$ (due to the coercivity of a), the space $D(A)$ is a Hilbert space with respect to the inner product

$$(u, v)_{D(A)} = \sum_{n=1}^{\infty} \lambda_n^2 (u, e_n)_H (e_n, v)_H = (Au, Av)_H.$$

Moreover, with this inner product, $A : D(A) \to H$ is a unitary mapping. In what follows, we will always take this inner product and the corresponding induced norm on $D(A)$.

Lemma 8.31 *Let* $u \in L_2((0, T), H)$ *be such that* $\sum_{n=1}^{\infty} \lambda_n^2 \int_0^T u_n(t)^2 \, dt < \infty$, *where* $u_n(t) = (u(t), e_n)_H$. *Then* $u \in L_2((0, T), D(A))$ *and*

$$\|u\|_{L_2((0,T),D(A))}^2 = \sum_{n=1}^{\infty} \lambda_n^2 \int_0^T |u_n(t)|^2 \, dt.$$

Proof It follows from the monotone convergence theorem that series

$$\sum_{n=1}^{\infty} \lambda_n^2 u_n(t)^2 < \infty$$

converges for all $t \in (0, \infty)$, after altering u on a set of measure zero if necessary. Thus $u(t) \in D(A)$ for all $t \in (0, \infty)$. The sequence $\hat{e}_n := \frac{1}{\lambda_n} e_n$, $n \in \mathbb{N}$, forms an orthonormal basis of $D(A)$, and $(u(t), \hat{e}_n)_{D(A)} = \lambda_n u_n(t)$. The claim now follows from Lemma 8.28.   □

We can now prove the following theorem on existence and uniqueness of solutions for the inhomogeneous evolution problem.

Theorem 8.32 *Let* $0 < T \leq \infty$, $f \in L_2((0, T), H)$ *and* $u_0 \in V$. *Then there exists a unique* $u \in L_2((0, T), D(A)) \cap H^1((0, T), H)$ *which solves the initial value problem*

$$\dot{u}(t) + Au(t) = f(t) \qquad \text{for almost every } t \in (0, T), \tag{8.46a}$$

$$u(0) = u_0. \tag{8.46b}$$

Moreover, the following estimates hold:

$$\|u\|^2_{L_2((0,T),H)} \le \frac{1}{2\lambda_1}\|u_0\|^2_H + \frac{1}{\lambda_1^2}\|f\|^2_{L_2((0,T),H)}, \tag{8.47}$$

$$\|\dot{u}\|^2_{L_2((0,T),H)} \le \frac{1}{2}a(u_0) + 4\|f\|^2_{L_2((0,T),H)}, \tag{8.48}$$

$$\|u\|^2_{L_2((0,T),D(A))} \le \frac{1}{2}a(u_0) + \|f\|^2_{L_2((0,T),H)}. \tag{8.49}$$

Note that $L_2((0,T),D(A))$ is well defined since $D(A)$ is a Hilbert space. The initial condition (8.46b) makes sense because $u \in C([0,T],H)$. It is remarkable that the solution u has *maximal regularity*: both functions $\dot{u}$ and Au are in $L_2((0,T),H)$.

Proof (of Theorem 8.32) Uniqueness: Let u be a solution of (8.46) and consider, for $n \in \mathbb{N}$, the function $u_n(t) := (u(t), e_n)_H$. Then $u_n \in H^1(0,T)$ and $\dot{u}_n(t) + \lambda_n u_n(t) = f_n(t) := (f(t), e_n)_H$, $u_n(0) = (u_0, e_n)_H$. We then deduce, using elementary properties of linear differential equations, that

$$u_n(t) = e^{-\lambda_n t}\left\{(u_0, e_n)_H + \int_0^t e^{\lambda_n s} f_n(s)\,ds\right\} \tag{8.50}$$

(see Exercise 5.10). This establishes uniqueness; it will also help us with the proof of existence.

Existence: We define u_n by (8.50), and prove using the above lemmata that $u(t) := \sum_{n=1}^\infty u_n(t)e_n$ is a solution of (8.46). In doing so, we will treat the u_0-term and the f-term separately.

1st case: $f \equiv 0$. Then $u_n(t) = e^{-\lambda_n t}(u_0, e_n)_H$, $\dot{u}_n(t) = -\lambda_n u_n(t)$, and so

$$\sum_{n=1}^\infty \int_0^T u_n(t)^2\,dt = \sum_{n=1}^\infty |(u_0, e_n)_H|^2 \int_0^T e^{-2\lambda_n t}\,dt \le \frac{1}{2\lambda_1}\|u_0\|^2_H.$$

By Lemma 8.28, there exists a unique $u \in L_2((0,T),H)$ for which holds that $u_n(t) = (u(t), e_n)_H$. For this function, we have $\|u\|^2_{L_2((0,T),H)} \le \frac{1}{2\lambda_1}\|u_0\|^2_H$. Since

$$\sum_{n=1}^\infty \int_0^T \dot{u}_n(t)^2\,dt = \sum_{n=1}^\infty \lambda_n^2 \int_0^T u_n(t)^2\,dt \le \sum_{n=1}^\infty \lambda_n^2 |(u_0, e_n)_H|^2 \int_0^T e^{-2\lambda_n t}\,dt$$

$$\le \frac{1}{2}\sum_{n=1}^\infty \lambda_n |(u_0, e_n)_H|^2 = \frac{1}{2}a(u_0),$$

we see that $u \in H^1((0,T),H)$ and $\|\dot{u}\|^2_{L_2((0,T),H)} \le \frac{1}{2}a(u_0)$. Since

$$\sum_{n=1}^\infty \int_0^T \lambda_n^2 u_n(t)^2\,dt \le \frac{1}{2}a(u_0)$$

(as follows from the above estimate), by Lemma 8.31 we also have that $u \in L_2((0,T),D(A))$ and $\|u\|^2_{L_2((0,T),D(A))} \le \frac{1}{2}a(u_0)$.

2nd case: $u_0 = 0$. Then $u_n(t) = \int_0^t e^{-\lambda_n(t-s)} f_n(s)\, ds$. We set $g_n(t) := e^{-\lambda_n t}$ for $t \geq 0$ and $g_n(t) = 0$ for $t < 0$; we likewise extend f_n by 0 on $(-\infty, 0] \cup (T, \infty)$. Then

$$(\mathcal{F} g_n)(s) = \frac{1}{\sqrt{2\pi}} \int_0^\infty e^{-ist} e^{-\lambda_n t}\, dt = \frac{1}{is + \lambda_n} \frac{1}{\sqrt{2\pi}}.$$

Since $u_n = f_n * g_n$, we have $\mathcal{F} u_n = \sqrt{2\pi}\, \mathcal{F} f_n \cdot \mathcal{F} g_n$. Now since the Fourier transform is isometric on $L_2(\mathbb{R}, \mathbb{C})$, it follows that

$$\int_0^T |u_n(t)|^2\, dt = \|\mathcal{F} u_n\|_{L_2(\mathbb{R})}^2 = 2\pi \|\mathcal{F} g_n \mathcal{F} f_n\|_{L_2(\mathbb{R})}^2$$

$$\leq \frac{1}{\lambda_n^2} \|\mathcal{F} f_n\|_{L_2(\mathbb{R})}^2 = \frac{1}{\lambda_n^2} \|f_n\|_{L_2(\mathbb{R})}^2. \tag{8.51}$$

Hence, on the one hand,

$$\sum_{n=1}^\infty \int_0^\infty |u_n(t)|^2\, dt \leq \frac{1}{\lambda_1^2} \sum_{n=1}^\infty \int_0^\infty |f_n|^2\, dt = \frac{1}{\lambda_1^2} \|f\|_{L_2((0,\infty),H)}^2,$$

and so by Lemma 8.28 there exists a unique $u \in L_2((0, \infty), H)$ such that $u_n = (u(\cdot), e_n)_H$ for all $n \in \mathbb{N}$, and we have the estimate

$$\|u\|_{L_2((0,\infty),H)}^2 \leq \frac{1}{\lambda_1^2} \|f\|_{L_2((0,\infty),H)}^2.$$

But on the other hand, by (8.51),

$$\sum_{n=1}^\infty \lambda_n^2 \int_0^\infty |u_n(t)|^2\, dt \leq \|f\|_{L_2((0,\infty),H)}^2,$$

and so, by Lemma 8.31, $u \in L_2((0, \infty), D(A))$, and we have the estimate

$$\|u\|_{L_2((0,\infty),D(A))}^2 \leq \|f\|_{L_2((0,\infty),H)}^2.$$

Finally, $\dot{u}_n(t) = \lambda_n u_n(t) + f_n(t)$. Hence

$$\sum_{n=1}^\infty \int_0^\infty \dot{u}_n(t)^2\, dt \leq \sum_{n=1}^\infty \int_0^\infty \left(2\lambda_n^2 u_n^2(t) + 2 f_n(t)^2\right) dt \leq 4 \|f\|_{L_2((0,T),H)}^2$$

by the above estimate. Thus (by Lemma 8.30) we get $u \in H^1((0, T), H)$ and $\|\dot{u}\|_{L_2((0,T),H)}^2 \leq 4 \|f\|_{L_2((0,T),H)}^2$. We thus finally obtain the estimates (8.47), (8.48) and (8.49), by adding the separate estimates for u_0 and f together.   □

In order to be able to prove an error estimate for the numerical solution of problem (8.46) via the finite element method, we require somewhat more regularity of the solutions. We can arrange this by making stronger regularity demands on f and u_0.

Theorem 8.33 *Let $0 < T < \infty$, let $u_0 \in D(A)$ be such that $Au_0 \in V$ and suppose that $f \in L_2((0,T), D(A))$.*
(a) Then the solution u of (8.46) is in $H^1((0,T), D(A))$ and we have

$$\|\dot{u}\|^2_{L_2((0,T),D(A))} \leq \frac{1}{2} a(Au_0) + 4\|f\|^2_{L_2((0,T),D(A))}.$$

(b) If in addition $f \in H^1((0,T), H)$, then u is also in $H^2((0,T), H)$, and

$$\|\ddot{u}\|^2_{L_2((0,T),H)} \leq a(Au_0) + 8\|f\|^2_{L_2((0,T),D(A))} + 2\|\dot{f}\|^2_{L_2((0,T),H)}.$$

Moreover, $u \in C^1([0,T], V)$.

Proof (a) Let $u_0 \in D(A)$ be such that $v_0 := Au_0 \in V$ and let $f \in L_2((0,T), D(A))$. Then $Af \in L_2((0,T),H)$. By Theorem 8.32 there exists a unique function $v \in L_2((0,T), D(A)) \cap H^1((0,T),H)$ such that $\dot{v} + Av = Af$, $v(0) = v_0$. Then $u = A^{-1}v$ is the solution of (8.46). By Theorem 8.32 we also have the estimate

$$\|\dot{v}\|^2_{L_2((0,T),H)} \leq \frac{1}{2} a(v_0) + 4\|Af\|^2_{L_2((0,T),H)}.$$

Since A^{-1} is an isometric isomorphism from H to $D(A)$ and $v \in H^1((0,T),H)$, it follows that also $u \in H^1((0,T), D(A))$ and $\dot{u} = A^{-1}\dot{v}$. Moreover,

$$\|\dot{u}\|^2_{L_2((0,T),D(A))} = \|\dot{v}\|_{L_2((0,T),H)} \leq \frac{1}{2} a(Au_0) + 4\|f\|^2_{L_2((0,T),D(A))}.$$

(b) Now assume additionally that $f \in H^1((0,T),H)$. Since $u = A^{-1}v$ and $\dot{v}+Av = Af$, we have $\dot{u} = A^{-1}\dot{v} = -v + f \in H^1((0,T),H)$. In addition, by the estimate in part (a),

$$\|\ddot{u}\|^2_{L_2((0,T),H)} \leq 2\|\dot{v}\|^2_{L_2((0,T),H)} + 2\|\dot{f}\|^2_{L_2((0,T),H)}$$
$$\leq a(Au_0) + 8\|Af\|^2_{L_2((0,T),H)} + 2\|\dot{f}\|^2_{L_2((0,T),H)},$$

which concludes the proof of the first assertion of b). The following interpolation result will show that $u \in C^1([0,T], V)$. □

For the missing assertion in the proof above we need an interpolation result for mixed Sobolev spaces. We add an integration by parts formula needed in the next section. We consider a Gelfand triple $V \hookrightarrow H \hookrightarrow V'$ as in Remark 4.51 (page 147). For that, let V, H be separable Hilbert spaces with inner products $(\cdot, \cdot)_H$, $(\cdot, \cdot)_V$ and induced norms $\|\cdot\|_H, \|\cdot\|_V$, such that V is continuously imbedded in H and also dense in H.[2] We identify H with a subspace of V'. For $a, b \in \mathbb{R}$, $a < b$, we consider the mixed Sobolev space

---

[2] It will turn out to be convenient to use this boldface notation here as we will later associate $D(A)$, V and H to the boldface written spaces.

$$\mathbb{W}(a, b) := L_2((a, b), V) \cap H^1((a, b), V').$$

Note that $H^1((a, b), V') \subset C([a, b], V')$.

Lemma 8.34 (Integration by parts) *(a)* $\mathbb{W}(a, b) \subset C([a, b], H)$
(b) For $u, v \in \mathbb{W}(a, b)$ we have

$$\int_a^b \langle \dot{u}(t), v(t) \rangle \, dt = (u(b), v(b))_H - (u(a), v(a))_H - \int_a^b \langle \dot{v}(t), u(t) \rangle \, dt.$$

Proof We refer to [25, Thm. XVIII.2] for a proof in the general case. Here we give a different proof for the special case where the imbedding of V in H is compact (as is the case in all our examples).

We assume that $\dim H = \infty$ and use the results and notations from Remark 4.51. Thus $(e_n)_{n \in \mathbb{N}}$ is an orthonormal basis of H, $V = \{u \in H : \sum_{n=1}^\infty \lambda_n (e_n, u)_H^2 < \infty\}$ with $\|u\|_V^2 = \sum_{n=1}^\infty \lambda_n (e_n, u)_H^2$, where $1 \le \lambda_1 \le \cdots \le \lambda_n \le \lambda_{n+1} \le \cdots$ and $\lim_{n \to \infty} \lambda_n = \infty$. Moreover, we may identify the dual space as $V' = \{y = (y_n)_{n \in \mathbb{N}} \subset \mathbb{R} : \sum_{n=1}^\infty \frac{y_n^2}{\lambda_n} < \infty\}$ with the norm $\|y\|_{V'}^2 = \sum_{n=1}^\infty \frac{y_n^2}{\lambda_n}$, where the duality is given by $\langle y, u \rangle = \sum_{n=1}^\infty y_n (u, e_n)_H$ for all $y \in V'$ and $u \in V$.

(a) Let $u \in \mathbb{W}(a, b) \subset C([a, b], V')$. Define $u_n(t) := (u(t), e_n)_H$. Since $e_n \in V$ we have $u_n \in H^1(a, b)$ and $\dot{u}_n(t) = (\dot{u}(t), e_n)_H$. Since $u \in L_2((a, b), V)$ there exists $t_0 \in (a, b)$ such that $u(t_0) \in V \subset H$. By Theorem 5.13 applied to $f = g = u_n$ one has for $t \in [a, b]$, using the Hölder and Young inequality,

$$u_n(t)^2 = u_n(t_0)^2 + 2 \int_{t_0}^t \dot{u}_n(s) \, u_n(s) \, ds = u_n(t_0)^2 + 2 \int_{t_0}^t \frac{\dot{u}_n(s)}{\sqrt{\lambda_n}} \sqrt{\lambda_n} \, u_n(s) \, ds$$

$$\le u_n(t_0)^2 + 2 \left(\int_{t_0}^t \frac{\dot{u}_n(s)^2}{\lambda_n} \, ds \right)^{1/2} \left(\int_{t_0}^t \lambda_n u_n(s)^2 \, ds \right)^{1/2}$$

$$\le u_n(t_0)^2 + \int_{t_0}^t \frac{\dot{u}_n(s)^2}{\lambda_n} \, ds + \int_{t_0}^t \lambda_n u_n(s)^2 \, ds.$$

Thus $\sum_{n=1}^\infty u_n(t)^2 < \infty$ and hence $u(t) \in H$ as well as

$$\|u(t)\|_H^2 \le \|u(t_0)\|_H^2 + \int_a^b \|\dot{u}(s)\|_{V'}^2 \, ds + \int_a^b \|u(s)\|_V^2 \, ds < \infty$$

for all $t \in [a, b]$.

We show that $u \in C([a, b], H)$. Since $(u(\cdot), e_n)_H \in C([a, b])$ for all $n \in \mathbb{N}$ and since the set $\{e_n : n \in \mathbb{N}\}$ is total in H, it follows from Lemma 4.34 that u is weakly continuous. It follows from the estimate above that

$$\left| \|u(t)\|_H^2 - \|u(t_0)\|_H^2 \right| \le \left| \int_{t_0}^t \|\dot{u}(s)\|_{V'}^2 \, ds \right| + \left| \int_{t_0}^t \|u(s)\|_V^2 \, ds \right| \longrightarrow 0$$

as $t \to t_0$ for all $t_0 \in [a, b]$.[3] Thus $\|u(\cdot)\|_{\mathbf{H}} \in C([a, b])$. Now it follows from Theorem 4.33 that $u \in C([a, b], \mathbf{H})$.

(b) Let $u, v \in \mathbb{W}(a, b)$ and set $u_n(t) := (u(t), e_n)_{\mathbf{H}}$, $v_n(t) := (v(t), e_n)_{\mathbf{H}}$. Then, $\langle \dot{u}(t), v(t) \rangle = \sum_{n=1}^{\infty} \dot{u}_n(t) v_n(t)$. Since

$$\sum_{n=1}^{\infty} |\dot{u}_n(t) v_n(t)| \le \left(\sum_{n=1}^{\infty} \frac{\dot{u}_n(t)^2}{\lambda_n} \right)^{1/2} \left(\sum_{n=1}^{\infty} \lambda_n v_n(t)^2 \right)^{1/2} \le \|\dot{u}(t)\|_{\mathbf{V}'}^2 + \|v(t)\|_{\mathbf{V}}^2$$

it follows that $\langle \dot{u}(\cdot), v(\cdot) \rangle \in L_1((a, b))$. Moreover, by Theorem 5.13,

$$\int_a^b \langle \dot{u}(t), v(t) \rangle \, dt = \int_a^b \sum_{n=1}^{\infty} \dot{u}_n(t) v_n(t) \, dt = \sum_{n=1}^{\infty} \int_a^b \dot{u}_n(t) v_n(t) \, dt$$

$$= \sum_{n=1}^{\infty} \left\{ u_n(b) v_n(b) - u_n(a) v_n(a) - \int_a^b \dot{v}_n(t) u_n(t) \, dt \right\}$$

$$= u(b) v(b) - u(a) v(a) - \int_a^b \dot{v}(t) u(t) \, dt,$$

which proves the lemma. $\qquad \square$

Proof (of the last assertion of Theorem 8.33) We finally have to show that $u \in C^1([0, T], V)$. Let $V = D(A)$, $\mathbf{H} = V$ and $V' = H$ (see Exercise 8.12). Then $\dot{u} \in H^1((0, T), H) \cap L_2((0, T), D(A)) = \mathbb{W}(0, T)$. Thus the claim follows from Lemma 8.34. $\qquad \square$

We now wish to apply these abstract results to the heat equation with Dirichlet boundary conditions. Let $\Omega \subset \mathbb{R}^d$ be a bounded open convex set, $H = L_2(\Omega)$, $V = H_0^1(\Omega)$, and $a(u, v) = \int_\Omega \nabla u \, \nabla v \, dx$ for all $u, v \in H_0^1(\Omega)$. By Theorem 6.86, the associated operator on $L_2(\Omega)$ is given by

$$D(A) = H^2(\Omega) \cap H_0^1(\Omega), \quad Av = -\Delta v.$$

We note in passing that $H^2(\Omega) \cap H_0^1(\Omega)$ is a closed subspace of $H^2(\Omega)$.

Theorem 8.35 *Let $0 < T \le \infty$ and $f \in L_2((0, T), L_2(\Omega))$, $u_0 \in H_0^1(\Omega)$. Then there exists a unique $u \in L_2((0, T), H^2(\Omega)) \cap H^1((0, T), L_2(\Omega))$ such that*

$$\dot{u}(t) = \Delta u(t) + f(t), \quad t \in (0, T), \tag{8.52a}$$

$$u(t) \in H_0^1(\Omega), \qquad t \in (0, T), \tag{8.52b}$$

$$u(0) = u_0. \tag{8.52c}$$

---

[3] We need the absolute values on the right-hand side as t might be smaller than t_0.

Note that $H^1((0,T), L_2(\Omega)) \subset C([0,T],L_2(\Omega))$, so that the initial condition (8.52c) makes sense. We interpret condition (8.52b) as a Dirichlet boundary condition, while (8.52a) states that u satisfies the inhomogeneous heat equation.

The proof of Theorem 8.35 follows directly from Theorem 8.32. For an error analysis in the context of the numerical treatment of the problem via finite elements we will require the following statement on regularity, which we can make under somewhat stronger assumptions on the initial value u_0 and the inhomogeneity f.

Theorem 8.36 *Let* $0 < T < \infty$. *Assume that there exists a constant* $c > 0$ *for which the following statements hold: Let* $f \in L_2((0,T), H^2(\Omega) \cap H_0^1(\Omega))$ *and suppose that* $u_0 \in H^2(\Omega) \cap H_0^1(\Omega)$, *with* $\Delta u_0 \in H_0^1(\Omega)$.

(a) Then the solution u *of (8.52) is in* $H^1((0,T), H^2(\Omega) \cap H_0^1(\Omega))$, *and*

$$\|\dot{u}\|^2_{L_2((0,T),H^2(\Omega)\cap H_0^1(\Omega))} \le c(\|\Delta u_0\|^2_{H^1(\Omega)} + \|f\|^2_{L_2((0,T),H^2(\Omega))}).$$

(b) If in addition $f \in H^1((0,T); L_2(\Omega))$, *then* $u \in H^2((0,T); L_2(\Omega))$ *and*

$$\|\ddot{u}\|^2_{L_2(0,T;L_2(\Omega))} \le c(\|\Delta u_0\|^2_{H^1(\Omega)} + \|f\|^2_{L_2((0,T),H^2(\Omega))} + \|\dot{f}\|^2_{L_2((0,T),L_2(\Omega))}).$$

Moreover, $u \in C^1([0,T], H_0^1(\Omega))$.

Theorem 8.36 is a direct consequence of Theorem 8.33 and the observation that $H^2(\Omega) \cap H_0^1(\Omega)$ is a closed subspace of $H^2(\Omega)$ and thus a Hilbert space.

8.6* Space/time variational formulations

The theorems of Banach–Nečas (Theorem 4.27) and Lions (Theorem 4.29), as generalizations of the Lax–Milgram theorem, permit us to give a variational formulation of time-dependent problems in space *and* time, and as such to give an alternative proof of their well-posedness. We will see later, in particular in the case of parabolic problems, that this formulation is advantageous for the numerical analysis of such problems.

Let us assume for the meantime that we have an elliptic problem in the spatial variables, that is, we consider a separable Hilbert space H (such as $H = L_2(\Omega)$) and a second separable Hilbert space V (such as $V = H_0^1(\Omega)$) which is continuously and densely imbedded in H. We denote elements of the dual space V' by ℓ and write

$$\langle \ell, v \rangle := \ell(v), \qquad v \in V.$$

As previously, we identify H with a subspace of V' by setting $\langle \ell, v \rangle := (\ell, v)_H$ for every $\ell \in H$. We now consider a coercive, but not necessarily symmetric, bilinear form $a : V \times V \to \mathbb{R}$. Then $(\mathcal{A}u)(v) := \langle \mathcal{A}u, v \rangle := a(u, v), u, v \in V$, defines a linear operator $\mathcal{A} : V \to V'$. (Here we use the notation $\mathcal{A} : V \to V'$ to distinguish this operator from $A : D(A) \to H$.) We now wish to treat the parabolic problem

$$\dot{u}(t) + \mathcal{A}u(t) = f(t), \tag{8.53a}$$

$$u(0) = 0, \tag{8.53b}$$

using a variational approach. Here we assume that $T > 0$ is a given end time and $f(t) \in V'$ for almost every $t \in I := (0, T)$ is a given external force, $f \in L_2(I, V')$. Then we have the following theorem on existence and uniqueness of solutions, which goes back to J.-L. Lions.

Theorem 8.37 *Given $f \in L_2(I, V')$, there exists exactly one solution $u \in H^1(I, V') \cap L_2(I, V)$ of (8.53).*

We recall that $H^1(I, V') \subset C([0, T], V')$, cf. Exercise 8.7, which means that $u(0)$ is well defined. For simplicity we will only consider the homogeneous initial condition $u(0) = 0$.

Proof (of Theorem 8.37) Let $f \in L_2(I, V')$. In order to prove existence of a solution we use the representation theorem of Lions (Theorem 4.29). For this we define $\mathbb{V} := L_2(I, V)$ and $\mathbb{W} := \{w \in C^1([0, T], V) : w(T) = 0\}$; then $\mathbb{V}$ is a Hilbert space with respect to the norm $\|v\|_{\mathbb{V}}^2 = \int_0^T \|v(t)\|_V^2 \, dt$, and $\mathbb{W} \subset \mathbb{V}$ is an inner product space for the same norm. Now define a bilinear form $b : \mathbb{V} \times \mathbb{W} \to \mathbb{R}$ by

$$b(v, w) := -\int_0^T (v(t), \dot{w}(t))_H \, dt + \int_0^T a(v(t), w(t)) \, dt; \tag{8.54}$$

then $b(\cdot, w) \in \mathbb{V}'$ for all $w \in \mathbb{W}$. In order to apply Theorem 4.29 we will show that condition (c) of Remark 4.30 holds. Given $w \in \mathbb{W}$, we apply the chain rule from Lemma 8.17 to obtain

$$b(w, w) = \int_0^T -\frac{1}{2}\frac{d}{dt} \|w(t)\|_H^2 \, dt + \int_0^T a(w(t), w(t)) \, dt$$
$$= \frac{1}{2}\|w(0)\|_H^2 + \int_0^T a(w(t), w(t)) \, dt \geq \alpha \|w\|_{\mathbb{W}}^2.$$

This shows that the assumptions of Theorem 4.29 are indeed satisfied. Since, given $f \in L_2(I, V')$, the expression $\langle \ell, w \rangle := \int_0^T \langle f(t), w(t) \rangle \, dt$ defines a continuous linear form $\ell \in \mathbb{W}'$, by Theorem 4.29 there exists some $u \in \mathbb{V}$ such that $b(u, w) = \langle \ell, w \rangle$ for all $w \in \mathbb{W}$.

Since $u \in \mathbb{V} = L_2(I, V)$ and $\mathcal{A} \in \mathcal{L}(V, V')$, we have $\mathcal{A}u \in L_2(I, V')$. For $\varphi \in \mathcal{D}(I)$ (see (6.2)) and $\psi \in V$ we can define a test function $w \in \mathbb{W}$ by $w(t) := \varphi(t)\psi$; for this function, we have

$$\int_0^T \langle f(t), w(t) \rangle \, dt = b(u, w) = -\int_0^T (u(t), \psi)_H \, \dot{\varphi}(t) \, dt + \int_0^T \langle \mathcal{A}u(t), \psi \rangle \, \varphi(t) \, dt.$$

Since $\psi \in V$ was arbitrary, it follows that

$$\int_0^T f(t) \, \varphi(t) \, dt = -\int_0^T u(t) \, \dot{\varphi}(t) \, dt + \int_0^T \mathcal{A}u(t) \, \varphi(t) \, dt$$

for all $\varphi \in \mathcal{D}(I)$. It now follows by definition (see 8.43) that $u \in H^1(I, V')$ and

$$\dot{u} + \mathcal{A}u = f. \tag{8.55}$$

In order to show that u is indeed a solution, we still need to show that $u(0) = 0$. To this end take $\varphi \in C^1([0, T])$ such that $\varphi(T) = 0$ and $\varphi(0) = 1$, and set $w(t) := \varphi(t)u(0)$, so that $w \in \mathbb{W}$. Replacing f by the expression on the left-hand side of (8.55) yields

$$\int_0^T \{-(u(t), \dot{w}(t))_H + a(u(t), w(t))\} \, dt = b(u, w) = \int_0^T \langle f(t), w(t) \rangle \, dt$$
$$= \int_0^T \langle \dot{u}(t) + \mathcal{A}u(t), w(t) \rangle \, dt = \int_0^T \langle \dot{u}(t), w(t) \rangle \, dt + \int_0^T a(u(t), w(t)) \, dt$$

$$= -(u(0), w(0))_H - \int_0^T \langle u(t), \dot{w}(t) \rangle \, dt + \int_0^T a(u(t), w(t)) \, dt,$$

since $w(T) = 0$. Hence $(u(0), w(0))_H = 0$. But since $w(0) = u(0)$, it follows that $u(0) = 0$, and we have proved existence.

To prove uniqueness, we suppose that $u \in H^1(I, V') \cap L_2(I, V)$ satisfies $\dot{u} + \mathcal{A}u = 0$ and $u(0) = 0$. Then

$$\int_0^\tau \{ \langle \dot{u}(t), u(t) \rangle + a(u(t), u(t)) \} \, dt = 0$$

for all $\tau \in [0, T]$. The coercivity of $a(\cdot, \cdot)$ implies in particular that $\int_0^\tau \langle \dot{u}(t), u(t) \rangle dt \leq 0$ for all $\tau \in [0, T]$. Part (b) of Lemma 8.34 now shows $\|u(\tau)\|_H = 0$ for all $\tau \in (0, T]$, whence $u = 0$. $\square$

Remark 8.38   (a) It follows from Theorem 4.29 that the solution u of (8.53) satisfies the estimate

$$\|u\|_{L_2(I,V)} \leq \frac{1}{\alpha} \|f\|_{L_2(I,V')},$$

where $\alpha > 0$ is the coercivity constant of $a(\cdot, \cdot)$. Observe that the imbedding constant c of Remark 4.30(c) is equal to 1 here.

(b) There is another approach which can be used to prove Theorem 8.37: one may take $\mathbb{V} := \{v \in H^1(I, V') \cap L_2(I, V) : v(0) = 0\}$, $\mathbb{W} = L_2(I, V)$ and

$$b(v, w) := \int_0^T \langle \dot{v}(t), w(t) \rangle \, dt + \int_0^T a(v(t), w(t)) \, dt. \tag{8.56}$$

In this case one simply "multiplies" the equation $\dot{u} + \mathcal{A}u = f$ by $w \in \mathbb{W}$ (without integrating by parts in the time variable).

It can be shown that the inf-sup condition for $b(\cdot, \cdot)$ (cf. Remark 8.39) and the assumptions of the theorem of Banach–Nečas (Theorem 4.27) all hold. The inf-sup condition is of considerable importance for the numerical treatment of the problem (cf. Theorem 9.42 on page 357) and allows an estimate on the norm of the operator $f \mapsto u$ in $L_2(I, V') \to \mathbb{V}$, to be contrasted with the norm in $L_2(I, V') \to L_2(I, V)$ as above. However, the surjectivity, the second condition in the theorem of Banach–Nečas, is not easy to show. One can either use Lions' theorem, as we have done, or else a Galerkin method as in [55, Thm. 5.1] (cf. Section 9.2.1 on page 330).

(c) One can go further and show that for every $f \in L_2(I, V')$ and every $u_0 \in H$ there exists a unique $u \in H^1(I, V') \cap L_2(I, V)$ such that $u' + \mathcal{A}u = f$, $u(0) = u_0$. One can even replace coercivity by the more general H-ellipticity condition (cf. (4.22)).

(d) The above analysis can also be performed in the same way on time-dependent, uniformly elliptic operators $\mathcal{A}(t)$. $\triangle$

Remark 8.39 Since the actual size of the inf-sup constant plays a major role in the numerical analysis of the problem, we shall provide a lower bound in terms of the bilinear form in (8.56), with $\mathbb{V}$ and $\mathbb{W}$ as in Remark 8.38(b), cf. [59, Prop. 1]. For this we will use a common technique, namely determining the supremizer in the sense of Remark 4.28(d).

So let $0 \neq v \in \mathbb{V}$. Denote by $\mathcal{A}^* : V \to V'$ the mapping given by the mapping $\langle \mathcal{A}^* u, w \rangle := a(w, u)$, (i.e., $\mathcal{A}^*$ is associated with the bilinear form a^* given by $a^*(u, w) := a(w, u)$). Then $\mathcal{A}^*$ is an isomorphism and so the mapping $L_2(I, V) \ni w \mapsto \mathcal{A}^* w \in L_2(I, V')$ is an isomorphism too. Thence, there exists a unique $z_v \in L_2(I, V)$ such that $\mathcal{A}^* z_v(t) = \dot{v}(t)$ for almost every $t \in I$. In particular, $a(\phi, z_v(t)) = (\dot{v}(t))(\phi) = \langle \dot{v}(t), \phi \rangle$ for all $\phi \in V$, and $\alpha \|z_v(t)\|_V \leq \|\dot{v}(t)\|_{V'} \leq C \|z_v(t)\|_V$ for almost every $t \in I$, where α and C are, respectively, the coercivity and continuity constants of the bilinear form $a(\cdot, \cdot)$. It follows that $z_v \in \mathbb{W}$. Now set $w := z_v + v \in \mathbb{W}$. Then

$$\|w\|_{\mathbb{W}}^2 = \int_0^T \|w(t)\|_V^2 \, dt = \int_0^T \|z_v(t) + v(t)\|_V^2 \, dt \leq 2 \int_0^T (\|z_v(t)\|_V^2 + \|v(t)\|_V^2) dt$$

$$\leq \frac{2}{\alpha} \int_0^T \|\dot{v}(t))\|_{V'}^2 \, dt + 2\|v\|_{L_2(I,V)}^2 = \frac{2}{\alpha} \|\dot{v}\|_{L_2(I,V')}^2 + 2\|v\|_{L_2(I,V)}^2$$

$$\leq 2 \max\{\alpha^{-1}, 1\} \|v\|_{\mathbb{V}}^2 < \infty.$$

Now we also have, by definition of z_v and Lemma 8.34,

$$b(v, w) = \int_0^T \{\langle \dot{v}(t) \, w(t)\rangle + a(v(t), w(t))\} dt$$

$$= \int_0^T \{\langle \dot{v}(t), v(t)\rangle + \langle \dot{v}(t), z_v(t)\rangle + a(v(t), z_v(t)) + a(v(t), v(t))\} \, dt$$

$$= \frac{1}{2} \|v(T)\|_H^2 + \int_0^T \{a(z_v(t), z_v(t)) + \langle \dot{v}(t), v(t)\rangle + a(v(t), v(t))\} \, dt$$

$$= \|v(T)\|_H^2 + \int_0^T \{a(z_v(t), z_v(t)) + a(v(t), v(t))\} \, dt.$$

The coercivity of $a(\cdot, \cdot)$ now implies that

$$b(v, w) \geq \alpha \int_0^T \{\|z_v(t)\|_V^2 + \|v(t)\|_V^2\} dt$$

$$\geq \alpha \min\{C^{-2}, 1\} \int_0^T \{\|\dot{v}(t)\|_{V'}^2 + \|v(t)\|_V^2\} dt = \alpha \min\{C^{-2}, 1\} \|v\|_{\mathbb{V}}^2$$

$$\geq \alpha \min\{C^{-2}, 1\} \frac{1}{\sqrt{2}} \min\{\sqrt{\alpha}, 1\} \|v\|_{\mathbb{V}} \|w\|_{\mathbb{W}}.$$

Hence $\tilde{\beta} := \frac{\alpha}{\sqrt{2}} \min\{C^{-2}, \sqrt{\alpha}, 1\}$ is a lower bound on the inf-sup constant β, that is, $\beta \geq \tilde{\beta} > 0$. In the case of the heat equation, with the norm $\|v\|_{\mathbb{V}}^2 := \|\dot{v}\|_{L_2(I,V')}^2 + \|v\|_{L_2(I,V)}^2 + \|v(T)\|_H^2$ and the energy norm $\|\phi\|_V^2 = a(\phi, \phi)$ on V, one can actually show that $\beta = 1$ [60], which is particularly convenient for numerical purposes.

In [55, Thm. 5.1] one can also find a bound on the inf-sup constant, which however degenerates as $T \to \infty$, whereas $\tilde{\beta}$ is independent of the endpoint T.   △

Remark 8.40 (Linear transport problems) Problems of the form $\dot{u} + \boldsymbol{b} \cdot \mathrm{grad}\, u + cu = f$ in $I \times \Omega$, equipped with suitable initial and boundary values, can also be treated using space/time variational methods as here. In [19] the well-posedness of a suitable variational formulation $u \in \mathbb{V}$ is proved, where $b(u, w) = f(w)$, $\mathbb{V} = L_2(I, L_2(\Omega))$, the bilinear form b is given by

$$b(v, w) := \int_0^T (v, -\dot{w} - \boldsymbol{b} \cdot \mathrm{grad}\, w + w(c - \nabla \cdot \boldsymbol{b}))_{L_2(\Omega)} \, dt$$

and $\mathbb{W}$ is chosen appropriately. This uses the theorem of Banach–Nečas, Theorem 4.27. More is shown: for certain special norms the inf-sup constant β equals 1.   △

Remark 8.41 (Wave equation) An analogous approach can be used to handle what is known as a *very weak* variational formulation of the wave equation $\ddot{u} - \Delta u = f$ in $I \times \Omega$ subject to suitable initial and boundary conditions. Here as well one can prove well-posedness and $\beta = 1$ (for suitable norms) if $\mathbb{V} := L_2(I, L_2(\Omega))$,

$$b(v, w) := \int_0^T (v, \ddot{w} - \Delta w)_{L_2(\Omega)} \, dt,$$

and the test space $\mathbb{W}$ is chosen appropriately, cf. [48, §9].   △

8.7* Comments on Chapter 8

In the present chapter we have described the evolution of physical systems. The essence of our principal theorems (Theorem 8.3 for the result in abstract form, with concrete realizations in Sections 8.2 and 8.3) is that given any initial state the evolution of the system is determined for all time. It is a far-reaching physical and philosophical question whether such a form of determinism correctly describes nature. Many fundamental physical laws can be formulated as evolution equations. Laplace in his works motivated such evolution equations by conceiving of an exterior intelligence as driving force, an intelligence which has become known as Laplace's demon. He wrote, "Une intelligence qui, à un instant donné, connaîtrait toutes les forces dont la nature est animée et la situation respective des êtres qui la compose embrasserait dans la même formule les mouvements des plus grands corps de l'univers et ceux du plus léger atome; rien ne serait incertain pour elle, et l'avenir, comme le passé, serait présent à ses yeux" [45, pages 32–33] ("An intelligence which at any given moment knew all forces which animate nature and the respective state of all beings which compose it, could express in the same formula the motions of the greatest heavenly bodies and those of the lightest atom; nothing would be uncertain for it; and the future, like the past, would be the present in its eyes.")

In our context, we know that such explicit formulae can only be derived in the simplest geometric situations; and we must content ourselves with numerical approximations, the topic of the next chapter. Regarding the philosophical background for the mathematics of evolution equations, we refer to the epilog in the book of Engel and Nagel [27].

8.8 Exercises

Exercise 8.1 Let $w(t) = \sum_{n=1}^{\infty} w_n(t) e_n$, where w_n is given by (8.33). Show by direct calculation that $a(w(t)) + \|\dot{w}(t)\|_H^2 = a(u_0) + \|u_1\|_H^2$ for all $t \geq 0$.

Exercise 8.2 (Separated variables) Let A be an operator on a real Hilbert space H and let $f \in D(A)$, $f \neq 0$.
(a) Show that there exists a unique solution $\dot{u}(t) = Au(t)$, $t \in (0, T)$ of the form $u(t) = v(t)f$, where $v \in C^1(0, T)$, $v \not\equiv 0$, if f is an eigenvector of A.
(b) Do the same for the equation $\ddot{u}(t) = Au(t)$, $t \in (0, T)$, where we now assume that $v \in C^2(0, T)$.

Exercise 8.3 Let Ω be a bounded C^1-domain and suppose that $b : \partial\Omega \to [0, \infty)$ is bounded and Borel measurable. Let $u_0 \in H^1(\Omega) \cap H^2_{\mathrm{loc}}(\Omega)$ be such that $\Delta u_0 \in L^2(\Omega)$ and $\frac{\partial u}{\partial \nu} + b(z)(Tu)(z) = 0$, and let $u_1 \in H^1(\Omega)$. Show that there exists a unique function $w \in C^2(\mathbb{R}_+, L_2(\Omega))$ such that $w(t) \in H^1(\Omega) \cap H^2_{\mathrm{loc}}(\Omega)$, $\Delta w(t) \in L_2(\Omega)$, $t \geq 0$ and

$$\ddot{w}(t) - \Delta w(t) = 0, \quad t \geq 0,$$
$$\frac{\partial w(t)}{\partial \nu} + bT(w(t)) = 0,$$
$$w(0) = u_0, \quad \dot{w}(0) = u_1.$$

Exercise 8.4 Let $b : \partial\Omega \to [0, \infty)$ be bounded and Borel measurable. Show that $\lambda_n^b \le \lambda_n^D$, $n \in \mathbb{N}$.

Exercise 8.5 Let $I \subset \mathbb{R}$ be an interval and H a separable Hilbert space.
(a) Let $u \in L_1(I, H)$. Show that there exists a unique element $\int_I u(t)\, dt$ of H such that $\int_I (u(t), v)_H\, dt = \left(\int_I u(t)\, dt, v\right)_H$ for all $v \in H$.
(b) Show that $\left\| \int_I u(t)\, dt \right\|_H \le \int_I \|u(t)\|_H\, dt$.
(c) Show that $L_1(I, H)$ is a vector space and that the mapping $u \mapsto \int_I u(t)\, dt :$ $L_1(I, H) \to H$ is linear.
(d) If $0 < T < \infty$, then $L_2((0, T), H) \subset L_1((0, T), H)$.

Exercise 8.6 Let H_1, H_2 be separable Hilbert spaces and let $B : H_1 \to H_2$ be continuous and linear. Show that $u \mapsto B \circ u$ defines a continuous linear mapping of $L_2((0, T), H_1)$ into $L_2((0, T), H_2)$, $0 < T \le \infty$, and that the mapping is unitary if B is unitary.

Exercise 8.7 Let H be a separable Hilbert space and $0 < T \le \infty$.
(a) Let $u \in H^1((0, T), H)$ be such that $\dot{u}(t) = 0$ almost everywhere. Show that there exists $x \in H$ such that $u(t) = x$ for almost every $t \in (0, T)$.
   Suggestion: Use the scalar result in Lemma 5.7.
(b) Let $v \in L_2((0, T), H)$ and $x_0 \in H$. Define $u : (0, T) \to H$ by $u(t) = x_0 + \int_0^t v(s)\, ds$. Show that $u \in H^1((0, T), H)$ and $u' = v$. Also show that $u \in C([0, T], H)$ if $T < \infty$.
(c) Let $u \in H^1((0, T), H)$. Show that $u \in C([0, T], H)$ and $u(t) = u(0) + \int_0^t u'(s)\, ds$ for all $t \in (0, T)$.
(d) Let $u \in H^1((0, T), H)$ be such that $\dot{u} \in C([0, T], H)$. Show that then $u \in C^1([0, T], H)$.

Exercise 8.8 Let $\{e_n : n \in \mathbb{N}\}$ be an orthonormal basis of the Hilbert space H and let $(\lambda_n)_{n \in \mathbb{N}}$ be a monotonically increasing sequence such that $0 \le \lambda_1$, $\lim_{n \to \infty} \lambda_n = \infty$. Define $T(t) : H \to H$, $t \ge 0$, by $T(t)x = \sum_{n=1}^\infty e^{-\lambda_n t}(x, e_n)_H e_n$. Show:
(a) $T(t) \in \mathcal{L}(H)$, $T(t + s) = T(t)T(s)$, $t, s \ge 0$, $T(0) = I$, $\lim_{h \to 0} T(t + h)x = T(t)x$ for all $x \in H$.
   Suggestion: consider the proof of Theorem 8.3 for the last property.
(b) The limit $\lim_{t \downarrow 0} \frac{1}{t}(T(t)x - x) =: Bx$ exists for each $x \in H$ if and only if $\sum_{n=1}^\infty \lambda_n^2 (x, e_n)_H^2 < \infty$.
   Suggestion: Let $\sum_{n=1}^\infty \lambda_n^2 (x, e_n)_H^2 < \infty$, and set $y = -\sum_{n=1}^\infty \lambda_n (x, e_n)_H e_n$. Show that $T(t)x - x = \int_0^t T(s)y\, ds$.

Exercise 8.9 Let H be a separable Hilbert space, $\lambda \in \mathbb{R}$, $T > 0$, $f \in L_2((0, T), H)$, and $x_0 \in H$. Show that there exists a unique function $u \in H^1((0, T), H)$ such that $\dot{u}(t) + \lambda u(t) = f(t)$ almost everywhere in $(0, T)$, $u(0) = x_0$, and that this function is given by $u(t) = e^{-\lambda t}\{x_0 + \int_0^t e^{\lambda s} f(s)\, ds\}$.

Exercise 8.10 Take the assumptions made at the beginning of Section 8.4.
(a) Show that $D_\infty(A) := \bigcap_{k \in \mathbb{N}} D(A^k)$ is dense in H.

Here $D(A^k) := \{u \in D(A^{k-1}) : A^{k-1}u \in D(A)\}$, $A^k := A(A^{k-1}u)$ for $k \geq 2$.
(b) Let $u_0, u_1 \in D_\infty(A)$ and $f = 0$. Show that the solution w of (8.39) is in $C^\infty([0,T], V)$.
(c) Let $m \in \mathbb{N}_0$, $f \in C^m([0,T], H)$, where $C^0([0,T], H) = C([0,T], H)$. Define $w_n(t) := \frac{1}{\sqrt{\lambda_n}} \int_0^t \sin(s\sqrt{\lambda_n}) f_n(t-s)\, ds$ if $\lambda_n > 0$ and $w_n(t) := \int_0^t s\, f_n(t-s)\, ds$ if $\lambda_n = 0$, where $f_n(t) = (f(t), e_n)_H$. Show that $\sum_{n=1}^\infty w_n^{(k)} e_n$ converges in V uniformly with respect to $t \in [0,T]$, $k = 0, \ldots, m$.
Deduce from this that $w(\cdot) = \sum_{n=1}^\infty w_n(\cdot) e_n \in C^m([0,T], V)$.
(d) Let $u_0 = u_1 = 0$ and $f \in C^2([0,T], H)$. Show that (8.39) has a unique solution w.
(e) Let $u_0, u_1 \in D_\infty(A)$ and $f \in C^\infty([0,T], H)$. Show that (8.39) has a unique solution w and that $w \in C^\infty([0,T], V)$.

Exercise 8.11 Let $\Omega \subset \mathbb{R}^d$ be bounded, open and convex, and suppose that $u_0 \in C^2(\overline{\Omega})$, $u_1 \in C^1(\overline{\Omega})$ are such that $u_{0|\partial\Omega} = u_{1|\partial\Omega} = 0$, and $f \in C^2([0,T], L_2(\Omega))$.
(a) Show that under these assumptions the problem

$$\ddot{w}(t) = \Delta w(t) + f(t), \qquad\qquad t \in [0,T],$$
$$w(t) \in H^2(\Omega) \cap H_0^1(\Omega), \qquad\qquad t \in [0,T],$$
$$w(0) = u_0, \quad \dot{w}(0) = u_1,$$

has a unique solution $w \in C^2([0,T], L_2(\Omega))$.
(b) Assume that $u_0 \in C^4(\overline{\Omega})$, $u_{0|\partial\Omega} = \Delta u_{0|\partial\Omega} = 0$ and $u_1 \in C^3(\overline{\Omega})$, $u_{1|\partial\Omega} = \Delta u_{1|\partial\Omega} = 0$ as well as $f \in C^3([0,T], L_2(\Omega))$. Show that the solution w of (a) is in $C^3([0,T], H_0^1(\Omega))$.
(c) Asume that $u_0 \in C^5(\overline{\Omega})$ such that $u_{0|\partial\Omega} = \Delta u_{0|\partial\Omega} = \Delta^2 u_{0|\partial\Omega} = 0$ and $u_1 \in C^4(\overline{\Omega})$ such that $u_{0|\partial\Omega} = \Delta u_{0|\partial\Omega} = 0$ as well as $f \in C^4([0,T], L_2(\Omega))$. Show that now $w \in C^4([0,T], H_0^1(\Omega))$.
(d) Assume that $u_0, u_1 \in C^\infty(\overline{\Omega})$ such that $\Delta^k u_{0|\partial\Omega} = \Delta^k u_{1|\partial\Omega} = 0$ for all $k \in \mathbb{N}_0$ and let $f \in C^\infty([0,T], L_2(\Omega))$. Show that in this case $w \in C^\infty([0,T], L_2(\Omega))$.

Exercise 8.12 Let H be a real Hilbert space with orthonormal basis $(e_n)_{n \in \mathbb{N}}$. Let $\lambda_n \in (0, \infty)$, $0 < \lambda_1 \leq \cdots \leq \lambda_n \leq \lambda_{n+1}$, $\lim_{n \to \infty} \lambda_n = \infty$. Consider the Hilbert spaces

$$V := \left\{ u \in H : \sum_{n=1}^\infty \lambda_n\, (u, e_n)_H^2 < \infty \right\}$$

with scalar product $(u, v)_V := \sum_{n=1}^\infty \lambda_n\, (u, e_n)_H\, (e_n, v)_H$ and

$$D := \left\{ u \in H : \sum_{n=1}^\infty \lambda_n^2\, (u, e_n)_H^2 < \infty \right\}$$

with scalar product $(u, v)_D := \sum\limits_{n=1}^{\infty} \lambda_n^2 (u, e_n)_H (e_n, v)_H$. Thus $D \hookrightarrow V \hookrightarrow H$.

Show that $H = D'$, where the duality is given by

$$\langle f, u \rangle = \sum_{n=1}^{\infty} \lambda_n (u, e_n)_H (e_n, f)_H$$

for all $u \in D$, $f \in H$. In particular $\|f\|_H = \|f\|_{D'}$.

Chapter 9
Numerical methods

Thus far we have met and modeled a number of partial differential equations, considered elementary solution methods for them and finally developed mathematical theories to handle them. Everything up until now has been "analytic" in nature, that is, we can perform everything on paper, and the solutions we have constructed satisfy their respective partial differential equations "exactly" in the desired sense, even if we first had to find a suitable notion of solution (classical, strong or weak) for each problem.

But such methods are sometimes not enough. For example, we have proved existence and uniqueness of a (weak) solution to the Poisson equation with inhomogeneous Dirichlet boundary conditions, but we can only find an explicit formula for the solution on a few very special domains. Even disks and rectangles were treated using Fourier series, which after all can only be calculated approximately. If one wants a representation of the solution in any other even slightly more complex situation, one needs approximation procedures to calculate it.

In such cases, the use of computers in combination with solution methods from *numerical* mathematics can help. *Finite differences* and *finite elements* are two very frequently used methods, which we shall introduce in this chapter.

Chapter overview

© The Author(s), under exclusive license to Springer Nature Switzerland AG 2023
W. Arendt, K. Urban, *Partial Differential Equations*, Graduate Texts in
Mathematics 294, https://doi.org/10.1007/978-3-031-13379-4_9

What are numerical methods?

Numerical methods have the goal of determining an *approximate solution* to a given problem, via the application of an algorithm. The Swiss mathematician Heinz Rutishauser is considered one of the pioneers of modern numerical mathematics, and also of computer science. In his words, *"numerical mathematics is concerned with finding a computational way to solve mathematically formulated problems"*. The desired accuracy of the approximation is often determined by the real problem being studied. Think, for example, of the solution of a partial differential equation which describes the statics of some structural element, say of a building, and the structural engineer requires that the approximate solution should differ from the exact solution by at most 5%. Such tolerances are also often determined by approximations or simplifications already contained in the model. In such cases it is sensible to determine an approximate solution in such a way that the error roughly corresponds to the error introduced by the model; anything beyond that could well be wasted effort.

At this juncture mathematicians are required, since a number of mathematical questions arise directly in the context of any such numerical method; in particular:

- How accurate is the approximate solution; can one estimate the error precisely? Of course the exact solution is unknown; nevertheless, the distance between it and the approximation needs to be estimated.
- How quickly can one obtain the approximate solution, and how much effort (computation time) is required to calculate it? This is known as the *complexity* of the method; if the complexity is low, then we speak of an *efficient* method.
- What is the effect of unavoidable errors (such as roundoff errors or inaccuracies in the measurement of the input quantities) on the result? If small input errors result in small output errors, then we call the algorithm *stable*. This is clearly closely related to the well-posedness of a problem, or to be more precise, to the continuous dependence on the data.
- How does the method behave when essential parameters such as the domain, the coefficients in the partial differential equation or the right-hand side are altered? Is the algorithm *robust* to such changes?

These are classical questions in numerical mathematics, for short, numerics. To answer them, one often falls back on exactly those functional analytic tools that were introduced in the previous chapters. Of course the actual construction of suitable numerical methods is another matter, as is their practical implementation, that is, their programming. Here one requires input from computer science and information technology, in areas such as compiler construction, algorithm development, computer architecture, software engineering and management, and so on. It would go well beyond the scope of this book to discuss these latter topics; our goal here is to discuss the mathematical aspects of the numerical methods we shall discuss. For their implementation we will use a programming language chosen more or less arbitrarily, or a corresponding numerical library.

9.1 Finite differences for elliptic problems

Difference methods (finite differences) are perhaps the simplest method available for solving partial differential equations numerically. They are especially suitable when the domain is rectangular. But they can also be used in combination with *discrete maximum principles* to yield existence results for classical solutions of partial differential equations, which can be derived as a consequence of the convergence of the method. Here we will introduce the method and its essential properties; but we will also point out the limits of its applicability.

We will first decribe the finite difference method (FDM) via a simple example in one space dimension (that is, a boundary value problem for an ordinary differential equation), as this case is significantly easier to understand, and then afterwards we consider the two-dimensional analog, that is, partial differential equations. In Section 9.4.1 we will then describe the solution of parabolic problems using FDMs and Section 9.5.1 is devoted to FDMs for the wave equation.

9.1.1 FDM: the one-dimensional case

We start by considering the boundary value problem

$$-u''(x) = f(x), \quad x \in (0, 1), \tag{9.1a}$$

$$u(0) = u(1) = 0. \tag{9.1b}$$

Before coming to the FDM we first wish to summarize a few known theoretical results for the problem (9.1) which we will require in this section. We will also write (9.1) in the form $Lu = f$, where L is the differential operator defined by (9.1); later, we will define a discrete approximation L_h of L. We will use the norm $\|v\|_\infty := \sup_{x \in [0,1]} |v(x)|$ for $v \in C([0, 1])$.

Boundary value problems

Existence and uniqueness of solutions of (9.1) were proved in Section 5.2.1; here we wish to write the solutions in terms of the corresponding *Green's function*.

Lemma 9.1 *The following statements hold.*
(a) For every $f \in C([0, 1])$ there exists exactly one solution $u \in C^2([0, 1])$ of (9.1). This is given by

$$u(x) = \int_0^1 G(x, s) f(s) \, ds, \tag{9.2}$$

where G is the Green's function

$$G(x, s) := \min\{x, s\}(1 - \max\{x, s\}) = \begin{cases} s(1 - x), & \text{if } 0 \le s \le x \le 1, \\ x(1 - s), & \text{if } 0 \le x \le s \le 1. \end{cases} \quad (9.3)$$

(b) Shift theorem: if $f \in C^m([0, 1])$, $m \ge 0$, then $u \in C^{m+2}([0, 1])$.
(c) Monotonicity: if $f \ge 0$, $f \in C([0, 1])$, then $u \ge 0$.
(d) Stability (maximum principle): for every $f \in C([0, 1])$, we have $\|u\|_\infty \le \frac{1}{8}\|f\|_\infty$.

Proof (a) If we set $F(s) := \int_0^s f(t) \, dt$, then we may write the solution of (9.1a) as $u(x) = c_1 + c_2 x - \int_0^x F(s) \, ds$, where $c_1, c_2 \in \mathbb{R}$ are constants of integration. We now invoke the boundary conditions (9.1b) to obtain $0 = u(0) = c_1$, $0 = u(1) = c_2 - \int_0^1 F(s) \, ds$. Next, by integration by parts, we have

$$\int_0^x F(s) \, ds = x F(x) - \int_0^x s f(s) \, ds = \int_0^x (x - s) f(s) \, ds$$

and thus

$$u(x) = c_1 + c_2 x - \int_0^x F(s) \, ds = x \int_0^1 (1 - s) f(s) \, ds - \int_0^x (x - s) f(s) \, ds$$

$$= \int_0^x [x(1 - s) - (x - s)] f(s) \, ds + \int_x^1 x(1 - s) f(s) \, ds$$

$$= \int_0^x s(1 - x) f(s) \, ds + \int_x^1 x(1 - s) f(s) \, ds = \int_0^1 G(x, s) f(s) \, ds,$$

that is, (9.2). This proves uniqueness of the solution $u \in C^2([0, 1])$ of (9.1)). It is easily checked that the function defined by (9.2) is indeed a solution.

(b) This follows directly from (9.2).

(c) Since $G(x, s) \ge 0$, this follows from (a).

(d) It also follows from the representation of u in (a) and the fact that $G \ge 0$ that

$$|u(x)| \le \int_0^1 G(x, s) |f(s)| \, ds \le \|f\|_\infty \int_0^1 G(x, s) \, ds$$

$$= \|f\|_\infty \left\{ \int_0^x s(1 - x) \, ds + \int_x^1 x(1 - s) \, ds \right\}$$

$$= \|f\|_\infty \left\{ (1 - x)\frac{1}{2}x^2 + x(\frac{1}{2} - x + \frac{1}{2}x^2) \right\} = \|f\|_\infty \left\{ \frac{1}{2}x(1 - x) \right\},$$

that is, $\|u\|_\infty \le \|f\|_\infty \max_{x \in [0, 1]} \left\{ \frac{1}{2}x(1 - x) \right\} = \frac{1}{8}\|f\|_\infty$.   □

Discretization

We now wish to solve (9.1) approximately. Of course, for the simple problem (9.1) we already know the exact solution, given by (9.2); thus we do not actually need any approximation. However, (9.1) is to be understood as a simple model example to help understand the FDM; the explicit solution formula will assist when determining the error.

We first replace the interval $[0, 1]$ (which consists of uncountably many points $x \in [0, 1]$) by a discrete (and finite) set of points, or *nodes*. The simplest way to do so is to distribute these points uniformly in $[0, 1]$, that is,

$$x_i := ih, \quad h := \frac{1}{N + 1}, \quad i = 0, \ldots, N + 1, \quad N \in \mathbb{N}. \tag{9.4}$$

We call $h > 0$ the *mesh size* and

$$\Omega_h := \{x_i : 0 \le i \le N + 1\}, \qquad \overset{\circ}{\Omega}_h := \Omega_h \setminus \{0, 1\}, \tag{9.5}$$

an *equidistant grid* (or *uniform mesh*), consisting of uniformly distributed nodes x_i. Clearly, one cannot consider derivatives of functions on such a discrete set as Ω_h; hence we replace derivatives by difference quotients. For example, we may replace the second derivative by the central difference quotient

$$\Delta_h v(x) := \frac{1}{h^2} \Big(v(x + h) - 2v(x) + v(x - h) \Big). \tag{9.6}$$

The following error estimate is well known in analysis, but since it will play an important role in what follows, we include the proof.

Lemma 9.2 *If $v \in C^4([0, 1])$, then for all $x \in [h, 1 - h]$ we have*

$$|v''(x) - \Delta_h v(x)| \le \frac{h^2}{12} \|v^{(4)}\|_\infty. \tag{9.7}$$

Proof We consider a Taylor expansion of $v(x \pm h)$ about x and obtain

$$v(x \pm h) = v(x) \pm hv'(x) + \frac{h^2}{2} v''(x) \pm \frac{h^3}{6} v'''(x) + \frac{h^4}{24} v^{(4)}(\xi_\pm),$$

where $\xi_+ \in (x, x + h)$ and $\xi_- \in (x - h, x)$, respectively. Summing the two expansions yields

$$v(x + h) + v(x - h) = 2v(x) + h^2 v''(x) + \frac{h^4}{24} \{v^{(4)}(\xi_+) + v^{(4)}(\xi_-)\},$$

from which the claim follows. $\qquad\qquad\qquad\qquad\qquad\qquad\qquad\qquad\qquad\qquad\square$

Observe that the restriction $x \in [h, 1 - h]$ is necessary in order to prevent $\Delta_h v$ from involving points outside the interval $[0, 1]$. The above estimate is thus only valid in the interior mesh $\mathring{\Omega}_h$.

Error analysis

The final step in the construction of our FDM consists of restricting (9.1a) to Ω_h, replacing the second derivative u'' by Δ_h and using the boundary conditions. This yields an approximation $u_h = (u_i)_{i=1,\dots,N} \in \mathbb{R}^N$, $u_i \approx u(x_i)$ via the following system of equations:

$$u_0 = u_{N+1} = 0, \tag{9.8a}$$

$$-u_{i+1} + 2u_i - u_{i-1} = h^2 f_i, \quad 1 \le i \le N, \tag{9.8b}$$

where $f_i := f(x_i)$. In matrix form this may be written as $A_h u_h = h^2 f_h =: \tilde{f}_h$, where $f_h := (f_i)_{i=1,\dots,N} \in \mathbb{R}^N$ and

$$A_h := \begin{bmatrix} 2 & -1 & & & 0 \\ -1 & 2 & \ddots & & \\ & \ddots & \ddots & \ddots & \\ & & \ddots & 2 & -1 \\ 0 & & & -1 & 2 \end{bmatrix} \in \mathbb{R}^{N \times N}. \tag{9.9}$$

For the analysis of the approximation obtained in this fashion, the following definitions will be useful.

Definition 9.3 (a) We call any mapping $w_h : \Omega_h \to \mathbb{R}$ a *grid function*. The set of all grid functions will be denoted by V_h, and we set $V_h^0 := \{w_h \in V_h : w_h(0) = w_h(1) = 0\}$.
(b) We define the *discrete operator* $L_h : V_h^0 \to V_h^0$ by

$$(L_h w_h)_i := -\Delta_h w_h(x_i), \ 1 \le i \le N, \qquad (L_h w_h)_0 := (L_h w_h)_{N+1} := 0,$$

for $w_h \in V_h^0$. $\triangle$

With this definition, (9.8) is equivalent to the problem

Find $u_h \in V_h^0$ such that $L_h u_h = f_h$, $\quad f_h := (f(x_i))_{1 \le i \le N}$.

For this problem, we define *discrete Green's functions* $G^k \in V_k^0$, $1 \le k \le N$, as solutions of the discrete problem

$$L_h G^k = \delta^k, \quad 1 \le k \le N, \tag{9.10}$$

where $\delta^k \in V_h^0$ is defined by

$$\delta^k(x_i) := \delta_{i,k} = \begin{cases} 1, & \text{if } i = k, \\ 0, & \text{otherwise,} \end{cases}$$

for all $1 \le i \le N$. We will now establish a relationship between the Green's function G from (9.3) and its discrete counterparts.

Lemma 9.4 *We have $G^k(x_i) = h\, G(x_i, x_k)$ for all $1 \le i, k \le N$.*

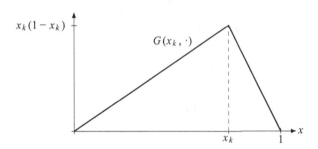

Fig. 9.1 Green's function as a function of its second argument, for fixed $x_k \in \Omega_h$.

Proof Fix a grid point $x_k \in \Omega_h$, then on the one hand we have, for all $1 \le i \le N$, $i \ne k$, $(L_h G(\cdot, x_k))(x_i) = G''(\cdot, x_k)(x_i) = 0$, since G, in accordance with (9.3), is a linear function of each of its arguments when the other is fixed, cf. Figure 9.1. On the other hand, for $i = k$, by definition

$$\begin{aligned}
(L_h G(\cdot, x_k))(x_k) &= \left(-\frac{1}{h^2}\right)\{G(x_{k+1}, x_k) - 2G(x_k, x_k) + G(x_{k-1}, x_k)\} \\
&= \left(-\frac{1}{h^2}\right)\{x_k(1 - x_{k+1}) - 2x_k(1 - x_k) + x_{k-1}(1 - x_k)\} \\
&= \left(-\frac{1}{h^2}\right)\{kh(N + 1 - k - 1)h - 2kh(N + 1 - k)h \\
&\quad + (k - 1)h(N + 1 - k)h\} \\
&= (-1)\{(N + 1 - k)(k - 2k + k - 1) + k\} = N + 1 = \frac{1}{h};
\end{aligned}$$

hence, by (9.10), $h(L_h G)(\cdot, x_k))(x_i) = \delta_{i,k} = (L_h G^k)_i$, for all $1 \le i \le N$, which proves the claim. □

We shall now investigate the questions of existence and uniqueness of solutions of the discrete problem; to this end, we will once again begin by defining a couple of relevant quantities.

Definition 9.5 Given $v_h, w_h \in V_h$, we define the *discrete inner product*

$$(v_h, w_h)_h := h \sum_{i=0}^{N+1} c_i v_i w_i, \tag{9.11}$$

where $c_0 := c_{N+1} := \frac{1}{2}, c_i := 1, i = 1, \ldots, N$ and $v_i := v_h(x_i), w_i := w_h(x_i)$, as well as the *discrete norm* $\|v_h\|_h := \sqrt{(v_h, v_h)_h}$. △

Clearly, $(v_h, w_h)_h$ corresponds to the composite trapezoidal rule for approximating the exact inner product $(v, w)_{L_2(0,1)} := \int_0^1 v(x) w(x) \, dx$ if v_h, w_h are interpolations of the functions $v, w \in C([0, 1])$ in Ω_h. That is, we have $(v_h, w_h)_h = T_h(v\, w)$ with the usual definition of the composite trapezoidal rule

$$T_h(v) := h\left(\frac{1}{2}v(0) + \sum_{i=1}^{N} v(x_i) + \frac{1}{2}v(1)\right).$$

The following stability analysis is also referred to as an *energy method*. The *discrete maximum norm* is defined by

$$\|v_h\|_{h,\infty} := \max_{0 \le i \le N+1} |v_h(x_i)|, \quad v_h \in V_h;$$

$\|v\|_{h,\infty}$ may be defined likewise for $v \in C([0, 1])$.

Lemma 9.6 *The following statements hold.*
(a) The operator L_h is symmetric: $(L_h v_h, w_h)_h = (v_h, L_h w_h)_h$ for all $v_h, w_h \in V_h^0$.
(b) The operator L_h is positive definite, that is, $(L_h v_h, v_h)_h \ge 0$ for all $v_h \in V_h^0$; and moreover $(L_h v_h, v_h)_h = 0$ if and only if $v_h = 0$.
(c) Discrete maximum principle: if $L_h u_h = f_h$, then $\|u_h\|_{h,\infty} \le \frac{1}{8}\|f_h\|_{h,\infty}$.
(d) The operator L_h is bounded: $(L_h v_h, w_h)_h \le 4\|v_h\|_h \|w_h\|_h$ for all $v_h, w_h \in V_h^0$.
(e) The eigenvectors and eigenvalues of L_h are given by $z_h^k = (z_j^k)_{j=0,\ldots,N+1}$, where $z_j^k = \sqrt{2h} \sin(jkh\pi)$ and $\lambda_k = \frac{4}{h^2} \sin^2(\frac{kh\pi}{2})$, $k = 1, \ldots, N$.

Remark 9.7 Together, these assertions guarantee the well-posedness of the discrete problem (9.8): (a) and (b) together imply the existence and uniqueness of solutions, (c) the continuous dependence of the solution on the data. In other words, (c) is a stability result. Observe that the norm in (c) is not induced by the inner product $(\cdot, \cdot)_h$; however, since all norms in finite-dimensional space are equivalent, it makes no difference. △

Proof (of Lemma 9.6) (a) Since $v_h, w_h \in V_h^0$, we have $v_0 = v_{N+1} = w_0 = w_{N+1} = 0$; and we also define $v_{-1} := w_{-1} := 0$. Then, by summation by parts,

$$(L_h v_h, w_h)_h = h \sum_{i=0}^{N+1} c_i \frac{1}{h^2}\left(-v_{i+1} + 2v_i - v_{i-1}\right) w_i$$

$$= \left(-\frac{1}{h}\right) \sum_{i=1}^{N} \Big((v_{i+1} - v_i) - (v_i - v_{i-1}) \Big) w_i$$

$$= \frac{1}{h} \sum_{i=0}^{N} (v_{i+1} - v_i)(w_{i+1} - w_i).$$

Again summing by parts, we obtain

$$(L_h v_h, w_h)_h = \left(-\frac{1}{h}\right) \sum_{i=1}^{N} v_i \Big((w_{i+1} - w_i) - (w_i - w_{i-1}) \Big) = (L_h w_h, v_h)_h.$$

(b) By (a) we have $(L_h v_h, v_h)_h = \frac{1}{h} \sum_{i=0}^{N} (v_{i+1} - v_i)^2 \geq 0$ for all $v_h \in V_h^0$. If $(L_h v_h, v_h)_h = 0$, then $v_i = v_{i+1}$ for all i. But since $v_0 = v_{N+1} = 0$ (as $v_h \in V_h^0$), it follows that $v_h \equiv 0$.

(c) Using the representation of $u_h = L_h^{-1} f_h$ in terms of the discrete Green's functions and the fact that $G(x, y) \geq 0$, we have

$$|u(x_k)| \leq \sum_{i=1}^{N} G^i(x_k) |f(x_i)|$$

$$\leq \|f\|_{h,\infty} \sum_{i=1}^{N} h\, G(x_k, x_i) = \|f\|_{h,\infty} \left(\frac{1}{2} x_k (1 - x_k) \right) \tag{9.12}$$

(cf. Exercise 9.2), which yields the claim.

(d) As in (a) we have $(L_h v_h, w_h)_h = \frac{1}{h} \sum_{i=0}^{N} (v_{i+1} - v_i)(w_{i+1} - w_i)$, where we have used the homogeneous boundary values. Now we apply Hölder's and Young's inequalities to obtain

$$(L_h v_h, w_h)_h \leq \frac{1}{h} \left(\sum_{i=0}^{N} (v_{i+1} - v_i)^2 \right)^{1/2} \left(\sum_{i=0}^{N} (w_{i+1} - w_i)^2 \right)^{1/2}$$

$$\leq 4 \left(\frac{1}{h} \sum_{i=0}^{N} v_i^2 \right)^{1/2} \left(\frac{1}{h} \sum_{i=0}^{N} w_i^2 \right)^{1/2} = 4 \|v_h\|_h \|w_h\|_h,$$

as claimed.

(e) By definition, we have

$$(L_h z_h^k)_j = \frac{1}{h^2} (-z_{j+1}^k + 2 z_j^k - z_{j-1}^k)$$

$$= h^{-3/2} \Big(-\sin\big(k(j+1)\pi h \big) + 2 \sin\big(k j \pi h \big) - \sin\big(k(j+1)\pi h \big) \Big).$$

Since $\sin\big(k(j \pm 1)\pi h \big) = \sin(k j \pi h) \cos(k \pi h) \pm \cos(k j \pi h) \sin(k \pi h)$, we have

$$(L_h z_h^k)_j = h^{-3/2} \sin(k j \pi h)(2 - 2\cos(k \pi h)) = h^{-3/2} \sin(k j \pi h) 4 \sin^2\left(\frac{k \pi h}{2}\right)$$

$$= \frac{4}{h^2} \sin^2\left(\frac{k \pi h}{2}\right) \sqrt{2h} \sin(k j \pi h) = \lambda_k (z_h^k)_j,$$

for all $j = 1, \ldots, N$, which completes the proof. $\qquad\qquad\square$

Theorem 9.8 (Discrete positivity) *Suppose that $f \in C([0, 1])$ satisfies $f(x) \geq 0$ for all $x \in [0, 1]$, and let u_h be the solution of (9.8). Then $u_i \geq 0$ for all $0 \leq i \leq N + 1$.*

Proof Since $L_h G^k = \delta^k$, we define $u_h \in V_h^0$ by $u_h := \sum_{k=1}^N f(x_k) G^k$. Then, firstly, we have $u_0 = u_{N+1} = 0$, which is the conclusion for $i \in \{0, N + 1\}$. Moreover, $L_h u_h = \sum_{k=1}^N f(x_k) L_h G^k = \sum_{k=1}^N f(x_k) \delta^k = f_h$, meaning that u_h, thus defined, is indeed the unique discrete solution. Since $G(x, y) \geq 0$ and $G^k(x_i) = h\, G(x_i, x_k) \geq 0$, the assumption that $f(x) \geq 0$ immediately implies $u_i \geq 0$ for all $1 \leq i \leq N$. $\quad\square$

Definition 9.9 Given $f \in C([0, 1])$, let $u \in C^2([0, 1])$ be the unique solution of $Lu = f$. Then the *local truncation error* $\tau_h \in V_h^0$ is defined by

$$\tau_h(x_i) := (L_h u)(x_i) - f(x_i), \quad 1 \leq i \leq N.$$

We call $e_h := u - u_h$ the *discretization error*. Finally, we say that a family $\{L_h\}_{h>0}$ of discrete operators $L_h : V_h^0 \to V_h^0$ is of *consistency order* $p \in \mathbb{N}$ if $\|\tau_h\|_{h,\infty} = O(h^p)$ as $h \to 0+$. $\qquad\qquad\triangle$

In our case, the above defined L_h is of consistency order 2 as we shall see next.

Lemma 9.10 *For all $f \in C^2([0, 1])$ we have $\|\tau_h\|_{h,\infty} \leq \frac{h^2}{12}\|f''\|_\infty$.*

Proof It follows from Lemma 9.1(b) that $u \in C^4([0, 1])$, and so, by Lemma 9.2,

$$\|\tau_h\|_{h,\infty} = \max_{0 \leq i \leq N+1} |(L_h u)(x_i) - f(x_i)| = \max_{0 \leq i \leq N+1} |-\Delta_h(x_i) + u''(x_i)|$$

$$\leq \frac{h^2}{12}\|u^{(4)}\|_\infty = \frac{h^2}{12}\|f''\|_\infty,$$

since $-u''(x) = f(x)$ for all $x \in (0, 1)$. $\qquad\qquad\square$

We will also need the following relationship between the local truncation error and the discretization error:

$$L_h e_h = L_h u - L_h u_h = L_h u - f_h = \tau_h. \qquad (9.13)$$

We can now formulate and prove a convergence theorem for the FDM for the problem (9.1).

Theorem 9.11 (Convergence of the FDM) *If $f \in C^2([0, 1])$, then*

$$\|u - u_h\|_{h,\infty} \leq \frac{h^2}{96}\|f''\|_\infty. \tag{9.14}$$

Proof It follows from the discrete maximum principle (Lemma 9.6(c)) that, for the equation $L_h u_h = f_h$, we have the estimate $\|u_h\|_{h,\infty} \leq \frac{1}{8}\|f_h\|_{h,\infty}$. If we apply this to (9.13), then we obtain $\|e_h\|_{h,\infty} \leq \frac{1}{8}\|\tau_h\|_{h,\infty}$. Combined with Lemma 9.10, this yields $\|e_h\|_{h,\infty} \leq \frac{1}{8}\|\tau_h\|_{h,\infty} \leq \frac{h^2}{96}\|f''\|_{h,\infty}$.   □

Notice in particular that the error estimate follows from the consistency and the discrete maximum principle; this gives us an indication of what will be necessary when considering other difference operators than Δ_h via the central difference quotient. One needs to prove these two properties, and then one obtains a corresponding convergence property.

Numerical solution and experiments

The above error analysis gives us the error arising from the discretization. However, this implicitly assumes that we have an exact solution of the system $L_h u_h = f_h$. In the one-dimensional case under consideration this is indeed possible (up to machine precision), since the matrix A_h in the linear system $A_h u_h = \tilde{f}_h$ has a special form, as we saw in (9.9). Indeed, A_h is a symmetric *tridiagonal matrix*, for which the entries off the main diagonal are constant. (We recall that a matrix is called *tridiagonal* if all entries are zero except those on the main diagonal and the two adjacent diagonals, above and below it.) A square matrix of size N which has at most c nonzero entries in each row is called *sparse*. Here $0 < c \ll N$ is a fixed (small) constant. For tridiagonal matrices we obviously have $c = 3$; for such matrices it is possible to derive a special Cholesky decomposition, which in turn leads to a recursion scheme, see e.g. [52, Ch. 3.7.1]. One can thus solve the linear system with *linear* cost, that is, with $O(N)$ operations, where $N \in \mathbb{N}$ is the dimension of the system.

Remark 9.12 (a) When working with non-equidistant grids one uses specially adapted difference quotients; in this case, the matrix A_h is generally no longer symmetric, but still tridiagonal.

(b) The tridiagonality is a consequence of the use of second-order difference operators. If one uses formulae of higher order, then the matrix will no longer have just three nontrivial diagonals, but rather a band structure.

(c) It is easy to see that in the case of Dirichlet boundary conditions it is sufficient to restrict to the treatment of homogeneous boundary conditions, cf. Section 6.8. Other boundary conditions (Neumann, Robin) need special treatment; one needs to adjust the difference quotients at the boundary. Variable coefficients may be treated with special FDMs. We refer to the numerical mathematics literature [16, 35, 46, 52].   △

To summarize, the advantage of the FDM in dimension one is its simplicity, especially as regards the actual implementation. The greatest disadvantage is perhaps the assumption of Theorem 9.11, that $f \in C^2([0, 1])$, which is a very restrictive demand.

We finish by describing a numerical experiment. Why an "experiment"?

On the one hand, the error analysis described above is of asymptotic nature, that is, the statements are valid in the regime $h \to 0+$, in other words, under the assumption that h is "sufficiently small". It could be the case that the method does not actually have the desired convergence properties for "reasonable" step sizes h (that is, which do not lead to unmanageably large systems). Whether this is actually the case, can often only be established through experiments.

Another point of interest is the constants in the error estimates. In the above analysis we could determine these explicitly ($\frac{1}{8}$ and $\frac{1}{96}$) thanks to the discrete Green's functions and explicit knowledge of the solutions. But in the case of variable coefficients, or in higher dimensions, this is often impossible; one can merely show that such constants exist. If one wishes to estimate their size, then this is generally only possible via numerical experiments. The same is true for the computational cost of a method, which in the theorems is expressed as an O-term. But the precise constants that these terms hide can have an enormous influence on runtime.

Additionally, when analyzing the method we always assumed that the linear systems which appear can always be solved exactly; the influence of roundoff errors, for example, was always neglected.

Finally, in the error analysis of a numerical method one always makes certain assumptions, which may not be satisfied in a given application. Even if one cannot actually apply the theory in such a case, there is at least a chance that the assumptions made were too restrictive and the method still delivers the desired result. This can only be established by suitable experimentation.

We now wish to describe such an experiment. In the case of FDMs the assumption that $f \in C^2([0, 1])$ is extremely restrictive for the solution of the boundary value problem; we wish to see to what extent the one-dimensional FDM also works for less regular functions. We will consider two different right-hand sides. Firstly, we choose the function $u_{\sin}(x) := \sin(2\pi x)$, which is clearly C^∞, and set $f_{\sin} := -u_{\sin}''$. For this example, we expect quadratic convergence ($O(h^2)$), since the assumptions of the above theorems are satisfied. We also define

$$f_\alpha(x) := \begin{cases} 1, & 0 \le x \le \alpha, \\ -1, & \alpha < x \le 1, \end{cases} \qquad (9.15)$$

which has a discontinuity at the point $\alpha \in (0, 1)$. Here we will take $\alpha = 0.5$; the solution is thus no longer C^2. One can show in an analogous fashion to above that the FDM still converges in this case, but only linearly ($O(h)$) and not quadratically; see Exercise 9.1.

In Figure 9.2 we give a log–log plot of the convergence history (that is, the error in dependence on the dimension of the linear system, the number of unknowns).

This permits us to read off the order of convergence as the (negative) gradient of the line, which is taken as a linear least squares fit to the data. As expected, we obtain quadratic convergence for $u_{\sin}$. For u_α, $\alpha = 0.5$, we obtain linear convergence, that is, a line with slope -1. We see that the error analysis is "sharp" in this case, that is, we obtain linear convergence and nothing more.

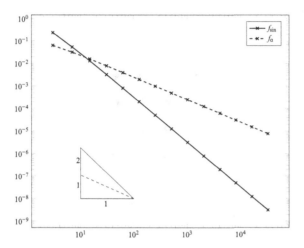

Fig. 9.2 Convergence history for the FDM, for the right-hand side f_α in (9.15) (dashed) and $f_{\sin}$ (solid line). In each case the error $\|u - u_h\|_{h,\infty}$ is plotted as a function of the number of mesh points N, as a log–log plot. The linear convergence for f_α and the quadratic convergence for $f_{\sin}$ are clearly recognizable.

This, and all further numerical examples in this chapter, were written in MATLAB (Mathworks®) and tested with version R2020a.

9.1.2 FDM: the two-dimensional case

We next consider the Dirichlet problem on the unit square,

$$-\Delta u(x) = f(x), \quad x \in \Omega := (0,1)^2, \qquad u(x) = 0, \quad x \in \Gamma := \partial\Omega. \qquad (9.16)$$

Discretization

We apply the same idea as in the one-dimensional case and define an equidistant grid (also called uniform mesh) analogous to (9.5)

$$\Omega_h := \{(x,y) \in \overline{\Omega} : x = kh, \quad y = \ell h, \quad 0 \le k, \ell \le N+1\} \qquad (9.17)$$

for $h := \frac{1}{N+1}$, $N = N_h \in \mathbb{N}$, as above. The boundary now consists of more than two points, namely

$$\partial\Omega_h := \Gamma \cap \Omega_h = \{(x, y) \in \Gamma : x = kh \text{ or } y = \ell h, \quad 0 \le k, \ell \le N + 1\} \quad (9.18)$$
$$= \{(x, y) \in \Gamma : x = kh \text{ and } y = \ell h, \quad 0 \le k, \ell \le N + 1\}, \quad (9.19)$$

cf. Figure 9.3. As in the one-dimensional case we define $\mathring{\Omega}_h := \Omega_h \setminus \partial\Omega_h$.

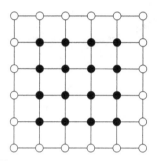

Fig. 9.3 Equidistant grid on $\overline{\Omega} = [0, 1]^2$. The interior points ($\mathring{\Omega}_h$) are represented by the solid black dots, $\partial\Omega_h$ by the empty white circles ($\circ$).

We will again approximate the derivatives, in this case second order partial derivatives, by the central difference quotients (9.6), that is,

$$\begin{aligned}
\Delta u(x, y) &= \frac{\partial^2}{\partial x^2} u(x, y) + \frac{\partial^2}{\partial y^2} u(x, y) \\
&\approx \frac{u(x + h, y) - 2u(x, y) + u(x - h, y)}{h^2} \\
&\quad + \frac{u(x, y + h) - 2u(x, y) + u(x, y - h)}{h^2} \\
&= \frac{u(x + h, y) + u(x - h, y) + u(x, y + h) + u(x, y - h) - 4u(x, y)}{h^2} \\
&=: \Delta_h(x, y). \quad (9.20)
\end{aligned}$$

Entirely analogously to Lemma 9.2, one can show using Taylor expansions that $\|\Delta u - \Delta_h u\|_\infty = O(h^2)$ as $h \to 0+$ if $u \in C^4(\overline{\Omega})$. One thus obtains consistency order 2, as before. With similar (albeit technically more involved) methods, one can prove a convergene theorem which states that on the square one has

$$\|u - u_h\|_{h,\infty} = O(h^2) \text{ as } h \to 0+, \text{ if } u \in C^4(\overline{\Omega}).$$

We now wish to show that the regularity assumption $u \in C^4(\overline{\Omega})$ is particularly restrictive in dimension two: even if the right-hand side is smooth, the solution may not have the required regularity.

Example 9.13 We consider the *L-shaped domain* $\Omega := (-1, 1)^2 \setminus \{[0, 1) \times (-1, 0]\} \subset \mathbb{R}^2$. In Example 6.88 we found a function $f \in C^\infty(\overline{\Omega})$ such that the solution of the Poisson problem $-\Delta u(x) = f(x)$, $x \in \Omega$ and $u(x) = 0$, $x \in \Gamma := \partial\Omega$, belongs to $H_0^1(\Omega) \cap H_{\text{loc}}^2(\Omega)$ but not $H^2(\Omega)$. In this example, $u \in C(\overline{\Omega})$ and $u_{|\Gamma} = 0$. Hence the L-shaped domain does *not* allow maximal H^2-regularity. In particular, there is *no* shift theorem as in Lemma 9.1 (b).   △

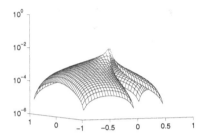 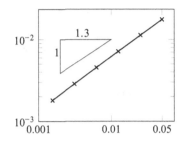

Fig. 9.4 Finite difference method on the L-shaped domain, error between approximate and exact solutions (left) and convergence history (right)

Now one could of course hope that the FDM nevertheless converges at the optimal rate since, for example, the analytic estimates are not optimal. We thus describe a numerical experiment with a given weak solution $u \in H^1(\Omega) \setminus \{C^2(\Omega) \cap C(\overline{\Omega})\}$, which we already met in Section 6.11:

$$u(x, y) = r^{2/3} \sin\left(\frac{2}{3}\varphi\right), \quad (x, y) = (r \cos \varphi, r \sin \varphi).$$

In Figure 9.4 we show the error for a uniform mesh of size $h = \frac{1}{80}$ (left) and the convergence history (right). Note that here we have displayed the mesh size h on the x-axis; thus the curve has positive slope (unlike in Figure 9.2). We first observe that the corners do in fact cause problems, as large errors occur in these points. We also note that the inverse of the slope of the line in the plot of the convergence history is approximately 1.3 (cf. the accompanying triangle), which gives the order of convergence. This is noteworthy for two reasons:

- the lack of H^2-regularity would appear to lead to a rate of convergence which is genuinely lower than 2;
- the FDM converges in this case even though the solution is not C^4; the rate of convergence is also superlinear.

A further problem with the FDM occurs if the domain has a more complicated geometry than one just involving right angles, as was the case with $(0, 1)^2$ and the L-shaped domain, cf. Figure 9.5.

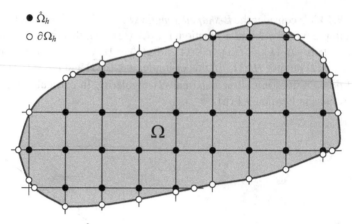

Fig. 9.5 Equidistant mesh $\mathring{\Omega}_h$ (solid dots) and boundary mesh $\partial\Omega_h$ (○) for a curvilinear domain $\Omega \subset \mathbb{R}^2$.

For curvilinear domains Ω the definitions (9.17) and (9.18) no longer make sense; instead, we use

$$\mathring{\Omega}_h := \{(x, y) \in \Omega : x = k \cdot h, \ y = \ell \cdot h, \ k, \ell \in \mathbb{Z}\},$$
$$\partial\Omega_h := \{(x, y) \in \Gamma : x = k \cdot h \ \text{or} \ y = \ell \cdot h, \ k, \ell \in \mathbb{Z}\}.$$

It is immediate that Ω_h does not represent the geometric structure of Ω well; the discrepancy between $\partial\Omega_h$ and Γ is even more pronounced.

Numerical solutions

We now wish to determine the numerical solutions. To this end, we first describe the linear system of equations $A_h u_h = f_h$. The exact form of the matrix $A_h \in \mathbb{R}^{N_h^2 \times N_h^2}$ will of course depend on the numbering of the mesh points. If one chooses what is known as the *lexicographic order* (cf. Figure 9.6)

$$z_k = (x_i, y_j), \quad x_i = ih, \quad y_j = jh, \quad k := (j - 1)N + i,$$

then the linear system corresponds to a *block tridiagonal matrix*

$$A_h = \begin{bmatrix} B_h & C_h & & 0 \\ C_h & B_h & \ddots & \\ & \ddots & \ddots & \ddots \\ & & \ddots & B_h & C_h \\ 0 & & & C_h & B_h \end{bmatrix} \in \mathbb{R}^{N_h^2 \times N_h^2}$$

with blocks

$$
B_h = \begin{bmatrix} 4 & -1 & & & 0 \\ -1 & 4 & \ddots & & \\ & \ddots & \ddots & \ddots & \\ & & \ddots & 4 & -1 \\ 0 & & & -1 & 4 \end{bmatrix} \in \mathbb{R}^{N_h \times N_h}, \quad C_h = \mathrm{diag}\,(-1) \in \mathbb{R}^{N_h \times N_h},
$$

right-hand side $f_h = (h^2 f(z_k))_{k=1,\ldots,N_h^2}$, $k = (j-1)N_h + i$, $1 \leq i, j \leq N_h$, and solution vector $u_h = (u(z_k))_{k=1,\ldots,N_h^2} \in \mathbb{R}^{N_h^2}$.

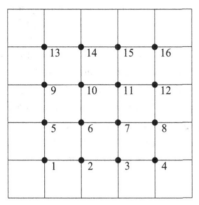

Fig. 9.6 Lexicographic ordering of the mesh points, $\Omega = (0,1)^2$, $N = 5$.

In this case it is not possible to derive a simple recursion formula for the Cholesky decomposition. But since A_h is sparse, symmetric and positive definite, it is possible, and indeed natural, to use an iterative numerical procedure to determine an approximate solution of the linear system. Starting with an initial value $u^{(0)}$, we can calculate a sequence of approximations $u_h^{(k)}$, $k = 1, 2, \ldots$ converging to the solution of the system $A_h u_h = f_h$ as $k \to \infty$. Since $A_h \in \mathbb{R}^{N_h \times N_h}$ (here $N_h = N_h^2$ in the above notation) is sparse and an iteration step requires a matrix-vector multiplication in just about every known iterative method, for the step from $u_h^{(k)}$ to $u_h^{(k+1)}$ the cost is linear, that is, one requires $O(N_h)$ operations. The total cost will, of course, depend on the total number of iterations, which may be measured as follows: fix an initial value $u_h^{(0)}$ and a corresponding initial error $\|u_h^{(0)} - u_h\|$ with respect to a suitable norm $\|\cdot\|$. We would like to know how many steps are necessary to reduce this initial error by a given factor $\varepsilon \in (0,1)$. This clearly describes the speed of convergence of the method; to obtain an estimate on this speed one may prove a bound of the form

$$
\|u_h^{(k+1)} - u_h\| \leq \rho_h^k \|u_h^{(k)} - u_h\|
$$

for some convergence factor $\rho_h \in (0,1)$. If this holds, then the iteration requires

$$k = \left\lceil \frac{\log \varepsilon}{\log \rho_h} \right\rceil$$

steps to reduce the error by the desired factor.

Now it may happen that this error reduction *degenerates* for small step sizes $h \to 0+$, that is, $\rho_h \to 1$. This would mean that the method converges more and more slowly as h becomes smaller. Since at the same time the dimension of the matrix grows as h becomes smaller, such a method may become rather inconvenient. It is thus desirable to apply a method which is *asymptotically optimal*, that is, $\rho_h \le \rho_0 < 1$, $h \to 0+$. In other words: the error reduction of an asymptotically optimal method is independent of the step size h, that is, the size N_h of the matrix.

The question of the construction of such asymptotically optimal methods was open for a long time. These days there are at least two known kinds of methods, namely the conjugate gradient method with what is known as the BPX preconditioner (named after Bramble, Pasciak and Xu, 1990 [17]), and the multigrid method (see for example [16, Ch. V]). Wavelet methods are another option, cf., e.g., [58].

As we have seen, the total cost depends decisively on the size of the matrix, that is, the number of mesh points. In the two dimensional case we have $N_h = N_h^2$. In dimension d one has N_h^d mesh points, which means that even using optimal solution methods with linear cost, i.e., $O(N_h^d)$ operations for a given error reduction, one eventually succumbs to the *curse of dimensionality*.

9.2 Finite elements for elliptic problems

The finite element method (FEM) was developed in the 1950s to deal with computations arising in automotive and aeronautical engineering, in particular for structural mechanics; it has since become a standard tool in many areas. The mathematical theory has come a long way, even if the FEM is still the subject of current research. Here we will give a short introduction which is directly connected to the analytical methods for partial differential equations treated in the previous chapters.

9.2.1 The Galerkin method

Unlike the finite difference method, the finite element method is based on the weak formulation of a given partial differential equation. As we will see, this has the advantage of permitting us to avoid the strong regularity assumptions necessary for the FDM.

We start with the variational formulation of a boundary value problem for an elliptic partial differential equation. Let H be a Hilbert space with inner product $(\cdot, \cdot)$, $V \hookrightarrow H$ a second Hilbert space continuously imbedded in H, and $a : V \times V \to \mathbb{R}$

a continuous, coercive bilinear form. By the Lax–Milgram theorem, Theorem 4.24, for each $f \in H$ there exists exactly one $u \in V$ such that

$$a(u, v) = (f, v), \quad v \in V. \tag{9.21}$$

Since V is in general infinite dimensional (e.g. $H_0^1(\Omega)$), (9.21) cannot be used directly for numerics. The solution is to consider *finite-dimensional* subspaces

$$V_h \subset V, \quad \dim(V_h) =: \mathcal{N}_h < \infty,$$

where the index "h" is used to suggest an analogy to mesh size. We will see later how one can construct V_h based on a net (or mesh) of mesh size h, whereby significantly more general nets than the classical FDM mesh will be allowed. This also enables us to avoid the second major disadvantage of FDM, namely the restriction to domains with simple geometry.

Given a finite-dimensional ("discrete") space $V_h \subset V$, we consider the following discrete problem: find $u_h \in V_h$ such that

$$a(u_h, \chi) = (f, \chi), \quad \chi \in V_h. \tag{9.22}$$

We call u_h the *discrete solution* of (9.21) in V_h. Since V_h is a subspace of V and we are considering the same bilinear form $a(\cdot, \cdot)$ as in (9.21), the Lax–Milgram theorem guarantees the existence and uniqueness of a solution $u_h \in V_h$ of (9.22), as well as its stability (continuous dependence on the data, here f). Thus the discrete problem is well posed. We will see in Section 9.2.2 that, upon construction a basis of V_h, (9.22) leads to a linear system of equations of dimension $\mathcal{N}_h = \dim V_h$.

As with FDMs we wish to investigate the error $u - u_h$. It turns out that one can make statements about this error under very weak assumptions.

Theorem 9.14 (Céa's lemma) *Suppose that the bilinear form $a : V \times V \to \mathbb{R}$ is continuous, that is, there exists $C > 0$ such that $a(u, v) \leq C\|u\|_V \|v\|_V$ for all $u, v \in V$, and coercive, that is, there exists $\alpha > 0$ such that $a(u, u) \geq \alpha\|u\|_V^2$ for all $u \in V$. Then, given $f \in H$, for the solutions u of (9.21) and u_h of (9.22) we have*

$$\|u - u_h\|_V \leq \frac{C}{\alpha} \inf_{\chi \in V_h} \|u - \chi\|_V =: \frac{C}{\alpha} \mathrm{dist}_V(u, V_h).$$

Proof Since $V_h \subset V$, in (9.21) we can, in particular, choose any $\chi \in V_h$ as a test function, and in doing so obtain $a(u, \chi) = (f, \chi)$ for all $\chi \in V_h$. We then substract (9.22) from this identity, which yields

$$a(u - u_h, \chi) = (f, \chi) - (f, \chi) = 0, \quad \chi \in V_h. \tag{9.23}$$

This equation is known as the *Galerkin orthogonality* relation for the error $e_h := u - u_h$, which states that this error is perpendicular to the test space V_h with respect to the bilinear form $a(\cdot, \cdot)$. Now by coercivity and the Galerkin orthogonality

$$\alpha \|u - u_h\|_V^2 \le a(u - u_h, u - u_h)$$
$$= a(u - u_h, u - \chi) + a(u - u_h, \chi - u_h) = a(u - u_h, u - \chi)$$

for arbitrary $\chi \in V_h$, since $\chi - u_h \in V_h$. Due to the continuity of $a(\cdot, \cdot)$ we also have

$$\alpha \|u - u_h\|_V^2 \le a(u - u_h, u - \chi) \le C\|u - u_h\|_V \cdot \|u - \chi\|_V;$$

dividing by $\|u - u_h\|_V$ and taking the infimum over all $\chi \in V_h$ now yields the claim.   □

The statement of Céa's lemma links numerical mathematics with approximation theory, since Theorem 9.14 states that the error $e_h = u - u_h$ is, up to the constant $\frac{C}{\alpha}$ which depends on the problem, essentially the *best approximation* of u in V_h. Thus the size of the error depends on the approximation property of the *trial space* V_h. In what follows, we will analyze the extent to which this is in fact the case.

9.2.2 Triangulation and approximation on triangles

In order to construct a concrete trial space V_h, we may consider a geometric subdivision of Ω, similar to Ω_h for the FDM, but substantially better. On this subdivision (mesh, grid, net) we may define functions (piecewise), which can then be chosen as basis functions for V_h; that is, the trial space will be the set of linear combinations of the basis functions.

We will restrict ourselves to open polygons $\Omega \subset \mathbb{R}^2$ in the plane; the geometric subdivision becomes more complicated in higher-dimensional spaces $\mathbb{R}^d$, $d > 2$. On domains whose boundary is not polygonal (but rather, for example, curvilinear), additional terms appear in the error estimates, which we will not consider here, cf. Section 9.3*.

Definition 9.15 We call a family of open sets $\mathcal{T} := \{T_i\}_{i=1}^N$, $N \in \mathbb{N}$, a *partition* of Ω if it has the following properties:
(a) $T_i \subset \Omega$ is open, $i = 1, \ldots, N$;
(b) $T_i \cap T_j = \emptyset$, $i \ne j$, $i, j = 1, \ldots, N$;
(c) $\bigcup_{i=1}^N \overline{T}_i = \overline{\Omega}$.
We call each $T \in \mathcal{T}$ an *element* of the partition, and speak of a *triangulation* if each $T \in \mathcal{T}$ is an open triangle.   △

In numerics one can also consider generalized triangles with curvilinear boundaries, or partitions into other geometric objects such as quadrilaterals; such partitions are also called triangulations. Here, however, we will only consider partitions into triangles. Later we will also demand certain "goodness" properties of the triangulations (cf. Definitions 9.24 and 9.26). On the individual triangles, we will exclusively consider affine functions (we also speak of *linear elements*). These will then be glued together to create continuous functions on Ω, which will yield the trial space V_h. By Céa's lemma (Theorem 9.14) we know that the quantity

$$\inf_{\chi \in V_h} \|u - \chi\|_{H^1(\Omega)} =: dist_{H^1(\Omega)}(u, V_h)$$

will be of central importance for the analysis of the error. We saw in Chapter 6 that the solution of the Poisson equation is in $H^2(\Omega)$ under suitable assumptions on Ω. We will investigate how well one can approximate arbitrary H^2-functions by piecewise affine functions (cf. Corollary 9.28).

9.2.3 Affine functions on triangles

Let $T \subset \mathbb{R}^2$ be an open triangle, to be fixed throughout. In keeping with common practice, we will say that a function $v : \overline{T} \to \mathbb{R}$ is *affine* if

$$v\big(\lambda x + (1 - \lambda)y\big) = \lambda\, v(x) + (1 - \lambda)\, v(y), \qquad x, y \in \overline{T},\ 0 \le \lambda \le 1. \tag{9.24}$$

We start with the following characterization.

Lemma 9.16 *The following statements hold.*
(a) The set $\mathcal{P}_1(T)$ of affine functions from $\overline{T}$ to $\mathbb{R}$ is a vector space of dimension three. Let t_1, t_2, t_3 be the three vertices of $\overline{T}$, then for each $i \in \{1, 2, 3\}$ there exists exactly one $v_i \in \mathcal{P}_1(T)$ such that $v_i(t_j) = \delta_{i,j}$, $i, j = 1, 2, 3$. The set $\{v_1, v_2, v_3\}$ forms a basis of $\mathcal{P}_1(T)$ known as its Lagrange basis.
(b) For each $v \in \mathcal{P}_1(T)$ there are uniquely determined coefficients $a, b, c \in \mathbb{R}$ such that

$$v(x) = a + bx_1 + cx_2, \qquad x = (x_1, x_2) \in \overline{T}. \tag{9.25}$$

Proof (a) Every $x \in \overline{T}$ has a unique expansion in terms of the *barycentric coordinates* (a_1, a_2, a_3), namely as

$$x = a_1 t_1 + a_2 t_2 + a_3 t_3, \qquad a_i \ge 0,\ i = 1, 2, 3,\ a_1 + a_2 + a_3 = 1, \tag{9.26}$$

cf. Exercise 9.13. Since for any $v \in \mathcal{P}_1(T)$ we have $v(x) = a_1 v(t_1) + a_2 v(t_2) + a_3 v(t_3)$, we see that v is uniquely determined by its values at the vertices t_1, t_2, t_3. Now, for each $i \in \{1, 2, 3\}$, define $v_i : \overline{T} \to \mathbb{R}$ by $v_i(x) := a_i$, with x as in (9.26). Then every $v \in \mathcal{P}_1(T)$ has a unique expansion $v = \alpha_1 v_1 + \alpha_2 v_2 + \alpha_3 v_3$, where $\alpha_j = v(t_j)$, $j = 1, 2, 3$. We have shown that $\{v_1, v_2, v_3\}$ is a basis of $\mathcal{P}_1(T)$.
(b) It is easy to see that the functions in (9.25) are affine. That they form a vector space E of dimension three, is the subject of Exercise 9.14. Thus $E = \mathcal{P}_1(T)$. $\square$

As an immediate consequence of the above lemma, we see that

$$\mathcal{P}_1(T) = \big\{p : \overline{T} \to \mathbb{R} : p(x) = a + bx_1 + cx_2,\quad a, b, c \in \mathbb{R},\quad x = (x_1, x_2)\big\}.$$

Affine functions can be defined between arbitrary convex sets. A function $f : \mathbb{R}^d \to \mathbb{R}^d$ is affine if and only if $f - f(0)$ is linear; that is, affine functions are translations of linear functions. Sometimes one also simply calls affine functions linear; this is where the name *linear elements* in numerical analysis comes from.

9.2.4 Norms on triangles

A first difficulty arises because when deriving error estimates for variational problems we need to control Sobolev norms, yet we have defined the Lagrange basis in terms of interpolation. The following auxiliary result is key to linking these concepts. Again let $T \subset \mathbb{R}^2$ be a fixed open triangle.

Lemma 9.17 *Let $T \subset \mathbb{R}^2$ be an open triangle with vertices t_1, t_2, t_3. Then*

$$\|\|v\|\|_{H^2(T)} := |v|_{H^2(T)} + \sum_{i=1}^{3} |v(t_i)|, \quad v \in H^2(T),$$

defines a norm which is equivalent to $\|\cdot\|_{H^2(T)}$, where

$$|v|_{H^2(T)}^2 = \int_T (|D_1^2 v(x)|^2 + 2|D_1 D_2 v(x)^2|^2 + |D_2^2 v(x)|^2)\, dx.$$

Proof Since the imbedding $H^2(T) \hookrightarrow C(\overline{T})$ is continuous (cf. Corollary 7.25), there exists a constant $c(T) > 0$ such that

$$\sup_{x \in \overline{T}} |v(x)| \le c(T) \|v\|_{H^2(T)}, \quad v \in H^2(T), \qquad (9.27)$$

whence $\|\|v\|\|_{H^2(T)} \le (1 + 3c(T)) \|v\|_{H^2(T)}$, which is the upper bound. In particular, $\|\cdot\|\|_{H^2(T)}$ is a continuous semi-norm on $H^2(T)$.

Let us assume for a contradiction that the lower bound $\|v\|_{H^2(T)} \le C\|\|v\|\|_{H^2(T)}$, $v \in H^2(T)$ is false for all $C > 0$. Then there exists a sequence $(v_k)_{k \in \mathbb{N}} \subset H^2(T)$ such that $\|v_k\|_{H^2(T)} = 1$ and $\|\|v_k\|\|_{H^2(T)} \le \frac{1}{k}$, $k \in \mathbb{N}$. By Theorem 7.22 the imbedding $H^2(T) \hookrightarrow H^1(T)$ is compact. Hence there exists a subsequence of $(v_k)_{k \in \mathbb{N}}$ which is convergent in $H^1(T)$; for simplicity we will denote this subsequence by $(v_k)_{k \in \mathbb{N}}$. Now $|v_k|_{H^2(T)} \le \|\|v_k\|\|_{H^2(T)} \le \frac{1}{k} \to 0$ as $k \to \infty$, and so

$$\|v_k - v_\ell\|_{H^2(T)}^2 = \|v_k - v_\ell\|_{H^1(T)}^2 + |v_k - v_\ell|_{H^2(T)}^2$$

$$\le \|v_k - v_\ell\|_{H^1(T)}^2 + \left(|v_k|_{H^2(T)} + |v_\ell|_{H^2(T)}\right)^2 \to 0$$

as $k, \ell \to \infty$. Hence $(v_k)_{k \in \mathbb{N}}$ is also a Cauchy sequence in $H^2(T)$ with limit $v \in H^2(T)$. For this limit, due to the continuity of $\|\|\cdot\|\|_{H^2(T)}$, on the one hand we have

$$\|\|v\|\|_{H^2(T)} = \lim_{k \to \infty} \|\|v_k\|\|_{H^2(T)} = 0, \qquad (9.28)$$

and on the other

$$\|v\|_{H^2(T)} = \lim_{k \to \infty} \|v_k\|_{H^2(T)} = 1. \tag{9.29}$$

Since $\||v|\|_{H^2(T)} = 0$ it follows that $|v|_{H^2(T)} = 0$ and hence $v \in \mathcal{P}_1(T)$ by the following lemma. But since $|v(t_i)| \le \||v|\|_{H^2(T)} = 0$, $i = 1, 2, 3$, we must have $v \equiv 0$. This is a contradiction to (9.29). $\qquad\square$

Lemma 9.18 *Let $T \subset \mathbb{R}^2$ be an open triangle and suppose that $v \in H^2(T)$ is such that $D_i D_j v = 0$ for all $i, j = 1, 2$. Then there exists an affine function $p \in \mathcal{P}_1(T)$ such that $p = v$ almost everywhere in T.*

Proof By Theorem 6.19 there exists a constant $c_j \in \mathbb{R}$ such that $D_j v = c_j$, $j = 1, 2$. Define $w(x) := v(x) - c_1 x_1 - c_2 x_2$, $x = (x_1, x_2) \in T$; then $D_j w = 0$, $j = 1, 2$. Hence there exists $c \in \mathbb{R}$ such that $w = c$, and so $v(x) = c + c_1 x_1 + c_2 x_2$ almost everywhere. $\qquad\square$

9.2.5 Transformation into a reference element

It is useful for both analytic and computational purposes to transform an arbitrary element T into a *reference element* $\hat{T}$. We will choose for our reference triangle the lower left half of the unit square

$$\hat{T} := \{(x_1, x_2) \in \mathbb{R}^2 : x_1, x_2 > 0, \ 0 < x_1 + x_2 < 1\},$$

as depicted in Figure 9.7. Again let $T \subset \mathbb{R}^2$ be a fixed open triangle. Then there exists an affine mapping $F : \hat{T} \to T$ which maps $\hat{T}$ bijectively onto T, that is,

$$F(\hat{x}) = b + B\hat{x}, \quad \hat{x} \in \hat{T}, \qquad F(\hat{T}) = T, \tag{9.30}$$

for an invertible matrix $B \in \mathbb{R}^{2 \times 2}$ and a point $b \in \overline{T}$. Clearly, $p \circ F \in \mathcal{P}_1(\hat{T})$ for all $p \in \mathcal{P}_1(T)$, that is, the set of affine functions is invariant under such affine transformations. This will allow us to reduce the proof of many statements about T to their counterparts on the reference element $\hat{T}$. We note that by Lemma 9.17 there exist constants $0 < \hat{c} \le \hat{C} < \infty$ such that

$$\hat{c} \, \||\hat{v}|\|_{H^2(\hat{T})} \le \|\hat{v}\|_{H^2(\hat{T})} \le \hat{C} \, \||\hat{v}|\|_{H^2(\hat{T})}, \qquad \hat{v} \in H^2(\hat{T}). \tag{9.31}$$

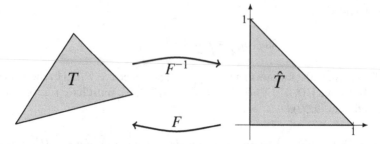

Fig. 9.7 Reduction to a reference triangle.

We will reduce the error analysis to one on the reference triangle $\hat{T}$. To this end, we need to investigate the effect of the mapping F and its inverse $F^{-1} : T \to \hat{T}$, given by $F^{-1}(x) = B^{-1}x - B^{-1}b$. Let $\|x\| := (x_1^2 + x_2^2)^{1/2}$, $x = (x_1, x_2) \in \mathbb{R}^2$, be the Euclidean norm and

$$\|A\| := \sup_{\|x\| \le 1} \|Ax\| = \sup_{\|x\|=1} \|Ax\| = \sup_{x \ne 0} \frac{\|Ax\|}{\|x\|}, \quad A \in \mathbb{R}^{2 \times 2},$$

the induced operator norm. Given the triangle T, we denote its *inradius* by ρ_T and its *circumradius* by r_T, cf. Figure 9.8 for the reference triangle $\hat{T}$. It is decisive for the error estimates we wish to prove that the norm of the transformation matrix B can be estimated by the circumradius, and the norm of its inverse by the inradius. This is the subject of the next lemma.

Lemma 9.19 *With the above notation we have* $\|B\| \le \frac{\sqrt{2}}{\sqrt{2}-1} r_T$ *and* $\|B^{-1}\| \le \frac{1}{\sqrt{2}\,\rho_T}$.

Proof Let $\hat{\rho}$ and $\hat{r}$ be the inradius and the circumradius, respectively, of the reference triangle $\hat{T}$, that is, $\hat{\rho} = 1 - \frac{1}{2}\sqrt{2}$ and $\hat{r} = \frac{1}{2}\sqrt{2}$, cf. Figure 9.8. Now let $\hat{x} \in \mathbb{R}^2$ be such that $\|\hat{x}\| = 2\hat{\rho}$. Then there exist two points $\hat{y}, \hat{z} \in \hat{T}$ such that $\hat{x} = \hat{y} - \hat{z}$, cf. Figure 9.9. Hence $F\hat{y}, F\hat{z} \in T$ and $\|B\hat{x}\| = \|F\hat{y} - F\hat{z}\| \le 2\,r_T$, and thus

$$\|B\| = \sup_{\|\hat{x}\|=2\hat{\rho}} \frac{\|B\hat{x}\|}{\|\hat{x}\|} \le \frac{r_T}{\hat{\rho}} = \frac{\sqrt{2}}{\sqrt{2}-1} r_T.$$

Conversely, if we begin with a point $x \in \mathbb{R}^2$ such that $\|x\| = 2\rho_T$, then we can find $y, z \in T$ such that $x = y - z$. Hence $\|B^{-1}x\| = \|F^{-1}y - F^{-1}z\| \le 2\hat{r}$. It now follows as above that

$$\|B\|^{-1} = \sup_{\|x\|=2\rho_T} \frac{\|B^{-1}x\|}{\|x\|} \le \frac{\hat{r}}{\rho_T} = \frac{1}{\sqrt{2}\,\rho_T}.$$

This proves the claim.   □

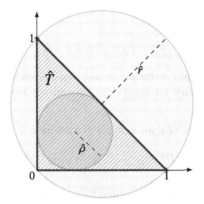

Fig. 9.8 Circumradius $\hat{r} = \frac{1}{2}\sqrt{2}$ and inradius $\hat{\rho} = 1 - \frac{1}{2}\sqrt{2} \approx 0.293$ of the reference triangle $\hat{T}$.

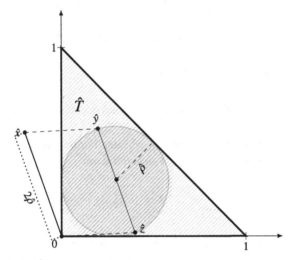

Fig. 9.9 Points $\hat{y}, \hat{z} \in \hat{T}$ such that $\hat{x} = \hat{y} - \hat{z}$.

In order to be able to reduce error estimates on T to ones on $\hat{T}$, we need to investigate the effects of the transformation on the Sobolev norms. We recall the definition of the function $F : \hat{T} \rightarrow T$ given by $F(\hat{x}) = B\hat{x} + b$ (Definition 9.30). Now, given $v : T \rightarrow \mathbb{R}$, we define the function $\hat{v} : \hat{T} \rightarrow \mathbb{R}$ by

$$\hat{v}(\hat{x}) := v(F(\hat{x})) = v(x),$$

where $x = F(\hat{x})$. Then

$$\frac{\partial}{\partial \hat{x}_i}\hat{v}(\hat{x}) = \frac{\partial}{\partial \hat{x}_i}v(F(\hat{x})) = \sum_{j=1}^{2}\frac{\partial}{\partial x_j}v(x)\frac{\partial}{\partial \hat{x}_i}(F(\hat{x}))_j = \sum_{j=1}^{2}\frac{\partial}{\partial x_j}v(x)\,B_{j,i},$$

where $B = (B_{i,j})_{i,j=1,2}$. We thus have

$$\nabla(v \circ F) = (\nabla v) \circ F \cdot B. \tag{9.32}$$

(Note that here the gradient is to be interpreted as a row vector, and on the right-hand side of (9.32) the dot stands for matrix multiplication.) This leads to the following estimate.

Lemma 9.20 *Given* $v \in H^1(T)$*, for* $\hat{v} := v \circ F$ *we have* $\hat{v} \in H^1(\hat{T})$ *and*

$$|\hat{v}|_{H^1(\hat{T})} \leq |\det B|^{-1/2}\|B\| \, |v|_{H^1(T)}, \quad |v|_{H^1(T)} \leq |\det B|^{1/2}\|B^{-1}\| \, |\hat{v}|_{H^1(\hat{T})}.$$

Proof Using (9.32) and the transformation formula for the change of variables from $\hat{T}$ to T we have

$$|\hat{v}|^2_{H^1(\hat{T})} = |v \circ F|^2_{H^1(\hat{T})} = \int_{\hat{T}} |\nabla(v \circ F)|^2 \, d\hat{x} = \int_{\hat{T}} |(\nabla v) \circ F \cdot B|^2 \, d\hat{x}$$

$$\leq \|B\|^2 \int_{\hat{T}} |(\nabla v) \circ F|^2 |\det B| \, d\hat{x} \cdot |\det B|^{-1}$$

$$= \|B\|^2 |\det B|^{-1} \int_T |\nabla v|^2 \, dx,$$

from which the first inequality follows. The second one may be shown analogously. □

We will need a similar estimate on $|\hat{v}|_{H^2(\hat{T})}$ later. For this we require an auxiliary result for matrices involving the *Hilbert–Schmidt norm*

$$\|A\|^2_{\mathrm{HS}} := \sum_{i,j=1}^n a_{ij}^2, \quad A := (a_{ij})_{i,j=1,\dots,d} \in \mathbb{R}^{d \times d}.$$

In numerics this is often also called the *Schur norm* or the *Frobenius norm*.

Lemma 9.21 *For all* $A, B \in \mathbb{R}^d$ *we have* $\|BA\|_{\mathrm{HS}} \leq \|B\| \cdot \|A\|_{\mathrm{HS}}$*, where* $\|\cdot\|$ *is the matrix norm induced by the Euclidean vector norm* $\|\cdot\|$*.*

Proof We have $\|A\|^2_{\mathrm{HS}} = \sum_{i=1}^d \|Ae_i\|^2$, where $e_i := (\delta_{1,i}, \dots, \delta_{d,i})^T \in \mathbb{R}^d$, cf. (1.42), and hence $\|BA\|^2_{\mathrm{HS}} = \sum_{i=1}^d \|BAe_i\|^2 \leq \|B\|^2 \sum_{i=1}^d \|Ae_i\|^2 = \|B\|^2 \cdot \|A\|^2_{\mathrm{HS}}$. This shows the claim. □

We can now give the following estimate, as announced above.

Lemma 9.22 *For all* $v \in H^2(T)$ *we have* $\hat{v} := v \circ F \in H^2(\hat{T})$*, and the inequality* $|\hat{v}|_{H^2(\hat{T})} \leq |\det B|^{-1/2}\|B\|^2 |v|_{H^2(T)}$ *holds.*

Proof We have

$$|\hat{v}|^2_{H^2(\hat{T})} = \int_{\hat{T}} \left\{ \left|\frac{\partial^2}{\partial \hat{x}_1^2} \hat{v}(\hat{x})\right|^2 + 2\left|\frac{\partial}{\partial \hat{x}_1} \frac{\partial}{\partial \hat{x}_2} \hat{v}(\hat{x})\right|^2 + \left|\frac{\partial^2}{\partial \hat{x}_2^2} \hat{v}(\hat{x})\right|^2 \right\} d\hat{x}.$$

Since $x = F\hat{x}$ and $\hat{v}(\hat{x}) = v(x)$, it follows that

$$
\begin{aligned}
\frac{\partial}{\partial \hat{x}_i}\frac{\partial}{\partial \hat{x}_j}\hat{v}(\hat{x}) &= \frac{\partial}{\partial \hat{x}_i}\left(\sum_{k=1}^{2}\frac{\partial}{\partial x_k}v(x)\,B_{k,j}\right) = \sum_{k=1}^{2}\frac{\partial}{\partial x_k}\left(\frac{\partial}{\partial \hat{x}_i}\hat{v}(\hat{x})\right)B_{k,j} \\
&= \sum_{k,\ell=1}^{2}\frac{\partial}{\partial x_k}\left(\frac{\partial}{\partial x_\ell}v(x)B_{\ell,i}\right)B_{k,j} = \sum_{k,\ell=1}^{2}B_{\ell,i}\frac{\partial}{\partial x_k}\frac{\partial}{\partial x_\ell}v(x)\,B_{k,j} \\
&= (B^T \mathcal{H}v(x)B)_{i,j}
\end{aligned}
$$

for $i,j = 1,2$, where $\mathcal{H}v(x) := \left(\frac{\partial}{\partial x_i}\frac{\partial}{\partial x_j}v(x)\right)_{i,j=1,2}$ is the Hessian. We thus have

$$
\begin{aligned}
|\hat{v}|_{H^2(\hat{T})}^2 &= \int_{\hat{T}}\|\hat{\mathcal{H}}\hat{v}(\hat{x})\|_{\mathrm{HS}}^2\,d\hat{x} = |\det B|^{-1}\int_T \|B^T \mathcal{H}v(x)B\|_{\mathrm{HS}}^2\,dx \\
&\leq |\det B|^{-1}\|B\|^4\int_T \|\mathcal{H}v(x)\|_{\mathrm{HS}}^2\,dx = |\det B|^{-1}\|B\|^4\,|v|_{H^2(T)}^2,
\end{aligned}
$$

whence the claim. $\qquad\qquad\square$

9.2.6 Interpolation for finite elements

As we have seen, Céa's lemma states that the norm of the error $\|u - u_h\|_{H^1(\Omega)}$ is, up to a constant, essentially the error of the best approximation in V_h to the solution u. Now, in order to be able to estimate this best approximation, we will construct an *interpolation operator* $I_h : C(\overline{\Omega}) \to V_h$. Since we are taking $\Omega \subset \mathbb{R}^2$ to be a polygon, we have $H^2(\Omega) \subset C(\overline{\Omega})$, cf. Corollary 7.25. Given any $u \in H^2(\Omega)$, $\|u - I_h u\|_{H^1(\Omega)}$ is then an upper bound on the error in the best approximation. We will define the interpolation operator piecewise on each $T \in \mathcal{T}$.

For every $v \in C(\overline{T})$ there exists a unique affine function $I_T v \in \mathcal{P}_1(T)$ such that $I_T v(t_i) = v(t_i)$, $i = 1,2,3$, where t_1, t_2, t_3 are, again, the vertices of T. In this way we can define a local interpolation operator

$$
I_T : C(\overline{T}) \to \mathcal{P}_1(T).
$$

We next wish to estimate the interpolation error for I_T in terms of the semi-norm $|\cdot|_{H^2(T)}$.

Theorem 9.23 *Let $T \subset \mathbb{R}^2$ be an open triangle with inradius ρ_T and circumradius r_T, and let $\hat{C}$ be as in (9.31). Then for all $v \in H^2(T)$ we have*

$$
\|v - I_T v\|_{L_2(T)} \leq 12\,\hat{C}\,r_T^2\,|v|_{H^2(T)},
$$

$$
|v - I_T v|_{H^1(T)} \leq 9\,\hat{C}\,\frac{r_T^2}{\rho_T}\,|v|_{H^2(T)},
$$

$$\|v - I_T v\|_{H^1(T)} \le 9\, \hat{C}\, r_T^2 \sqrt{2 + \rho_T^{-2}}\, |v|_{H^2(T)}.$$

Proof Let $F(\hat{x}) = B\hat{x} + b$ be the affine transformation mapping $\hat{T}$ onto T. If $\hat{t}_1, \hat{t}_2, \hat{t}_3$ are the vertices of $\hat{T}$, then $t_i := F(\hat{t}_i)$, $i = 1, 2, 3$, are the vertices of T. Now if $v \in H^2(T)$, then $\hat{v} := v \circ F \in H^2(\hat{T})$ and, moreover,

$$\widehat{I_T v} = I_{\hat{T}} \hat{v}, \tag{9.33}$$

as can be seen as follows: since both $I_T v : T \to \mathbb{R}$ and $F : \hat{T} \to T$ are affine, so is $\widehat{I_T v} = I_T v \circ F$ as the composition of affine functions. Hence both functions in (9.33) are affine. Since they agree at the three points $\hat{t}_i$, $i = 1, 2, 3$, they are identical. This shows (9.33). Now since $(I_{\hat{T}} \hat{v} - \hat{v})(\hat{t}_i) = 0$, $i = 1, 2, 3$, by (9.31), we have

$$\|I_{\hat{T}} \hat{v} - \hat{v}\|_{H^2(\hat{T})} \le \hat{C}\, \|\|I_{\hat{T}} \hat{v} - \hat{v}\|\|_{H^2(\hat{T})} = \hat{C}\, |I_{\hat{T}} \hat{v} - \hat{v}|_{H^2(\hat{T})} = \hat{C}\, |\hat{v}|_{H^2(\hat{T})}, \tag{9.34}$$

since $I_{\hat{T}} \hat{v} \in \mathcal{P}_1(\hat{T})$ and thus $|I_{\hat{T}} \hat{v}|_{H^2(\hat{T})} = 0$.

We next prove the L_2-estimate. Using the above estimate, we have

$$\|v - I_T v\|_{L_2(T)} = |\det B|^{1/2}\, \|\widehat{v - I_T v}\|_{L_2(\hat{T})} = |\det B|^{1/2}\, \|\hat{v} - I_{\hat{T}} \hat{v}\|_{L_2(\hat{T})}$$

$$\le |\det B|^{1/2}\, \|\hat{v} - I_{\hat{T}} \hat{v}\|_{H^2(\hat{T})} \le \hat{C}\, |\det B|^{1/2}\, |\hat{v}|_{H^2(\hat{T})}.$$

Now we apply Lemma 9.22 and obtain

$$\|v - I_T v\|_{L_2(T)} \le \hat{C}\, \|B\|^2\, |v|_{H^2(T)} \le \hat{C}\, \frac{2}{(\sqrt{2} - 1)^2}\, r_T^2\, |v|_{H^2(T)}$$

$$\le 12\, \hat{C}\, r_T^2\, |v|_{H^2(T)},$$

where we have also used Lemma 9.19. This establishes the first estimate. For the proof of the second inequality (involving the H^1-semi-norm) we use Lemma 9.20 and obtain

$$|v - I_T v|_{H^1(T)} \le |\det B|^{1/2}\, \|B^{-1}\|\, |\hat{v} - I_{\hat{T}} \hat{v}|_{H^1(\hat{T})}$$

$$\le |\det B|^{1/2}\, \|B^{-1}\|\, \|\hat{v} - I_{\hat{T}} \hat{v}\|_{H^2(\hat{T})}$$

$$\le \hat{C}\, |\det B|^{1/2}\, \|B^{-1}\|\, |\hat{v}|_{H^2(\hat{T})} \le \hat{C}\, \|B\|^2 \|B^{-1}\|\, |v|_{H^2(T)},$$

where we have also used (9.34) and Lemma 9.22. Combined with Lemma 9.19, this yields

$$|v - I_T v|_{H^1(T)} \le \hat{C}\left(\frac{\sqrt{2}}{\sqrt{2} - 1} r_T\right)^2 \frac{1}{\sqrt{2}\, \rho_T}\, |v|_{H^2(T)}$$

$$= \hat{C}\, \frac{\sqrt{2}}{(\sqrt{2} - 1)^2}\, \frac{r_T^2}{\rho_T}\, |v|_{H^2(T)} \le 9\, \hat{C}\, \frac{r_T^2}{\rho_T}\, |v|_{H^2(T)}.$$

We finally turn to the H^1-norm. The above estimate implies

$$\|v - I_T v\|^2_{H^1(T)} = \|v - I_T v\|^2_{L_2(T)} + |v - I_T v|^2_{H^1(T)}$$

$$\leq \left(144\ \hat{C}^2 r_T^4 + 81\ \hat{C}^2 \frac{r_T^4}{\rho_T^2}\right) |v|^2_{H^2(T)}$$

$$\leq 81\ \hat{C}^2 r_T^4 \left(2 + \rho_T^{-2}\right) |v|^2_{H^2(T)},$$

and thus the claim.   □

9.2.7 Finite element spaces

Up until now we have only considered a single, fixed triangle $T \subset \mathbb{R}^2$. We now return to the triangulation of a polygon $\Omega \subset \mathbb{R}^2$ introduced in Section 9.2.2; an example is depicted in Figure 9.10. On Ω we define the vector space

$$V_{\mathcal{T}} := \{v \in C_0(\Omega) : v_{|\overline{T}} \in \mathcal{P}_1(T), T \in \mathcal{T}\}$$

of globally continuous, piecewise affine functions which satisfy homogeneous Dirichlet boundary conditions. Here, as before, $C_0(\Omega) = \{u \in C(\overline{\Omega}) : u_{|\partial\Omega} = 0\}$. Note that $V_{\mathcal{T}}$ is a subspace of $H^1_0(\Omega)$ (see Exercise 7.6 and Theorem 6.30). We call the elements of $V_{\mathcal{T}}$ *linear finite elements*, as the functions in $V_{\mathcal{T}}$ are piecewise affine.

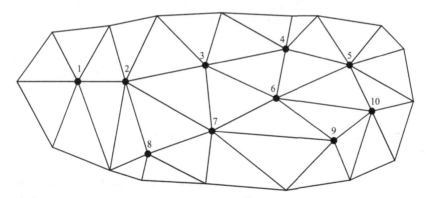

Fig. 9.10 Triangulation of a polygon Ω. The interior nodes are numbered arbitrarily.

We next wish to determine the dimension of the space $V_{\mathcal{T}}$; this will also be the dimension of the linear system of equations which we will have to solve in order to determine the numerical solution in $V_{\mathcal{T}}$. Here we will also impose an additional "goodness" property on the triangulation.

Definition 9.24 A triangulation $\mathcal{T}$ will be called *admissible* if the following conditions are satisfied:

(i) If, given $T_i, T_j \in \mathcal{T}$, $\overline{T}_i \cap \overline{T}_j$ consists of exactly one point, then this point is a vertex of both T_i and T_j.

(ii) If, given $T_i, T_j \in \mathcal{T}$, $i \neq j$, $\overline{T}_i \cap \overline{T}_j$ consists of more than one point, then $\overline{T}_i \cap \overline{T}_j$ is an edge of both T_i and T_j. △

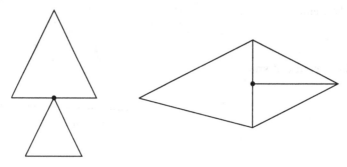

Fig. 9.11 Non-admissible partitions. On the left, condition (i) of Definition 9.24 is violated, on the right condition (ii).

Figure 9.11 should help to clarify this definition, as it describes two non-admissible partitions: conditions (i) and (ii) prevent hanging nodes. Let $\mathcal{T}$ be an admissible triangulation. If $z \in \overline{\Omega}$ is a vertex of two distinct triangles in $\mathcal{T}$, then we call z a *node*. We will denote by $\overset{\circ}{\mathcal{T}}$ the set of interior nodes, i.e., all nodes lying in Ω, and by $|\overset{\circ}{\mathcal{T}}|$ the number of interior nodes, cf. Figure 9.10. Using $\overset{\circ}{\mathcal{T}}$ we can now construct a basis of $V_{\mathcal{T}}$, as follows.

Lemma 9.25 *Let $\mathcal{T}$ be an admissible triangulation of a polygon $\Omega \subset \mathbb{R}^2$. Then for every $t_i \in \overset{\circ}{\mathcal{T}} = \{t_1, \ldots, t_N\}$, $N = |\overset{\circ}{\mathcal{T}}|$, there exists a unique $\varphi_i \in V_{\mathcal{T}}$ such that*

$$\varphi_i(t_j) = \delta_{i,j}, \qquad j = 1, \ldots, N. \tag{9.35}$$

The functions $\{\varphi_1, \ldots, \varphi_N\}$ form a basis of $V_{\mathcal{T}}$, known as the Lagrange basis *or the* nodal basis *of $V_{\mathcal{T}}$.*

Proof Fix $i \in \{1, \ldots, N\}$ and let $T \in \mathcal{T}$ be a triangle which has t_i as a vertex. There exists a unique $f_T \in \mathcal{P}_1(T)$ such that $f_T(t_i) = 1$ but f_T vanishes in the other two vertices of T. Now let $S \in \mathcal{T}$ be such that $\overline{T} \cap \overline{S} \neq \emptyset$, $S \neq T$. We will show that $f_T = f_S$ on $\overline{T} \cap \overline{S}$. This will mean that $\varphi_i(x) := f_T(x)$ is well defined for all $x \in \overline{T}$, and so $\varphi_i \in V_{\mathcal{T}}$ will be the desired function which satisfies (9.35). Since the triangulation is admissible, $\overline{T} \cap \overline{S}$ is either a node or a common edge, as hanging nodes have been ruled out. If $\overline{T} \cap \overline{S} = \{z\}$ is a common vertex of T and S, then $f_T(z) = f_S(z)$. In the second case we may write $\overline{T} \cap \overline{S} = \{\lambda x + (1 - \lambda)y : 0 \leq \lambda \leq 1\}$, where x and y are two common vertices of T and S. Then we have, for all $0 \leq \lambda \leq 1$,

$$f_T(\lambda x + (1 - \lambda)y) = \lambda f_T(x) + (1 - \lambda)f_T(y) = \lambda f_S(x) + (1 - \lambda)f_S(y)$$
$$= f_S(\lambda + (1 - \lambda)y),$$

meaning that $f_S = f_T$ on $\overline{T} \cap \overline{S}$. This proves (9.35).

We still need to show that $\{\varphi_1, \ldots, \varphi_N\}$ is a basis of V_T. By (9.35) it is clear that these functions are linearly independent. Now take any $u \in V_T$, then $v := u - \sum_{i=1}^{N} u(t_i)\varphi_i \in V_T$ and we have $v(t_j) = 0$ for all $j = 1, \ldots, N$. Since $v \in V_T$ vanishes on the boundary $\partial\Omega$ of Ω, it follows that $v(x) = 0$ for every vertex x of every triangle $T \in \mathcal{T}$. Since $V_{|\overline{T}} \in \mathcal{P}_1(T)$, we thus have $v_{|\overline{T}} = 0$ for all $T \in \mathcal{T}$, and so $v \equiv 0$. This means that u admits the expansion $u = \sum_{i=1}^{N} u(t_i)\varphi_i$. □

Using the Lagrange basis we can now define an interpolation operator

$$I_T : C_0(\Omega) \to V_T \quad \text{as follows:} \quad I_T u := \sum_{i=1}^{N} u(t_i)\varphi_i, \ u \in C_0(\Omega). \qquad (9.36)$$

That is, $I_T u$ is the uniquely determined element of V_T which coincides with u in all nodes. The operator I_T is linear and continuous, and $I_T^2 = I_T$, that is, it is a continuous projection of $C_0(\Omega)$ onto V_T. We also call I_T the *Clément operator*. We now wish to study convergence properties of such triangulations. We will consider a family $\{\mathcal{T}_h\}_{h>0}$ of regular triangulations of the polygon Ω; the notation $\mathcal{T}_h$ is used to indicate the assumption, which we will always take, that

$$r_T \leq h \quad \text{for all } T \in \mathcal{T}_h.$$

We will show that under suitable assumptions

$$\|v - I_T v\|_{L_2(\Omega)} = O(h^2) \quad \text{as} \quad h \to 0+, \text{ for all } v \in H^2(\Omega).$$

Of course we will need to eliminate the possibility that the triangles may degenerate for small h; such a case could occur if the ratio of the longest to the shortest size should blow up as $h \downarrow 0$. The following definition will prevent this from happening.

Definition 9.26 A family $\{\mathcal{T}_h\}_{h>0}$ of triangulations is said to be *quasi-uniform* if $r_T \leq h$ for all $T \in \mathcal{T}_h$ and there exists some $\kappa \geq 1$ such that

$$\frac{r_T}{\rho_T} \leq \kappa$$

for all $T \in \mathcal{T}_h$ and all $h > 0$. △

The condition of quasi-uniformity imposes that the ratio of the circumradius to the inradius is uniformly bounded by κ, for all $T \in \mathcal{T}_h$ and all $h > 0$. Thus narrow, drawn-out triangles are ruled out as $h \to 0+$. For every $h > 0$ we now consider the finite element space

$$V_h := \{v \in C_0(\Omega) : v_{|\overline{T}} \in \mathcal{P}_1(T) \text{ for all } T \in \mathcal{T}_h\}, \qquad (9.37)$$

as well as the Clément operator $I_h : C_0(\Omega) \to V_h$ defined above. For these we have the following error estimates.

Theorem 9.27 *Let $\{\mathcal{T}_h\}_{h>0}$ be a quasi-uniform family of admissible triangulations of a polygon $\Omega \subset \mathbb{R}^2$. Then the following error estimates hold:*

$$\|v - I_h v\|_{L_2(\Omega)} \leq 12\,\hat{C}\,h^2\,|v|_{H^2(\Omega)}, \qquad \|v - I_h v\|_{H^1(\Omega)} \leq 16\,\hat{C}\,\kappa\,h\,|v|_{H^2(\Omega)},$$

for all $v \in H^2(\Omega)$ and all $0 < h \leq 1$, where $\hat{C}$ is as in (9.31).

Proof By assumption, $r_T \leq h$ for all $T \in \mathcal{T}_h$. Furthermore,

$$\|v - I_h v\|_{L_2(\Omega)}^2 = \sum_{T \in \mathcal{T}_h} \|v - I_h v\|_{L_2(T)}^2$$

$$\leq 144\,\hat{C}^2\,h^4 \sum_{T \in \mathcal{T}_h} |v|_{H^2(T)}^2 = 144\,\hat{C}^2\,h^4\,|v|_{H^2(\Omega)}^2$$

by Theorem 9.23. The proof of the second inequality is entirely analogous, using that $\rho_T \leq r_T \leq h \leq 1$; indeed, we have that $81(2 + \rho_T^{-2}) \leq 243\,\rho_T^{-2}$ and so, by Theorem 9.23,

$$\|v - I_h v\|_{H^1(\Omega)}^2 = \sum_{T \in \mathcal{T}_h} \|v - I_h v\|_{H^1(T)}^2 \leq 81\,\hat{C}^2 \sum_{T \in \mathcal{T}_h} r_T^4(2 + \rho_T^{-2})\,|v|_{H^2(T)}^2$$

$$\leq 243\,\hat{C}^2 \sum_{T \in \mathcal{T}_h} r_T^2\,\frac{r_T^2}{\rho_T^2}\,|v|_{H^2(T)}^2 \leq 243\,\hat{C}^2\,h^2\,\kappa^2 \sum_{T \in \mathcal{T}_h} |v|_{H^2(T)}^2$$

$$= 243\,\hat{C}^2\,h^2\,\kappa^2\,|v|_{H^2(\Omega)}^2,$$

whence the second estimate.   □

As a consequence of Theorem 9.27 we have the following estimates, which will be important in conjunction with Céa's lemma.

Corollary 9.28 *Under the assumptions of Theorem 9.27, the following estimates hold for all $v \in H^2(\Omega)$ and $0 < h \leq 1$:*

$$\inf_{\chi \in V_h} \|v - \chi\|_{L_2(\Omega)} \leq 12\,\hat{C}\,h^2\,|v|_{H^2(\Omega)},$$

$$\inf_{\chi \in V_h} \|v - \chi\|_{H^1(\Omega)} \leq 16\,\hat{C}\,\kappa\,h\,|v|_{H^2(\Omega)}.$$

9.2.8 The Poisson problem on polygons

The above analysis yielded estimates on the error in the best approximation in V_h of an arbitrary function $v \in H^2(\Omega)$. We now wish to apply these estimates to the solution u of the elliptic boundary value problem (9.21). For this, we need to specify the bilinear form in question, which in turn will determine the constants C and α in Céa's lemma (Theorem 9.14). The final step will be to replace the norm of the unknown solution by an expression which only depends on known data of the problem.

We will again restrict ourselves to the Poisson problem $-\Delta u = f$ in a convex polygon $\Omega \subset \mathbb{R}^2$ with homogeneous Dirichlet boundary conditions $u = 0$ on $\partial\Omega$. So let $\Omega \subset \mathbb{R}^2$ be a convex polygon. By Theorem 6.86 the Poisson problem is H^2-*regular*, that is, for every $f \in L_2(\Omega)$ there exists a unique $u \in H_0^1(\Omega) \cap H^2(\Omega)$ such that $-\Delta u = f$. In what follows, we will call this u the solution of the Poisson problem (with right-hand side f). Its H^2-norm can be controlled in the following sense.

Theorem 9.29 *There exists a constant $c_2 > 0$, which depends only on Ω, such that*

$$\|u\|_{H^2(\Omega)} \leq c_2 \|f\|_{L_2(\Omega)}$$

for all $f \in L_2(\Omega)$, where u is the solution of the Poisson problem with right-hand side f.

Proof The mapping $A^{-1} : L_2(\Omega) \to H^2(\Omega)$ which maps each $f \in L_2(\Omega)$ to the unique solution of the Poisson problem, is clearly linear (see Theorem 8.7 for the notation A^{-1}). It follows from the closed graph theorem that A^{-1} is continuous. Thus $\|u\|_{H^2(\Omega)} = \|A^{-1}f\|_{H^2(\Omega)} \leq \|A^{-1}\|_{\mathcal{L}(L_2(\Omega),H^2(\Omega))}\|f\|_{L_2(\Omega)}$, which was exactly the claim with $c_2 := \|A^{-1}\|_{\mathcal{L}(L_2(\Omega),H^2(\Omega))}$. □

The bilinear form associated with the Poisson problem is given by $a(u,v) = \int_\Omega \nabla u \nabla v \, dx$, $u,v \in H_0^1(\Omega)$.

Lemma 9.30 *For the bilinear form associated with the Poisson problem we have, for all $u,v \in H_0^1(\Omega)$,*

$$|a(u,v)| \leq \|u\|_{H^1(\Omega)} \|v\|_{H^1(\Omega)}, \qquad a(u,u) \geq \alpha\|u\|_{H^1(\Omega)}^2$$

for some constant $\alpha > 0$.

Proof The first inequality is trivial, the second follows from the Poincaré inequality (Theorem 6.32). □

The first estimate states that in Céa's lemma (Theorem 9.14) we may take the constant C to be $C = 1$. We now have all the necessary ingredients for the proof of the following central theorem, which delivers linear convergence of the error with respect to the H^1-norm.

Theorem 9.31 *Let $\{\mathcal{T}_h\}_{h>0}$ be a quasi-uniform family of admissible triangulations of a convex polygon $\Omega \subset \mathbb{R}^2$. Given $f \in L_2(\Omega)$, let u be the solution of the Poisson problem and let $u_h \in V_h$ be the discrete solutions via linear elements in accordance* with (9.22). *Then there exists a constant $c > 0$, which depends only on κ and Ω, such that, for all $0 < h \leq 1$,*

$$\|u - u_h\|_{H^1(\Omega)} \leq c\,h\|f\|_{L_2(\Omega)}.$$

Proof By Theorem 9.14 (Céa's lemma), Theorem 9.29 and Corollary 9.28, we have

$$\|u - u_h\|_{H^1(\Omega)} \leq \frac{1}{\alpha} \inf_{\chi \in V_h} \|u - \chi\|_{H^1(\Omega)} \leq \frac{1}{\alpha} 16\,\hat{C}\,\kappa\,h\,|u|_{H^2(\Omega)}$$

$$\leq \frac{1}{\alpha} 16\,\hat{C}\,\kappa\,h\,\|u\|_{H^2(\Omega)} \leq \frac{1}{\alpha} 16\,c_2\,\hat{C}\,\kappa\,h\,\|f\|_{L_2(\Omega)}$$

and thus the claim, with $c = \frac{1}{\alpha} 16\,c_2\,\hat{C}\,\kappa$. $\qquad\qquad\qquad\qquad\qquad\qquad\square$

An L_2-estimate

In this section we will show how one can obtain L_2-estimates in the concrete case of the Poisson problem. It turns out that linear convergence with respect to the H^1-norm in Theorem 9.31 leads to quadratic convergence in the L_2-norm. The general result is due to Aubin (1967) and Nitsche (1968) and is valid for general (not necessarily symmetric) forms. The essential step consists of considering a so-called *dual problem* to the actual variational problem of interest, which is as follows: given $g \in L_2(\Omega)$, let $w \in H_0^1(\Omega)$ the solution of

$$a(v, w) = (g, v), \quad v \in H_0^1(\Omega).$$

In the case of a symmetric form $a(\cdot, \cdot)$ and the right-hand side considered here the dual problem has exactly the same form as the original Poisson problem; as such, here the designation "dual problem" is superfluous (but not in the case of non-symmetric or even nonlinear problems, which we will not be considering here), but perhaps useful to highlight the key step in the proof (see Exercise 9.16 for a more general statement). This is how we gain an order of convergence in the passage from the H^1-norm to the L_2-norm.

Theorem 9.32 (Aubin–Nitsche, special case) *Under the assumptions of Theorem 9.31 we have, for all $0 < h \leq 1$,*

$$\|u - u_h\|_{L_2(\Omega)} \leq c^2\,h^2\,\|f\|_{L_2(\Omega)},$$

where c is the constant from Theorem 9.31.

Proof Fix $0 < h \leq 1$. By the theorem of Riesz–Fréchet there exists an $g \in L_2(\Omega)$ such that $\|g\|_{L_2(\Omega)} \leq 1$ and

$$\|u - u_h\|_{L_2(\Omega)} = (u - u_h, g)_{L_2(\Omega)}.$$

We now consider the solution $w \in H_0^1(\Omega)$ of the dual problem, here the Poisson problem with right-hand side g. Using the Galerkin orthogonality $a(u - u_h, \chi) = 0$ for all $\chi \in V_h$, we obtain

$$\|u - u_h\|_{L_2(\Omega)} = (u - u_h, g)_{L_2(\Omega)} = a(u - u_h, w) = a(u - u_h, w - \chi)$$

for arbitrary $\chi \in V_h$, where in the second step we chose the test function $v = u - u_h$ in the dual problem, that is, $(g, u - u_h) = a(u - u_h, w)$. The continuity of the bilinear form $a(\cdot, \cdot)$ and the error estimate of Theorem 9.31 yield

$$\|u - u_h\|_{L_2(\Omega)} \leq \|u - u_h\|_{H^1(\Omega)} \|w - \chi\|_{H^1(\Omega)} \leq c\, h \,\|f\|_{L_2(\Omega)} \|w - \chi\|_{H^1(\Omega)}.$$

Since $\chi \in V_h$ is arbitrary, we may choose $\chi = w_h$, the discrete solution of the dual problem. Again applying Theorem 9.31, we obtain

$$\|u - u_h\|_{L_2(\Omega)} \leq c\, h \,\|f\|_{L_2(\Omega)} \,c\, h \,\|g\|_{L_2(\Omega)} \leq c^2\, h^2 \,\|f\|_{L_2(\Omega)},$$

whence the claim. $\square$

We thus obtain the error estimate $\|u - u_h\|_{L_2(\Omega)} = O(h^2)$, $h \to 0+$, that is, the same order as with the FDM (which we had proved there for the discrete maximum norm $\|\cdot\|_{h,\infty}$). However, the extremely restrictive assumption $u \in C^4(\overline{\Omega})$ needed for the FDM is not needed for the FEM. The significantly greater flexibility with regards to the geometry of the domain Ω is an additional advantage.

A suboptimal L_∞-estimate

Up until now, all estimates have been formulated in terms of Lebesgue norms. To finish we wish to prove an estimate (of first order) with respect to the supremum norm

$$\|v\|_{C(\overline{\Omega})} := \sup_{x \in \overline{\Omega}} |v(x)|, \qquad v \in C(\overline{\Omega}).$$

We speak here of a "suboptimal" estimate since with rather more effort one could prove a higher order of convergence (cf. Section 9.3*). For the desired estimate we require what is known as an *inverse estimate*, for which we will however need a further assumption on the triangulation.

Definition 9.33 A family $\{\mathcal{T}_h\}_{h>0}$ of triangulations is called *uniform* if $r_T \leq h$ for all $T \in \mathcal{T}_h$ and there exists some $\kappa \geq 1$ such that $\frac{h}{\rho_T} \leq \kappa$ for all $T \in \mathcal{T}_h$ and all $h > 0$.

$\triangle$

Since $r_T \leq h$ it is clear that the requirement of uniformity is stronger than the quasi-uniformity of Definition 9.26. We can now introduce and prove the announced inverse estimate.

Lemma 9.34 (Inverse estimate) *Let* $\{\mathcal{T}_h\}_{h>0}$ *be a uniform family of admissible triangulations of a polygon* $\Omega \subset \mathbb{R}^2$. *Then the inequality* $\|\chi\|_{C(\overline{\Omega})} \leq 3\frac{\kappa}{h}\|\chi\|_{L_2(\Omega)}$ *holds for all* $\chi \in V_h$, *cf.* (9.37).

Proof Let $T \in \mathcal{T}_h$, then $\chi_{|\overline{T}} \in \mathcal{P}_1(T)$, so that $\chi_{|T}^2$ is a quadratic polynomial. We now use a quadrature formula on T which is exact for cubic polynomials, cf. [57, Ch. 8.8, T_n: 3-3]. In order to be able to formulate this, we will need the following notation: let t_i, $i = 1, 2, 3$ be the vertices T, $m_{i,j}$, $i < j$, $i = 1, 2$, $j = 2, 3$, the midpoints of the edges, and s the center of mass of T. Then we have, for $\chi \in V_h$,

$$\|\chi\|_{L_2(T)}^2 = \int_T \chi(x)^2 \, dx = \frac{|T|}{60}\left(3 \sum_{i=1}^3 \chi(t_i)^2 + 8 \sum_{i<j} \chi(m_{i,j})^2 + 27\chi(s)^2\right)$$

$$\geq \frac{\pi}{20}\rho_T^2 \left(\max_{i=1,2,3} |\chi(t_i)|\right)^2,$$

since $|T| \geq \pi\rho_T^2$. We obtain $\|\chi\|_{C(\overline{T})} = \max_{i=1,2,3} |\chi(t_i)|$ by $\chi_{|\overline{T}} \in \mathcal{P}_1(T)$; hence we have shown that $\|\chi\|_{C(\overline{T})} \leq (\frac{20}{\pi})^{1/2}\frac{1}{\rho_T}\|\chi\|_{L_2(T)} \leq \frac{3}{\rho_T}\|\chi\|_{L_2(T)}$. Since $\|\chi\|_{C(\overline{\Omega})} = \sup_{T \in \mathcal{T}_h} \|\chi\|_{C(\overline{T})}$, the claim follows. $\qquad\square$

We now come to the suboptimal L_∞-estimate.

Theorem 9.35 *Let* $\{\mathcal{T}_h\}_{h>0}$ *be a uniform family of admissible triangulations of a polygon* $\Omega \subset \mathbb{R}^2$. *Given* $f \in L_2(\Omega)$, *let* u *be the solution of the Poisson problem and* $u_h \in V_h$ *the discrete solution via linear elements in accordance with* (9.22). *Then there exists a constant* $C > 0$, *which depends only on the domain* Ω, *such that* $\|u - u_h\|_{C(\overline{\Omega})} \leq C\kappa\,h\,\|f\|_{L_2(\Omega)}$. *Here* κ *is the constant from Definition 9.33.*

Proof Since $F_T : \hat{T} \to T$ is bijective, we have, for every $T \in \mathcal{T}_h$,

$$\|u - I_T u\|_{C(\overline{T})} = \|\hat{u} - I_{\hat{T}}\hat{u}\|_{C(\overline{T})} \leq c_1 |\hat{u} - I_{\hat{T}}\hat{u}|_{H^2(\hat{T})} = c_1 |\hat{u}|_{H^2(\hat{T})},$$

where c_1 is the imbedding constant of $H^2(\hat{T}) \hookrightarrow C(\overline{\hat{T}})$ and where we use that the second derivatives of $I_{\hat{T}}\hat{u}$ vanish. Now, by Lemma 9.22 and Lemma 9.19,

$$|\hat{u}|_{H^2(\hat{T})} \leq |\det B_T|^{-1/2} \|B_T\|^2 |u|_{H^2(T)}$$

$$\leq |\det B_T|^{-1/2} \frac{2}{(\sqrt{2}-1)^2} r_T^2 |u|_{H^2(T)}.$$

Next, since $|\hat{T}| = \frac{1}{2}$,

$$| \det B_T | = 2 \int_{\hat{T}} | \det B_T | \, d\hat{x} = 2 \int_T dx = 2 \, |T| \geq 2\,\pi\,\rho_T^2 \geq 2\,\pi\,\kappa^{-2} r_T^2,$$

whence, by Theorem 9.29 and the fact that $r_T \leq h$ and $\dfrac{\sqrt{2}}{(\sqrt{2}-1)^2 \sqrt{\pi}} \leq 5$,

$$|\hat{u}|_{H^2(\hat{T})} \leq 5\,\kappa\,h\,|u|_{H^2(T)} \leq 5\,\kappa\,h\,|u|_{H^2(\Omega)} \leq 5\,c_2\,\kappa\,h\,\|f\|_{L_2(\Omega)}.$$

Taking the maximum over u, we obtain the following estimate for the mapping $I_h : C_0(\Omega) \rightarrow V_h$: $\|u - I_h u\|_{C(\overline{\Omega})} \leq 5 c_1 c_2 \kappa h \|f\|_{L_2(\Omega)}$. In fact, we observe that $u_h - I_h u$ belongs to V_h and so the inverse estimate of Lemma 9.34 holds. We apply Theorem 9.27 to the term $\|u_h - I_h u\|_{L_2(\Omega)}$ and the quadratic estimate from Theorem 9.32 to $\|u - u_h\|_{L_2(\Omega)}$ to obtain

$$\|u - u_h\|_{C(\overline{\Omega})} \leq \|u - I_h u\|_{C(\overline{\Omega})} + \|u_h - I_h u\|_{C(\overline{\Omega})}$$
$$\leq 5 c_1 c_2 \kappa h \|f\|_{L_2(\Omega)} + 3\frac{\kappa}{h}\|u_h - I_h u\|_{L_2(\Omega)}$$
$$\leq 5 c_1 c_2 \kappa h \|f\|_{L_2(\Omega)} + 3\frac{\kappa}{h}\|u - I_h u\|_{L_2(\Omega)} + 3\frac{\kappa}{h}\|u - u_h\|_{L_2(\Omega)}$$
$$\leq 5 c_1 c_2 \kappa h \|f\|_{L_2(\Omega)} + 3\frac{\kappa}{h} 12\,\hat{C}\,h^2\|u\|_{H^2(\Omega)} + 3\frac{\kappa}{h}h^2 c^2 \|f\|_{L_2(\Omega)}$$
$$\leq h\kappa\,(5 c_1 c_2 + 36\,\hat{C} c_2 + 3 c^2)\,\|f\|_{L_2(\Omega)},$$

which proves the claim for $C := 5 c_1 c_2 + 36\,\hat{C} c_2 + 3 c^2$. $\qquad\square$

We record a further property of uniform triangulations for later use.

Remark 9.36 If $\{\mathcal{T}_h\}_{h>0}$ is a uniform family of admissible triangulations, then the number of elements to which each node belongs can be controlled independently of h. In fact, since $\frac{r_T}{\rho_T} \leq \frac{h}{\rho_T} \leq \kappa$ is bounded independently of h, the smallest angle of each triangle is likewise bounded from below independently of the mesh size h. $\quad\triangle$

With this we can now prove a further variant of the inverse estimate. This often goes by the name of *Bernstein's inequality* in the approximation theory literature.

Theorem 9.37 (Inverse estimate) *Let $\{\mathcal{T}_h\}_{h>0}$ be a uniform family of admissible triangulations of a polygon $\Omega \subset \mathbb{R}^2$. Then there exists a constant $C_{\mathrm{inv}} > 0$ such that $\|\nabla \chi\|^2_{L_2(\Omega)} \leq C_{\mathrm{inv}} h^{-2} \|\chi\|^2_{L_2(\Omega)}$ for all $\chi \in V_h$.*

Proof By Remark 9.36 it suffices to show the claim in the case $\Omega = T$; the rest follows by summing over the elements. To treat $\Omega = T$, in turn we may reduce everything to the reference element $\hat{T}$, on which we use barycentric coordinates. Then every $\hat{\chi}$ on $\hat{T}$ may be written as $\hat{\chi} = \sum_{i=1}^3 \alpha_i \hat{\varphi}_i$, where $\hat{\varphi}_1(\hat{x}) := 1 - \hat{x}_1 - \hat{x}_2$, $\hat{\varphi}_2(\hat{x}) := \hat{x}_1$, $\hat{\varphi}_3(\hat{x}) := \hat{x}_2$ and $\alpha := (\alpha_i)_{i=1,2,3}$. An elementary calculation shows that $\|\hat{\nabla} \hat{\chi}\|^2_{L_2(\hat{T})} = ((\alpha_2 - \alpha_1)^2 + (\alpha_3 - \alpha_1)^2)\|\chi_{\hat{T}}\|^2_{L_2(\hat{T})} = \frac{1}{2}(2\alpha_1^2 + \alpha_2^2 + \alpha_3^2 - \alpha_1\alpha_2 - \alpha_1\alpha_3) \leq \frac{3}{2}|\alpha|^2$. On the other hand, another elementary calculation (if necessary using Maple®, cf. Chapter 10) shows that $\|\hat{\varphi}_i\|^2_{L_2(\hat{T})} = \frac{1}{12}$ and $(\hat{\varphi}_i, \hat{\varphi}_j)_{L_2(\hat{T})} = \frac{1}{24}$ for $i \neq j$. Thus $\|\hat{\chi}\|^2_{L_2(\hat{T})} = \frac{1}{12}(\alpha_1^2 + \alpha_2^2 + \alpha_3^2 + \alpha_1\alpha_2 + \alpha_1\alpha_3 + \alpha_2\alpha_3) = \frac{1}{36}(\alpha_1 + \alpha_2 + \alpha_3)^2 + \frac{1}{18}|\alpha|^2$, and so $\|\hat{\nabla}\hat{\chi}\|^2_{L_2(\hat{T})} \leq \frac{1}{27}\|\hat{\chi}\|^2_{L_2(\hat{T})}$. We now apply Lemma 9.19 and, setting $C_{\text{inv}} := \frac{1}{54}$, obtain the estimate

$$\|\nabla\chi\|^2_{L_2(T)} \leq |\det B| \, \|B^{-1}\|^2 \, \|\hat{\nabla}\hat{\chi}\|^2_{L_2(\hat{T})} \leq C_{\text{inv}} h^{-2} |\det B| \, \|\hat{\chi}\|^2_{L_2(\hat{T})}$$
$$= C_{\text{inv}} h^{-2} \|\chi\|^2_{L_2(T)},$$

which proves the claim. $\square$

Corollary 9.38 (Inverse Cauchy–Schwarz inequality) *Under the assumptions of Theorem 9.37, for the bilinear form $a(\cdot, \cdot)$ associated with the Poisson problem the inequality $a(u_h, v_h) \leq C_{\text{inv}} h^{-2} \|u_h\|_{L_2(\Omega)} \|v_h\|_{L_2(\Omega)}$ holds for all $u_v, v_h \in V_h$.*

Proof By the Cauchy–Schwarz inequality and the inverse estimate, we have

$$a(u_h, v_h) = (\nabla u_h, \nabla v_h)_{L_2(\Omega)} \leq \|\nabla u_h\|_{L_2(\Omega)} \|\nabla u_h\|_{L_2(\Omega)}$$
$$\leq C_{\text{inv}} h^{-2} \|u_h\|_{L_2(\Omega)} \|v_h\|_{L_2(\Omega)},$$

for all $u_h, v_h \in V_h$, which completes the proof. $\square$

9.2.9 The stiffness matrix and the linear system of equations

We shall now describe how one can solve the discrete problem (9.22)

$$a(u_h, \chi) = (f, \chi), \quad \chi \in V_h$$

numerically, where $a(\cdot, \cdot)$ is a symmetric, coercive and continuous bilinear form. Here V_h is a finite-dimensional Hilbert space with basis

$$\Phi_h := \{\varphi_1, \ldots, \varphi_{N_h}\} \subset V_h.$$

We seek a solution of the discrete problem in the form of a linear combination of these basis elements,

$$u_h = \sum_{i=1}^{N_h} u_i \varphi_i \in V_h.$$

We form a vector from the coordinates $\boldsymbol{u}_h := (u_i)_{1 \leq i \leq N_h} \in \mathbb{R}^{N_h}$ and use the basis functions φ_j, $j = 1, \ldots, N_h$, of V_h as test functions in (9.22). Thus the discrete problem is equivalent to

$$(f, \varphi_j)_{L_2(\Omega)} = a(u_h, \varphi_j) = \sum_{i=1}^{N_h} a(\varphi_i, \varphi_j) u_i = (A_h u_h)_j, \quad j = 1, \ldots, N_h,$$

where $A_h := (a(\varphi_i, \varphi_j))_{1 \le i,j \le N_h}$ is the *stiffness matrix*. As with the FEM itself, the name has its origins in structural mechanics, where the entries of A_h can be interpreted as describing the stiffness of the elements being analyzed.

In order to calculate A_h one needs to calculate inner products (normally integrals of products of derivatives), often numerically. There are many strategies available to do this, and to construct the complete matrix efficiently, which can be found in textbooks on the numerics of partial differential equations, e.g. [16, Ch. II.8].

Since $a(\cdot, \cdot)$ is symmetric and coercive, the matrix A_h is symmetric and positive definite (s.p.d.). In the context of finite elements, A_h is additionally sparse, since *supp* φ_i overlaps with *supp* φ_j for very few $j \ne i$, and only such entries $a(\varphi_i, \varphi_j)$ can be distinct from zero, as long as $a(\cdot, \cdot)$ is local, as in the Poisson problem. Thus, just as with the FDM, one can bring efficient iterative solution procedures (such as the BPX preconditioned conjugate gradient method, the multigrid method and wavelet methods, as mentioned above) to bear to solve the linear system of equations.

9.2.10 Numerical experiments

Just as for the FDM, we now wish to describe several numerical experiments, in order to complement the theoretical results and also demonstrate some interesting quantitative effects.

One-dimensional examples

We start with the same one-dimensional example we saw in the context of the FDM. In one dimension the elements T are open subintervals of $\Omega = (0, 1)$. To keep things simple, suppose that the partition is *equidistant*, that is, $h := \frac{1}{N_h}$, $N_h > 1$, and $x_i := ih$, $0 \le i \le N_h$. Then we obtain

$$\mathcal{T}_h = \{T_i\}_{i=1}^{N_h}, \quad T_i := (x_{i-1}, x_i), \quad 0 = x_0 < x_1 < \cdots < x_{N_h-1} < x_{N_h} = 1.$$

For this partition we determine a basis of the space V_h consisting of piecewise linear continuous functions. Clearly, V_h is identical to the spline space generated by linear B-splines with respect to $\mathcal{T}_h$. These functions, which we shall call *hat functions* but are also known as *triangle functions* and *tent functions*, are given by

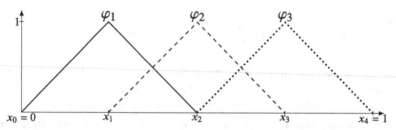

Fig. 9.12 Hat functions φ_1, φ_2, φ_3 with three interior nodes.

$$\varphi_i(x) := \begin{cases} \frac{1}{h}(x - x_{i-1}), \ x \in (x_{i-1}, x_i), \\[2mm] \frac{1}{h}(x_{i+1} - x), \ x \in (x_i, x_{i+1}), \\[2mm] 0, \qquad\qquad \text{otherwise,} \end{cases}$$

for $1 \le i \le N_h - 1$. Figure 9.12 shows an example for $N_h = 4$ and three functions φ_1, φ_2, φ_3. Clearly the space V_h has dimension equal to the number of interior nodes of the mesh, $\mathcal{N}_h = \dim(V_h) = N_h - 1$, where N_h is the number of elements (that is, subintervals). Hence the resulting linear system of equations has dimension $\mathcal{N}_h = N_h - 1$.

One can check easily (cf. Exercise 9.3) that in the one-dimensional uniform partition case the stiffness matrix corresponds exactly to the matrix A_h from the FDM. However, the right-hand sides are different: we have

$$f_i^{\text{FDM}} = h^2 f(x_i), \qquad f_i^{\text{FEM}} = (f, \varphi_i)_{L_2(\Omega)} = \int_0^1 f(x)\,\varphi_i(x)\,dx.$$

For our experiments, on the one hand we again consider the function $u_{\sin}(x) := \sin(2\pi x) \in C^\infty(0, 1)$, and on the other the solution of the two point boundary value problem with discontinuous right-hand side f_α in (9.15). Since f_α is piecewise constant (and in particular in L_2), the corresponding solution u_α belongs to $H^2(0, 1)$. We thus expect quadratic convergence with respect to $\|\cdot\|_{L_2(\Omega)}$ in both cases. This is exactly what Figure 9.13 shows us: we obtain two parallel lines with the same slope, indicating quadratic convergence in both cases (here we have chosen $\|\cdot\|_{h,\infty}$ in order to facilitate the comparison with the FDM). We also see that the curve for u_α is below the other, so that the errors are *quantitatively* smaller even though the *qualitative* behavior is the same. The reason for this is that the right-hand side of the error estimate of Corollary 9.32 contains the L_2-norm of the right-hand side, and this norm is smaller for u_α than for $u_{\sin}$, that is, $\|f_\alpha\|_{L_2(0,1)} < \|f_{\sin}\|_{L_2(0,1)}$. This explains the quantitative difference.

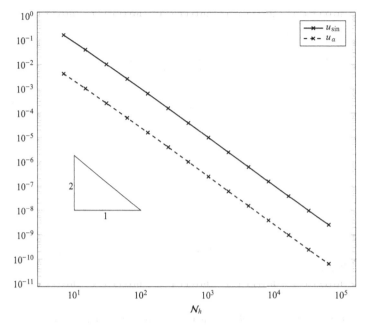

Fig. 9.13 Convergence history for the FEM for the right-hand sides f_α (dashed line) and $f_{\sin}$ (solid line) in (9.15). In both cases the error $\|u - u_h\|_{h,\infty}$ is plotted against the number of degrees of freedom N_h in a log-log scale. The quadratic convergence is clearly visible for both examples.

Two-dimensional examples

We now consider three examples in two space dimensions, as depicted in Figure 9.14. The first is a square, for which we have already determined the exact solution in the course of the book (Section 3.3.1); we thus know the analytic solution. This is not the case for the other two examples. For the L-shaped domain we know that due to the inward-pointing corner (the domain is not convex) we must expect a lower degree of regularity of solutions and hence a lower order of convergence. The third example is a "general" convex polygon, which will show the advantages of the FEM. In this case we have H^2-regularity. The triangulations were generated with the MATLAB command `createpde` under specification of a maximal mesh size h. As above, in each case we will use piecewise linear elements for $h = 2^{-k}$, $k = 1, \ldots, 7$.

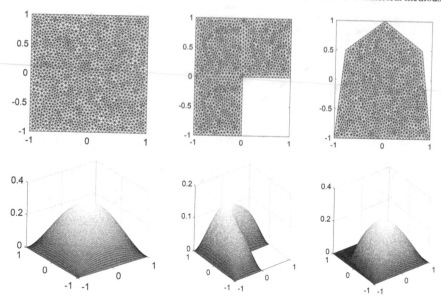

Fig. 9.14 Three illustrative examples for finite elements in 2D, the square (left), an L-shaped domain (center) and a convex polygon (right). The respective FEM solutions are shown in the second row.

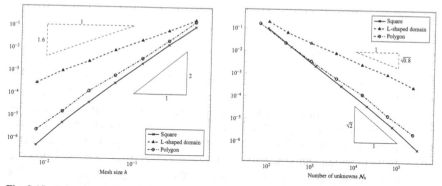

Fig. 9.15 Convergence history for the three two-dimensional finite elements examples, left w.r.t. mesh size h, right w.r.t. number of unknowns N_h.

In each case we have numerically determined a "reference solution" on a very fine mesh, and then determined the error to this solution. In all numerical examples the Lebesgue norms were calculated with the help of quadrature formulae of corresponding orders. The errors and the orders of convergence are given in Figure 9.15. On the left-hand side we see the error as a function of the maximal mesh size h in a log-log plot. The orders of convergence of ca. 2 for the square and polygon and ca. 1.6 for the L-shaped domain, as predicted by the theory, can be seen clearly. We see in particular that the FEM admits the same order of convergence on general convex polygons as on the square. On the L-shaped domain we obtain a lower order of convergence; since we know that the solution is in H^1 but not in H^2, this was to be expected. It is also evident that the lack of regularity of the solution does indeed result in a reduction of the rate of convergence. In this sense the convergence estimates described above appear to be sharp.

On the right-hand side of Figure 9.15 we have plotted the error against the number of unknowns $\mathcal{N}_h$. This is interesting for two reasons: for a given maximal mesh size h the number of elements varies, since on the one hand the domains are different and on the other the generation of the meshes results in a different number of nodes. If we had a uniform mesh, then we would have $\mathcal{N}_h \sim h^2$ – we can recognize the corresponding rate of convergence despite the differences in the mesh. The second reason is that $\mathcal{N}_h$ also essentially determines the computational cost, and hence the runtime.

9.3* Extensions and generalizations

9.3.1 The Petrov–Galerkin method

When treating the Galerkin method in Section 9.2.1 we were acting on the assumption that the trial and test spaces of the bilinear forms were identical. We now wish to consider problems of the type (4.12), that is, given $f \in W'$,

$$\text{find } u \in V \text{ such that } b(u, w) = f(w) \text{ for all } w \in W, \tag{9.38}$$

where $b : V \times W \to \mathbb{R}$ is a continuous bilinear form and the trial space V and the test space W are now possibly distinct. In this case one may consider finite-dimensional subspaces of the form

$$V_h \subset V, \quad W_h \subset W, \quad 0 \neq \dim(V_h) = \dim(W_h) = \mathcal{N}_h < \infty, \quad h > 0. \tag{9.39}$$

Here we are supposing for simplicity that trial and test spaces have the same dimension, which will result in a linear system with the same number of variables as equations to be solved. If one does not wish to assume this, then one obtains a least squares problem. In the theorem of Banach–Nečas, Theorem 4.27, the inf-sup-condition for $b|_{V_h \times W_h}$ guarantees the existence and uniqueness of solutions of the discrete problem, that is, that there exists exactly one $u_h \in V_h$ such that

$$b(u_h, w_h) = f(w_h), \quad w_h \in W_h. \tag{9.40}$$

However, here we can quickly recognize an essential difference from the numerical approximation of coercive problems using the Galerkin method. In the latter case the discrete problem "inherits" the coercivity constant from the variational problem; we do not need to distinguish between α and a potential α_h. This is not the case with the inf-sup condition, since there is no reason for the supremum with respect to $w \in W$ to be attained in W_h. This means that the spaces V_h and W_h need to be compatible with each other, in the sense that the inf-sup-condition for $b|_{V_h \times W_h}$ needs to be uniform in $h > 0$.

Definition 9.39 The spaces $V_h \subset V$ and $W_h \subset W$ in (9.39) are said to satisfy a *Ladyshenskaya–Babuška–Brezzi (LBB) condition* with respect to the bilinear form $b : V \times W \to \mathbb{R}$ if there exists a constant $\beta > 0$ such that

$$\inf_{v_h \in V_h} \sup_{w_h \in W_h} \frac{b(v_h, w_h)}{\|w_h\|_W \|v_h\|_V} \geq \beta \tag{9.41}$$

for all $h > 0$.   △

Note that the constant β in the LBB condition is not permitted to depend on h. It can be a very delicate problem to construct discrete spaces V_h and W_h in such a way that the LBB condition is satisfied; we refer to [16]. We also observe that the constant β in (9.41) need not be the inf-sup constant of the bilinear form b (if the latter exists, which is often assumed); strictly speaking one should introduce an additional symbol for this constant. However, by considering the minimum of the two constants, one can survive with just one letter β for the two conditions.

Remark 9.40 So far the LBB condition does not imply that b satisfies an inf-sup condition (in fact, the spaces V_h, W_h could be independent of $h > 0$). But we will see below that LBB does in fact imply an inf-sup condition for b if we assume that the spaces V_h and W_h approximate V and W, respectively, as $h \to 0+$ (see (9.46), (9.47) below).   △

We now wish to investigate in which form an analog of Theorem 9.14 (Céa's lemma) holds; for this, we assume the LBB condition. Let $f \in W'$. It follows from the theorem of Banach–Nečas, Theorem 4.27, that (9.40) has a unique solution u_h for all $h > 0$. Assume that $u \in V$ is a solution[1] of the continuous problem (9.38). Then Galerkin orthogonality holds, that is,

$$b(u - u_h, w_h) = 0 \quad \text{for all } w_h \in W_h. \tag{9.42}$$

Now let $v_h \in V_h$ be arbitrary; then by the triangle inequality, the LBB condition and Galerkin orthogonality $(b(u_h - v_h, w_h) = b(u - v_h, w_h))$, we have

$$\|u - u_h\|_V \leq \|u - v_h\|_V + \|u_h - v_h\|_V \leq \|u - v_h\|_V + \frac{1}{\beta} \sup_{w_h \in W_h} \frac{b(u_h - v_h, w_h)}{\|w_h\|_W}$$

$$= \|u - v_h\|_V + \frac{1}{\beta} \sup_{w_h \in W_h} \frac{b(u - v_h, w_h)}{\|w_h\|_W} \leq \left(1 + \frac{C}{\beta}\right) \|u - v_h\|_V.$$

Since $v_h \in V_h$ was arbitrary, we have thus shown that

$$\|u - u_h\|_V \leq \left(1 + \frac{C}{\beta}\right) \inf_{v_h \in V_h} \|u - v_h\|_V.$$

As with Céa's lemma, the Petrov–Galerkin approximation is thus, up to a multiplicative constant, as good as the best approximation to u in the trial space V_h. However, the constant is significantly worse – and, as we shall see, not optimal. We will next prove a result from [61], that the "1+" can be eliminated. To this end we first require a result of T. Kato from [40].

Lemma 9.41 (Kato, 1960) *Let H be a Hilbert space and suppose that $P : H \to H$ is idempotent, i.e., $P^2 = P$. If $P \neq 0$ and $P \neq I$, then*

$$\|P\| = \|I - P\|. \tag{9.43}$$

---

[1] Note that here we do not need to assume uniqueness of solutions, cf. the end of the section. Existence and uniqueness of solutions are guaranteed by the theorem of Banach–Nečas, Theorem 4.27 (page 133) in the case that the inf-sup condition (4.11) and the surjectivity condition (4.13) are satisfied.

Proof (1) First suppose that $\dim(H) = 2$, so that necessarily $rank(P) = rank(I - P) = 1$. Hence there exist $a, b, c, d \in H$ such that $Pv = (b, v)_H a$ and $(I - P)v = (d, v)_H c$ for all $v \in H$, and $(a, b)_H = (c, d)_H = 1$. Hence $\|P\| = \|a\|_H \|b\|_H$ and $\|I - P\| = \|c\|_H \|d\|_H$. Moreover, for all $v \in H$ we have $v = Pv + (I - P)v = (b, v)_H a + (d, v)_H c$. Now we substitute b and d for v and obtain, respectively, $b = \|b\|_H^2 a + (b, d)_H c$ and $d = (b, d)_H a + \|d\|_H^2 c$. We also have $1 = (a, b)_H = \|a\|_H^2 \|b\|_H^2 + (a, c)_H (b, d)_H$ and $1 = (c, d)_H = (a, c)_H (b, d)_H + \|c\|_H^2 \|d\|_H^2$, that is

$$\|a\|_H^2 \|b\|_H^2 = 1 - (a, c)_H (b, d)_H = \|c\|_H^2 \|d\|_H^2. \tag{9.44}$$

In particular, $\|P\| = \|I - P\|$, which yields the claim when $\dim(H) = 2$.

(2) Now let $\dim(H) > 2$. We show that $\|I - P\| \leq \|P\|$. Choose $x \in H$ such that $\|x\|_H = 1$. Set $X := span\{x, Px\}$; then the space X, of dimension at most 2, is invariant under both P and $I - P$. If $\dim(X) = 1$, then $x = Px$ and hence $(I - P)x = 0$. In the other case, $\dim(X) = 2$, by (1) we have $\|(I - P)x\|_H \leq \|(I - P)_{|X}\| \|x\|_H = \|P_{|X}\| \leq \|P\|$. In both cases we have $\|(I - P)x\|_H \leq \|Px\|_H$, that is, $\|I - P\| \leq \|P\|$. Exchanging the roles of P and $I - P$ yields $\|P\| \leq \|I - P\|$. This proves the claim. $\qquad\square$

We are now able to prove the optimal estimate announced above.

Theorem 9.42 (Xu, Zikatanov, 2003) *Let* $b : V \times W \to \mathbb{R}$ *be a continuous bilinear form with continuity constant* $0 < C < \infty$*, for which the LBB condition (9.41) holds for some constant* $\beta > 0$*. Given* $u \in V$ *let* $u_h \in V_h$ *be the solution of* $b(u_h, w_h) = b(u, w_h)$ *for all* $w_h \in W_h$*. Then* u *and* u_h *satisfy the estimate*

$$\|u - u_h\|_V \leq \frac{C}{\beta} \inf_{v_h \in V_h} \|u - v_h\|_V. \tag{9.45}$$

Proof Define the mapping $P_h : V \to V_h$ by $u \mapsto P_h u := u_h$. Now P_h is clearly idempotent. The LBB condition (9.41) implies that $P_h \neq 0$ for all $h > 0$. If $I - P_h = 0$, then (9.45) is trivial. Thus we may assume that $I - P_h \neq 0$. In this case, by Lemma 9.41, $\|P_h\| = \|I - P_h\|$. Now let $v_h \in V_h$ be arbitrary. Since $(I - P_h)v_h = 0$ it follows that $\|u - u_h\|_V = \|(I - P_h)(u - v_h)\|_V \leq \|I - P_h\| \|u - v_h\|_V = \|P_h\| \|u - v_h\|_V$. Furthermore, by definition of P_h, for all $v \in V$

$$\|P_h v\|_V \leq \frac{1}{\beta} \sup_{w_h \in W_h} \frac{b(P_h v, w_h)}{\|w_h\|_W} = \frac{1}{\beta} \sup_{w_h \in W_h} \frac{b(v, w_h)}{\|w_h\|_W} \leq \frac{C}{\beta} \|v\|_V.$$

Thus $\|P_h\| \leq \frac{C}{\beta}$, which shows the claim. $\qquad\square$

So far the finite-dimensional subspaces V_h and W_h were quite arbitrary. We now want to assume that they approximate V and W, respectively, as $h \to 0+$. In fact, if

$$\text{for all } v \in V \text{ we have } dist(v, V_h) \to 0 \quad \text{as} \quad h \to 0+, \tag{9.46}$$

then for a given $f \in V'$ and a solution u of (9.38) the solution u_h of the discrete problem (9.40) converges to u as $h \to 0+$. In particular, this implies uniqueness of the solutions of (9.38). If, in addition, we assume that

$$\text{for all } w \in W \text{ we have } dist(w, W_h) \to 0 \quad \text{as} \quad h \to 0+, \tag{9.47}$$

the LBB condition also implies existence for the continuous problem (9.38), as we will now see.

Theorem 9.43 *Let* $f \in W'$*. Assume the LBB condition, along with (9.46) and (9.47). Then there exist unique solutions* u *of (9.38) and* u_h *of (9.40). Moreover, (9.45) holds, and* $\lim_{h \to 0} u_h = u$*.*

Proof The LBB condition implies that

$$\|u_h\|_V \le \frac{1}{\beta} \sup_{w_h \in W_h} \frac{b(u_h, w_h)}{\|w_h\|_W} = \frac{1}{\beta} \sup_{w_h \in W_h} \frac{|f(w_h)|}{\|w_h\|_W} \le \frac{1}{\beta} \|f\|_{W'}$$

for all $h > 0$. Hence $(u_h)_{h>0}$ is bounded. By Theorem 4.35 there exist some $u \in V$ and a sequence $(h_k)_{k \in \mathbb{N}}$ such that $h_k > 0$, $\lim_{k \to \infty} h_k = 0$, and $u_{h_k} \rightharpoonup u$ weakly in V as $k \to \infty$. Now let $w \in W$, then by (9.47) there exists a sequence $(w_k)_{k \in \mathbb{N}}$, $w_k \in W_{h_k}$, such that $\lim_{k \to \infty} \|w - w_k\|_W = 0$. Hence, by the Galerkin orthogonality,

$$b(u, w) = \lim_{k \to \infty} b(u_{h_k}, w) = \lim_{k \to \infty} b(u_{h_k}, w - w_k + w_k) = \lim_{k \to \infty} b(u_{h_k}, w_k)$$
$$= \lim_{k \to \infty} f(w_k) = f(w).$$

This shows that u is a solution. Now Theorem 9.42 shows that (9.45) holds for u; hence also $u = \lim_{h \to 0} u_h$, since we are assuming (9.46). This implies uniqueness. $\square$

Remark 9.44 (a) The converse of Theorem 9.43 also holds: if for each $f \in W'$ there exist unique solutions u_h of (9.40) and u of (9.38), respectively, and $\lim_{h \to 0} u_h = u$, then the LBB condition holds. See [8, Thm. 2.4].

(b) If $V = W$ and b is coercive, then LBB holds for arbitrary appoximating subspaces, provided that $V_h = W_h$ for all $h > 0$. This continues to hold if (9.40) is well posed and the form is only *essentially coercive*, i.e., if $u_n \rightharpoonup 0$ weakly in V and $b(u_n, u_n) \to 0$ always implies that $\|u_n\|_V \to 0$.

The converse is also true: if the form is not essentially coercive but (9.40) is well posed, then there exist $V_h = W_h$ which satisfy (9.46), but for which the LBB condition is violated, [8, Thm. 5.2]. An example of an essentially coercive form can be found by replacing the Laplacian in the Poisson problem from Section 9.2.8 by $\Delta - \lambda I$ for arbitrary $\lambda \in \mathbb{R}$ (this is also known as the *Helmholtz problem*). If λ is not an eigenvalue of the Dirichlet Laplacian, then the continuous problem is well posed. In this case, the finite element convergence theorems, Theorems 9.31 and 9.32, continue to hold, [8, Thm. 7.6].

(c) When computing an approximate solution u_h of u numerically, one usually assumes the well-posedness of (9.38). The main point of focus then typically consists of constructing LBB stable discretizations V_h and W_h. In practice this often requires knowledge of analytic properties of the continuous problem, and an approach adapted specifically to the problem. $\triangle$

9.3.2 Further extensions

Higher orders of convergence. When dealing with convergence questions we have always restricted ourselves to *linear* finite elements, the reason being that this leads to quadratic convergence in $L_2(\Omega)$, as long as the solution of the partial differential equation is in $H^2(\Omega)$. We had seen that the solutions have this degree of regularity (under certain assumptions) in Section 6.11. If one wishes for a higher order of convergence, then one needs to use elements of higher order and, correspondingly, make stronger regularity assumptions. Both the construction of elements of higher order and the associated convergence analysis are technically more demanding; upon performing these steps, one may obtain estimates of the form

$$\|u - u_h\|_{H^1(\Omega)} \le c \cdot h^k |u|_{H^{k+1}(\Omega)}, \quad 0 < h \le 1, \quad u \in H^{k+1}(\Omega).$$

More information on such results can be found in books on the numerics of partial differential equations, for example [16, 35, 46, 52].

General L_2-estimates. The above L_2 error estimate can also be derived for general variational problems. For this the following theorem is required, which may be obtained via a formalization of the proof of Theorem 9.32, cf. [16, Thm. II.7.6].

Theorem 9.45 (Aubin–Nitsche lemma) *Let $V \hookrightarrow H$ be a continuously imbedded and dense subspace of a Hilbert space H, and $V_h \subset V$ a finite-dimensional subspace of V. Let $a(\cdot, \cdot)$ be bilinear, continuous and coercive. Then*

$$\|u - u_h\|_H \leq C \|u - u_h\|_V \sup_{g \in H} \left\{ \frac{1}{\|g\|_H} \inf_{\chi \in V_h} \|\varphi_g - \chi\|_V \right\},$$

where $C > 0$ is the continuity constant of $a(\cdot, \cdot)$, and, for given $g \in H$, φ_g, is the solution of the dual problem $a(v, \varphi_g) = (g, v)$ for all $v \in V$.

L_∞-estimates. We have proved an error estimate which is "suboptimal" with respect to the supremum norm. With considerably more effort it can be shown that the following estimate holds for H^2-regular problems in two-dimensional domains, cf., e.g., [21, Thm. 3.3.7]:

$$\|u - u_h\|_{C(\overline{\Omega})} \leq c\, h^2 \, |\log h|^{3/2} \|\nabla^2 u\|_{C(\overline{\Omega})}, \quad u \in C^2(\overline{\Omega}).$$

More general elements. One can of course consider geometric partitions of Ω not just into triangles but also, for example, quadrilaterals (or in the three-dimensional case tetrahedra and hexahedra). Quadrilaterals have the advantage that one can easily construct piecewise polynomials via tensor products; this results in simple elements and efficient computations. On the other hand, triangles are more flexible when it comes to partitioning arbitrary polygons.

More general equations. Instead of the Poisson problem one could also consider convection-diffusion-reaction equations (as in Theorem 5.23, for example). This results in a non-symmetric system of equations, which needs to be solved using other numerical methods. In addition, in the convection-dominated case numerical instabilities can arise which requires special discretizations. Neumann and Robin boundary conditions must likewise be discretized suitably. There are other problems of elliptic type, such as those listed in Table 2.1 (page 47), for which one can define suitable finite elements; however, this generally requires problem-specific constructions.

Non-polygonal domains. We will close this section with two remarks on what happens when Ω is not a polygon. In this case Ω can, obviously, no longer be partitioned exactly into triangles; at the boundary there is a deviation. There are (at least) two strategies which may be adopted here; the first involves considering "triangles" with curvilinear boundary, the other involves approximating Ω by a polygon Ω_h.

If $\partial\Omega$ is smooth, then one can use what are known as *isoparametric* elements at the boundary, for which one side is curved. Of course, in this case the transformation onto the reference element $\hat{T}$ will no longer be affine, which means that both the analysis and the numerical implementation will be significantly more involved.

If Ω is convex and smooth, then the Poisson problem is H^2-regular. If one chooses a quasi-uniform family of admissible triangulations, in such a way that the vertices of the inscribed polygonal domain of Ω lie on the boundary $\partial\Omega$, then one can show that the geometric error which arises by approximating Ω by Ω_h, is of order of magnitude $O(h^2)$. This means that in this case one has the same order of convergence as in the polygonal case, that is, $O(h)$ in H^1 and $O(h^2)$ in L_2, [35, §8.6]. Finally, if Ω is smooth but not convex, then the variational problem is still H^2-regular (see Remark 6.87) and one can still prove the optimal order of convergence, albeit by other methods. Indeed, since in the non-convex case the vertices of a triangle may lie outside Ω, it follows that $V_h \not\subset V$. In this case we may speak of a *non-conforming* partition. For such partitions other techniques are needed for the error estimates; see for example [16, Ch. III].

9.4 Parabolic problems

We now wish to describe two numerical methods for parabolic equations. Here as well we will restrict ourselves to a limited number of aspects of these methods and refer any readers who wish for more information to [46, 52], for example. For these problems we will require a discretization in both space *and* time. For the spatial discretization, as above we will consider finite differences and finite elements.

9.4.1 Finite differences

As in the elliptic case, we will start with the finite difference method (FDM). For this we will again consider the initial-boundary value problem for the heat equation on the interval $[0, 1]$, that is,

$$u_t = u_{xx} \qquad\qquad \text{in } (0, T) \times (0, 1), \qquad\qquad (9.48a)$$
$$u(t, 0) = u(t, 1) = 0, \qquad \text{for } t \in [0, T], \qquad\qquad (9.48b)$$
$$u(0, x) = u_0(x), \qquad\qquad x \in [0, 1]. \qquad\qquad (9.48c)$$

Here $u_0 \in C([0, 1])$ should satisfy $u_0(0) = u_0(1) = 0$. By Corollary 3.37 problem (9.48) has a unique solution $u \in C^\infty((0, T] \times [0, 1]) \cap C([0, T] \times [0, 1])$. For the following estimates we will need more regularity as $t \downarrow 0$. We will thus assume throughout this section that

$$u_0 \in C^4([0, 1]) \text{ with } u_0^{(m)}(0) = u_0^{(m)}(1) = 0 \text{ for } m = 0, 2, 4. \qquad (9.49)$$

In this case, we know by Theorem 3.38 that $u \in C^{2,4}([0, T] \times [0, 1])$, where as above $[0, T]$, $T > 0$, is the time interval under consideration. In order to construct a FDM for (9.48) we clearly need a mesh in both space and time. For simplicity we choose equidistant mesh sizes in both variables,

$$\Delta t = \frac{T}{N}, \quad h = \frac{1}{M}, \quad M, N \in \mathbb{N}.$$

Definition 9.46 (Difference operators) For brevity of notation we introduce the following difference operators:
(a) *Forward difference* with respect to the *time* variable:
  $D_{\Delta t}^+ v(t, x) := \frac{1}{\Delta t}\big(v(t + \Delta t, x) - v(t, x)\big)$;
(b) *Backward difference* with respect to the *time* variable:
  $D_{\Delta t}^- v(t, x) := \frac{1}{\Delta t}\big(v(t, x) - v(t - \Delta t, x)\big)$;
(c) *Symmetric difference* of second order with respect to the *space* variable:
  $D_h^2 v(t, x) := \frac{1}{h^2}\big(v(t, x + h) - 2v(t, x) + v(t, x - h)\big).$ △

Using the FDM we will now determine an approximation

$$U_i^k \approx u(t^k, x_i), \quad t^k := k\,\Delta t, \quad x_i := i\,h, \quad k = 0, \ldots, N, \, i = 0, \ldots, M,$$

of the exact solution u of (9.48).

The explicit Euler method

To this end we discretize the time derivative $\dot{u}$ using the forward difference operator $D_{\Delta t}^+$ and arrive at the *explicit Euler method*:

$$D_{\Delta t}^+ U_i^k = D_h^2 U_i^k, \qquad 1 \le i \le M - 1, \, 0 \le k \le N, \tag{9.50a}$$

$$U_0^k = U_M^k = 0, \qquad 0 \le k \le N, \tag{9.50b}$$

$$U_i^0 = u_{0,i} = u_0(x_i), 0 \le i \le M. \tag{9.50c}$$

This system of equations for the unknowns $U_i^k \in \mathbb{R}$ can be easily solved. Firstly, (9.50a) reads

$$\Delta t^{-1} (U_i^{k+1} - U_i^k) = h^{-2} (U_{i+1}^k - 2U_i^k + U_{i-1}^k);$$

if we introduce the abbreviation

$$\lambda := \frac{\Delta t}{h^2},$$

then (9.50a) becomes

$$U_i^{k+1} = \lambda\, U_{i+1}^k + (1 - 2\lambda)U_i^k + \lambda\, U_{i-1}^k =: (E_\lambda U^k)_i,$$

complemented by the boundary conditions $U_0^{k+1} = U_M^{k+1} = 0$. We can now recognize the origin of the name "explicit". Starting with the known initial condition (9.50c) we can calculate $U^{k+1} := (U_i^{k+1})_{0 \le i \le M} \in \mathbb{R}^{M+1}$ directly from U^k, that is, U^{k+1} is given explicitly in terms of U^k. In order to analyze (9.50) we (re-)introduce the discrete maximum norm (in the space variables):

$$\|U^k\|_{h,\infty} := \max_{0 \le j \le M} |U_j^k| \qquad \text{(recall } h = \tfrac{1}{M}\text{)}.$$

We suppose for the meantime that

$$0 < \lambda = \frac{\Delta t}{h^2} \le \frac{1}{2}, \quad \text{that is, } \Delta t \le \frac{1}{2}\,h^2. \tag{9.51}$$

Then $1 - 2\lambda \ge 0$, and it follows that

$$\|E_\lambda U^k\|_{h,\infty} = \|U^{k+1}\|_{h,\infty} \le \lambda\|U^k\|_{h,\infty} + |1 - 2\lambda|\,\|U^k\|_{h,\infty} + \lambda\|U^k\|_{h,\infty}$$

$$= \|U^k\|_{h,\infty} \le \cdots \le \|U^0\|_{h,\infty} \le \|u_0\|_\infty.$$

Hence the operator E_λ is bounded (i.e. stable), with $\|E_\lambda\| \leq 1$, in the case where $\lambda \leq \frac{1}{2}$.

Remark 9.47 It is not hard to see that the condition $\lambda \leq \frac{1}{2}$ is necessary for the stability of the operator. To show this, choose the initial condition u_0 so that $U_i^0 = (-1)^i \sin(i\,h\,\pi), 0 \leq i \leq M$. Then

$$
\begin{aligned}
U_i^1 &= \lambda U_{i+1}^0 + (1 - 2\lambda)U_i^0 + \lambda U_{i-1}^0 \\
&= (-1)^i \{-\lambda \sin((i+1)\,h\,\pi) + (1 - 2\lambda)\sin(i\,h\,\pi) - \lambda \sin((i-1)\,h\,\pi)\} \\
&= (-1)^i \sin(i\,h\,\pi)\{-\lambda \cos(h\,\pi) + (1 - 2\lambda) - \lambda \cos(h\,\pi)\} \\
&\quad + (-1)^i \{-\lambda \cos(i\,h\,\pi)\sin(h\,\pi) + \lambda \cos(i\,h\,\pi)\sin(h\,\pi)\} \\
&= (1 - 2\lambda - 2\lambda \cos(h\,\pi))U_i^0
\end{aligned}
$$

and, by induction $U_i^k = (1 - 2\lambda - 2\lambda \cos(h\,\pi))^k U_i^0$. Now, for every $\lambda > \frac{1}{2}$ we can choose h so small that $|1 - 2\lambda - 2\lambda \cos(h\,\pi)| = 2\lambda(\cos(h\,\pi) + 1) - 1 > 1$, and so U^k diverges as $k \to \infty$. △

Remark 9.48 The condition (9.51) expresses the relationship between the respective step sizes in time and space. Such a condition arises in general in the context of numerical methods for time-dependent problems, not only parabolic ones (see Section 9.5 below for the wave equation, a hyperbolic problem). It is often called a *Courant–Friedrichs–Lewy (CFL) condition*. In the above case λ is also called the *CFL number*. △

Analogously to the FDM for elliptic problems we define the *local discretization error* as follows:

$$
\tau_i^k := D_{\Delta t}^+ u_i^k - D_h^2 u_i^k, \qquad u_i^k := u(t^k, x_i). \tag{9.52}
$$

For the analysis of the error τ^k we will require the following seminorm for functions $v \in C^4([0, 1])$:

$$
|v|_{C^4} := \sup_{x \in (0,1)} |v_{xxxx}(x)| = \sup_{x \in (0,1)} \left| \frac{d^4}{dx^4}v(x) \right| = \sup_{x \in (0,1)} |v^{(4)}(x)|. \tag{9.53}
$$

We now have the following consistency estimate.

Theorem 9.49 *Let $\lambda \leq \frac{1}{2}$, suppose that (9.49) holds, and suppose that $u \in C^{2,4}([0, T] \times [0, 1])$ is the unique solution of (9.48). Then $\|\tau^k\|_{h,\infty} \leq c\,h^2\,|u_0|_{C^4}$ for some constant $c > 0$.*

Proof Performing a Taylor expansion of the exact solution $u(t, x)$, we have that

$$u_i^{k+1} = u(t^{k+1}, x_i) = u(t^k, x_i) + (t^{k+1} - t^k)\dot{u}(t^k, x_i) + \frac{1}{2}(t^{k+1} - t^k)^2 \ddot{u}(\sigma^k, x_i)$$

$$= u_i^k + \Delta t\, \dot{u}(t^k, x_i) + \frac{1}{2}(\Delta t)^2\, \ddot{u}(\sigma^k, x_i)$$

for a suitable $\sigma^k \in (t^k, t^{k+1})$, and analogously

$$|u_{xx}(t^k, x_i) - D_h^2 u(t^k, x_i)| \le \frac{1}{12} h^2 \max_{x \in (x_{i-1}, x_{i+1})} |u_{xxxx}(t^k, x)|$$

for some $\xi_i \in (x_i, x_{i+1})$. Since $u_t = u_{xx}$, it follows that

$$\tau_i^k = D_{\Delta t}^+ u_i^k - D_h^2 u_i^k - (u_t(t^k, x_i) - u_{xx}(t^k, x_i))$$

$$= (D_{\Delta t}^+ u(t^k, x_i) - u_t(t^k, x_i)) - (D_h^2 u(t^k, x_i) - u_{xx}(t^k, x_i))$$

$$= \frac{1}{2} \Delta t\, u_{tt}(\sigma^k, x_i) - (D_h^2 u(t^k, x_i) - u_{xx}(t^k, x_i)).$$

Since $u \in C^{2,4}([0, T] \times [0, 1])$, we have $u_{tt} = (u_{xx})_t = (u_t)_{xx} = u_{xxxx}$ and hence, since $\Delta t \le \frac{1}{2} h^2$.

$$|\tau_i^k| \le \frac{1}{2} \Delta t \max_{t \in (t^k, t^{k+1})} |u_{tt}(t, x_i)| + \frac{1}{12} h^2 \max_{x \in (x_{i-1}, x_{i+1})} |u_{xxxx}(t^k, x)|$$

$$\le c\, h^2 \max_{t \in (t^k, t^{k+1})} \max_{x \in (x_{i-1}, x_{i+1})} |u_{xxxx}(t, x)|.$$

Finally, we use the smoothing property of parabolic equations, which in particular states that $|u(t, \cdot)|_{C^4} \le |u_0|_{C^4}$ for all $t \in [0, T]$, as follows immediately from (8.13). Hence $\|\tau^k\|_{h,\infty} \le c\, h^2 |u_0|_{C^4}$, which was the claim. $\square$

With the help of the above consistency estimate we can now turn to the principal convergence theorem.

Theorem 9.50 *Let $\lambda \le \frac{1}{2}$, suppose that (9.49) holds, let $u \in C^{2,4}([0, T] \times [0, 1])$ be the unique solution of (9.48), write $u^k := u(t^k, \cdot)$, and take U^k to be the discrete solution of (9.50). Then there exists a constant $c > 0$ such that*

$$\|U^k - u^k\|_{h,\infty} \le c\, t_k\, h^2 |u_0|_{C^4}$$

for all $0 \le k \le N$.

Proof We define the error $e^k := U^k - u^k$. Since U^k is the discrete solution of (9.50), it follows that $D_{\Delta t}^+ e_i^k - D_h^2 e_i^k = (D_{\Delta t}^+ - D_h^2)U_i^k - (D_{\Delta t}^+ - D_h^2)u_i^k = -\tau_i^k$, that is, $(\Delta t)^{-1}(e_i^{k+1} - e_i^k) = D_h^2 e_i^k - \tau_i^k$, and thus

$$e_i^{k+1} = e_i^k + \Delta t\, D_h^2 e_i^k - \Delta t\, \tau_i^k = \lambda e_{i-1}^k + (1 - 2\lambda)e_i^k + \lambda e_{i+1}^k - \Delta t\, \tau_i^k$$

$$= (E_\lambda e^k)_i - \Delta t\, \tau_i^k.$$

It follows by induction that

$$e^{k+1} = E_\lambda e^k - \Delta t\, \tau^k = E_\lambda^{k+1} e^0 - \Delta t \sum_{\ell=0}^{k} E_\lambda^{k-\ell} \tau^\ell.$$

Since $e^0 = U^0 - u^0 = 0$, by Theorem 9.49 and the fact that $\|E_\lambda\| \le 1$,

$$\|e^k\|_{h,\infty} \le \Delta t \sum_{\ell=0}^{k-1} \|E_\lambda^{k-\ell}\tau^\ell\|_{h,\infty} \le \Delta t \sum_{\ell=0}^{k-1} \|\tau^\ell\|_{h,\infty}$$

$$\le c\,k\,\Delta t\, h^2\, |u_0|_{C^4} = c\,t_k\,h^2\,|u_0|_{C^4},$$

whence the claim. $\square$

Remark 9.51 With the help of Fourier transforms and what is known as *von Neumann stability theory* one can obtain L_2-estimates on the errors, in particular $\|U^k - u^k\|_{L_2(0,1)} \le c\,t_k\,h^2\,|u_0|_{H^4(0,1)}$, $u_0 \in H^4(0,1)$, cf. [46, Ch. 9]. $\triangle$

The implicit Euler method

A major disadvantage of the explicit Euler method is the stability condition $\Delta t \le \frac{1}{2}h^2$ (CFL), which in practice often leads to excessively small step sizes. Replacing (9.50a) by

$$D_{\Delta t}^- U_i^{k+1} = D_h^2 U_i^{k+1}, \quad 1 \le i \le M-1,\ 0 \le k \le N-1, \tag{9.54}$$

leads to the *implicit Euler method*, which in full reads as follows $\Delta t^{-1}\,(U_i^{k+1} - U_i^k) = h^2\,(U_{i+1}^{k+1} - 2U_i^{k+1} + U_{i-1}^{k+1})$, that is,

$$(1+2\lambda)U_i^{k+1} - \lambda U_{i+1}^{k+1} - \lambda U_{i-1}^{k+1} = U_i^k. \tag{9.55}$$

We thus obtain a linear system of equations involving a tridiagonal matrix whose main diagonal entries are all $1 + 2\lambda$, and the entries of the respective off-diagonals above and below it are all $-\lambda$. In this case, U^{k+1} is only *implicitly* given in terms of U^k, even if the linear system may be simply solved in $O(M)$ operations. We may write (9.55) in matrix form as

$$B_\lambda U^{k+1} = U^k, \quad B_\lambda \in \mathbb{R}^{M \times M},$$

where B_λ is the symmetric tridiagonal matrix just described. Thus, for an appropriate index i_0, we get by (9.55)

$$\|U^{k+1}\|_{h,\infty} = |U_{i_0}^{k+1}| = \frac{1}{1+2\lambda}\left|U_{i_0}^k + \lambda U_{i_0+1}^{k+1} + \lambda U_{i_0-1}^{k+1}\right|$$

$$\le \frac{1}{1+2\lambda}\left(|U_{i_0}^k| + \lambda|U_{i_0+1}^{k+1}| + \lambda|U_{i_0-1}^{k+1}|\right)$$

$$\leq \frac{2\lambda}{1+2\lambda}\|U^{k+1}\|_{h,\infty} + \frac{1}{1+2\lambda}\|U^k\|_{h,\infty}.$$

Hence, we get $\|U^{k+1}\|_{h,\infty} \leq \|U^k\|_{h,\infty}$ for all $\lambda \geq 0$, which means that the method is stable without any restriction on λ. A corresponding convergence result may be obtained analogously to the one above; the error estimate in this case reads

$$\|U^k - u^k\|_{h,\infty} \leq c\, t_k (h^2 + \Delta t) \max_{0 \leq t \leq t^k} |u(t, \cdot)|_{C^4},$$

meaning the method is of first order with respect to the time variable. If one wishes for a higher order in time, one may consider, for example, the Crank–Nicolson method or a Runge–Kutta method. We will not go into details here.

Numerical experiments

We will now illustrate the significance of the CFL condition by means of a numerical experiment for the one-dimensional heat equation on a rod $\Omega = (0,1)$ with end time $T = 1$, right-hand side $f \equiv 0$ and initial condition $u_0(x) := x(1-x)$. As t grows, $u(t,x)$ tends to 0. We choose different step sizes in space and time, $\Delta t = \frac{T}{N}$, $h = \frac{1}{M}$. The results of this experiment are depicted in Figure 9.16. In the left and the center the *explicit* Euler method was applied; we see how sensitively the method reacts to the CFL number $\lambda = \Delta t / h^2$. For $\lambda = 0.5$ we see (center) that the method converges stably, whereas if $\lambda = 0.506$ then large oscillations emerge (left, note the range up to 10^7). On the right we see that the *implicit* Euler method also converges for very course step sizes (here $\lambda = 45 \gg 0.5$).

9.4.2 Finite elements

Let $\Omega \subset \mathbb{R}^2$ be a convex polygon and let $0 < T < \infty$. We know from Theorem 8.35 that the inhomogeneous heat equation is well posed: that is, if $u_0 \in H_0^1(\Omega)$ and $f \in C([0,T], L_2(\Omega))$, then there exists a unique solution $u \in L_2((0,T), H^2(\Omega)) \cap H^1((0,T), L_2(\Omega))$ of the problem

$$\dot{u}(t) - \Delta u(t) = f(t), \qquad t \in (0,T), \tag{9.56a}$$

$$u(t) \in H_0^1(\Omega), \quad t \in (0,T), \tag{9.56b}$$

$$u(0) = u_0. \tag{9.56c}$$

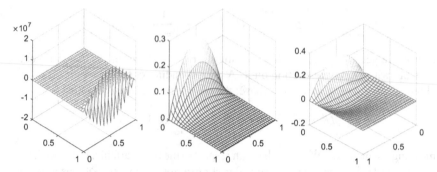

Fig. 9.16 Numerical solution of the heat equation via the FDM for the spatial variables, $M = 30$. Left: explicit Euler method with $N = 1780$ (CFL $\lambda = 0.506$); center: explicit Euler method with $N = 1800$ ($\lambda = 0.5$); right: implicit Euler method, $N = 20$ ($\lambda = 45$).

In terms of the associated bilinear form $a : H_0^1(\Omega) \times H_0^1(\Omega) \to \mathbb{R}$, $a(v, w) := (\nabla v, \nabla w)_{L_2(\Omega)}$, the solution u satisfies the equation

$$(\dot{u}(t), v)_{L_2(\Omega)} + a(u(t), v) = (f(t), v)_{L_2(\Omega)} \tag{9.57}$$

for almost all $t \in (0, T)$ and all $v \in H_0^1(\Omega)$.

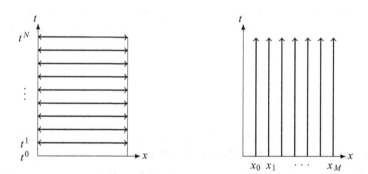

Fig. 9.17 Horizontal (left) and vertical (right) method of lines.

One possible strategy is to begin by discretizing in only *one* of the two variables, viz. time t and space x; this is referred to as *semi-discretization*. First discretizing in time and then in space is known as the *horizontal method of lines*, cf. Figure 9.17 (left). In this case, for each point in time t^k, $k = 1, \ldots, N$, one obtains a boundary value problem. These N problems can be solved in parallel. Here we will describe the *vertical method of lines*, whereby one first discretizes the spatial domain Ω, for example using the FEM, cf. Figure 9.17 (right). This leads to a system of ordinary differential equations. Let $(\mathcal{T}_h)_{0<h\leq 1}$ be a quasi-uniform family of admissible triangulations of Ω. We suppose that $\mathcal{T}_h$ consists of N_h triangles (we are again assuming that Ω is a polygon). Suppose that the corresponding finite element space $V_h \subset H_0^1(\Omega)$ is generated by the Lagrange basis

$$\Phi_h = \{\varphi_1^h, \ldots, \varphi_{N_h}^h\}, \quad \dim V_h = N_h.$$

Then the *spatial semi-discrete problem* is as follows: determine a function $u_h \in C^1([0,T], V_h)$ such that

$$u_h(0) = u_{0,h}, \tag{9.58a}$$

$$(\dot{u}_h(t), \chi)_{L_2(\Omega)} + a(u_h(t), \chi) = (f(t), \chi)_{L_2(\Omega)}, \qquad \chi \in V_h, \ t > 0, \tag{9.58b}$$

where $u_{0,h} \in V_h$ is an approximation of the initial condition u_0. Problem (9.58) has a unique solution $u_h \in C^1([0,T], V_h)$. This can be shown by the classical Picard–Lindelöf theorem on existence and uniqueness of solutions to ordinary differential equations or by applying Theorem 8.22 to $a|_{V_h \times V_h}$ and $P_h f \in C([0,T], V_h)$, where P_h is the orthogonal projection of H to V_h. See also Exercise 9.18. Subsequently, we will discretize the time in terms of finite difference approximations. This is the reason why we need here that f is continuous in time.

Before we turn to the analysis of (9.58), we first wish to describe the form of the semi-discrete problem more precisely. We are clearly searching for some

$$u_h(t, x) = \sum_{i=1}^{N_h} \alpha_i(t)\varphi_i^h(x) \in V_h$$

with time-dependent coefficients $\alpha_i(t)$, $t \geq 0$. Then, setting $\chi = \varphi_j^h$, (9.58b) reads

$$\sum_{i=1}^{N_h} \dot{\alpha}_i(t)(\varphi_i^h, \varphi_j^h)_{L_2(\Omega)} + \sum_{i=1}^{N_h} \alpha_i(t)\,a(\varphi_i^h, \varphi_j^h) = (f(t), \varphi_j^h)_{L_2(\Omega)}, \quad 1 \leq j \leq N_h,$$

and the initial conditions $\alpha_i(0) = \gamma_i$ are given by $u_{0,h} = \sum_{i=1}^{N_h} \gamma_i \varphi_i^h$. We define the *stiffness matrix* $A_h := (a(\varphi_i^h, \varphi_j^h))_{i,j=1,\ldots,N_h} \in \mathbb{R}^{N_h \times N_h}$, take the right-hand side $b_h(t) := (f(t), \varphi_j^h)_{L_2(\Omega)})_{j=1,\ldots,N_h} \in \mathbb{R}^{N_h}$ and introduce what is known as the *mass matrix* $M_h := ((\varphi_i^h, \varphi_j^h)_{L_2(\Omega)})_{i,j=1,\ldots,N_h} \in \mathbb{R}^{N_h \times N_h}$; then our problem may be represented in matrix form as

$$M_h \dot{\alpha}_h(t) + A_h \alpha_h(t) = b_h(t), \quad t > 0, \ \alpha_h(0) = \gamma. \tag{9.59}$$

We thus have a system of ordinary differential equations of dimension N_h. We will come to the discretization of this system somewhat later; we first observe that A_h and M_h are symmetric and positive definite, and so in particular M_h is regular. Hence (9.59) reads

$$\dot{\alpha}_h(t) + M_h^{-1} A_h \alpha_h(t) = M_h^{-1} b_h(t), \quad t > 0, \ \alpha_h(0) = \gamma.$$

This initial value problem has a unique solution $\alpha_h \in C^1([0,T];\mathbb{R}^n)$ by the classical Picard–Lindelöf theorem; see also Exercise 9.18. We next perform a stability analysis.

Lemma 9.52 *We have the stability estimate*

$$\|u_h(t)\|_{L_2(\Omega)} \le \|u_{0,h}\|_{L_2(\Omega)} + \int_0^t \|f(s)\|_{L_2(\Omega)}\,ds, \quad t \in [0,T]. \tag{9.60}$$

Proof We take $v_h = u_h(t)$ as a test function in (9.58b) and obtain

$$(\dot{u}_h(t), u_h(t))_{L_2(\Omega)} \le (\dot{u}_h(t), u_h(t))_{L_2(\Omega)} + a(u_h(t), u_h(t))$$
$$= (f(t), u_h(t))_{L_2(\Omega)}.$$

For $\varepsilon > 0$ let $g_\varepsilon(t) := (\|u_h(t)\|^2_{L_2(\Omega)} + \varepsilon)^{1/2}$. Then, $g_\varepsilon \in C^1([0,T])$ and by Lemma 8.17 and the Cauchy–Schwarz inequality we have

$$g_\varepsilon'(t) = \frac{(\dot{u}_h(t), u_h(t))_{L_2(\Omega)}}{g_\varepsilon(t)} \le \frac{(f(t), u_h(t))_{L_2(\Omega)}}{g_\varepsilon(t)} \le \|f(t)\|_{L_2(\Omega)} \frac{\|u_h(t)\|_{L_2(\Omega)}}{g_\varepsilon(t)}$$
$$\le \|f(t)\|_{L_2(\Omega)}$$

Thus,

$$g_\varepsilon(t) = g_\varepsilon(0) + \int_0^t g_\varepsilon'(s)\,ds \le g_\varepsilon(0) + \int_0^t \|(f(s)\|_{L_2(\Omega)}\,ds$$

Letting $\varepsilon \searrow 0$ gives the desired inequality.   □

The following operator will be extremely useful in the error analysis. We call $R_h : H_0^1(\Omega) \to V_h$, defined by

$$a(R_h v, \chi) = a(v, \chi), \quad v \in H_0^1(\Omega),\ \chi \in V_h, \tag{9.61}$$

the *Ritz projection*. This is clearly the orthogonal projection onto V_h with respect to the energy inner product $a(u,v) := (\nabla u, \nabla v)_{L_2(\Omega)}$ (for the equivalence of the induced norm, see Corollary 6.33). This projection also appeared in the proof of Theorem 9.42. The error analysis will require the following estimates. As with the elliptic equations we will restrict ourselves to the case of linear elements, that is, to V_h as in (9.37).

Theorem 9.53 *Let Ω be a convex polygon. For the space V_h of linear finite elements as in (9.37) we have*
(a) $\|R_h v - v\|_{L_2(\Omega)} \le Ch\|v\|_{H^1(\Omega)}$, $v \in H_0^1(\Omega)$,
(b) $\|R_h v - v\|_{L_2(\Omega)} \le Ch^2\|v\|_{H^2(\Omega)}$, $v \in H_0^1(\Omega) \cap H^2(\Omega)$,
where the constant $C > 0$ depends only on Ω and κ.

Proof (a) Let $v \in H_0^1(\Omega)$ and choose $g \in L_2(\Omega)$, $\|g\|_{L_2(\Omega)} = 1$, such that $\|R_h v - v\|_{L_2(\Omega)} = (R_h v - v, g)_{L_2(\Omega)}$. Given this g, choose $w \in H_0^1(\Omega) \cap H^2(\Omega)$ for which $-\Delta w = g$, and let $w_h \in V_h$ be the corresponding discrete solution, i.e. $a(w_h, \chi) = (g, \chi)_{L_2(\Omega)}$, $\chi \in V_h$. By the definition of R_h in (9.61) we have $a(v - R_h v, \chi) = 0$, $v \in H_0^1(\Omega)$, $\chi \in V_h$, that is, the Galerkin orthogonality. Hence, using $-\Delta u = g$ and the test function $\chi = R_h v - v$,

$$\|R_h v - v\|_{L_2(\Omega)} = (R_h v - v, g)_{L_2(\Omega)} = a(R_h v - v, w) = a(R_h v - v, w - w_h)$$
$$\leq |R_h v - v|_{H^1(\Omega)} \|w - w_h\|_{H^1(\Omega)} \leq c_1 h |R_h v - v|_{H^1(\Omega)},$$

where $c_1 > 0$ is the constant in the error estimate of Theorem 9.31. Finally, $I - R_h$ is a contraction with respect to $|\cdot|_{H^1(\Omega)}$ (as an orthogonal projection with respect to $a(\cdot, \cdot)$); hence $|R_h v - v|_{H^1(\Omega)} \leq |v|_{H^1(\Omega)}$. This yields the claim.

(b) Now let $v \in H_0^1(\Omega) \cap H^2(\Omega)$, set $f := -\Delta v$, and consider the corresponding discrete solution $v_h \in V_h$, that is, $a(v_h, \chi) = (f, \chi)_{L_2(\Omega)}$, $\chi \in V_h$. Since R_h is the a-orthogonal projection of $H_0^1(\Omega)$ onto V_h, it follows that

$$|R_h v - v|_{H^1(\Omega)} = |R_h(v - v_h) - (v - v_h)|_{H^1(\Omega)} \leq |v_h - v|_{H^1(\Omega)}$$
$$\leq c\, h \|f\|_{L_2(\Omega)} \leq c\, c_2\, h \|v\|_{H^2(\Omega)},$$

where c is the constant from Theorem 9.31 and c_2 the one of Theorem 6.86. Combining this with the estimate from part (a), we have $\|R_h v - v\|_{L_2(\Omega)} \leq c_1 h |R_h v - v|_{H^1(\Omega)} \leq c\, c_1\, c_2\, h^2 \|v\|_{H^2(\Omega)}^2$. $\qquad\square$

We can now prove the desired error estimate for the semi-discrete approximation.

Theorem 9.54 *Let $0 < T < \infty$ and let Ω be a convex polygon. Moreover, let $f \in L_2((0, T), H_0^1(\Omega) \cap H^2(\Omega)) \cap H^1((0, T), L_2(\Omega))$, and let $u_0 \in H_0^1(\Omega) \cap H^2(\Omega)$ such that $\Delta u_0 \in H_0^1(\Omega)$. Finally, let $u_{0,h} \in V_h$ and suppose that u_h is the solution of (9.58). Then the following error estimate holds for the exact solution u of (9.56):*

$$\|u_h(t) - u(t)\|_{L_2(\Omega)} \leq \|u_{0,h} - u_0\|_{L_2(\Omega)}$$
$$+ c\, h^2 \left\{ \|u_0\|_{H^2(\Omega)} + \sqrt{t}\left(\|\Delta u_0\|_{H^1(\Omega)} + \|f\|_{L_2((0,t), H^2(\Omega))}\right) \right\}$$

for all $t \in [0, T]$ and a constant $c > 0$ depending only on Ω and κ.

Proof Recall from Theorem 8.36 that by our assumptions we have that $u \in H^1((0, T), H^2(\Omega) \cap H_0^1(\Omega)) \cap H^2((0, T), L_2(\Omega)) \cap C^1([0, T], H_0^1(\Omega))$. In particular $u \in C([0, T], H^2(\Omega) \cap H_0^1(\Omega))$. We rewrite the error $u_h - u$ using the Ritz projection as

$$u_h - u = (u_h - R_h u) + (R_h u - u) =: \theta_h + \rho_h. \tag{9.62}$$

For the second term in (9.62) we may use the error estimate of Theorem 9.53 (b) for R_h; let $t \in [0, T]$, since $u(t) \in H_0^1(\Omega) \cap H^2(\Omega)$ there exists a constant $c > 0$ such that

$$\|\rho_h(t)\|_{L_2(\Omega)} = \|R_h u(t) - u(t)\|_{L_2(\Omega)}$$

$$\leq c\,h^2 \|u(t)\|_{H^2(\Omega)} = c\,h^2 \left\| u(0) + \int_0^t \dot{u}(s)\,ds \right\|_{H^2(\Omega)}$$

$$\leq c\,h^2 \left(\|u_0\|_{H^2(\Omega)} + \int_0^t \|\dot{u}(s)\|_{H^2(\Omega)}\,ds \right)$$

$$\leq c\,h^2 (\|u_0\|_{H^2(\Omega)} + \sqrt{t}\|\dot{u}\|_{L_2((0,t),H^2(\Omega))}),$$

where we have used Theorem 8.29 (b) and Hölder's inequality in the last step. We now use Theorem 8.36 (a) (for $T = t$) to obtain

$$\|\dot{u}\|_{L_2((0,t),H^2(\Omega))} \leq c\,(\|\Delta u_0\|_{H^1(\Omega)}^2 + \|f\|_{L_2((0,t),H^2(\Omega))}^2)^{1/2}$$

$$\leq c\,(\|\Delta u_0\|_{H^1(\Omega)} + \|f\|_{L_2((0,t),H^2(\Omega))}),$$

whence (with a possibly different constant c)

$$\|\rho_h(t)\|_{L_2(\Omega)} \leq c\,h^2 \left(\|u_0\|_{H^2(\Omega)} + \sqrt{t}(\|\Delta u_0\|_{H^1(\Omega)} + \|f\|_{L_2((0,t),H^2(\Omega))}) \right).$$

For the first term $\theta_h = u_h - R_h u$ in (9.62) we consider a related initial boundary problem. Namely, for $\chi \in V_h$ we have

$$(\dot{\theta}_h(t), \chi)_{L_2(\Omega)} + a(\theta_h(t), \chi) =$$

$$= (\dot{u}_h(t), \chi)_{L_2(\Omega)} + a(u_h(t), \chi) - \left(\frac{d}{dt} R_h u(t), \chi \right)_{L_2(\Omega)} - a(R_h u(t), \chi)$$

$$= (f(t), \chi)_{L_2(\Omega)} - (R_h \dot{u}(t), \chi)_{L_2(\Omega)} - a(u(t), \chi)$$

$$= (\dot{u}(t) - R_h \dot{u}(t), \chi)_{L_2(\Omega)}$$

by properties of the Ritz projection and since u is a solution of (9.56). Due to Theorem 8.36, we have $\dot{u} \in C([0, T]; H_0^1(\Omega))$, which allows us to apply R_h to $\dot{u}$. But this equation says that

$$(\dot{\theta}_h(t), \chi)_{L_2(\Omega)} + a(\theta_h(t), \chi) = -(\dot{\rho}_h(t), \chi)_{L_2(\Omega)}, \quad \chi \in V_h. \tag{9.63}$$

Hence the two terms in (9.62) are related. Since $u \in C^1([0, T], H_0^1(\Omega))$, we get $\rho_h \in C^1([0, T], H_0^1(\Omega))$. Thus, we can apply the stability estimate (9.60) to (9.63) to obtain

$$\|\theta_h(t)\|_{L_2(\Omega)} \leq \|\theta_h(0)\|_{L_2(\Omega)} + \int_0^t \|\dot{\rho}_h(s)\|_{L_2(\Omega)}\,ds.$$

To deal with the term in the integral, we apply the error estimate in Theorem 9.53 (b) for the Ritz projection. Since $\dot{u}(t) \in H_0^1(\Omega) \cap H^2(\Omega)$, Theorem 9.53 (b) yields for all $t \in [0, T]$

$$\|\dot{\rho}_h(t)\|_{L_2(\Omega)} = \|R_h \dot{u}(t) - \dot{u}(t)\|_{L_2(\Omega)} \le c h^2 \|\dot{u}(t)\|_{H^2(\Omega)}.$$

We also have by our assumption that $u_0 \in H_0^1(\Omega) \cap H^2(\Omega)$, so that again Theorem 9.53 (b) yields

$$\begin{aligned}
\|\theta_h(0)\|_{L_2(\Omega)} &= \|u_h(0) - R_h u(0)\|_{L_2(\Omega)} = \|u_{0,h} - R_h u_0\|_{L_2(\Omega)} \\
&\le \|u_{0,h} - u_0\|_{L_2(\Omega)} + \|u_0 - R_h u_0\|_{L_2(\Omega)} \\
&\le \|u_{0,h} - u_0\|_{L_2(\Omega)} + c h^2 \|u_0\|_{H^2(\Omega)}.
\end{aligned}$$

The claim now follows upon putting all this together.   □

The above proof also yields a result which can be found, e.g., in [46, Thm. 10.4] or in slightly different form in [34, Thm. 5.10].

Corollary 9.55 *Under the assumptions of Theorem 9.54 we have for all $t \in [0, T]$*

$$\begin{aligned}
\|u_h(t) - u(t)\|_{L_2(\Omega)} &\le \|u_{0,h} - u_0\|_{L_2(\Omega)} \\
&\quad + c h^2 \left\{ \|u_0\|_{H^2(\Omega)} + \int_0^t \|\dot{u}(s)\|_{H^2(\Omega)} \, ds \right\}.
\end{aligned}$$

Remark 9.56 Both estimates derived above involve $\|u_{0,h} - u_0\|_{L_2(\Omega)}$, the initial error. So far, we did not specify the choice of the initial value $u_{0,h}$. At least three appropriate choices are useful: (i) Interpolation using the Clément operator in (9.36), (ii) the Ritz projection (9.61) or (iii) the orthogonal projection i.e., $(u_{0,h}, \chi)_{L_2(\Omega)} = (u_0, \chi)_{L_2(\Omega)}$ for all $\chi \in V_h$. Recalling Theorem 9.27, Theorem 9.53 or standard estimates for the orthogonal projection (also known as Bramble-Hilbert lemma, see e.g. [16, Ch. II,§6]) we have in all three cases that there exists a constant $c > 0$ such that

$$\|u_{0,h} - u_0\|_{L_2(\Omega)} \le c h^2 \|u_0\|_{H^2(\Omega)}, \quad u_0 \in H_0^1(\Omega) \cap H^2(\Omega). \tag{9.64}$$

This means that the term $\|u_{0,h} - u_0\|_{L_2(\Omega)}$ is dominated by the remaining quantities in the estimates in Theorem 9.54 or Corollary 9.55. Hence, we can eliminate the initial error from the estimates as long as $u_{0,h}$ is chosen to satisfy (9.64).   △

The implicit Euler method

The final step is a discretization of the time variable in the initial value problem for the ordinary differential equation given in (9.59). We already saw with the FDM that the explicit Euler method is only stable under restrictive assumptions on the step size in time and space; here we will limit ourselves to the implicit Euler method and a fixed step size Δt, that is, $t^k = k \Delta t$, $k = 0, ..., N = \frac{T}{\Delta t}$. In this case the time discretization of (9.58) reads

$$(D_{\Delta t}^- U^k, \chi)_{L_2(\Omega)} + a(U^k, \chi) = (f(t^k), \chi)_{L_2(\Omega)}, \qquad \chi \in V_h, \qquad (9.65a)$$

$$U^0 = u_{0,h}. \qquad (9.65b)$$

Here $U^k \in H_0^1(\Omega)$, $k = 1, ..., N$ and the difference operators are defined as follows (see also Definition 9.46).

Definition 9.57 (Difference operators) For a sequence $(U^k)_{k=0,...,N}$ of vectors we define (a) the *forward difference* as $D_{\Delta t}^+ U^k := \frac{1}{\Delta t}(U^{k+1} - U^k)$, $k = 0, ..., N - 1$, and (b) the *backward difference* by $D_{\Delta t}^- U^k := \frac{1}{\Delta t}(U^k - U^{k-1})$, $k = 1, ..., N$.   △

Thus (9.65a) reads for for $k = 1, ..., N$,

$$(U^k, \chi)_{L_2(\Omega)} + \Delta t\, a(U^k, \chi) = (U^{k-1} + \Delta t\, f(t^k), \chi)_{L_2(\Omega)}, \; \chi \in V_h. \qquad (9.66)$$

Here we suppose that $f \in C([0,T], L_2(\Omega))$. We seek U^k of the form $U^k(x) = \sum_{i=1}^{N_h} \alpha_i^k \varphi_i(x) \in V_h$, in which case (9.66) becomes

$$\sum_{i=1}^{N_h} \alpha_i^k (\varphi_i^h, \varphi_j^h)_{L_2(\Omega)} + \Delta t \sum_{i=1}^{N_h} \alpha_i^k a(\varphi_i^h, \varphi_j^h) =$$

$$= \sum_{i=1}^{N_h} \alpha_i^{k-1} (\varphi_i^h, \varphi_j^h)_{L_2(\Omega)} + \Delta t\, (f(t^k), \varphi_j^h)_{L_2(\Omega)}$$

for $j = 1, \ldots, N_h$. In matrix form this reads

$$(M_h + \Delta t\, A_h)\alpha_h^k = M_h \alpha_h^{k-1} + \Delta t\, b_h(t^k),$$

where $\alpha_h^k = (\alpha_i^k)_{1 \le i \le N_h} \in \mathbb{R}^{N_h}$, M_h and A_h are the mass and stiffness matrices, respectively, and the right-hand side $b_h(t^k)$ is in $\mathbb{R}^{N_h}$. Since $M_h + \Delta t\, A_h$ is s.p.d. (and thus of course regular), (9.65) has a unique solution. We start with a stability result.

Lemma 9.58 *For all $k = 1, ..., N$, we have the estimate*

$$\|U^k\|_{L_2(\Omega)} \le \|U^0\|_{L_2(\Omega)} + \Delta t \sum_{j=1}^k \|f(t^j)\|_{L_2(\Omega)}.$$

Proof Choose $\chi = U^k$ in (9.66), then

$$\|U^k\|_{L_2(\Omega)}^2 + \Delta t\, a(U^k, U^k) = (U^{k-1}, U^k)_{L_2(\Omega)} + \Delta t\, (f(t^k), U^k)_{L_2(\Omega)}$$

$$\le \|U^k\|_{L_2(\Omega)} \big(\|U^{k-1}\|_{L_2(\Omega)} + \Delta t\, \|f(t^k)\|_{L_2(\Omega)}\big).$$

Since $a(U^k, U^k) \ge 0$, it follows that

$$\|U^k\|_{L_2(\Omega)} \le \|U^{k-1}\|_{L_2(\Omega)} + \Delta t\, \|f(t^k)\|_{L_2(\Omega)}.$$

We now apply this inequality recursively to obtain

$$\|U^k\|_{L_2(\Omega)} \leq \|U^{k-2}\|_{L_2(\Omega)} + \Delta t \left(\|f(t^{k-1})\|_{L_2(\Omega)} + \|f(t^k)\|_{L_2(\Omega)}\right)$$

$$\leq \cdots \leq \|U^0\|_{L_2(\Omega)} + \Delta t \sum_{j=1}^{k} \|f(t^j)\|_{L_2(\Omega)},$$

as claimed. $\qquad\square$

As with the FDM we see that the stability result for the implicit Euler method holds with no restrictions on Δt or h. We now come to the error estimate in this case. Of course, this estimate depends on the solution. But given f and u_0 it says that the error is quadratic in h, but only linear in Δt.

Theorem 9.59 *Let $0 < T < \infty$ and let $\Omega \subset \mathbb{R}^d$ be open, bounded and convex. Let $u_0 \in H_0^1(\Omega) \cap H^2(\Omega)$ be such that $\Delta u_0 \in H_0^1(\Omega)$, and let $f \in L_2((0,T), H^2(\Omega)) \cap H^1((0,T), L_2(\Omega))$ be such that $f(t) \in H_0^1(\Omega)$ for all $t \in [0,T]$. Finally, let u be the corresponding solution of (9.56) and $u_{0,h} \in V_h$ satisfy (9.64). There exists a constant depending only on Ω such that for all $k = 1, ..., N$*

$$\|U^k - u(t^k)\|_{L_2(\Omega)} \leq c\, h^2 \left\{\|u_0\|_{H^2(\Omega)} + \int_0^{t^k} \|\dot{u}(s)\|_{H^2(\Omega)}\, ds\right\}$$

$$+ c\, \Delta t \int_0^{t^k} \|\ddot{u}(s)\|_{L_2(\Omega)}\, ds.$$

Proof The assumptions on f imply that $f \in C([0,T], H_0^1(\Omega))$. By Theorem 8.36, we have that $u \in H^1((0,T), H^2(\Omega)) \cap H^2((0,T), L_2(\Omega))$, in particular we obtain $u \in C([0,T], H^2(\Omega))$. Moreover, $u \in C^1([0,T], H_0^1(\Omega))$ by the last assertion of Theorem 8.36. Analogously to the proof of Theorem 9.54, we set for $k = 1, ..., N$

$$U^k - u(t^k) = (U^k - R_h u(t^k)) + (R_h u(t^k) - u(t^k)) =: \theta_h^k + \rho_h^k$$

and estimate each of the latter two terms separately. Firstly, by Theorem 9.53,

$$\|\rho_h^k\|_{L_2(\Omega)} = \|R_h u(t^k) - u(t^k)\|_{L_2(\Omega)} \leq c\, h^2 \|u(t^k)\|_{H^2(\Omega)}$$

$$= c\, h^2 \left\|u_0 + \int_0^{t^k} \dot{u}(s)\, ds\right\|_{H^2(\Omega)}$$

$$\leq c\, h^2 \left\{\|u_0\|_{H^2(\Omega)} + \int_0^{t^k} \|\dot{u}(s)\|_{H^2(\Omega)}\, ds\right\}.$$

We shall derive an initial value problem for θ_h^k, just as we did earlier. We have, for all $\chi \in V_h$,

$$(D_{\Delta t}^-\theta_h^k, \chi)_{L_2(\Omega)} + a(\theta_h^k, \chi) =$$
$$= (D_{\Delta t}^- U^k, \chi)_{L_2(\Omega)} + a(U^k, \chi) - (D_{\Delta t}^- R_h u(t^k), \chi)_{L_2(\Omega)} - a(R_h u(t^k), \chi)$$
$$= (f(t^k), \chi)_{L_2(\Omega)} - (D_{\Delta t}^- R_h u(t^k), \chi)_{L_2(\Omega)} - a(u(t^k), \chi)$$
$$= (\dot{u}(t^k) - D_{\Delta t}^- R_h u(t^k), \chi)_{L_2(\Omega)} \quad =: -(\omega^k, \chi)_{L_2(\Omega)},$$

where the right-hand side is

$$\omega^k = R_h D_{\Delta t}^- u(t^k) - \dot{u}(t^k) = (R_h - I)D_{\Delta t}^- u(t^k) + \left(D_{\Delta t}^- u(t^k) - \dot{u}(t^k)\right)$$
$$=: \omega_1^k + \omega_2^k.$$

We proceed as we did previously, this time using Lemma 9.58:

$$\|\theta_h^k\|_{L_2(\Omega)} \le \|\theta_h^0\|_{L_2(\Omega)} + \Delta t \sum_{j=1}^k \|\omega^j\|_{L_2(\Omega)}$$

$$\le \|\theta_h^0\|_{L_2(\Omega)} + \Delta t \sum_{j=1}^k \|\omega_1^j\|_{L_2(\Omega)} + \Delta t \sum_{j=1}^k \|\omega_2^j\|_{L_2(\Omega)}.$$

In order to estimate the term involving ω_1^j we use the identity

$$\omega_1^j = (R_h - I)D_{\Delta t}^- u(t^j) = (R_h - I)\frac{1}{\Delta t} \int_{t^{j-1}}^{t^j} \dot{u}(s)\, ds$$
$$= \frac{1}{\Delta t} \int_{t^{j-1}}^{t^j} (R_h - I)\dot{u}(s)\, ds,$$

that is,

$$\Delta t \sum_{j=1}^k \|\omega_1^j\|_{L_2(\Omega)} \le \sum_{j=1}^k \int_{t^{j-1}}^{t^j} \|(R_h - I)\dot{u}(s)\|_{L_2(\Omega)}\, ds$$
$$= \int_0^{t^k} \|(R_h - I)\dot{u}(s)\|_{L_2(\Omega)}\, ds \le c\, h^2 \int_0^{t^k} \|\dot{u}(s)\|_{H^2(\Omega)}\, ds$$

by Theorem 9.53. For the second term we use that $u \in H^2((0,T), L_2(\Omega))$ and Theorem 8.29

$$\omega_2^j = \frac{u(t^j) - u(t^{j-1})}{\Delta t} - \dot{u}(t^j) = -\frac{1}{\Delta t}\int_{t^{j-1}}^{t^j} (s - t^{j-1})\ddot{u}(s)\, ds.$$

It follows that

$$\Delta t \sum_{j=1}^k \|\omega_2^j\|_{L_2(\Omega)} \le \sum_{j=1}^k \left\|\int_{t^{j-1}}^{t^j} (s - t^{j-1})\ddot{u}(s)\, ds\right\|_{L_2(\Omega)}$$

$$\leq \Delta t \int_0^{t^k} \|\ddot{u}(s)\|_{L_2(\Omega)}\, ds.$$

From these estimates we obtain

$$\|\theta_h^k\|_{L_2(\Omega)} \leq \|\theta_h^0\|_{L_2(\Omega)} + c\, h^2 \int_0^{t^k} \|\dot{u}(s)\|_{H^2(\Omega)}\, ds + \Delta t \int_0^{t^k} \|\ddot{u}(s)\|_{L_2(\Omega)}\, ds.$$

Finally $\|\theta_h^0\|_{L_2(\Omega)} = \|U^0 - R_h u(0)\|_{L_2(\Omega)} = \|u_{0,h} - R_h u_0\|_{L_2(\Omega)} \leq \|u_{0,h} - u_0\|_{L_2(\Omega)} + \|u_0 - R_h u_0\|_{L_2(\Omega)} \leq c\, h^2 \|u_0\|_{H^2(\Omega)}$ by (9.64) and Theorem 9.53. Combining all these estimates yields the claim. $\qquad\square$

The Crank–Nicolson method

The estimate of Theorem 9.59 reveals an essential disadvantage of the Euler method: it is of second order in the space variable (h^2) but only first order in time (Δt). As already mentioned (albeit not in detail) in Section 9.4.1 (FDM) this disadvantage can be overcome using the Crank–Nicolson method. We are now going to describe this method, which can also analogously be used with the FDM in space. To this end, we expand the semi-discrete equation symmetrically about the point $t^{k-1/2} := (k - \frac{1}{2})\Delta t$ and obtain

$$(D_{\Delta t}^- U^k, \chi)_{L_2(\Omega)} + a(\tfrac{1}{2}(U^k + U^{k-1}), \chi) = (f(t^{k-1/2}), \chi)_{L_2(\Omega)}, \quad \chi \in V_h, \qquad (9.67a)$$

$$U^0 = u_{0,h}, \qquad (9.67b)$$

or, for $k = 1, ..., N$,

$$(U^k, \chi)_{L_2(\Omega)} + \tfrac{\Delta t}{2} a(U^k, \chi) = (U^{k-1} + \Delta t\, f(t^{k-1/2}), \chi)_{L_2(\Omega)} - \tfrac{1}{2} a(U^{k-1}, \chi) \quad (9.68)$$

for all $\chi \in V_h$. In matrix form this reads

$$(M_h + \tfrac{\Delta t}{2} A_h)\alpha_h^k = (M_h - \tfrac{1}{2} A_h)\alpha_h^{k-1} + \Delta t\, b_h(t^{k-1/2}),$$

where $\alpha_h^k = (\alpha_i^k)_{1 \leq i \leq N_h} \in \mathbb{R}^{N_h}$ is the unknown vector, M_h and A_h are, again, mass and stiffness matrices, respectively, and the right-hand side $b_h(t^{k-1/2}) \in \mathbb{R}^{N_h}$, analogously to the Euler method. One can obtain a stability result analogous to Lemma 9.58 upon using $\chi = U^k + U^{k-1}$ as a test function in (9.67). Indeed, firstly, $\Delta t\, (D_{\Delta t}^- U^k, U^k + U^{k-1})_{L_2(\Omega)} = \|U^k\|_{L_2(\Omega)}^2 - \|U^{k-1}\|_{L_2(\Omega)}^2 = (\|U^k\|_{L_2(\Omega)} - \|U^{k-1}\|_{L_2(\Omega)})(\|U^k\|_{L_2(\Omega)} + \|U^{k-1}\|_{L_2(\Omega)})$. It now follows from (9.67) that

$$\frac{1}{\Delta t}(\|U^k\|_{L_2(\Omega)} - \|U^{k-1}\|_{L_2(\Omega)})(\|U^k\|_{L_2(\Omega)} + \|U^{k-1}\|_{L_2(\Omega)}) \leq$$

$$\leq \frac{1}{\Delta t}(\|U^k\|_{L_2(\Omega)} - \|U^{k-1}\|_{L_2(\Omega)})(\|U^k\|_{L_2(\Omega)} + \|U^{k-1}\|_{L_2(\Omega)})$$

$$+ \tfrac{1}{2}a(U^k + U^{k-1}, U^k + U^{k-1})$$
$$= (f(t^{k-1/2}), U^k + U^{k-1})_{L_2(\Omega)} \leq \|f(t^{k-1/2})\|_{L_2(\Omega)} \|U^k + U^{k-1}\|_{L_2(\Omega)}$$
$$\leq \|f(t^{k-1/2})\|_{L_2(\Omega)}(\|U^k\|_{L_2(\Omega)} + \|U^{k-1}\|_{L_2(\Omega)}),$$

that is, $\frac{1}{\Delta t}(\|U^k\|_{L_2(\Omega)} - \|U^{k-1}\|_{L_2(\Omega)}) \leq \|f(t^{k-1/2})\|_{L_2(\Omega)}$. This is equivalent to the inequality $\|U^k\|_{L_2(\Omega)} \leq \|U^{k-1}\|_{L_2(\Omega)} + \Delta t \|f(t^{k-1/2})\|_{L_2(\Omega)}$, and so

$$\|U^k\|_{L_2(\Omega)} \leq \|U^0\|_{L_2(\Omega)} + \Delta t \sum_{j=1}^{k} \|f(t^{j-1/2})\|_{L_2(\Omega)}.$$

The following error estimate can be proved analogously to Theorem 9.59, cf. [46].

Theorem 9.60 Let $u_0 \in H_0^1(\Omega) \cap H^2(\Omega)$ and f be such that the solution u of (9.56) satisfies the condition $u \in H^2((0,T), H^2(\Omega)) \cap H^3((0,T), L_2(\Omega))$. Assume that $u_{0,h} \in V_h$ satisfies (9.64). Then there exists a constant $c > 0$ depending only on Ω such that for all $k = 0, ..., N$

$$\|U^k - u(t^k, \cdot)\|_{L_2(\Omega)} \leq c\, h^2 \left\{ \|u_0\|_{H^2(\Omega)} + \int_0^{t^k} \|\dot{u}\|_{H^2(\Omega)}\, ds \right\}$$

$$+ c\,(\Delta t)^2 \int_0^{t^k} (\|\ddot{u}(s)\|_{L_2(\Omega)} + \|\Delta\ddot{u}(s)\|_{L_2(\Omega)})\, ds.$$

Numerical experiments. We conclude this section with an experiment related to the Crank–Nicolson method. Here we will choose the data of the initial-boundary value problem in such a way that we can specify an exact solution and, correspondingly, determine the right-hand side exactly. This means we can calculate all the errors exactly and plot these against the number of unknowns, N_h (cf. Figure 9.18). Here we are also choosing $\Delta t = h$. The respective linear and quadratic orders, which arise exclusively from the discretization in time, are clearly recognizable.

9.4.3* Error estimates via space/time variational formulations

We showed in Section 8.6* that we can interpret parabolic initial-boundary value problems as variational problems of the form

$$\text{find } u \in \mathbb{V} \text{ such that } b(u, w) = f(w), \qquad \forall w \in \mathbb{W}, \tag{9.69}$$

with distinct trial and test spaces

$$\mathbb{V} := H_{(0)}^1(I,V') \cap L_2(I,V) = \{v \in H^1(I, V') \cap L_2(I, V):\ v(0) = 0\} \text{ and } \mathbb{W} := L_2(I,V).$$

Here $I = (0,T)$ for some $T > 0$ and V, H are given separable Hilbert spaces such that V is continuously and densely imbedded in H. Let $a : V \times V \to \mathbb{R}$ be a continuous, coercive bilinear

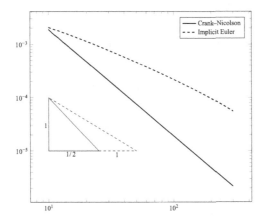

Fig. 9.18 Convergence history for the Crank–Nicolson and implicit Euler methods for the heat equation.

form and let $b : \mathbb{V} \times \mathbb{W} \to \mathbb{R}$ be given by (8.56). We saw in Section 8.6* that this space/time bilinear form is continuous (with constant C) and that (9.69) is well posed. Moreover, in Remark 8.39 it was proved that an inf-sup condition is satisfied for some constant β; both constants are independent of T. Hence we can use the Xu–Zikatanov theorem, Theorem 9.42, to derive an error estimate for a corresponding Petrov–Galerkin approximation. We now introduce a special discretization and show that this is equivalent to the Crank–Nicolson method. This means that we can use the space/time formulation to give another error estimate (which will turn out to be sharper than the one above).

Let $\mathbb{V}_\delta \subset \mathbb{V}$, $\mathbb{W}_\delta \subset \mathbb{W}$ be finite-dimensional spaces and let $u_\delta \in \mathbb{V}_\delta$ be a solution of the discrete problem

$$b(u_\delta, w_\delta) = f(w_\delta), \qquad \forall w_\delta \in \mathbb{W}_\delta, \tag{9.70}$$

where we restrict ourselves to the case $H = L_2(\Omega)$, $V = H_0^1(\Omega)$. Let $\mathbb{V}_\delta := S_{\Delta t} \otimes V_h$, $\mathbb{W}_\delta := Q_{\Delta t} \otimes V_h$,

$$\delta := (\Delta t, h), \tag{9.71}$$

where $S_{\Delta t}$ and V_h are finite element spaces spanned by piecewise linear functions and $Q_{\Delta t}$ a finite element space spanned by piecewise constant functions with respect to triangulations $\mathcal{T}_h^{\text{space}}$ in space and $\mathcal{T}_{\Delta t}^{\text{time}} \equiv \{t^{k-1} \equiv (k-1)\Delta t < t \leq k\,\Delta t \equiv t^k, 1 \leq k \leq K\}$ in time, $\Delta t := T/K$, as appropriate.

We have $S_{\Delta t} = span\,\{\sigma^1, \ldots, \sigma^K\}$, where σ^k is the (interpolating) hat function with respect to the nodes t^{k-1}, t^k and t^{k+1} (truncated to $[0, T]$ when $k = K$), and also $Q_{\Delta t} = span\{\tau^1, \ldots, \tau^K\}$, where $\tau^k = \chi_{I^k}$ is the characteristic function on $I^k := (t^{k-1}, t^k)$. Finally, take $V_h = span\,\{\phi_1, \ldots, \phi_{n_h}\}$ to be the nodal basis with respect to $\mathcal{T}_h^{\text{space}}$. Given functions $v_\delta = \sum_{k=1}^K \sum_{i=1}^{n_h} v_i^k \sigma^k \otimes \phi_i \in \mathbb{V}_\delta$ and $w_\delta = \sum_{\ell=1}^K \sum_{j=1}^{n_h} w_j^\ell \tau^k \otimes \phi_j$ (with coefficients v_i^k and w_j^ℓ, respectively), we obtain

$$b(v_\delta, w_\delta) = \int_I \left\{ \langle \dot{v}_\delta(t), w_\delta(t) \rangle_{V' \times V} + a(v_\delta(t), w_\delta(t)) \right\} dt$$

$$= \sum_{k,\ell=1}^K \sum_{i,j=1}^{n_h} v_k^i w_\ell^j \left\{ (\dot{\sigma}^k, \tau^\ell)_{L_2(I)} (\phi_i, \phi_j)_H + (\sigma^k, \tau^\ell)_{L_2(I)} a(\phi_i, \phi_j) \right\}$$

$$= v_\delta^T B_\delta w_\delta,$$

where

$$B_\delta := N^{\text{time}}_{\Delta t} \otimes M^{\text{space}}_h + M^{\text{time}}_{\Delta t} \otimes A^{\text{space}}_h \qquad (9.72)$$

and $M^{\text{space}}_h := [(\phi_i, \phi_j)_{L_2(\Omega)}]_{i,j=1,\dots,n_h}$, $M^{\text{time}}_{\Delta t} := [(\sigma^k, \tau^\ell)_{L_2(I)}]_{k,\ell=1,\dots,K}$ are the respective mass matrices with respect to time and space, and the stiffness matrices are given by $N^{\text{time}}_{\Delta t} := [(\dot{\sigma}^k, \tau^\ell)_{L_2(I)}]_{k,\ell=1,\dots,K}$ and $A^{\text{space}}_h := [a(\phi_i, \phi_j)]_{i,j=1,\dots,n_h}$. For our special choice of spaces we obtain, denoting by $\delta_{k,\ell}$ the discrete Kronecker delta,

$$(\dot{\sigma}^k, \tau^\ell)_{L_2(I)} = \delta_{k,\ell} - \delta_{k+1,\ell}, \qquad (\sigma^k, \tau^\ell)_{L_2(I)} = \frac{\Delta t}{2}(\delta_{k,\ell} + \delta_{k+1,\ell}),$$

$$b(v_\delta, \tau^\ell \otimes \phi_j) = \sum_{i=1}^{n_h} \left[(v_i^\ell - v_i^{\ell-1})(\phi_i, \phi_j)_H + \frac{\Delta t}{2}(v_i^\ell + v_i^{\ell-1}) a(\phi_i, \phi_j) \right]$$

$$= \Delta t \left[M^{\text{space}}_h \frac{1}{\Delta t}(v^\ell - v^{\ell-1}) + A^{\text{space}}_h v^{\ell-1/2} \right],$$

where $v^\ell := (v_i^\ell)_{i=1,\dots,n_h}$, $v_i^{\ell-1/2} := \frac{1}{2}(v_i^\ell + v_i^{\ell-1})$, and correspondingly for $v^{\ell-1/2}$. Now we apply the trapezoidal rule to approximate the integral with respect to the time variable on the right-hand side,

$$f(\tau^\ell \otimes \phi_j) = \int_0^T \langle f(t), \tau^\ell \otimes \phi_j \rangle_{V' \times V} \, dt$$

$$\approx \frac{\Delta t}{2} \langle f(t^{\ell-1}) + f(t^\ell), \phi_j \rangle_{V' \times V} = \frac{\Delta t}{2}(f^{\ell-1} + f^\ell)_j = \Delta t \, f_j^{\ell-1/2},$$

where $f^\ell := ((\langle f(t^\ell), \phi_j \rangle_{V' \times V})_{j=1,\dots,n_h}$. Now we can write (9.70) as

$$\frac{1}{\Delta t} M^{\text{space}}_h (v^\ell - v^{\ell-1}) + A^{\text{space}}_h v^{\ell-1/2} = f^{\ell-1/2}, \qquad v^0 := 0, \qquad (9.73)$$

which is exactly the Crank–Nicolson method introduced above. We can thus use the space/time variational formulation to derive error estimates for the Crank–Nicolson method.

Remark 9.61 We can also prove the inf-sup condition. To this end we consider a slightly modified norm: given $v \in V$, we set $\bar{v}^k := (\Delta t)^{-1} \int_{I^k} v(t) \, dt \in V$ and $\bar{v} := \sum_{k=1}^K \chi_{I^k} \otimes \bar{v}^i \in L_2(I, V)$, as well as $\|v\|^2_{V,\delta} := \|\dot{v}\|^2_{L_2(I,V')} + \|\bar{v}\|^2_{L_2(I,V)} + \|v(T)\|^2_H$. This averaging in time is in fact the "natural" norm for the analysis of the Crank–Nicolson method. For the inf-sup and continuity constants (recall the definition of δ in (9.71))

$$\beta_\delta := \inf_{v_\delta \in \mathbb{V}_\delta} \sup_{w_\delta \in \mathbb{W}_\delta} \frac{b(v_\delta, w_\delta)}{\|v_\delta\|_{V,\delta} \|w_\delta\|_{\mathbb{W}}}, \qquad \gamma_\delta := \sup_{v_\delta \in \mathbb{V}_\delta} \sup_{w_\delta \in \mathbb{W}_\delta} \frac{b(v_\delta, w_\delta)}{\|v_\delta\|_{V,\delta} \|w_\delta\|_{\mathbb{W}}},$$

we have $\beta_\delta = \gamma_\delta = 1$, as long as $a(\cdot, \cdot)$ is assumed to be symmetric. We use the energy norm $\|\phi\|^2_V := a(\phi, \phi)$, $\phi \in V$, [60]. A comprehensive treatment of stable space/time discretizations for parabolic problems can be found in [4]. $\triangle$

We now come to the error estimate announced above. For this, we suppose Ω to be convex. We recall that $\mathbb{V}_\delta = S_{\Delta t} \otimes V_h$ is a tensor product, and that both $S_{\Delta t}$ and V_h are finite element spaces. Both admit corresponding Ritz projections as in (9.61); we shall call these $R_{\Delta t} : H^1_{(0)}(I) \to S_{\Delta t}$ and $R_h : H^1_0(\Omega) \to V_h$, respectively. Then by Theorem 9.53,

$$\|R_{\Delta t}\sigma - \sigma\|_{L_2(I)} \le C^2 (\Delta t)^2 \|\sigma\|_{H^2(I)}, \qquad \sigma \in H^1_{(0)}(I) \cap H^2(I), \qquad (9.74a)$$

$$\|R_h \phi - \phi\|_{L_2(\Omega)} \le C^2 h^2 \|\phi\|_{H^2(\Omega)}, \qquad \phi \in H^1_0(\Omega) \cap H^2(\Omega). \qquad (9.74b)$$

We now define the space/time Ritz projection $\mathbb{R}_\delta : V \to V_\delta$ by $\mathbb{R}_\delta := R_{\Delta t} \otimes R_h$. Then we have, for all $v \in V$,

$$\mathbb{R}_\delta v - v = (R_{\Delta t} \otimes R_h)v - (I \otimes R_h)v + (I \otimes R_h)v - v$$
$$= [(R_{\Delta t} - I) \otimes R_h]v + [I \otimes (R_h - I)]v,$$

and we obtain the estimates $\|\mathbb{R}_\delta v - v\|_{L_2(I,L_2(\Omega))} \leq C((\Delta t)^2 + h^2) \|v\|_{H^2(I,H^2(\Omega))}$ for all $v \in V \cap H^2(I,H^2(\Omega))$. Upon applying Theorems 9.42 and 9.45 (Xu–Zikatanov and the Aubin–Nitsche lemma, respectively), we obtain the following result.

Theorem 9.62 *Let u_0 and f be such that the solution u of (9.56) satisfies $u \in H^2(I,H^2(\Omega))$. There exists a constant $c > 0$ depending only on Ω such that*

$$\|u - u_\delta\|_{L_2(I,L_2(\Omega))} \leq c((\Delta t)^2 + h^2) \|u\|_{H^2(I,H^2(\Omega))},$$
$$\|u - u_\delta\|_V \leq c(\Delta t + h) \|u\|_{H^2(I,H^2(\Omega))}$$

(with $\delta = (\Delta t, h)$ as in (9.71)).

This estimate is better than the one of Theorem 9.60 in at least two respects. Firstly, it is stated in terms of the "natural" norm, that is, the norm which matches the statement of the problem; secondly, we obtain the same order under weaker regularity assumptions. Extensive numerical experiments concerning the required regularity as well as the efficiency of space-time discretizations compared with standard time-marching schemes are described in [37].

9.5 The wave equation

In this section we will describe elementary numerical methods for the solution of the wave equation; we refer to Section 8.4 for the corresponding theory. As we did for the heat equation, we will apply a semi-discretization utilizing the vertical method of lines (Figure 9.17), which will lead to a system of initial value problems for ordinary differential equations. We will then discretize the second time derivative using the central difference quotients $D_{\Delta t}^2 := D_{\Delta t}^- D_{\Delta t}^+$ with respect to an equidistant time discretization of mesh size $\Delta t = \frac{T}{N}$ and nodes $t^k := k \Delta t$, $k = 0, \ldots, N$ in the time interval $[0, T]$. For the space discretization we will consider finite differences and finite elements, as above.

9.5.1 Finite differences

To keep things simple, we will consider the case of one space dimension, that is, the initial-boundary value problem

$$u_{tt} - c^2 u_{xx} = f \qquad \text{in } (0, T) \times (0, 1), \qquad (9.75a)$$
$$u(t, 0) = u(t, 1) = 0 \qquad \text{for all } t \in (0, T), \qquad (9.75b)$$
$$u(0, x) = u_0(x), u_t(0, x) = u_1(x) \qquad \text{for all } x \in (0, 1), \qquad (9.75c)$$

for given continuous functions $u_0, u_1 : [0, 1] \to \mathbb{R}$ and $f : [0, T] \times [0, 1] \to \mathbb{R}$ (we assume in particular that $u_0(0) = u_0(1) = 0$). Uniqueness was established in Theorem 3.6 (page 54); that is, there exists at most one solution $u \in C^2([0, T] \times [0, 1])$ of (9.75). In the subsequent convergence analysis of various numerical schemes we sometimes require higher regularity of the solution u of (9.75), which can often be ensured by assuming appropriate properties of the data u_0, u_1 and f.

An explicit scheme: The leapfrog method

If we also approximate the second derivative in the space variable using the central difference quotient at time t^{k-1}, on an equidistant mesh of size $h := \frac{1}{M}$, $M \in \mathbb{N}$, then we obtain the discrete scheme

$$(\Delta t)^{-2}(U_i^k - 2U_i^{k-1} + U_i^{k-2}) - c^2 h^{-2}(U_{i-1}^{k-1} - 2U_i^{k-1} + U_{i+1}^{k-1}) = f_i^{k-1} \qquad (9.76)$$

(where $f_i^k := f(t^k, x_i)$) for determining an approximation U_i^k to $u_i^k := u(t^k, x_i)$. We still need the initial conditions for $k = 0$ and $k = 1$. A canonical choice for the initial value is $U_i^0 := u_0(x_i)$. Since we are using central difference quotients of second order, we also require an approximation of second order to u_i^1. This can be obtained via a Taylor expansion

$$u(t^1, x_i) = u(0, x_i) + \Delta t\, u_t(0, x_i) + \frac{(\Delta t)^2}{2} u_{tt}(0, x_i) + O((\Delta t)^3)$$

$$= u_0(x_i) + \Delta t\, u_1(x_i) + \frac{(\Delta t)^2}{2} \left(c^2 u_{xx}(0, x_i) + f(0, x_i)\right) + O((\Delta t)^3);$$

then, setting

$$\lambda := \frac{\Delta t}{h},$$

we have the desired $O((\Delta t)^2)$-approximation given by

$$U_i^1 := u_0(x_i) + \Delta t\, u_1(x_i) + \frac{c^2}{2} \lambda^2 \{u_0(x_{i-1}) - 2u_0(x_i) + u_0(x_{i+1})\}$$

$$+ \frac{(\Delta t)^2}{2} f_i^0. \qquad (9.77)$$

As for elliptic and parabolic problems, here we have defined the *local discretization error* by

$$\tau_i^k := f(t^k, x_i) - D_{\Delta t}^2 u_i^k + c^2 D_h^2 u_i^k, \qquad \tau^k := (\tau_i^k)_{i=1,\dots,M-1}. \qquad (9.78)$$

Theorem 9.63 (Consistency) *Suppose the CFL condition* $\lambda \leq c^{-1}$ *holds, and let* $u \in C^{4,4}([0, T] \times [0, 1])$ *be the unique solution of* (9.75) *and* U *the solution of* (9.76) *with* (9.77) *and* $U_i^0 := u_0(x_i)$. *Then* $\|\tau^k\|_{h,\infty} \leq \frac{c^2}{12} h^2 |u|_{C^4}$.

Proof By assumption,

$$
\begin{aligned}
\tau_i^k &= (\ddot{u}(t^k, x_i) - c^2 u_{xx}(t^k, x_i)) - (D_{\Delta t}^2 u_i^k - c^2 D_h^2 u_i^k) \\
&= (\ddot{u}(t^k, x_i) - D_{\Delta t}^2 u_i^k) - c^2 (u_{xx}(t^k, x_i) - D_h^2 u_i^k) \\
&= \frac{(\Delta t)^2}{12} u_{tttt}(\sigma^k, x_i) - c^2 \frac{h^2}{12} u_{xxxx}(t^k, \xi_i)
\end{aligned}
$$

for some points $\sigma^k \in (t^{k-1}, t^{k+1})$ and $\xi_i \in (x_{i-1}, x_{i+1})$ (cf. the proof of Theorem 9.49). Since $u_{tttt} = c^2 u_{xxtt} = c^4 u_{xxxx}$, we also have that

$$
\begin{aligned}
|\tau_i^k| &\leq \frac{c^2}{12} |c^2 (\Delta t)^2 - h^2| \max_{(t,x) \in (t^{k-1}, t^{k+1}) \times (x_{i-1}, x_{i+1})} |u_{xxxx}(t, x)| \\
&\leq \frac{c^2}{12} |1 - c\lambda| h^2 |u|_{C^4}.
\end{aligned}
$$

The claim now follows immediately, since $1 - c\lambda \leq 1$ if $\lambda \leq c^{-1}$.   □

Note that $\tau_i^k = 0$ in the case $\lambda = c^{-1}$, meaning that the method is exact in this case. The name "leapfrog method" is derived from the image depicted in Figure 9.19 (left). In the literature the name is also used to describe central difference methods in space and time for first order problems. In this case, the central point (t^k, x_i) would be missing from Figure 9.19 (left).

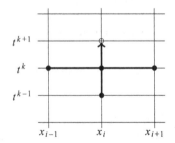 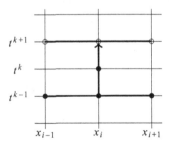

Fig. 9.19 Left: an explanation for the name "leapfrog" method. The value at (t^{k+1}, x_i) is determined from the other four. Right: difference stencil for the implicit method with $\theta = 1/2$.

We shall now prove that the leapfrog method converges; for this we will impose a number of assumptions on the data. To simplify the presentation we will use the same notation as in Section 9.1.1. Setting $U^k := (U_i^k)_{i=1,\ldots,M-1}$, we denote the difference operator in the space variable by (see Definition 9.3)

$$c^2 (L_h U^k)_i = -\left(\frac{c}{h}\right)^2 (U^k_{i-1} - 2U^k_i + U^k_{i+1}), \qquad i = 1, \ldots, M-1.$$

Then (9.76) reads $(\Delta t)^{-2}(U^k - 2U^{k-1} + U^{k-2}) + c^2 L_h U^{k-1} = f^k$, where $f^k :=$ $(f^k_i)_{i=1,\ldots,M-1}$. Setting $\mathcal{U}^{\Delta t} := (U^k)_{k=0,\ldots,N}$, the leapfrog method (9.76) reads

$$(\mathcal{L}^{\Delta t} \mathcal{U}^{\Delta t})_k := U^k - (2I - (\Delta t)^2 L_h)U^{k-1} + U^{k-2} = (\Delta t)^2 f^k. \qquad (9.79)$$

Recall from Lemma 9.6 that L_h is bounded, symmetric and positive definite with respect to the discrete inner product $(\cdot, \cdot)_h$ of Definition 9.5. We will perform the stability analysis with respect to the (discrete) L_2-norm in time, i.e.,

$$\|\mathcal{U}^{\Delta t}\|^2_{L_2} := \sum_{k=0}^{N} \|U^k\|^2_h.$$

Theorem 9.64 (Stability) *Suppose that the CFL condition $\lambda = \frac{\Delta t}{h} \le c^{-1}$ is satisfied. Then the leapfrog method is L_2-stable for the homogeneous problem $(f \equiv 0)$ in the following sense: if $\mathcal{L}^{\Delta t} \mathcal{U}^{\Delta t} = 0$, then there exists a constant $C > 0$ independent of Δt such that $\|\mathcal{U}^{\Delta t}\|_{L_2} \le C(\|U^0\|_h + \|U^1\|_h)$.*

Proof By Lemma 9.6, there exists an orthonormal basis $\{e^1, \ldots, e^{M-1}\}$ of eigenvectors of L_h corresponding to the positive eigenvalues $\mu_i = \frac{4}{h^2} \sin^2(\frac{i\pi h}{2})$, $i = 1, \ldots, M-1$. We may expand each U^k in this basis, i.e., $U^k = \sum_{i=1}^{M-1} \alpha^k_i e^i$, so that $\mathcal{L}^{\Delta t} \mathcal{U}^{\Delta t} = 0$ can be rewritten as

$$\alpha^k_i - (2 - c^2(\Delta t)^2 \mu_i)\alpha^{k-1}_i + \alpha^{k-2}_i = 0, \quad k = 2, \ldots, N, \; i = 1, \ldots, M-1.$$

For every i, this is a second order linear difference equation complemented by the initial conditions $\alpha^0_i = \alpha_{i,0}$ and $\alpha^1_i = \alpha_{i,1}$. The usual *ansatz* to solve such equations is $\alpha^k_i = c_{i,1}(\zeta_{1,i})^k + c_{i,2}(\zeta_{2,i})^k$ for $\zeta_{i,1/2}$ and $c_{i,1/2}$ to be determined. This leads to the characteristic equation $\varrho_i(\zeta) := \zeta^2 - \sigma_i \zeta + 1 = 0$, where $\sigma_i := 2 - c^2(\Delta t)^2 \mu_i$, with roots $\zeta_{i,1/2} = \frac{\sigma_i}{2} \pm (\frac{\sigma_i^2}{4} - 1)^{1/2}$. Finally, $c_{i,1}, c_{i,2} \in \mathbb{R}$ are determined in such a way as to satisfy the initial conditions, namely

$$c_{i,1} = \frac{\alpha^1_i - \alpha^0_i \zeta_{i,2}}{\zeta_{i,1} - \zeta_{i,2}}, \quad c_{i,2} = \frac{\alpha^0_i \zeta_{i,1} - \alpha^1_i}{\zeta_{i,1} - \zeta_{i,2}}.$$

Moreover, it is readily seen that $c^2_{i,1} + c^2_{i,2} \le C_0\{(\alpha^0_i)^2 + (\alpha^1_i)^2\}$ for some constant $C_0 > 0$. This leads to

$$\|\mathcal{L}^{\Delta t} \mathcal{U}^{\Delta t}\|^2_{L_2} = \sum_{k=0}^{N} \|U^k\|^2_h = \sum_{k=0}^{N} \left\| \sum_{i=1}^{M-1} \alpha^k_i e^i \right\|^2_h = \sum_{k=0}^{N} \sum_{i=1}^{M-1} |\alpha^k_i|^2$$

$$= \sum_{k=0}^{N} \sum_{i=1}^{M-1} |c_{i,1} \zeta_{i,1}^k + c_{i,2} \zeta_{i,2}^k|^2$$

$$\leq 2 \sum_{k=0}^{N} \sum_{i=1}^{M-1} \left(c_{i,1}^2 |\zeta_{i,1}|^{2k} + c_{i,2}^2 |\zeta_{i,2}^k|^{2k} \right).$$

This term is bounded as $N \to \infty$ ($\Delta t \to 0$) if and only if $|\zeta_{i,1/2}| \leq 1$, which holds if $|\sigma_i| \leq 2$, i.e., $|2 - c^2 (\Delta t)^2 \mu_i| \leq 2$. By Lemma 9.6 (e), we have $0 < \mu_i \leq \frac{4}{h^2}$, so that the CFL condition in fact ensures that $|\zeta_{i,1/2}| \leq 1$. This implies that there is a constant $\hat{C} > 0$ independent of N such that

$$\|\mathcal{L}^{\Delta t} \mathcal{U}^{\Delta t}\|_{L_2}^2 \leq 2 \hat{C} \sum_{i=1}^{M-1} \left(c_{i,1}^2 + c_{i,2}^2 \right) \leq 2 \hat{C} C_0 \sum_{i=1}^{M-1} \left((\alpha_i^0)^2 + (\alpha_i^1)^2 \right)$$

$$= 2 \hat{C} C_0 (\|U^0\|_h^2 + \|U^1\|_h^2),$$

as claimed. □

We can now formulate the convergence theorem announced earlier. In order to do so, we introduce some further useful notation. We set $\mathcal{F}^{\Delta t} := ((\Delta t)^2 f^k)_{k=0,\dots,N}$, $u^k := (u(t^k, x_i))_{i=1,\dots,M-1}$, $\mathcal{U}^{\Delta t} := (u^k)_{k=0,\dots,N}$, $\tau^k := (\tau_i^k)_{i=1,\dots,M-1}$ and $\mathcal{T}^{\Delta t} := (\tau^k)_{k=0,\dots,N}$. Then $\mathcal{L}^{\Delta t} \mathcal{U}^{\Delta t} = \mathcal{F}^{\Delta t}$ and $\mathcal{T}^{\Delta t} = \mathcal{F}^{\Delta t} - \mathcal{L}^{\Delta t} \mathcal{U}^{\Delta t}$.

Theorem 9.65 (Convergence) *Suppose that the CFL condition* $\lambda = \frac{\Delta t}{h} \leq c^{-1}$ *holds, and let* $u \in C^{4,4}([0, T] \times [0, 1])$ *be the unique solution of* (9.75) *and* U *the unique solution of* (9.76) *with* (9.77) *and* $U_i^0 := u_0(x_i)$. *Then the leapfrog method converges; more precisely, there exists a constant* $C(u, f) > 0$ *such that*

$$\max_{k=0,\dots,N} \|U^k - u^k\|_h \leq C(u, f)((\Delta t)^2 + h^2).$$

Proof Given $u^k := (u(t^k, x_i))_{i=1,\dots,M-1}$, we set $E^k := U^k - u^k$ and $\mathcal{E}^{\Delta t} := (E^k)_{k=0,\dots,N}$. Then $E^0 = 0$ and we get

$$E^1 = U^1 - u^1 = u_0 + \Delta t \, u^1 + \frac{c^2}{2} \lambda^2 L_h u_0 + \frac{(\Delta t)^2}{2} f^0$$

$$- u_0 - \Delta t \, u^1 - \frac{(\Delta t)^2}{2} (c^2 u_{xx}^0 + f^0) + O((\Delta t)^3)$$

$$= O((\Delta t)^2 + h^2) + O((\Delta t)^3).$$

Next, $\mathcal{L}^{\Delta t} \mathcal{E}^{\Delta t} = \mathcal{L}^{\Delta t} \mathcal{U}^{\Delta t} - \mathcal{L}^{\Delta t} \mathcal{U}^{\Delta t} = \mathcal{T}^{\Delta t}$, where $\|\mathcal{T}^{\Delta t}\|_{L_2} = O((\Delta t)^2 + h^2)$ by Theorem 9.63 under the given regularity assumptions. Theorem 9.64 now yields $\|\mathcal{E}^{\Delta t}\|_{L_2} = O((\Delta t)^2 + h^2)(\|E^0\|_h + \|E^1\|_h)$, where the relevant constant only depends on u and f. □

Theorem 9.65 shows that the convergence is in fact a corollary of stability and consistency. This principle is also considered one of the fundamental theorems in the analysis of finite differences for the numerics of (ordinary) differential equations. The theorem is valid in much greater generality than what was stated here; for example, often convergence on the one hand and stability and consistency on the other are actually *equivalent*; and goes by the name *Lax–Richtmyer theorem*.

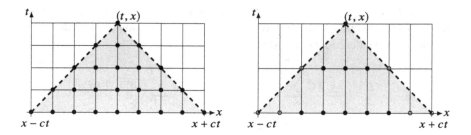

Fig. 9.20 Numerical domain of dependence for the wave equation: on the left the CFL condition is satisfied and the domain of dependence is contained in the numerical domain of dependence. On the right the step size is too large: the mesh points ∘ are in the domain of dependence but are not used to approximate $u(t, x)$; rather, only the • points are.

In Figure 9.20 we elucidate the meaning of the CFL condition. This condition states that the domain of dependence of the wave equation (cf. Figure 3.1 on page 52) must contain all the mesh points which are required to calculate the approximation at a point (t, x) (this is also called the *numerical domain of dependence*). On the left-hand side the CFL condition is satisfied, on the right-hand side it is not.

An implicit method

Just as with the heat equation, the CFL condition represents a restriction on the possible choice of step size for space and time, even if it is less restrictive for the wave equation than for the heat equation. We recall that implicit methods (such as the implicit Euler method and the Crank–Nicolson method) for the heat equation, are stable with respect to the choice of step size; we say they are *unconditionally stable*. We shall describe a very common such implicit scheme for the wave equation. For this, we take $\theta \in [0, \frac{1}{2}]$ to be a parameter which we shall specify more precisely later, and define

$$\tilde{f}_\theta^k := (1 - 2\theta)f^{k-1} + \theta(f^k + f^{k-2}),$$

so that

$$D_{\Delta t}^2 U^k + \theta L_h(U^k + U^{k-2}) + (1 - 2\theta)L_h U^{k-1} = \tilde{f}_\theta^k. \tag{9.80}$$

When $\theta = 0$, this obviously corresponds to the explicit leapfrog method considered above. Here, we define the *local discretization error*, analogously to the explicit case, by

$$\tau_\theta^k := \tilde{f}_\theta^k - D_{\Delta t}^+ u^k - \theta L_h(u^k + u^{k-2}) - (1 - 2\theta)L_h u^{k-1}. \qquad (9.81)$$

Theorem 9.66 (Consistency) *Let $u \in C^{4,4}([0, T] \times [0, 1])$ be the unique solution of (9.75) and U the solution of (9.80) with U_i^1 given by (9.77) and $U_i^0 := u_0(x_i)$. Then $\|\tau_\theta^k\|_{h,\infty} = O((\Delta t)^2 + h^2)$ for all $\theta \in [0, \frac{1}{2}]$.*

Proof Firstly, a Taylor expansion shows that $L_h u^k - f^k = -c^2 u_{xx}^k + O(h^2) - f^k = -\ddot{u}^k + O(h^2)$. Hence

$$\begin{aligned}
\tau_\theta^k &= \theta(f^k + f^{k-2}) + (1 - 2\theta)f^{k-1} \\
&\quad - D_{\Delta t}^+ u^k - \theta L_h(u^k + u^{k-2}) - (1 - 2\theta)L_h u^{k-1} \\
&= (1 - 2\theta)\ddot{u}^{k-1} + O(h^2) - \ddot{u}^{k-1} + O((\Delta t)^2) + \theta(\ddot{u}^k + \ddot{u}^{k-2}) \\
&= \theta\,(\ddot{u}^k + \ddot{u}^{k-2} - 2\ddot{u}^{k-1}) + O((\Delta t)^2 + h)^2) \\
&= \theta\,|u|_{C^4}\, O((\Delta t)^2) + O((\Delta t)^2 + h)^2) = O((\Delta t)^2 + h)^2),
\end{aligned}$$

whence the claim.   □

The proof of stability is similar to the one in the explicit case.

Theorem 9.67 (Stability) *The implicit method (9.80) is L_2-stable for the homogeneous problem in the sense of Theorem 9.64*
(a) for all $1/4 \le \theta \le 1/2$;
(b) for all $0 \le \theta < 1/4$, if the CFL condition $\lambda = \frac{\Delta t}{h} \le (c\sqrt{1 - 4\theta})^{-1}$ holds.
That is, in both of these cases, for the solution $\mathcal{U}^{\Delta t}$ of the homogeneous problem, there exists a constant $C > 0$ independent of Δt such that $\|\mathcal{U}^{\Delta t}\|_{L_2} \le C(\|U^0\|_h + \|U^1\|_h)$.

Proof We follow the lines of the proof of Theorem 9.64 and arrive at the difference equation $\alpha_i^k - \sigma_i \alpha_i^{k-1} - \alpha_i^{k-2} = 0$, where now $\sigma_i := \frac{2-(\Delta t)^2(1-2\theta)\mu_i}{1+(\Delta t)^2\theta\mu_i}$, for $k = 2, ..., N$, $i = 1, ..., M - 1$. As in the proof of Theorem 9.64 we need to show that $|\sigma_i| \le 2$. The condition $\sigma_i \le 2$ is equivalent to $2 - (\Delta t)^2(1 - 2\theta)\mu_i \le 2 + 2(\Delta t)^2\theta\mu_i$, that is, $0 \le (\Delta t)^2\mu_i(2\theta + 1 - 2\theta) = (\Delta t)^2\mu_i$, which is always satisfied since L_h is positive definite. For the mirror condition $\sigma_i \ge -2$ we have to distinguish between the two cases (a) and (b).

(a) $0 \le \theta < 1/4$: $\sigma_i \ge -2$ means that $2 - (\Delta t)^2(1 - 2\theta)\mu_i \ge -2 - 2(\Delta t)^2\theta\mu_i$, that is, $4 \ge (\Delta t)^2\mu_i(1 - 4\theta)$. But since $\mu_i \in (0, 4c^2 h^{-2}]$ by Lemma 9.6 (e), this condition is satisfied if $4 \ge (1 - 4\theta)4c^2\lambda^2$, i.e. if $\lambda \le (c\sqrt{1 - 4\theta})^{-1}$.

(b) $1/4 \le \theta \le 1/2$: as above, if $\sigma_i \ge -2$, then $4 \ge (\Delta t)^2\mu_i(1 - 4\theta)$. But since $\theta \ge 1/4$, we have $1 - 4\theta \le 0$, so that the condition is always satisfied.   □

Remark 9.68 Note that in the explicit case $\theta = 0$ we obtain exactly the same CFL condition as in the above analysis. For $\theta \geq 1/4$ the implicit method is *unconditionally stable*. When $\theta = 1/4$ we obtain the Crank–Nicolson method.   △

We now obtain the following statement, whose proof is completely analogous to the one in the explicit case, see Theorem 9.65. This is in the spirit of the Lax–Richtmyer theorem since we can again obtain the convergence of the method from the relevant statements on consistency and stability.

Theorem 9.69 (Convergence) *Let* $u \in C^{4,4}([0,T] \times [0,1])$ *for the solution of (9.75). Let* U *be the solution of (9.80) with (9.77) and* $U_i^0 := u_0(x_i)$. *Then the method converges*
(a) for all $1/4 \leq \theta \leq 1/2$;
(b) for all $0 \leq \theta < 1/4$, *if the condition* $\lambda \leq (c\sqrt{1 - 4\theta})^{-1}$ *holds.*
That is, in both cases there exists a constant $C(u, f) > 0$ *such that*

$$\max_{k=0,\ldots,N} \|U^k - u^k\|_h \leq C(u, f)((\Delta t)^2 + h^2).$$

Numerical experiments

We shall now describe a numerical experiment related to the stability of the time discretization. We consider a one-dimensional example with a smooth solution in form of a wave, cf. Figure 9.21. We choose $h = \frac{1}{50}$ and $\Delta t = \frac{1}{100}$, that is, $\lambda = 0.5$. For the special choice $c = 2.025$ the CFL condition is (just) still satisfied, as one can see from the left column of Figure 9.21. The errors start to oscillate at the end of the time interval, but they are still within the tolerance specified by the step size. If one should alter these numbers, even only slightly, then the error would quickly grow beyond all bounds. However, in the case of the implicit method, things are much more stable. Even with $\theta = 0.01$ the error is significantly smoother; the condition $\lambda \leq (c\sqrt{1 - 4\theta})^{-1}$ of Theorem 9.67 is satisfied here. In this case, however, the error can still rapidly become large if one alters the data. This is to be contrasted with what happens in the unconditionally stable case $\theta = 0.5$, which is very robust to changes in the data. We thus see that the investigation of stability undertaken above yields a sharp result, in the sense that even in the case of a small violation of the conditions numerical instabilities emerge.

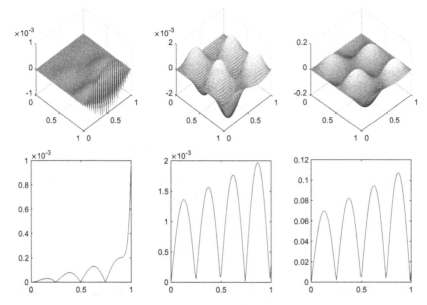

Fig. 9.21 The wave equation via finite differences. Left column: the explicit method ($\theta = 0$, leapfrog), middle column: $\theta = 0.01$, right column: $\theta = 0.5$, $h = \frac{1}{50}$, $\Delta t = \frac{1}{100}$. Upper row: error function in the space and time variables, lower row: maximal error in the space variable per timestep (note the differing scales on the y-axis).

Upon considering the error plots more carefully one can see that the errors in the two implicit methods are *quantitatively* larger than in the explicit case – but still within the predicted tolerance. This may be explained by the additional error introduced when solving the linear system of equations in the implicit case.

9.5.2 Finite elements

Next we will describe Galerkin methods using finite elements for the discretization of the space variables in the case of the wave equation on $\Omega \subset \mathbb{R}^d$, i.e.,

$$\ddot{u}(t) - c^2 \Delta u(t) = f(t) \qquad \text{in } (0, T) \times \Omega, \tag{9.82a}$$

$$u(t) \in H_0^1(\Omega) \cap H^2(\Omega) \qquad \text{for all } t \in (0, T), \tag{9.82b}$$

$$u(0) = u_0, u_t(0) = u_1, \tag{9.82c}$$

for given $u_0 \in H^2(\Omega) \cap H_0^1(\Omega)$, $u_1 \in H_0^1(\Omega)$ and $f \in C([0, T], H_0^1(\Omega))$. In this section, we shall always assume that $\Omega \subset \mathbb{R}^2$ is a convex polygon and $0 < T < \infty$. We know from Theorem 8.22 that there exists a unique solution $u \in C^2([0, T], L_2(\Omega))$ of (9.82), which we call the *exact solution*.

The first part of the following numerical analysis is based in part on [46, §13] and [34, Ch. 5]. We will use the notation from Section 9.4.2. The spatial semi-discrete analog of (9.58) for the wave equation reads

$$u_h(0) = u_{0,h}, \quad \dot{u}_h(0) = u_{1,h}, \tag{9.83a}$$

$$(\ddot{u}_h(t), \chi)_{L_2(\Omega)} + a(u_h(t), \chi) = (f(t), \chi)_{L_2(\Omega)}, \qquad \chi \in V_h, \ t > 0, \tag{9.83b}$$

where $u_{0,h}, u_{1,h} \in V_h$ are suitable approximations of the initial conditions u_0 and u_1, respectively. Here, we restrict ourselves to spaces V_h generated by linear Lagrange elements on a family $\{\mathcal{T}_h\}_{h>0}$ of admissible and quasi-uniform triangulations. The matrix representation reads

$$M_h \ddot{\alpha}_h(t) + A_h \alpha_h(t) = b_h(t), \ t > 0, \quad \alpha_h(0) = \gamma_0, \ \dot{\alpha}_h(0) = \gamma_1 \tag{9.84}$$

for the right-hand side b_h, where M_h and A_h are, respectively, the mass and stiffness matrices, and we have the finite element representation of the initial values

$$u_{\ell,h} = \sum_{i=1}^{N_h} \gamma_{\ell,i} \varphi_i^h, \qquad \gamma_\ell = (\gamma_{\ell,i})_{i=1,\dots,N_h}, \quad \ell = 0, 1.$$

Since (9.84) is an initial value problem of a system of second order linear ordinary differential equations with constant coefficients, the classical Picard–Lindelöf theorem yields existence and uniqueness of a solution α_h.

We again use the *Ritz projection* R_h introduced in (9.61) and recall the error estimate of Theorem 9.53; this will give us the error estimate in the semi-discrete case. For this, we use the following discrete weighted norm,

$$\|\varphi\|_{1,h} := \sqrt{\|\varphi\|_{L_2(\Omega)}^2 + h^2 |\varphi|_{H^1(\Omega)}^2}, \qquad \varphi \in H^1(\Omega),$$

and restrict ourselves to the case $a(\varphi, \psi) = (\nabla\varphi, \nabla\psi)_{L_2(\Omega)}$, that is, $a(\varphi) = |\varphi|_{H^1(\Omega)}^2$, see (4.5), page 129.

Theorem 9.70 *Let* $\{\mathcal{T}_h\}_{0<h\leq h_0}$ *for some fixed* $h_0 > 0$ *be a family of admissible triangulations of a convex polygon* $\Omega \subset \mathbb{R}^2$ *and let* u *be the exact solution of* (9.82) *such that* $u \in C^2([0,T], H^2(\Omega) \cap H_0^1(\Omega))$. *If* $u_{0,h}, u_{1,h} \in V_h$, *and* u_h *is the solution of* (9.83), *then*

$$\|u_h(t) - u(t)\|_{1,h} + \|\dot{u}_h(t) - \dot{u}(t)\|_{L_2(\Omega)} \leq$$
$$\leq c \left(|u_{0,h} - R_h u_0|_{H^1(\Omega)} + \|u_{1,h} - R_h u_1\|_{L_2(\Omega)} \right)$$
$$+ c\, e^t\, h^2 \left\{ \|u(t)\|_{H^2(\Omega)} + \|\dot{u}(t)\|_{H^2(\Omega)} + \|\ddot{u}\|_{L_2((0,t),H^2(\Omega))} \right\}$$

for all $t \geq 0$, *where the constant* $c > 0$ *only depends on* Ω *and* κ.

Proof We again use the Ritz projection to rewrite the error $u_h - u$ as $u_h - u = (u_h - R_h u) + (R_h u - u) =: \theta_h + \rho_h$, and start by estimating the second term ρ_h. Let $t \in (0, T)$. By Theorem 9.53 and its proof one has $\|\rho_h(t)\|_{L_2(\Omega)} \leq c\, h^2 \|u(t)\|_{H^2(\Omega)}$; $|\rho_h(t)|_{H^1(\Omega)} \leq c\, h \|u(t)\|_{H^2(\Omega)}$ and $\|\dot{\rho}_h(t)\|_{L_2(\Omega)} \leq c\, h^2 \|\dot{u}(t)\|_{H^2(\Omega)}$. Putting these together, for the second term we thus obtain

$$\|\rho_h(t)\|_{1,h} + \|\dot{\rho}_h(t)\|_{L_2(\Omega)} \le ch^2(\|u(t)\|_{H^2(\Omega)} + \|\dot{u}(t)\|_{H^2(\Omega)}). \tag{9.85}$$

Now, as in the proof of Theorem 9.54, we derive an initial value problem for the first term $\theta_h = u_h - R_h u$. Namely, for all $\chi \in V_h$ we have

$$
\begin{aligned}
(\ddot{\theta}_h(t), \chi)_{L_2(\Omega)} + a(\theta_h(t), \chi) &= \\
&= (\ddot{u}_h(t), \chi)_{L_2(\Omega)} + a(u_h(t), \chi) - (\tfrac{d^2}{dt^2} R_h u(t), \chi)_{L_2(\Omega)} - a(R_h u(t), \chi) \\
&= (f(t), \chi)_{L_2(\Omega)} - (R_h \ddot{u}(t), \chi)_{L_2(\Omega)} - a(u(t), \chi) \\
&= (\ddot{u}(t) - R_h \ddot{u}(t), \chi)_{L_2(\Omega)} = -(\ddot{\rho}_h(t), \chi)_{L_2(\Omega)}.
\end{aligned} \tag{9.86}
$$

We now investigate the influence of the error coming from the inhomogeneous right-hand side and the inhomogeneous initial conditions separately (this is also known as the *superposition principle*): we set $\theta_h = \vartheta_h + \zeta_h$ to be such that

$$
\begin{aligned}
(\ddot{\vartheta}_h(t), \chi)_{L_2(\Omega)} + a(\vartheta_h(t), \chi) &= 0, && \forall \chi \in V_h, t > 0, \\
\vartheta_h(0) = \theta_h(0), \quad \dot{\vartheta}_h(0) &= \dot{\theta}_h(0), \\
(\ddot{\zeta}_h(t), \chi)_{L_2(\Omega)} + a(\zeta_h(t), \chi) &= -(\ddot{\rho}_h(t), \chi)_{L_2(\Omega)}, && \forall \chi \in V_h, t > 0, \\
\zeta_h(0) = 0, \quad \dot{\zeta}_h(0) &= 0.
\end{aligned}
$$

Just as in Theorem 8.16, in the case of a homogeneous right-hand side the solution satisfies conservation of energy, that is,

$$
\begin{aligned}
|\vartheta_h(t)|^2_{H^1(\Omega)} + \|\dot{\vartheta}_h(t)\|^2_{L_2(\Omega)} &= |\theta_h(0)|^2_{H^1(\Omega)} + \|\dot{\theta}_h(0)\|^2_{L_2(\Omega)} \\
&= |u_h(0) - R_h u(0)|^2_{H^1(\Omega)} + \|\dot{u}_h(0) - R_h \dot{u}(0)\|^2_{L_2(\Omega)} \\
&= |u_{0,h} - R_h u_0|^2_{H^1(\Omega)} + \|u_{1,h} - R_h u_1\|^2_{L_2(\Omega)}
\end{aligned} \tag{9.87}
$$

for all $t \ge 0$. Now, by Theorem 6.32 (the Poincaré inequality), the estimate $\|\vartheta_h(t)\|_{L_2(\Omega)} \le c\,|\vartheta_h(t)|_{H^1(\Omega)}$ holds, and we obtain, possibly for a different constant c, that for $0 < h \le 1$,

$$\|\vartheta_h(t)\|_{1,h} + \|\vartheta_h(t)\|_{L_2(\Omega)} \le c(|u_{0,h} - R_h u_0|_{H^1(\Omega)} + \|u_{1,h} - R_h u_1\|_{L_2(\Omega)}). \tag{9.88}$$

It remains to obtain an estimate on ζ_h, i.e., homogeneous initial conditions but an inhomogeneous right-hand side. We use $\chi = \dot{\zeta}_h$ as a test function in the differential equation for ζ_h to obtain

$$
\begin{aligned}
\frac{1}{2} \frac{d}{dt} \left(\|\dot{\zeta}_h\|^2_{L_2(\Omega)} + |\zeta_h(t)|^2_{H^1(\Omega)} \right) &= (\ddot{\zeta}_h(t), \dot{\zeta}_h)_{L_2(\Omega)} + a(\zeta_h(t), \dot{\zeta}_h(t)) \\
&= -(\ddot{\rho}_h(t), \dot{\zeta}_h(t))_{L_2(\Omega)} \le \|\ddot{\rho}_h(t)\|_{L_2(\Omega)} \|\dot{\zeta}_h(t)\|_{L_2(\Omega)} \\
&\le \frac{1}{2} (\|\ddot{\rho}_h(t)\|^2_{L_2(\Omega)} + \|\dot{\zeta}_h(t)\|^2_{L_2(\Omega)}),
\end{aligned}
$$

where we have also used the Cauchy–Schwarz inequality and Young's inequality. We now integrate this inequality over the interval $[0, t]$ and use that $\zeta_h(0) = \dot{\zeta}_h(0) = 0$, to obtain the estimate

$$\|\dot{\zeta}_h(t)\|^2_{L_2(\Omega)} + |\zeta_h(t)|^2_{H^1(\Omega)} \le \int_0^t \|\ddot{\rho}_h(s)\|^2_{L_2(\Omega)}\, ds + \int_0^t \|\dot{\zeta}_h(s)\|^2_{L_2(\Omega)}\, ds.$$

We next apply Gronwall's lemma (Exercise 2.7) to the function y defined as $y(t) := \|\dot{\zeta}_h(t)\|^2_{L_2(\Omega)} + |\zeta_h(t)|^2_{H^1(\Omega)}$ and obtain

$$\|\dot{\zeta}_h(t)\|^2_{L_2(\Omega)} + |\zeta_h(t)|^2_{H^1(\Omega)} \le e^t \int_0^t \|\ddot{\rho}_h(s)\|^2_{L_2(\Omega)}\, ds$$

$$= e^t \int_0^t \|R_h \ddot{u}(s) - \ddot{u}(s)\|^2_{L_2(\Omega)}\, ds$$

$$\le C^2 h^4 e^t \int_0^t \|\ddot{u}(s)\|^2_{H^2(\Omega)}\, ds = C^2 h^4 e^t \|\ddot{u}\|^2_{L_2((0,t), H^2(\Omega))},$$

also using Theorem 9.53. Here, as well, the Poincaré inequality (Theorem 6.32) yields $\|\zeta_h(t)\|_{L_2(\Omega)} \le c\, |\zeta_h(t)|_{H^1(\Omega)}$, and so, again upon taking a larger constant c if necessary,

$$\|\zeta_h(t)\|_{1,h} + \|\zeta_h(t)\|_{L_2(\Omega)} \le c h^2 e^t \|\ddot{u}\|_{L_2((0,t), H^2(\Omega))}. \tag{9.89}$$

The estimates (9.85), (9.88) and (9.89) together yield the claim. $\square$

Now, in order to obtain an estimate on the terms involving the Ritz projection, as was the case for the heat equation we need to impose stronger regularity assumptions on the initial values. Indeed, as in the proof of Theorem 9.54 we have the estimates

$$|u_{0,h} - R_h u_0|_{H^1(\Omega)} \le |u_{0,h} - u_0|_{H^1(\Omega)} + c h^2 \|u_0\|_{H^3(\Omega)},\ u_0 \in H^3(\Omega), \tag{9.90a}$$

$$\|u_{1,h} - R_h u_1\|_{L_2(\Omega)} \le \|u_{1,h} - u_1\|_{L_2(\Omega)} + c h^2 \|u_1\|_{H^2(\Omega)},\ u_1 \in H^2(\Omega), \tag{9.90b}$$

meaning that we obtain quadratic convergence if we take suitable approximations of the initial conditions in the spatial variables, cf. also Section 9.5.2.

It remains to describe the time discretization, usually performed using finite differences; in most textbooks this is only briefly treated, if at all. Our approach is based on [49].

The leapfrog method

As with finite differences, the leapfrog method comes about by using the central difference quotient of second order on an equidistant mesh in the time variable, that is, one seeks an approximation $U^k = \sum_{i=1}^{N_h} \alpha_i^k \varphi_i$ such that for $f^k := f(t^k)$

$$U^0 = u_{0,h}, \quad U^1 = u_{1,h}, \tag{9.91a}$$

$$(D^2_{\Delta t} U^k, \chi)_{L_2(\Omega)} + a(U^k, \chi) = (f^k, \chi)_{L_2(\Omega)}, \quad \chi \in V_h, \ k \geq 1. \tag{9.91b}$$

Here $u_{0,h}$ and $u_{1,h}$ are approximations of u_0 and u_1, respectively, which will be specified later, see Section 9.5.2. In matrix form this method reads

$$M_h \alpha_h^k = (\Delta t)^2 (f_h^{k-1} - A_h \alpha_h^{k-1}) + M_h (2\alpha_h^{k-1} - \alpha_h^{k-2}) \tag{9.92}$$

for $k \geq 2$. We recall that the leapfrog method is explicit in the case of a finite difference discretization in the spatial variables. In (9.92), in each step a linear system of equations with mass matrix M_h needs to be solved; thus the method is actually implicit – however, such systems involving the mass matrix are actually relatively simple to handle thanks to their good conditioning.

In addition to consistency and stability, numerical methods for the wave equation need to satisfy a further condition, namely conservation of energy, which we saw in Theorem 8.16. It would not make sense for a numerical method to violate this fundamental, physically motivated property of the equation. However, it turns out that the notion of "energy" depends on the discretization; this is a consequence of the fact that the terms in the definition of the energy $\mathbb{E}(t) := \|\dot{u}(t)\|^2_{L_2(\Omega)} + a(u(t), u(t))$ need to be discretized.

Definition 9.71 (Energy of the leapfrog method) Let U^k be the iterate of the leapfrog method (9.91) at timestep k. Then the *discrete energy of the leapfrog method* for the wave equation is defined by

$$\mathbb{E}_{LF}^k := \tfrac{1}{2} \|D^+_{\Delta t} U^k\|^2_{L_2(\Omega)} + \tfrac{1}{2} a(U^k, U^{k+1}),$$

cf. Definition 9.46. △

In order to analyze the leapfrog method we require a special form of the CFL condition; for this we require the constant C_{inv} from the inverse estimate of Theorem 9.37. We say that the leapfrog method satisfies a CFL condition if the time step size Δt is chosen so small that $C_{\mathrm{inv}} \frac{(\Delta t)^2}{h^2} < 2$; more precisely:

Definition 9.72 The leapfrog method satisfies a *CFL condition* if there exists some $\lambda \in (0, 2)$ such that

$$C_{\mathrm{inv}} \frac{(\Delta t)^2}{h^2} \leq 2 - \lambda \tag{9.93}$$

holds. △

Lemma 9.73 *Let* $\{\mathcal{T}_h\}_{h>0}$ *be a uniform family of admissible triangulations of a polygon* $\Omega \subset \mathbb{R}^2$, *and suppose that* Δt *is chosen in such a way that the CFL condition* (9.93) *is satisfied. Then* $\mathbb{E}_{LF}^k \geq \tfrac{1}{4} \|D^+_{\Delta t} U^k\|^2_{L_2(\Omega)} + \tfrac{1}{4} [a(U^{k+1}, U^{k+1}) + a(U^k, U^k)] \geq 0.$

Proof By the inverse Cauchy–Schwarz inequality (Corollary 9.38),

$$2\,a(U^k, U^{k+1}) = a(U^{k+1}, U^{k+1}) + a(U^k, U^k) - a(U^{k+1} - U^k, U^{k+1} - U^k)$$

$$= a(U^{k+1}, U^{k+1}) + a(U^k, U^k) - (\Delta t)^2\, a(D_{\Delta t}^+ U^k, D_{\Delta t}^+ U^k)$$

$$\geq a(U^{k+1}, U^{k+1}) + a(U^k, U^k) - C_{\mathrm{inv}} \frac{(\Delta t)^2}{h^2} \|D_{\Delta t}^+ U^k\|_{L_2(\Omega)}^2.$$

It follows that

$$\mathbb{E}_{\mathrm{LF}}^k = \tfrac{1}{2} \|D_{\Delta t}^+ U^k\|_{L_2(\Omega)}^2 + \tfrac{1}{2} a(U^k, U^{k+1})$$

$$\geq \left(\frac{1}{2} - \frac{1}{4} C_{\mathrm{inv}} \frac{(\Delta t)^2}{h^2} \right) \|D_{\Delta t}^+ U^k\|_{L_2(\Omega)}^2 + \frac{1}{4} \Big(a(U^{k+1}, U^{k+1}) + a(U^k, U^k) \Big).$$

Since $\frac{1}{2} - \frac{1}{4} C_{\mathrm{inv}} \frac{(\Delta t)^2}{h^2} \geq \frac{1}{2} - \frac{1}{4}(2 - \lambda) = \frac{\lambda}{4}$ by the CFL condition (9.93), the claim follows.   □

We can now prove that the leapfrog method satisfies conservation of energy.

Theorem 9.74 (Conservation of energy) *Let $\{\mathcal{T}_h\}_{h>0}$ be a uniform family of admissible triangulations of a polygon $\Omega \subset \mathbb{R}^2$ and suppose Δt is chosen in such a way that the CFL condition (9.93) is satisfied. If $f \equiv 0$, then $\mathbb{E}_{\mathrm{LF}}^k = \mathbb{E}_{\mathrm{LF}}^0$, and if $f \neq 0$, then*

$$\sqrt{\mathbb{E}_{\mathrm{LF}}^k} \leq \sqrt{\mathbb{E}_{\mathrm{LF}}^0} + \sum_{j=1}^{k} \frac{\Delta t}{\sqrt{\lambda}} \|f^j\|_{L_2(\Omega)}.$$

Proof We use $\chi = U^{k+1} - U^{k-1} = (U^{k+1} - U^k) + (U^k - U^{k-1})$ as a test function in (9.91b) and obtain $(D_{\Delta t}^2 U^k, U^{k+1} - U^{k-1})_{L_2(\Omega)} + a(U^k, U^{k+1} - U^{k-1}) = (f^k, U^{k+1} - U^{k-1})_{L_2(\Omega)}$. We consider each of these terms separately. Firstly, $(D_{\Delta t}^2 U^k, U^{k+1} - U^{k-1})_{L_2(\Omega)} = \frac{1}{(\Delta t)^2}((U^{k+1} - U^k) - (U^k - U^{k-1}), (U^{k+1} - U^k) + (U^k - U^{k-1}))_{L_2(\Omega)} = \|D_{\Delta t}^+ U^k\|_{L_2(\Omega)}^2 - \|D_{\Delta t}^+ U^{k-1}\|_{L_2(\Omega)}^2$. It follows that

$$2(\mathbb{E}_{\mathrm{LF}}^k - \mathbb{E}_{\mathrm{LF}}^{k-1}) = \|D_{\Delta t}^+ U^k\|_{L_2(\Omega)}^2 + a(U^k, U^{k+1})$$

$$- \|D_{\Delta t}^+ U^{k-1}\|_{L_2(\Omega)}^2 - a(U^{k-1}, U^k)$$

$$= (f^k, U^{k+1} - U^{k-1})_{L_2(\Omega)}.$$

If $f \equiv 0$, then the first part of the claim now follows by induction. For the case $f \neq 0$ we have, adding and subtracting U^k,

$$2(\mathbb{E}_{\mathrm{LF}}^k - \mathbb{E}_{\mathrm{LF}}^{k-1}) = (f^k, U^{k+1} - U^k)_{L_2(\Omega)} + (f^k, U^k - U^{k-1})_{L_2(\Omega)}$$

$$= \Delta t\,(f^k, D_{\Delta t}^+ U^k)_{L_2(\Omega)} + \Delta t\,(f^k, D_{\Delta t}^+ U^{k-1})_{L_2(\Omega)}$$

$$\leq \Delta t\, \|f^k\|_{L_2(\Omega)} (\|D_{\Delta t}^+ U^k\|_{L_2(\Omega)} + \|D_{\Delta t}^+ U^{k-1}\|_{L_2(\Omega)})$$

$$\leq 2 \frac{\Delta t}{\sqrt{\lambda}} \|f^k\|_{L_2(\Omega)} ((\mathbb{E}_{\mathrm{LF}}^k)^{1/2} + (\mathbb{E}_{\mathrm{LF}}^{k-1})^{1/2}),$$

where in the last step we used Lemma 9.73. We assume for the meantime that $E_{LF}^k \neq 0$; then

$$(\mathbb{E}_{LF}^k)^{1/2} - (\mathbb{E}_{LF}^{k-1})^{1/2} = \frac{\mathbb{E}_{LF}^k - \mathbb{E}_{LF}^{k-1}}{(\mathbb{E}_{LF}^k)^{1/2} + (\mathbb{E}_{LF}^{k-1})^{1/2}} \leq \frac{\Delta t}{\sqrt{\lambda}} \|f^k\|_{L_2(\Omega)}.$$

Observe that this inequality is still valid if $E_{LF}^k = 0$, since $\mathbb{E}_{LF}^{k-1} \geq 0$. Hence, in both cases, $(\mathbb{E}_{LF}^k)^{1/2} \leq (\mathbb{E}_{LF}^{k-1})^{1/2} + \frac{\Delta t}{\sqrt{\lambda}} \|f^k\|_{L_2(\Omega)}$. An induction argument now yields the claim. $\qquad \square$

We recall that the conservation of energy in Theorem 8.16 was the key to the uniqueness statement in Theorem 8.15. It should thus not come as a particular surprise that the discrete form of conservation of energy yields the stability of the leapfrog method.

Theorem 9.75 (Stability of the leapfrog method) *Let $\{\mathcal{T}_h\}_{h>0}$ be a uniform family of admissible triangulations of a polygon $\Omega \subset \mathbb{R}^2$ and suppose Δt is chosen in such a way that the CFL condition (9.93) is satisfied. Then the leapfrog method is stable in the sense that there exists a constant $c > 0$ independent of h and Δt such that for $k = 0, \ldots, N - 1$*

$$\|D_{\Delta t}^+ U^k\|_{L_2(\Omega)} + \|U^{k+1}\|_{H^1(\Omega)} \leq \qquad (9.94)$$

$$\leq c \left(\|D_{\Delta t}^+ U^0\|_{L_2(\Omega)} + \|U^0\|_{H^1(\Omega)} + \|U^1\|_{H^1(\Omega)} + \sum_{j=1}^k \Delta t \|f^j\|_{L_2(\Omega)} \right).$$

Proof Using the coercivity of the bilinear form a and the energy estimate of Lemma 9.73, we have

$$\|D_{\Delta t}^+ U^k\|_{L_2(\Omega)} + \|U^{k+1}\|_{H^1(\Omega)} \leq \|D_{\Delta t}^+ U^k\|_{L_2(\Omega)} + \alpha^{-1/2} \sqrt{a(U^{k+1}, U^{k+1})}$$

$$\leq \sqrt{\frac{4}{\lambda} \mathbb{E}_{LF}^k} + 2\alpha^{-1/2} \sqrt{\mathbb{E}_{LF}^k} \leq C_1 \sqrt{\mathbb{E}_{LF}^k}$$

for some constant $C_1 > 0$ independent of Δt and h. We now use Young's inequality in the form $2a(\phi, \psi) \leq a(\phi, \phi) + a(\psi, \psi)$ and estimate, using the continuity constant C of a,

$$\mathbb{E}_{LF}^0 = \frac{1}{2} \|D_{\Delta t}^+ U^0\|_{L_2(\Omega)}^2 + \frac{1}{2} a(U^0, U^1)$$

$$\leq \frac{1}{2} \|D_{\Delta t}^+ U^0\|_{L_2(\Omega)}^2 + \frac{1}{4} a(U^0, U^0) + \frac{1}{4} a(U^1, U^1)$$

$$\leq \frac{1}{2} \|D_{\Delta t}^+ U^0\|_{L_2(\Omega)}^2 + \frac{C}{4} (\|U^0\|_{H^1(\Omega)}^2 + \|U^1\|_{H^1(\Omega)}^2).$$

Theorem 9.74 now yields the claim,

$$\sqrt{\mathbb{E}_{LF}^k} \leq \sqrt{\mathbb{E}_{LF}^0} + 2 \sum_{j=1}^{k} \frac{\Delta t}{\sqrt{\lambda}} \|f^j\|_{L_2(\Omega)}$$

$$\leq c\left(\|D_{\Delta t}^+ U^0\|_{L_2(\Omega)} + \|U^0\|_{H^1(\Omega)} + \|U^1\|_{H^1(\Omega)} + \sum_{j=1}^{k} \Delta t \|f^j\|_{L_2(\Omega)}\right)$$

for a suitable constant $c > 0$. $\qquad\square$

In the next theorem we need the exact solution to be in $C^4([0,T], H_0^1(\Omega))$. Theorem 8.23 gives conditions on the data u_0, u_1 and f which imply this degree of regularity.

Theorem 9.76 (Convergence of the leapfrog method) *Let $\{\mathcal{T}_h\}_{0 < h \leq h_0}$ for some fixed $h_0 > 0$ be a uniform family of admissible triangulations of a convex polygon $\Omega \subset \mathbb{R}^2$ and choose Δt in such a way that the CFL condition (9.93) is satisfied. Let u be the solution of (9.82) and assume that $u \in C^4([0,T], H_0^1(\Omega))$. Let $\{U^k\}_{k=0,\ldots,N}$ be the approximate solutions yielded by the leapfrog method (9.91). Then there exists a constant $C > 0$ independent of Δt and h such that for all $k = 0, \ldots, N-1$*

$$\|D_{\Delta t}^+(U^k - u(t^k))\|_{L_2(\Omega)} + \|U^{k+1} - u(t^{k+1})\|_{L_2(\Omega)} + \|U^k - u(t^k)\|_{L_2(\Omega)} \leq$$

$$\leq C\left(\|D_{\Delta t}^+(U^0 - R_h u_0)\|_{L_2(\Omega)} + \|U^0 - R_h u_0\|_{H^1(\Omega)} + \|U^1 - R_h u_1\|_{H^1(\Omega)}\right.$$

$$\left. + T\|u - R_h u\|_{C^2([0,T], L_2(\Omega))} + T(\Delta t)^2 \|u^{(4)}\|_{C([0,T], H^1(\Omega))}\right). \tag{9.95}$$

Proof As in the previous convergence proofs, we split up the error: $U^k - u(t^k) = [U^k - R_h u(t^k)] + [R_h u(t^k) - u(t^k)] =: \theta_h^k + \rho_h(t^k)$. To deal with the first part, we derive a discretized differential equation for it: for all $\chi \in V_h$, using the definition of the leapfrog method and properties of the Ritz projection, we have

$$(D_{\Delta t}^2 \theta_h^k, \chi)_{L_2(\Omega)} + a(\theta_h^k, \chi)$$
$$= (D_{\Delta t}^2 U^k, \chi)_{L_2(\Omega)} + a(U^k, \chi) - (D_{\Delta t}^2 R_h u(t^k), \chi)_{L_2(\Omega)} - a(R_h u(t^k), \chi)$$
$$= (f^k, \chi)_{L_2(\Omega)} - (D_{\Delta t}^2 R_h u(t^k), \chi)_{L_2(\Omega)} - a(u(t^k), \chi)$$
$$= (\ddot{u}(t^k) - D_{\Delta t}^2 u(t^k) + D_{\Delta t}^2 u(t^k) - D_{\Delta t}^2 R_h u(t^k), \chi)_{L_2(\Omega)}$$
$$= (\sigma(t^k) - D_{\Delta t}^2 \rho_h(t^k), \chi)_{L_2(\Omega)}, \tag{9.96}$$

where $\sigma(t^k) := \ddot{u}(t^k) - D_{\Delta t}^2 u(t^k)$ is the cut-off error of the time discretization. We next estimate the terms on the right-hand side of the differential equation (9.96) individually. Firstly, a Taylor expansion yields the estimate (cf. Exercise 9.19)

$$\|\sigma(t^k)\|_{L_2(\Omega)} \leq \frac{(\Delta t)^2}{12} \|u^{(4)}\|_{C([0,T], L_2(\Omega))}.$$

We estimate the second term in a similar fashion:

$$\|D^2_{\Delta t}\rho_h(t^k)\|_{L_2(\Omega)} \leq \|\ddot{\rho}_h(t^k)\|_{L_2(\Omega)} + \frac{(\Delta t)^2}{12}\|\rho_h^{(4)}\|_{C([0,T],L_2(\Omega))}$$

$$= \|\ddot{\rho}_h(t^k)\|_{L_2(\Omega)} + \frac{(\Delta t)^2}{12}\|R_h u^{(4)} - u^{(4)}\|_{C([0,T],L_2(\Omega))}$$

$$\leq \|u - R_h u\|_{C^2([0,T],L_2(\Omega))} + C h\frac{(\Delta t)^2}{12}\|u^{(4)}\|_{C([0,T],H^1(\Omega))},$$

where in the last step we used the boundedness of the Ritz projection, cf. Theorem 9.53. We now apply the stability estimate of Theorem 9.75 to equation (9.96) for θ_h^k (and observe that $\|\theta_h^{k+1}\|_{L_2(\Omega)} \leq c\|\theta_h^{k+1}\|_{H^1(\Omega)}$):

$$\|D^+_{\Delta t}\theta_h^k\|_{L_2(\Omega)} + \|\theta_h^{k+1}\|_{L_2(\Omega)} \leq \tag{9.97}$$

$$\leq c\Big(\|D^+_{\Delta t}\theta_h^0\|_{L_2(\Omega)} + \|\theta_h^0\|_{H^1(\Omega)} + \|\theta_h^1\|_{H^1(\Omega)}$$

$$+ \sum_{j=1}^{k}\Delta t\,\|\sigma(t^j) - D^2_{\Delta t}\rho_h(t^j)\|_{L_2(\Omega)}\Big)$$

and

$$\sum_{j=1}^{k}\Delta t\,\|\sigma(t^j) - D^2_{\Delta t}\rho_h(t^j)\|_{L_2(\Omega)} \leq$$

$$\leq \Delta t\sum_{j=1}^{N}(\|\sigma(t^j)\|_{L_2(\Omega)} + \|D^2_{\Delta t}\rho_h(t^j)\|_{L_2(\Omega)})$$

$$\leq \Delta t\sum_{j=1}^{N}\Big\{\frac{(\Delta t)^2}{12}\|u^{(4)}\|_{C([0,T],L_2(\Omega))} + \|u - R_h u\|_{C^2([0,T],L_2(\Omega))}$$

$$+ C\frac{h(\Delta t)^2}{12}\|u^{(4)}\|_{C([0,T],H^1(\Omega))}\Big\}$$

$$\leq T\max\{1, Ch\}\Big\{\|u - R_h u\|_{C^2([0,T],L_2(\Omega))} + (\Delta t)^2\|u^{(4)}\|_{C([0,T],H^1(\Omega))}\Big\}. \tag{9.98}$$

The estimates (9.97) and (9.98) still hold, albeit possibly for a larger constant, if we add $\|\theta_h^k\|_{L_2(\Omega)}$ to the left-hand side of (9.97). It remains to estimate the terms in the second part of the error, $\rho_h(t^k) = R_h u(t^k) - u(t^k)$. A Taylor expansion about $t^k + \frac{\Delta t}{2}$ yields $\|D^+_{\Delta t}\rho_h(t^k)\|_{L_2(\Omega)} \leq \|\dot{\rho}_h(t^k + \frac{\Delta t}{2})\|_{L_2(\Omega)} + \frac{(\Delta t)^2}{24}\|\rho_h^{(3)}\|_{C([0,T],L_2(\Omega))}$ (cf. Exercise 9.19), and so

$$\|D^+_{\Delta t}\rho_h(t^k)\|_{L_2(\Omega)} + \|\rho_h(t^{k+1})\|_{L_2(\Omega)} + \|\rho_h(t^k)\|_{L_2(\Omega)} \leq$$

$$\leq \|\dot{\rho}_h\|_{C([0,T],L_2(\Omega))} + \frac{(\Delta t)^2}{24}\|\rho_h^{(3)}\|_{C([0,T],L_2(\Omega))} + 2\|u - R_h u\|_{C([0,T],L_2(\Omega))}$$

$$\leq 3\|u - R_h u\|_{C^1([0,T],L_2(\Omega))} + Ch\frac{(\Delta t)^2}{24}\|u^{(3)}\|_{C([0,T],H^1(\Omega))}.$$

Combining the individual estimates on all the terms considered above yields

$$|D^+_{\Delta t}(U^k - u(t^k))\|_{L_2(\Omega)} + \|U^{k+1} - u(t^{k+1})\|_{L_2(\Omega)} + \|U^k - u(t^k)\|_{L_2(\Omega)} \le$$

$$\le |D^+_{\Delta t}\theta^k_h\|_{L_2(\Omega)} + \|\theta^{k+1}_h\|_{L_2(\Omega)} + \|\theta^k_h\|_{L_2(\Omega)}$$

$$+ |D^+_{\Delta t}\rho_h(t^k)\|_{L_2(\Omega)} + \|\rho_h(t^{k+1})\|_{L_2(\Omega)} + \|\rho_h(t^k)\|_{L_2(\Omega)}$$

$$\le c\Big(\|D^+_{\Delta t}\theta^0_h\|_{L_2(\Omega)} + \|\theta^0_h\|_{H^1(\Omega)} + \|\theta^1_h\|_{H^1(\Omega)}$$

$$+ T \max\{1, Ch\}\Big\{\|\rho_h\|_{C^2([0,T],L_2(\Omega))} + (\Delta t)^2\|u^{(4)}\|_{C([0,T],H^1(\Omega))}\Big\}\Big)$$

$$+ 3\|\rho_h\|_{C^1([0,T],L_2(\Omega))} + Ch\frac{(\Delta t)^2}{24}\|u^{(3)}\|_{C([0,T],H^1(\Omega))},$$

which results in the claimed estimate.   □

Remark 9.77 The above proof can also be carried out for the stronger norm $\|\theta^\ell_h\|_{H^1(\Omega)}$ in place of $\|\theta^\ell_h\|_{L_2(\Omega)}$; in this case the regularity assumptions and a few norms on the right-hand side need to be adjusted.   △

Corollary 9.78 *For linear finite elements, under the assumptions of Theorem 9.76, one has for $k = 0, ..., N - 1$*

$$\|D^+_{\Delta t}(U^k - u(t^k))\|_{L_2(\Omega)} + \|U^{k+1} - u(t^{k+1})\|_{L_2(\Omega)} + \|U^k - u(t^k)\|_{L_2(\Omega)} \le$$

$$\le C\Big(\|D^+_{\Delta t}(U^0 - R_h u_0)\|_{L_2(\Omega)} + \|U^0 - R_h u_0\|_{H^1(\Omega)}$$

$$+ \|U^1 - R_h u_1\|_{H^1(\Omega)} + T(h^2 + (\Delta t)^2)\|u\|_{C^4([0,T],H^2(\Omega))}\Big), \qquad (9.99)$$

if $u \in C^4([0, T], H^2(\Omega) \cap H^1_0(\Omega))$, where u is the solution of the wave equation.

Proof The claim follows from Theorem 9.76 using Theorem 9.53.   □

We thus obtain a method of order $O(h^2 + (\Delta t)^2)$, if the initial values admit a corresponding approximation. We can also see the dependence of the results on the length T of the time interval.

The Crank–Nicolson method

The leapfrog method has the advantage of being efficient in each timestep; however, it once again presupposes a CFL condition (9.93). To finish, we shall now describe the Crank–Nicolson method in this case and show that this method is unconditionally stably convergent. The method reads

$$U^0 = u_{0,h}, \quad U^1 = u_{1,h}, \tag{9.100a}$$

$$(D_{\Delta t}^2 U^k, \chi)_{L_2(\Omega)} + \frac{1}{4}a(U^{k+1} + 2U^k + U^{k-1}, \chi) = \tag{9.100b}$$

$$= \frac{1}{4}(f^{k+1} + 2f^k + f^{k-1}, \chi)_{L_2(\Omega)}, \quad \chi \in V_h, \ k \geq 1,$$

where, again, $u_{0,h}$ and $u_{1,h}$ are approximations of u_0 and u_1, respectively, to be specified later (see page 401). In matrix form this method reads for $k \geq 1$

$$\left(M_h + \frac{(\Delta t)^2}{4}A_h\right)\alpha_h^{k+1} = \frac{(\Delta t)^2}{4}(f_h^{k+1} + 2f_h^k + f_h^{k-1}) \tag{9.101}$$

$$- \frac{(\Delta t)^2}{4}A_h(2\alpha_h^k + \alpha_h^{k-1}) + M_h(2\alpha_h^k - \alpha_h^{k-1}).$$

We proceed in the same way as for the leapfrog method: we first define a suitable notion of energy and show that this energy is conserved; we then prove stability and so, finally, convergence of the method.

Definition 9.79 (Energy of the Crank–Nicolson method) Let U^k be the iterate of the Crank–Nicolson method (9.100) at timestep k. Then the *discrete energy of the Crank–Nicolson method* for the wave equation is defined by

$$\mathbb{E}_{CN}^k := \frac{1}{2}\|D_{\Delta t}^+ U^k\|_{L_2(\Omega)}^2 + \frac{1}{2}a(U^{k+1/2}, U^{k+1/2}),$$

where $U^{k+1/2} := \frac{1}{2}(U^{k+1} + U^k)$. △

We now show that the Crank–Nicolson method conserves this energy, without any CFL condition and under weaker assumptions on the finite element mesh.

Theorem 9.80 (Energy conservation) *Let $\{\mathcal{T}_h\}_{h>0}$ a family of admissible triangulations of a polygon $\Omega \subset \mathbb{R}^2$. Then the Crank–Nicolson method is energy conserving: if $f \equiv 0$, then $\mathbb{E}_{CN}^k = \mathbb{E}_{CN}^0$; if $f \neq 0$, then for $k = 1, ..., N$*

$$\sqrt{\mathbb{E}_{CN}^k} \leq \sqrt{\mathbb{E}_{CN}^0} + \frac{\Delta t}{4\sqrt{2}}\sum_{j=1}^k \|f^{j+1} + 2f^j + f^{j-1}\|_{L_2(\Omega)}.$$

Proof We use $\chi = U^{k+1} - U^{k-1} = (U^{k+1} - U^k) + (U^k - U^{k-1}) = (U^{k+1} + U^k) - (U^k + U^{k-1}) \in V_h$ as a test function in (9.100b) to obtain the relation

$$(D_{\Delta t}^2 U^k, (U^{k+1} - U^k) + (U^k - U^{k-1}))_{L_2(\Omega)} =$$

$$= \|D_{\Delta t}^+ U^k\|_{L_2(\Omega)}^2 - \|D_{\Delta t}^+ U^{k-1}\|_{L_2(\Omega)}^2$$

for the first term in (9.100b), and the relation $\frac{1}{4}a(U^{k+1} + 2U^k + U^{k-1}, (U^{k+1} + U^k) - (U^k + U^{k-1})) = a(U^{k+1/2}, U^{k+1/2}) - a(U^{k-1/2}, U^{k-1/2})$ for the second term. Hence the left-hand side of (9.100b) equals $2(\mathbb{E}_{CN}^k - \mathbb{E}_{CN}^{k-1})$, which in turn is equal to $\frac{1}{4}(f^{k+1} + 2f^k + f^{k-1}, (U^{k+1} - U^k) + (U^k - U^{k-1}))_{L_2(\Omega)}$ by (9.100b). In the case $f \equiv 0$ this yields $E_{CN}^k = \mathbb{E}_{CN}^{k-1}$, and the claim follows by induction. If $f \neq 0$, then

$$2(\mathbb{E}_{CN}^k - \mathbb{E}_{CN}^{k-1}) = \tfrac{1}{4}(f^{k+1} + 2f^k + f^{k-1}, (U^{k+1} - U^k) + (U^k - U^{k-1}))_{L_2(\Omega)}$$
$$\leq \|\tfrac{1}{4}(f^{k+1} + 2f^k + f^{k-1})\|_{L_2(\Omega)} \Delta t(\|D_{\Delta t}^+ U^k\|_{L_2(\Omega)} + \|D_{\Delta t}^+ U^{k-1}\|_{L_2(\Omega)})$$
$$\leq \tfrac{\sqrt{2}}{4} \Delta t \|f^{k+1} + 2f^k + f^{k-1}\|_{L_2(\Omega)} \left(\sqrt{\mathbb{E}_{CN}^k} + \sqrt{\mathbb{E}_{CN}^{k-1}}\right).$$

The rest of the proof proceeds analogously to the proof of Theorem 9.74.   □

Theorem 9.81 (Stability of the Crank–Nicolson method) *Let $\{\mathcal{T}_h\}_{h>0}$ be a family of admissible triangulations of a polygon $\Omega \subset \mathbb{R}^2$. Then the Crank–Nicolson method is stable, that is, there exists a constant $c > 0$ independent of h and Δt such that for $k = 0, ..., N - 1$*

$$\|D_{\Delta t}^+ U^k\|_{L_2(\Omega)} + \|U^{k+1/2}\|_{H^1(\Omega)} \leq \tag{9.102}$$

$$\leq c\left(\|D_{\Delta t}^+ U^0\|_{L_2(\Omega)} + \|U^{1/2}\|_{H^1(\Omega)} + \frac{\Delta t}{4} \sum_{j=1}^k \|f^{j+1} + 2f^j + f^{j-1}\|_{L_2(\Omega)}\right).$$

Proof Using the coercivity of the bilinear form a and the energy estimate of Theorem 9.80, we have

$$\|D_{\Delta t}^+ U^k\|_{L_2(\Omega)} + \|U^{k+1/2}\|_{H^1(\Omega)} \leq$$

$$\leq \|D_{\Delta t}^+ U^k\|_{L_2(\Omega)} + \alpha^{-1/2}\sqrt{a(U^{k+1/2}, U^{k+1/2})}$$

$$\leq 2 \max\{1, \alpha^{-1/2}\}\sqrt{\mathbb{E}_{CN}^k}$$

$$\leq 2 \max\{1, \alpha^{-1/2}\}\left\{\sqrt{\mathbb{E}_{CN}^0} + \frac{\Delta t}{4\sqrt{2}} \sum_{j=1}^k \|f^{j+1} + 2f^j + f^{j-1}\|_{L_2(\Omega)}\right\}$$

$$\leq c\left(\|D_{\Delta t}^+ U^0\|_{L_2(\Omega)} + \|U^{1/2}\|_{H^1(\Omega)} + \frac{\Delta t}{4} \sum_{j=1}^k \|f^{j+1} + 2f^j + f^{j-1}\|_{L_2(\Omega)}\right)$$

for a constant $c > 0$ independent of Δt and h, where we have also used Young's inequality, as before.   □

Of course, we can also apply the triangle inequality to (9.102) to obtain the further estimate

$$\|D_{\Delta t}^+ U^k\|_{L_2(\Omega)} + \|U^{k+1/2}\|_{H^1(\Omega)} \leq$$

$$\leq c\left(\|D_{\Delta t}^+ U^0\|_{L_2(\Omega)} + \|U^{1/2}\|_{H^1(\Omega)} + \Delta t \sum_{j=0}^{k+1} \|f^j\|_{L_2(\Omega)}\right). \tag{9.102'}$$

We again refer to Theorem 8.23 for conditions on the given data which imply the regularity of the exact solution required in the next statement.

Theorem 9.82 (Convergence of the Crank–Nicolson method) *Let* $\{\mathcal{T}_h\}_{0<h\leq h_0}$ *for some fixed* $h_0 > 0$ *be a family of admissible triangulations of a convex polygon* $\Omega \subset \mathbb{R}^2$. *Suppose that the exact solution* u *of* (9.82) *satisfies* $u \in C^4([0,T], H_0^1(\Omega))$, *and let* $\{U^k\}_{k=0,\ldots,N}$ *be the approximate solutions obtained via the Crank–Nicolson method* (9.100). *Then there exists a constant* $c > 0$ *independent of* Δt *and* h *such that for* $k = 0,\ldots,N-1$, *setting* $t^{k+1/2} := t^k + \frac{\Delta t}{2}$

$$\|D_{\Delta t}^+(U^k - u(t^k))\|_{L_2(\Omega)} + \|U^{k+1/2} - u(t^{k+1/2})\|_{L_2(\Omega)} \leq$$

$$\leq c\Big(\|D_{\Delta t}^+(U^0 - R_h u_0)\|_{L_2(\Omega)} + \|U^{1/2} - R_h u(t^{1/2})\|_{H^1(\Omega)}$$

$$+ T\|u - R_h u\|_{C^2([0,T],L_2(\Omega))} + T(\Delta t)^2 \|u^{(4)}\|_{C([0,T],H^1(\Omega))}\Big). \tag{9.103}$$

Proof The proof is very similar to the one of Theorem 9.76, the convergence result for the leapfrog method. We just sketch the main steps and again split up the error term: $U^k - u(t^k) = [U^k - R_h u(t^k)] + [R_h u(t^k) - u(t^k)] =: \theta_h^k + \rho_h(t^k)$. We insert the first part θ_h^k into the left-hand side of the Crank–Nicolson method, then by properties of the Ritz projections, for all $\chi \in V_h$ we have

$$(D_{\Delta t}^2 \theta_h^k, \chi)_{L_2(\Omega)} + \tfrac{1}{4}a(\theta_h^{k+1} + 2\theta_h^k + \theta_h^{k-1}, \chi) =$$

$$= (D_{\Delta t}^2 U^k, \chi)_{L_2(\Omega)} + \tfrac{1}{4}a(U^{k+1} + 2U^k + U^{k-1}, \chi)$$

$$- (D_{\Delta t}^2 R_h u(t^k), \chi)_{L_2(\Omega)} - \tfrac{1}{4}a(u(t^{k+1}) + 2u(t^k) + u(t^{k-1}), \chi)$$

$$= \tfrac{1}{4}(f^{k+1} + 2f^k + f^{k-1}, \chi)_{L_2(\Omega)} - (D_{\Delta t}^2 R_h u(t^k), \chi)_{L_2(\Omega)}$$

$$+ \tfrac{1}{4}(\ddot{u}(t^{k+1}) + 2\ddot{u}(t^k) + \ddot{u}(t^{k-1}), \chi)_{L_2(\Omega)} - \tfrac{1}{4}(f^{k+1} + 2f^k + f^{k-1}, \chi)_{L_2(\Omega)}$$

$$= (\ddot{u}(t^k) - D_{\Delta t}^2 u(t^k) + D_{\Delta t}^2 u(t^k) - D_{\Delta t}^2 R_h u(t^k), \chi)_{L_2(\Omega)}$$

$$+ \tfrac{1}{4}(\ddot{u}(t^{k+1}) - 2\ddot{u}(t^k) + \ddot{u}(t^{k-1}), \chi)_{L_2(\Omega)}$$

$$= (\sigma(t^k) - D_{\Delta t}^2 \rho_h(t^k) + \tfrac{\Delta t}{4}(D_{\Delta t}^+ \ddot{u}(t^k) - D_{\Delta t}^- \ddot{u}(t^k)), \chi)_{L_2(\Omega)}, \tag{9.104}$$

where $\sigma(t^k) := \ddot{u}(t^k) - D_{\Delta t}^2 u(t^k)$ is the cut-off error of the time discretization. We estimate the first two terms in (9.104) as above, that is, we have the bounds

$$\|\sigma(t^k)\|_{L_2(\Omega)} \leq \frac{(\Delta t)^2}{12}\|u^{(4)}\|_{C([0,T],L_2(\Omega))} \quad \text{and}$$

$$\|D_{\Delta t}^2 \rho_h(t^k)\|_{L_2(\Omega)} \leq \|u - R_h u\|_{C^2([0,T],L_2(\Omega))} + Ch\frac{(\Delta t)^2}{12}\|u^{(4)}\|_{C([0,T],H^1(\Omega))}.$$

The new term which appears here and not in the leapfrog method can be controlled using another Taylor expansion,

$$\|D_{\Delta t}^+ \ddot{u}(t^k) - D_{\Delta t}^- \ddot{u}(t^k)\|_{L_2(\Omega)} \leq C\Delta t \|u^{(4)}\|_{C([0,T],L_2(\Omega))}.$$

We now apply the stability estimate (9.102) to (9.104) for θ_h^k, setting $\theta_h^{k+1/2} := \frac{1}{2}(\theta_h^{k+1} + \theta_h^k)$ similar to Definition 9.79:

$$
\|D_{\Delta t}^+ \theta_h^k\|_{L_2(\Omega)} + \|\theta_h^{k+1/2}\|_{L_2(\Omega)} \leq c\Big(\|D_{\Delta t}^+ \theta_h^0\|_{L_2(\Omega)} + \|\theta_h^{1/2}\|_{H^1(\Omega)}
$$
$$
+ \Delta t \sum_{j=1}^{k} \|\sigma(t^j) - D_{\Delta t}^2 \rho_h(t^j) + \tfrac{\Delta t}{4}(D_{\Delta t}^+ \ddot{u}(t^j) - D_{\Delta t}^- \ddot{u}(t^j)\|_{L_2(\Omega)} \Big)
$$
$$
\leq c\Big(\|D_{\Delta t}^+ \theta_h^0\|_{L_2(\Omega)} + \|\theta_h^{1/2}\|_{H^1(\Omega)}
$$
$$
+ T\Big\{ \|u - R_h u\|_{C^2([0,T], L_2(\Omega))} + (\Delta t)^2 \|u^{(4)}\|_{C([0,T], H^1(\Omega))} \Big\} \Big).
$$

Next, we estimate the terms in the second part of the error, $\rho_h(t^k) = R_h u(t^k) - u(t^k)$. As with the proof of convergence of the leapfrog method, a Taylor expansion about $t^{k+1/2} = t^k + \frac{\Delta t}{2}$ (cf. Exercise 9.19) yields (by Theorem 9.53, p. 368)

$$
\|D_{\Delta t}^+ \rho_h(t^k)\|_{L_2(\Omega)} + \|\rho_h(t^{k+1/2})\|_{L_2(\Omega)} \leq
$$
$$
\leq \|\dot{\rho}_h\|_{C([0,T], L_2(\Omega))} + \frac{(\Delta t)^2}{24} \|\rho_h^{(3)}\|_{C([0,T], L_2(\Omega))} + \|\rho_h(t^{k+1/2})\|_{L_2(\Omega)}
$$
$$
\leq 2\|u - R_h u\|_{C^1([0,T], L_2(\Omega))} + C h \frac{(\Delta t)^2}{24} \|u^{(3)}\|_{C([0,T], H^1(\Omega))}.
$$

Finally, abbreviating $u^{k+1/2} := \frac{1}{2}\big(u(t^{k+1}) + u(t^k)\big)$, we use a Taylor expansion of $\|U^{k+1/2} - u(t)\|_{L_2(\Omega)}$ about $t^{k+1/2}$ (cf. Exercise 9.19) to obtain the estimate $\|U^{k+1/2} - u(t^{k+1/2})\|_{L_2(\Omega)} \leq \|U^{k+1/2} - u^{k+1/2}\|_{L_2(\Omega)} + \frac{(\Delta t)^2}{24}\|u^{(2)}\|_{C([0,T], L_2(\Omega))}$. Then, $U^{k+1/2} - u^{k+1/2} = \theta_h^{k+1/2} + \frac{1}{2}(\rho_h(t^{k+1}) + \rho_h(t^k))$ and combining all the above estimates yields the claim. $\square$

Corollary 9.83 *For linear finite elements on a uniform triangulation, under the assumptions of Theorem 9.82, one has for $k = 0, ..., N - 1$*

$$
\|D_{\Delta t}^+ (U^k - u(t^k))\|_{L_2(\Omega)} + \|U^{k+1/2} - u(t^{k+1/2})\|_{L_2(\Omega)} \leq \qquad (9.105)
$$
$$
\leq C\Big(\|D_{\Delta t}^+(U^0 - R_h u_0)\|_{L_2(\Omega)} + \|U^{1/2} - R_h u(t^{1/2})\|_{H^1(\Omega)}
$$
$$
+ T(h^2 + (\Delta t)^2)\|u\|_{C^4([0,T], H^2(\Omega))} \Big),
$$

if $u \in C^4([0, T], H^2(\Omega) \cap H_0^1(\Omega))$, where u is the solution of the wave equation.

Proof The estimate follows by (9.103) and the properties of the Ritz projector in Theorem 9.53. $\square$

Initialization

As for finite differences (see, e.g., (9.77)), for the numerical implementation of both the Crank–Nicolson and the leapfrog methods using finite elements for the space variables we still require approximations of the initial values, $U_0 = u_{0,h}$ and $U_1 = u_{1,h}$ in order to bound the first terms on the right-hand side of (9.90). The process of finding these is sometimes known as *initialization*. The error estimates which we have derived and proved in both cases, still contain terms which depend on $U^0 - R_h u_0$ and $U^1 - R_h u_1$, respectively, which in particular involve the Ritz projection from (9.61). This motivates the following result. We refer once more to Theorem 8.23 for conditions on the given data which imply the required regularity of the exact solution.

Theorem 9.84 (Initialization by elliptic projection) *Let $\{\mathcal{T}_h\}_{0<h\leq h_0}$ for some fixed $h_0 > 0$ be a uniform family of admissible triangulations of a convex polygon $\Omega \subset \mathbb{R}^2$. Let u be the exact solution of (9.82) and assume that $u \in C^3([0,T], H_0^1(\Omega))$. Define*

$$U_0 = u_{0,h} := R_h u_0, \quad U_1 = u_{1,h} := R_h \left(u_0 + \Delta t\, u_1 + \frac{(\Delta t)^2}{2} \ddot{u}(0) \right),$$

where u_0 and u_1 are as in (9.100). Then there exists a constant $C = C(u,T) > 0$ independent of h and $\Delta t \leq 1$ such that for $u^{k+1/2} := \frac{1}{2}(u(t^{k+1}) + u(t^k))$

$$\|D_{\Delta t}^+(U^0 - R_h u_0)\|_{L_2(\Omega)} + \|U^{1/2} - R_h u^{1/2}\|_{H^1(\Omega)} \leq C\,(\Delta t)^2.$$

Proof Firstly, by the triangle inequality,

$$\|U^{1/2} - R_h u^{1/2}\|_{H^1(\Omega)} \leq \tfrac{1}{2}\|U^0 - R_h u_0\|_{H^1(\Omega)} + \tfrac{1}{2}\|U^1 - R_h u(t^1)\|_{H^1(\Omega)}$$
$$= \tfrac{1}{2}\|U^1 - R_h u(t^1)\|_{H^1(\Omega)}$$

by assumption. Using the definition (9.61) of the Ritz projection and the coercivity and boundedness of the bilinear form $a(\cdot,\cdot)$, we obtain that, for any $v \in H_0^1(\Omega)$,

$$\alpha\,\|R_h v\|_{H^1(\Omega)}^2 \leq a(R_h v, R_h v) = a(R_h v, v) = a(v,v) \leq C\,\|v\|_{H^1(\Omega)}^2,$$

that is, $\|R_h v\|_{H^1(\Omega)} \leq \sqrt{\frac{C}{\alpha}}\|v\|_{H^1(\Omega)}$. Similar to above we now use a Taylor expansion of $\|u(t^1)\|_{H^1(\Omega)} = \|u(\Delta t)\|_{H^1(\Omega)}$ about 0 to obtain (cf. Exercise 9.19)

$$\|u(\Delta t) - u(0) - \Delta t\,\dot{u}(0) - \tfrac{(\Delta t)^2}{2}\ddot{u}(0)\|_{H^1(\Omega)} \leq \tfrac{(\Delta t)^3}{6}\|u\|_{C^3([0,T];H^1(\Omega))}$$

and thus, recalling that $u(0) = u_0$ and $\dot{u}(0) = u_1$

$$\|U^1 - R_h u(t^1)\|_{H^1(\Omega)} = \|u_{1,h} - R_h u(\Delta t)\|_{H^1(\Omega)}$$
$$= \left\| R_h\left(u_0 + \Delta t\, u_1 + \tfrac{(\Delta t)^2}{2}\ddot{u}(0) \right) - R_h u(\Delta t) \right\|_{H^1(\Omega)}$$

$$\leq \sqrt{\frac{C}{\alpha}} \left\| u_0 + \Delta t\, u_1 + \frac{(\Delta t)^2}{2} \ddot{u}(0) - u(\Delta t) \right\|_{H^1(\Omega)}$$

$$\leq \frac{\sqrt{C}}{6\sqrt{\alpha}} (\Delta t)^3 \|u\|_{C^3([0,T];H^1(\Omega))}.$$

Finally, it follows from our choice of $u_{0,h}$ that

$$\|D_{\Delta t}^+(U^0 - R_h u_0)\|_{L_2(\Omega)} = \frac{1}{\Delta t} \|(U^1 - R_h u_1) - (U^0 - R_h u_0)\|_{L_2(\Omega)}$$

$$= \frac{1}{\Delta t} \|U^1 - R_h u_1\|_{L_2(\Omega)} \leq \frac{1}{\Delta t} \|U^1 - R_h u_1\|_{H^1(\Omega)}$$

$$\leq \frac{\sqrt{C}}{6\sqrt{\alpha}} (\Delta t)^2 \|u\|_{C^3([0,T];H^1(\Omega))},$$

which yields the claim for $C(u,T) := \frac{\sqrt{C}}{4\sqrt{\alpha}} \|u\|_{C^3([0,T];H^1(\Omega))}$.   □

Remark 9.85 Theorem 9.84 clearly yields the optimal convergence rates, if we use the approximations of the initial values in Theorem 9.84 for the leapfrog and Crank–Nicolson methods. However, calculating the Ritz projections requires the use of the bilinear form $a(\cdot, \cdot)$ and thus the derivatives of u_0, u_1 and $\ddot{u}(0)$ with respect to the *space* variables, which is numerically costly. In practice the desired accuracy can often be obtained by recourse to simple nodal interpolations.   △

Numerical experiments

To finish, we will present the results of a few numerical experiments; these are essentially from [36]. These experiments are concerned with two questions: (1) how sharp is the CFL condition (9.93) in the case of the leapfrog method? We wish to investigate this question of stability quantitatively; (2) how strict are the regularity requirements in the proofs of convergence? We saw that strong assumptions were necessary on the unknown solution (and also on the initial conditions) in order to prove our convergence results. We wish to investigate the behavior of the methods when these assumptions are not satisfied.

In addition to the methods considered above, viz. the leapfrog and the Crank–Nicolson methods, we will also implement the implicit Euler method. As we already saw when investigating the heat equation, it should not come as a surprise that this method is only of first order in time, since the partial derivative in the time variable is approximated by a forward difference quotient of first order. On the other hand, we can expect stability of this method without an additional CFL condition.

The Case of One Space Dimension

We start with the case $\Omega = (0, 1)$, and consider the initial-boundary value problem (9.75) with $T = 1$ and $f \equiv 0$. If u_0 is at least piecewise continuous and u_1 integrable, then we can use the solution formula of d'Alembert, see (3.7), to obtain the exact solution and thus compute the (exact) error in the numerical approximation.

Stability analysis. We first investigate the CFL condition (9.93) for the leapfrog method. To this end we choose $u_0(x) := \sin(2\pi x)$, $u_1 \equiv 0$ and $\Delta t = h$. We make two choices for the wave speed c in (9.75): in the first case, $c = 0.5$, the CFL condition is satisfied, in the other case, $c = 1$, it is not. The results are depicted in Figure 9.22. The convergence of first and second order, respectively, when the CFL condition is satisfied, is clearly recognizable (left). On the right-hand side we see that Crank–Nicolson and implict Euler are stable, whereas the leapfrog method diverges for coarse step sizes. We obtain similar results in the case $u_0 \equiv 0$ und $u_1(x) := \sin(2\pi x)$.

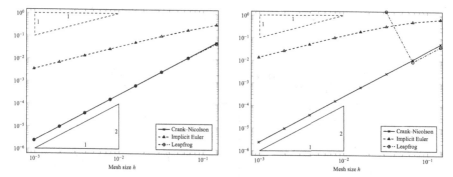

Fig. 9.22 Rate of convergence for FEM for the wave equation in one space dimension; comparison of the leapfrog, Crank–Nicolson and implicit Euler methods. Left: $c = 0.5$, the CFL condition is satisfied (the curves for Crank–Nicolson and leapfrog almost coincide), right: $c = 1$, the CFL condition is violated.

Regularity assumptions. In the case of one space dimension and homogeneous right-hand side, the smoothness of the solution depends exclusively on the initial displacement u_0 and velocity u_1. We will consider various cases, listed in Table 9.1. The cases 1 and 2 are exactly the ones which we considered in detail above.

The results of the convergence analysis are depicted in Table 9.2. We first observe that the implicit Euler method is at least quantitatively influenced by the CFL condition; the convergence rate drops to 0.8. In order to make clearer the effects of regularity in question, the table is ordered in such a way that the convergence rates for leapfrog and Crank–Nicolson are in descending order – interestingly this order is equal in the two cases. We see that discontinuous and non-continuously differentiable initial displacements have the largest negative influence on the convergence rates. The methods of second order (Crank–Nicolson and leapfrog) actually converge more slowly than the first order method (Euler). The behavior in the case of discontinuous initial velocity is comparable to that in the case of non-continuously differentiable initial displacements: we only see convergence of first order.

Case	u_0	u_1
1 smooth, CFL	$\sin(2\pi x) \in C^\infty(\Omega)$	$\equiv 0 \in C^\infty(\Omega)$
2 smooth, no CFL	$\sin(2\pi x) \in C^\infty(\Omega)$	$\equiv 0 \in C^\infty(\Omega)$
3 $u_0 \in C^0(\Omega) \setminus C^1(\Omega)$	$\begin{cases} x, & x < 0.5, \\ 1-x, & \text{otherwise.} \end{cases}$	$\equiv 0$
4 $u_0 \notin C^0(\Omega)$	$\begin{cases} 1, & x \in [0.4, 0.6], \\ 0, & \text{otherwise.} \end{cases}$	$\equiv 0$
5 $u_1 \in C^0(\Omega) \setminus C^1(\Omega)$	$\equiv 0$	$\begin{cases} x, & x < 0.5, \\ 1-x, & \text{otherwise.} \end{cases}$
6 $u_1 \notin C^0(\Omega)$	$\equiv 0$	$\begin{cases} 1, & x \in [0.4, 0.6], \\ 0, & \text{otherwise.} \end{cases}$

Table 9.1 Test cases for finite elements for the 1D wave equation.

Case/Conv. Rate	Leapfrog	Crank–Nicolson	Implicit Euler
1 Smooth, CFL	2.05	2.07	0.95
2 Smooth, no CFL	—	2.07	0.80
5 $u_1 \in C^0 \setminus C^1$	1.76	1.80	0.95
6 $u_1 \notin C^0$	1.06	1.05	0.83
3 $u_0 \in C^0 \setminus C^1$	1.02	0.99	0.99
4 $u_0 \notin C^0$	0.38	0.33	0.72

Table 9.2 Convergence rates of the various time discretization methods for finite elements for the 1D wave equation (sorted by convergence rate of leapfrog and Crank–Nicolson).

We wish to study the case of a discontinuous initial displacement (case 4) more closely, cf. Figure 9.23. On the left-hand side we see the initial displacement and its numerical approximation (which is the same for all methods). We can already see oscillations, which arise through the finite difference approximation of the discontinuous function u_0. In the case of the leapfrog method, even for moderate step sizes these oscillations build up so much that after just a few timesteps the approximation falls outside the machine accuracy. For this reason the leapfrog approximation can no longer be recognized on the right-hand side. There, the exact solution and the approximate solutions are depicted at time $t = 0.75$. What we can recognize is the differing behavior of the other two methods: the Crank–Nicolson method produces oscillations (which do not build up as much as with the leapfrog method, due to the stability of the former), whereas the implicit Euler method "blurs" the solution, that is, produces a very smooth (and poor) approximation of the discontinuous solution. The (poor) convergence behavior is depicted in Figure 9.24.

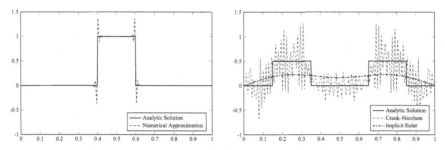

Fig. 9.23 Evolution of the analytic solution and numerical approximations of the 1D wave equation in time, case 4 ($h = \frac{1}{140}$, $\Delta t = \frac{1}{500}$, $T = 0.75$). Left: initial time $t = 0$; right: time $t = 0.75$.

The Case of Two Space Dimensions

As with the heat equation, the two-dimensional case offers the opportunity of investigating how the geometry of the domain $\Omega \subset \mathbb{R}^2$ affects the rate of convergence of the methods. For this, as in Figure 9.14 (page 354), we will compare the cases of a square, a convex polygon and the L-shaped domain. Since there is no closed formula for the solution in these cases, we have calculated a reference solution numerically on a very fine mesh using the implicit Euler method with very small timestep.

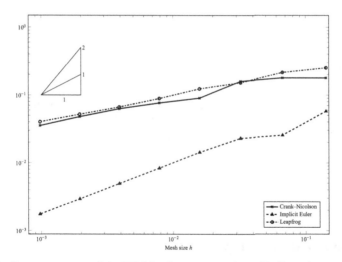

Fig. 9.24 Convergence rate of the FEM for the wave equation with discontinuous initial displacement u_0 in one space dimension; comparison of leapfrog, Crank–Nicolson and implicit Euler methods.

We again test diverse choices of u_0, u_1 and the right-hand side f, in particular to generate cases where the condition $u \in C^4([0,T], H_0^1(\Omega) \cap H^2(\Omega))$ is either satisfied or violated. The results are summarized in Figure 9.25. We choose functions for u_0, u_1 and f of successively lower regularity in each case (in the nine pictures in Figure 9.25, we have in each case, from left to right on the horizontal axis, C^∞, C^1, $C^0 \setminus C^1$, discontinuous). In all cases the influence of the geometry of the domain is considerably weaker than the influence of insufficient regularity of the data. This also demonstrates the flexibility of the finite element method compared with finite differences. In the case of a discontinuous right-hand side f the influence of the domain is actually the opposite of what we know for elliptic and parabolic problems; however, this will be far more a consequence of the specific choice of the data here, in the sense that this particular f will be particularly well suited to the L-shaped domain.

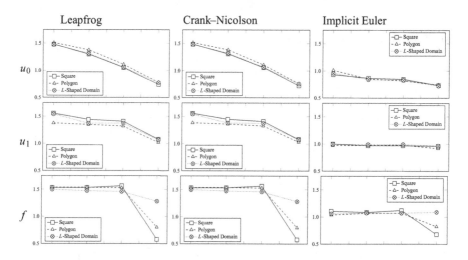

Fig. 9.25 Convergence rates for the finite element method for the wave equation in two space dimensions; comparison of leapfrog, Crank–Nicolson and implicit Euler methods for various geometries. Upper row: different initial displacements, middle row: different initial velocities, lower row: different right-hand sides, in each case with decreasing regularity from left to right.

9.6* Comments on Chapter 9

The finite element method goes back to Alexander Hrennikoff (1941) and Richard Courant (1942). While Hrennikoff discretized the domains using a rectangular mesh, similar to FDM, Courant used triangles. After studies in Wrocław (at the time Breslau in Prussia), Zurich and Göttingen, Courant obtained his doctorate in 1910 under David Hilbert; his thesis was entitled *Über die Anwendung des Dirichlet'schen Prinzipes auf die Probleme der konformen Abbildung* (on the application of the Dirichlet principle to problems of conformal mapping). In the same year he married Nerina Runge, the daughter of the Göttingen mathematics professor Carl Runge (known among other things for the Runge–Kutta method). He obtained his *habilitation* in 1912, before being drafted into the army to fight in the First World War, where he was wounded. He was a professor in Göttingen from 1920 until 1933, when he left Germany, having been dismissed from his position by the Nazis owing both to his Jewish heritage and his membership in the German Social Democratic Party. After a year in Cambridge, Courant took up a professorship at New York University in 1935. There, from 1935 until 1958 he directed the institute which would come to bear his name in 1964, and which to this day is one of the leading research institutes globally in applied mathematics. Among his successors as director of the Courant Institute we can find such names as Louis Nirenberg and Peter Lax.

In 1928, together with K. Friedrichs and H. Lewy, Courant published his famous work *Über die partiellen Differenzengleichungen der mathematischen Physik* (on the partial differential equations of mathematical physics) in the *Mathematische Annalen*, which already contains the stability condition later known as the CFL condition (cf. Remark 9.48). Friedrichs had obtained his doctorate in 1925 in Göttingen under Courant, writing his dissertation on *Die Randwert- und Eigenwertprobleme aus der Theorie der elastischen Platten* (boundary value and eigenvalue problems from the theory of elastic plates). He accompanied Courant to New York and was, in turn, the doctoral supervisor of Peter Lax in 1949. Robert Richtmyer worked with John von Neumann, among others, and developed numerical methods for the solution of complex problems, investigated in turn by Stanislaw Ulam. The resulting method is known today as the *Monte Carlo method*. In 1953 Richtmyer moved to the Courant Institute in New York, where he published a joint paper with Peter Lax on the Lax–Richtmyer equivalence theorem in 1956.

When developing the finite element method, Courant built on earlier work of Rayleigh, Ritz and Galerkin, by constructing the finite-dimensional spaces in the Galerkin method with the help of triangulations, as described above. As suitable as the finite element method is for treating elliptic equations of second order (as we have seen), its later success was only made possible by the development of sufficiently powerful computers. This is because the generation of a triangulation and setting up the linear systems of equations (for which integrals need to be calculated numerically) require significantly more computing power than is the case with the FDM. This also explains the statement of Lothar Collatz in 1950, when the FDM seemed unchallenged: *the difference method is a generally applicable method for boundary value problems. It is easy to set up and even for coarse mesh sizes generally yields, for relatively little calculation, a description of the solution function which is often sufficient for technical purposes. In particular, in partial differential equations there are areas where the difference method is the only practicable method and where other methods only capture the boundary conditions with difficulty, if they are capable of doing so at all.*[2] How much the world has changed!

Philippe Clément achieved a breakthrough in the convergence analysis of finite elements in 1975 with the eponymous interpolation operator and corresponding error estimates, [22]. It had been known since the doctoral thesis of Jean Céa in 1964 that the error in the finite element approximation was, up to a multiplicative constant, essentially the error of the best approximation in the trial space. In order to estimate the error of the best approximation, one uses the Clément operator, whose real brilliance consists in linking Sobolev regularity and interpolation.

---

[2] "Das Differenzenverfahren ist ein bei Randwertaufgaben allgemein anwendbares Verfahren. Es ist leicht aufstellbar und liefert bei groben Maschenweiten im Allgemeinen bei relativ kurzer Rechnung einen für technische Zwecke oft ausreichenden Überblick über die Lösungsfunktion. Insbesondere gibt es bei partiellen Differenzialgleichungen Bereiche, bei denen das Differenzenverfahren das einzig praktisch brauchbare Verfahren ist und bei denen andere Verfahren die Randbedingungen nur schwer oder gar nicht zu erfassen vermögen."

9.7 Exercises

Exercise 9.1 Let f_α be as in (9.15) and let $u_\alpha \in C^1([0, 1])$ be the solution of $Lu_\alpha = f_\alpha$ in accordance with (9.1). Define L_h in terms of the central difference quotient on an equidistant mesh and show that for the solution $u_{\alpha,h}$ of $L_h u_{\alpha,h} = f_\alpha$ we have $\|u_\alpha - u_{\alpha,h}\|_{h,\infty} = O(h)$ as $h \to 0+$.

Exercise 9.2 For $x_i := ih$, $h = \frac{1}{N+1}$, $0 \le i \le N + 1$, show (9.12)), i.e. show that $\sum_{k=1}^{N} G(x_i, x_k) = \frac{1}{2h} x_i(1 - x_i)$ for the Green's function G from (9.3).

Exercise 9.3 Show that the stiffness matrix of the Dirichlet problem in one space dimension is identical to the system matrix of the FDM with central difference quotients. Here the mesh is equidistant in both cases.

Exercise 9.4 Let T be a triangle with vertices t_1, t_2, t_3. Show that for each $i \in \{1, 2, 3\}$ there exists exactly one $v_i \in \mathcal{P}_1(T)$ such that $v_i(t_j) = \delta_{i,j}$, $j = 1, 2, 3$.

Exercise 9.5 Consider the initial value problem $y'(t) = -\frac{1}{y(t)}\sqrt{1 - y(t)^2}$, $y(0) = 1$ with solution $y(t) = \sqrt{1 - t^2}$, $0 \le t < 1$. Why does the explicit Euler method yield the solution $y \equiv 1$ independently of the step size?

Exercise 9.6 Let $\Omega = (a, b) \subset \mathbb{R}$, $-\infty < a < b < \infty$ and $\mathcal{P} := \bigcup_{p \in \mathbb{N}} \mathcal{P}_p(\Omega)$, where $\mathcal{P}_p(\Omega)$ denotes the space of polynomials on Ω of degree no greater than $p \in \mathbb{N}$. Show that $\mathcal{P}$ is a normed space for the norm $\|\cdot\|_\infty$, but not a Banach space.

Exercise 9.7 Consider a variant i_h of the Clément interpolation of $L_1(\Omega)$ in the space of linear finite elements $V_h = X_h^{1,0}$ (without homogeneous Dirichlet boundary conditions) on quasi-uniform meshes $\{\mathcal{T}_h\}$, defined as follows. For every node t_i we consider the functional $\pi_i : L_1(\Omega) \to \mathbb{R}$ given by

$$\pi_i(u) = \left(\int_{\omega_i} u\phi_i \, dx\right)\left(\int_{\omega_i} \phi_i \, dx\right)^{-1},$$

with node basis function ϕ_i and $\omega_i := supp\phi_i$. The interpolation is then defined by $i_h u = \sum_{x_i \in \mathcal{N}(\mathcal{T}_h)} \pi_i(u)\phi_i$, where $\mathcal{N}(\mathcal{T}_h)$ denotes the set of nodes of $\mathcal{T}_h$. Prove the following estimates:
(a) If $u \ge 0$, then $i_h u \ge 0$.
(b) $\|u - \pi_i(u)\|_{L_2(\omega_i)} \le c h_i \|\nabla u\|_{L_2(\omega_i)}$, where $h_i = \max_{T \subset \omega_i} h_T$.
(c) $\|u - i_h u\|_{L_2(T)} \le ch\|\nabla u\|_{L_2(\tilde{\omega}_T)}$, where $\tilde{\omega}_T := \bigcup\{K \in \mathcal{T}_h : \overline{K} \cap \overline{T} \ne \emptyset\}$.
(d) $\|u - i_h u\|_{L_2(\Omega)} \le ch\|\nabla u\|_{L_2(\Omega)}$.
(e) $\|u - i_h u\|_{H^{-1}(\Omega)} \le ch^2\|\nabla u\|_{L_2(\Omega)}$, where the dual space $H^{-1}(\Omega)$ of $H_0^1(\Omega)$ is equipped with the norm $\|v\|_{H^{-1}(\Omega)} = \sup_{\phi \in H_0^1(\Omega), \phi \ne 0} \frac{(v, \phi)}{\|\nabla \phi\|_{L_2(\Omega)}}$.

Exercise 9.8 Let $T \subset \mathbb{R}^2$ be a parallelogram. Prove that there exists an affine transformation $\sigma : \hat{T} = (0,1)^2 \to T$. What does the transformation that maps $\hat{T}$ onto a general quadrilateral look like?

Exercise 9.9 Let the boundary value problem

$$-u''(x) = f(x), \quad x \in (0,1),$$
$$u(0) = a, \quad u(1) = b$$

be given, for fixed constants a and b and a continuous function $f : [0,1] \to \mathbb{R}$. Discretizing using finite differences with uniform mesh size $h = 1/(N+1)$, $N \in \mathbb{N}$, yields the linear system of equations $A_h u_h = f_h$ as in (9.8). Show that the eigenvalues of A_h are given by $\lambda_j = \frac{4}{h^2} \sin^2\left(\frac{j\pi h}{2}\right)$, $j = 1, 2, \ldots, N$, whose associated orthonormal eigenvectores are $v_j = \sqrt{2h}[\sin(ij\pi h)]_{i=1}^{N}$, $j = 1, 2, \ldots, N$.

Exercise 9.10 In 1870 Weierstrass found the following counterexample to the Dirichlet principle, which consists of the following minimization problem for continuous functions which does not have a solution: find a function u which minimizes the functional

$$J(u) = \int_{-1}^{1} [xu'(x)]^2 \, dx$$

among all functions $u \in C^1([-1,1])$ such that $u(-1) = 0$ and $u(1) = 1$. Show using the sequence of functions $\{u_n\}$ defined by

$$u_n(x) = \frac{1}{2} + \frac{1}{2} \frac{\arctan(nx)}{\arctan n}, \quad n = 1, 2, \ldots$$

that this problem does, indeed, not have a solution.

Exercise 9.11 Let $\Omega \subset \mathbb{R}^2$ be a simply connected polygon. Show that for any triangulation of Ω the number of triangles plus the number of nodes minus the number of edges always equals 1. Why does this not hold if Ω is multiply connected?

Exercise 9.12 Show that for any family of triangles the quasi-uniformity condition $\frac{r_T}{\rho_T} \leq \kappa$, $T \in \mathcal{T}_h$, $h > 0$ is equivalent to the existence of a strictly positive global lower bound on the smallest interior angle of all triangles.

Exercise 9.13 Let $T \subset \mathbb{R}^2$ be an open triangle with vertices t_1, t_2, t_3. Show that every $x \in T$ has a unique representation of the form $x = a_1 t_1 + a_2 t_2 + a_3 t_3$, $a_1 + a_2 + a_3 = 1$, $a_i \geq 0$, $i = 1, 2, 3$.

Exercise 9.14 Let $T \subset \mathbb{R}^2$ be an open triangle and $E := \{v : T \to \mathbb{R} : \exists\, a, b, c \in \mathbb{R} \text{ such that } v(x) = a + bx_1 + cx_2, \ x = (x_1, x_2) \in \overline{T}\}$. Show that $\dim E = 3$.

Exercise 9.15 Let $A = (a_{i,j})_{i,j=1,...,M} \in \mathbb{R}^{(M-1)\times(M-1)}$, $M \geq 2$, be a tridiagonal matrix such that $a_{i,i} = 2$ for $i = 1, \ldots, M - 1$ and $a_{i,i+1} = a_{i+1,i} = -1$ for $1 \leq i \leq M - 2$. Show that, for all $k = 1, \ldots, M - 1$,

$$v^k := \left(\sin\left(ik\frac{\pi}{M}\right)\right)_{i=1,...,M-1}, \quad \mu_k := \left(2 - 2\cos\left(k\frac{\pi}{M}\right)\right) = 4\sin^2\left(\frac{k}{2}\frac{\pi}{M}\right),$$

are, respectively, the eigenvalues and eigenvectors of A.

Exercise 9.16 (Convergence rate of the Galerkin approximation) Let V and H be real Hilbert spaces such that $V \hookrightarrow H$ and let $a : V \times V \to \mathbb{R}$ be bilinear, continuous and coercive. Set $D(A) := \{u \in V : \exists f \in H \text{ such that } a(u, v) = (f, v)_H \text{ for all } v \in V\}$. For $0 < h \leq 1$ let $V_h \subset V$ be a finite-dimensional subspace of V. Finally, suppose that for each $u \in D(A)$ there exists a constant $c_u > 0$ such that $dist_V(u, V_h) := \inf\{\|u - \chi\|_V : \chi \in V_h\} \leq c_u \, \varphi(h)$ for all $0 < h \leq 1$, where $\varphi : (0, 1] \to (0, \infty)$ is some function such that $\lim_{h \downarrow 0} \varphi(h) = 0$.

(a) Prove the following statement: for $f \in H$, let $u \in V$, $u_h \in V_h$ be the solution of the problem $a(u, v) = (f, v)_H$, $v \in V$ and the solution of the approximating problem $a(u_h, \chi) = (f, \chi)_H$, $\chi \in V_h$, respectively. Then there exists a constant $c_1 > 0$ such that $\|u - u_h\|_V \leq c_1 \, \varphi(h) \|f\|_H$ for all $0 < h \leq 1$.

(b) Set $D(A^*) := \{w \in V : \exists g \in H \text{ such that } a(v, w) = (v, g)_H \text{ for all } v \in V\}$, and suppose that for every $w \in D(A^*)$ there exists a constant $c_w^* > 0$ such that $dist_V(w, V_h) \leq c_w^* \, \varphi(h), 0 < h \leq 1$. Show that then there exists a constant $c_2 > 0$ such that $\|u - u_h\|_H \leq c_2 \, \varphi(h)^2 \|f\|_H, 0 < h \leq 1$.

(c) Let $H = L_2(\Omega)$, $\Omega \subset \mathbb{R}^2$ be a polygon, let $(\mathcal{T}_h)_{h>0}$ be a quasi-uniform family of admissible triangulations, and let V_h be given by (9.37). Further suppose that the form a is such that $D(A) \subset H^2(\Omega)$ and $D(A^*) \subset H^2(\Omega)$. Show that under these assumptions there exists a constant $c_3 > 0$ such that $\|u - u_h\|_{L_2(\Omega)} \leq c_3 \, h^2 \|f\|_{L_2(\Omega)}, 0 < h \leq 1$.

Suggestion: (a) Use Céa's lemma and the uniform boundedness principle [2]. (b) Apply (a) to the dual problem, as in the proof of Theorem 9.32(a).

Exercise 9.17 (Crank–Nicolson method) Prove Theorem 9.60.

Exercise 9.18 Let $A \in \mathbb{R}^{n\times n}$, $g \in L_2((0, T), \mathbb{R}^n)$ and $u_0 \in \mathbb{R}^n$. Show that the problem

$$\dot{u}(t) + Au(t) = g(t), \ t \in (0, T), \qquad u(0) = u_0,$$

has a unique solution $u \in H^1((0, T), \mathbb{R}^n)$ given by the formula $u(t) = u_0 + \int_0^t e^{-(t-s)A} g(s) \, ds$.

Exercise 9.19 (Taylor expansion)
Let X be a Banach space, $f \in C^{n+1}([a, b], X)$ and $t_0 \in (a, b)$. One can show as in the case $X = \mathbb{R}$ that $f(t) = T_n(t) + R_n(t)$ with

$$T_n(t) = \sum_{k=0}^{n} \frac{f^{(k)}(t_0)}{k!}(t - t_0)^k, \qquad R_n(t) = \int_{t_0}^{t} \frac{(t-s)^n}{n!} f^{(n+1)}(s)\, ds.$$

(a) Show that $\|R_n(t)\|_X \le \dfrac{|t - t_0|^{n+1}}{(n+1)!}\|f^{(n+1)}\|_{C([a,b],X)}$.

(b) Let $n = 2$. Show that $\|R_n(t)\|_X \le \dfrac{(b-a)^3}{48}\|f^{(3)}\|_{C([a,b],X)}$.

Suggestion: Choose $t_0 = \frac{1}{2}(a + b)$.

(c) If $\dim X = 1$, then it is well-known that for each $t \in [a, b]$ there exists a τ between t and t_0 such that $R_n(t) = \frac{1}{(n+1)!}(t - t_0)^{n+1} f^{(n+1)}(\tau)$. Show that this is no longer true if $\dim X \ge 2$.

Suggestion: Choose $n = 0$ and $X = \mathbb{R}^2$.

Chapter 10
Maple®, or why computers can sometimes help

Maple® is a program developed and distributed by the company Maplesoft (also in cooperation with universities and research institutes). It is a computer algebra system, that is, a program which undertakes mathematical manipulations *exactly* according to given rules; in particular, it does not commit roundoff errors as is the case with numerical approximation methods.

Among other advantages, Maple® thus enables one to perform complicated calculations on the computer, saving time and effort and reducing potential sources of error. Of course, Maple® is not the only product on the market for computer algebra systems; for example, *MuPAD* and *Mathematica* offer similar features. The following remarks on the use of computer algebra systems are based on the syntax of Maple®; the statements about strategies in general as well as the advantages and limitations of such tools are, however, equally valid for all such programs. Most such computer algebra systems also include some numerical methods; for Maple® this is the case from Version 12 on.

In this section a number of basic technics will be introduced; however, we cannot, nor do we wish to, lay claim to giving a complete treatise on the solution of the partial differential equations presented here using Maple®.

Chapter overview

© The Author(s), under exclusive license to Springer Nature Switzerland AG 2023
W. Arendt, K. Urban, *Partial Differential Equations*, Graduate Texts in
Mathematics 294, https://doi.org/10.1007/978-3-031-13379-4_10

10.1 Maple®

We will, in particular, use the package *PDEtools*, originally developed at the University of Waterloo (Ontario, Canada) by Edgardo S. Cheb-Terrab among others, together with Waterloo Maple Inc. Starting with the version R5, PDEtools is an integral part of Maple®, meaning in particular that very comprehensive online documentation is available. This also means that the commands from PDEtools can be called up directly without the need for a library.

In what follows, we will show a few examples of *Maple® worksheets*, all of which were generated using Version 12 and tested up to Version 17. The program commands are always given after the symbol ">". The output from Maple® appears in normal font (not typewriter style) and is additionally centered; such output can be, and was, generated by Maple® using the LaTeXexport command. At the head of each of the following Maple® commands is the file name. The corresponding Maple® worksheets (whence the file extension "mw") are available on the website of the book.

10.1.1 Elementary examples

We start with two elementary examples, for the linear transport equation and the wave equation. For the linear transport equation, we first consider Worksheet 10.1.

File: LinTransport.mw

```
>  restart:
>  infolevel[pdsolve]:=5:
>  eq := diff(u(t,x),t)+c*diff(u(t,x),x) = 0;
```

$$eq := \frac{\partial}{\partial t} u(t, x) + c \frac{\partial}{\partial x} u(t, x) = 0$$

```
>  pdsolve(eq);
```
Checking arguments ...
First set of solution methods (general or quase general solution)
Second set of solution methods (complete solutions)
Trying methods for first order PDEs
Second set of solution methods successful

Maple®-Worksheet 10.1: Linear transport equation.

The commands used are largely self-explanatory. Partial derivatives can be declared using the command `diff`, so that `LinTransport` defines the homogeneous linear transport equation. This partial differential equation is then solved (analytically) using `pdsolve`, a command from PDEtools. Here we have generated further output using the command infolevel=5. We see that various solution strategies (and heuristics) are tested internally. Of course, we obtain the same general solution as in Example 2.2 with $c = a$. Freely choosable solution components are always indicated in Maple® by an underscore "_", thus _F1 is the initial value function $u_0 \in C^1(\mathbb{R})$ given in Example 2.2.

For our second example we consider the wave equation in Worksheet 10.2. Here, as well, the Maple® commands used should be self-explanatory. We recover the general solution formula of d'Alembert (3.3) with the two freely choosable functions _F1 and _F2 (labeled there as φ and ψ).

File: WaveEqn.mw

```
>  restart:
>  eq:=diff(u(t,x),t,t) - diff(u(t,x),x,x)=0;
```

$$eq := \frac{\partial^2}{\partial t^2} u(t, x) - \frac{\partial^2}{\partial x^2} u(t, x) = 0$$

```
>  pdsolve(eq);
```

$$u(t, x) = \_F1(x + t) + \_F2(x - t)$$

Maple®-Worksheet 10.2: Wave equation.

Of course Maple® cannot determine such general solution formulae "by itself"; it is programmed so as to test whether, given a differential equation, internally known solution methods or heuristics can be used. If a solution can be determined, then it is returned as output, if not, then not. Thus Maple® can do exactly as much as it has been taught to do.

10.1.2 Solutions via Fourier transforms

Here we consider the initial value problem for the heat equation, cf. Example 3.60. We wish to solve this problem with the help of Fourier transforms; for this, we consider the Maple® worksheet HeatExFourier (Worksheet 10.3 on page 416). We are already familiar with the definition of the differential equation; the definition of the initial conditions is, again, largely self-explanatory. Note that at this point in time the function $f(x)$ is still unknown. Here we assume asymptotic boundary conditions in space, that is, that the solution decays as $x \to \pm\infty$. Only afterwards (under point 1.) do we define the initial value $f(x)$ as the characteristic function $\mathbb{1}_{(0,1)}$ of the unit interval. In order to be able to use the integral transformations that Maple® can perform, we need to load the corresponding package.

Under 2. we first express the function f as a linear combination of Heaviside functions, which are known to Maple®. The reason for doing so is that Maple® also

File: HeatExFourier.mw

```
>   restart;
>   eq:=diff(u(t,x),t)-a^2*diff(u(t,x),x,x)=0;
>   ICs:=u(0,x)=f(x);
```

$$eq := \frac{\partial}{\partial t} u(t,x) - a^2 \frac{\partial^2}{\partial x^2} u(t,x) = 0$$

$$ICs := u(0,x) = f(x)$$

— 1. ————————————————————

```
>   a:=1;
>   f:=x->piecewise(x<0,0,x<1,1,0):
>   'f(x)'=f(x);
```

$$a := 1$$

$$f(x) = \begin{cases} 0 & x<0 \\ 1 & x<1 \\ 0 & otherwise \end{cases}$$

```
>   with(inttrans):
```

— 2. ————————————————————

```
>   f:=unapply(convert(f(x),Heaviside),x);
```

$$f := x \mapsto Heaviside(x) - Heaviside(-1+x)$$

```
>   assume(t>0);
```

— 3. ————————————————————

```
>   fourier(eq,x,w);
```

$$w^2 fourier(u(t,x),x,w) + \frac{\partial}{\partial t} fourier(u(t,x),x,w) = 0$$

— 4. ————————————————————

```
>   ode:=subs(fourier(u(t,x),x,w)=s(t),%);
```

$$ode := w^2 s(t) + \frac{d}{dt} s(t) = 0$$

— 5. ————————————————————

```
>   dsolve({ode,s(0)=fourier(f(x),x,w)},s(t));
```

$$s(t) = \frac{i\left(e^{-iw}-1\right)e^{-w^2 t}}{w}$$

— 6. ————————————————————

```
>   sol:=invfourier(rhs(%),w,x);
```

$$sol := -1/2 \operatorname{erf}\left(1/2 \frac{-1+x}{\sqrt{t}}\right) + 1/2 \operatorname{erf}\left(1/2 \frac{x}{\sqrt{t}}\right)$$

```
>   plot3d(sol,x=-1..2,t=0..1,orientation=[50,40],
>   numpoints=2000,axes=framed,color="gray");
```

— 7. ————————————————————

```
>   pdetest(u(t,x)=sol,eq);
```

$$0$$

Maple®-Worksheet 10.3: Fourier transformation for the heat equation.

knows the Fourier transform of the Heaviside function and can thus easily deal with such a linear combination of them. This also illustrates how one sometimes needs to apply small "tricks" to help Maple® out.

In 3. we transform the differential equation with the help of the Fourier transform, and 4. serves to write the ordinary differential equation conveniently in terms of the unknown function $s(t)$. This ordinary differential equation can be solved using the command `dsolve` (cf. 5.), and in the case one obtains a closed formula for the solution. In 6. we apply the inverse Fourier transformation. We then express the solution `sol` in terms of the function `erf`, where

$$\mathrm{erf}(x) := \frac{2}{\sqrt{\pi}} \int_0^x e^{-t^2}\, dt$$

is the *error function*, related but not identical to the distribution function of the standard normal distribution. Here again, we can see that one already needs to know in advance how to use Maple®. The integral `erf` does not have a closed representation, meaning that Maple® cannot express the solution (as an inverse Fourier transform) in terms of a simple formula. However, if one already knows what the solution could look like, then one can try to express `sol` in terms of the corresponding functions, as has been done successfully in this case. We next depict the solution thus obtained graphically, cf. Figure 10.1 (left). Finally, in 7. we check the solution by inserting it into the differential equation and obtaining the error value 0, as desired.

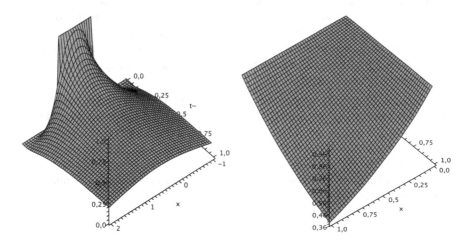

Fig. 10.1 Maple® solution of $u_t = u_{xx}$ with $u_0 = \chi_{[0,1)}$ using Fourier transforms (left) and Maple® solution of the inhomogeneous heat equation using Laplace transforms (right).

10.1.3 Laplace transforms

For our next example we consider the initial-boundary value problem (3.109) for the inhomogeneous heat equation on the interval $[0, 1]$, that is,

$$u_t - u_{xx} = f(t, x) := -(t^2 + x)e^{-tx}, \quad (t, x) \in \mathbb{R}^+ \times (0, 1),$$
$$u(0, x) = 1, \qquad\qquad\qquad\qquad x \in (0, 1),$$
$$u(t, 0) = a(t) := 1, \quad u(t, 1) = b(t) := e^{-t}, \qquad t > 0,$$

cf. Example 3.66. The corresponding commands are listed in Worksheet 10.4 (named HeatExLaplace, page 419). The declaration of the differential equation and the boundary and initial conditions should now be clear. In 1. we then transform the equation using Laplace transforms and substitute in the initial conditions. In 2. and 3. we set up the boundary value problem for the ordinary differential equation, and solve it in 4. using dsolve. After the inverse transformation in 5., we wish to represent the solution in terms of trigonometric functions, which does indeed lead to a simple form of the solution, represented graphically in Figure 10.1 (right). Finally, in 6. we check the result by inserting the solution into the original equation.

10.1.4 It can also be done numerically

As mentioned at the beginning of the chapter, in Version 12 the set of numerical solution methods available in Maple® was expanded. We now wish to demonstrate this briefly using the same example as just now, in Worksheet 10.5. The only difference is in the syntax of the additional option numeric in pdsolve. There is no recognizable difference to the function depicted in Figure 10.1 (right), the solution found using Laplace transforms.

Nevertheless, several warnings are called for. This example certainly shows how simply one can obtain a numerical solution using Maple®; however, we do not know which numerical method is used internally by Maple® to do so. We have already seen that this depends strongly on the particular problem. Secondly, at least up until now Maple® is known for not being optimally efficient in terms of calculation time. If short calculation times are important, then it is advisable to use a specialist numerics program instead.

10.1.5 Calculating function values

To finish, we return again to the solution formula for the Black–Scholes equation from Sections 1.5 and 3.5, where we derived a closed formula for the solution (albeit in terms of the distribution function $\mathcal{N}$ of the standard normal distribution, whose

File: HeatExLaplace.mw

```
>   restart;
>   f := (t,x) -> (-1)*(t^2+x)*exp(-t*x);
>   eq:=diff(u(t,x),t)-diff(u(t,x),x,x)=f(t,x);
>   a := t->1;
>   b := t->exp(-t);
>   ICs:=u(0,x)=1;
>   BCs:=u(t,0)=a(t), u(t,1)=b(t);
```

$$f := (t, x) \mapsto -\left(t^2 + x\right) e^{-tx}$$

$$eq := \frac{\partial}{\partial t} u(t, x) - \frac{\partial^2}{\partial x^2} u(t, x) = -\left(t^2 + x\right) e^{-tx}$$

$$a := t \mapsto 1$$

$$b := t \mapsto e^{-t}$$

$$ICs := u(0, x) = 1$$

$$BCs := u(t, 0) = 1, \, u(t, 1) = e^{-t}$$

```
>   with(inttrans):
```
— 1. ————————————————————————————————
```
>   laplace(eq,t,s):
>   subs(ICs,%);
```

$$s\,laplace(u(t, x), t, s) - 1 - \frac{\partial^2}{\partial x^2} laplace(u(t, x), t, s) = -2\,(s + x)^{-3} - \frac{x}{s + x}$$

— 2. ————————————————————————————————
```
>   ode:=subs(laplace(u(t,x),t,s)=U(x),%);
```

$$ode := sU(x) - 1 - \frac{d^2}{dx^2} U(x) = -2\,(s + x)^{-3} - \frac{x}{s + x}$$

— 3. ————————————————————————————————
```
>   BC1 := U(0) = laplace(a(t),t,s): subs(BCs, %);
>   BC2 := U(1) = laplace(b(t),t,s): subs(BCs, %);
```

$$U(0) = s^{-1}$$

$$U(1) = (1 + s)^{-1}$$

— 4. ————————————————————————————————
```
>   dsolve({ode, BC1, BC2},U(x));
```

$$U(x) = (s + x)^{-1}$$

— 5. ————————————————————————————————
```
>   invlaplace(rhs(%),s,t):
>   sol:=collect(map(simplify,expand(%)),trig);
```

$$sol := \left(e^{tx}\right)^{-1}$$

```
>   plot3d(sol,x=0..1,t=0..1,orientation=[50,40],
>   numpoints=2000,axes=framed,color="gray");
```
— 6. ————————————————————————————————
```
>   pdetest(u(t,x)=sol,eq);
```

$$0$$

Maple®-Worksheet 10.4: Laplace transformation for solving the inhomogeneous heat equation.

```
File: HeatNumerical.mw

  >   restart:
  >   f := (t,x) -> (-1)*(t^2+x)*exp(-t*x);
```
$$f := (t, x) \mapsto -\left(t^2 + x\right) e^{-tx}$$
```
  >   eq := diff(u(t,x),t) - diff(u(t,x),x,x) = f(t,x);
```
$$eq := \frac{\partial}{\partial t} u(t, x) - \frac{\partial^2}{\partial x^2} u(t, x) = -\left(t^2 + x\right) e^{-tx}$$
```
  >   cond := u(0,x)=1, u(t,0)=1, u(t,1)=exp(-t);
```
$$cond := u(0, x) = 1, u(t, 0) = 1, u(t, 1) = e^{-t}$$
```
  >   infolevel['pdsolve/numeric']:=3:
  >   Sol := pdsolve(eq,{cond},numeric);
```
$Sol :=$ module $()$ export $plot, plot3d, animate, value, settings; \dots$ endmodule
```
  >   Sol:-plot3d(t=0..1,x=0..1.0,orientation=[50,40],
  >   numpoints=2000,axes=box,axes=framed,color="gray");
```

Maple®-Worksheet 10.5: Numerical solution of the inhomogeneous heat equation.

```
File: BlackScholes.mw

  >   restart:
  >   with(Statistics):

  >   # Density of the standard normal distribution
  >   N := proc(d)
  >   CDF(Normal(0,1),d);
  >   end proc:

  >   # Formula for the value of a European call
  >   BlackScholesCall:=proc(S,K,T,r,sigma)
  >   local d1,d2;
  >   d1:=(ln(S/K)+(r+sigma^2/2)*T)/(sigma*sqrt(T));
  >   d2:=d1-sigma*sqrt(T);
  >   S*N(d1)-K*exp(-r*T)*N(d2);
  >   end proc:

  >   r     := 0.05:
  >   sigma := 0.5:
  >   S     := 100.0:
  >   T     := 1.0:
  >   K     := 100.0:

  >   BlackScholesCall(S,K,T,r,sigma);
                       21.79260422
```

Maple®-Worksheet 10.6: Calculating function values for the solution of the Black–Scholes equation.

values cannot be calculated exactly). We wish to show how one can obtain these values using Maple®, that is, one is essentially using Maple® as a "high-powered, intelligent calculator". For this, in Worksheet 10.6 we see two procedures; we have used Maple® much like a programming language. The first procedure returns the values of the function N. Since this can only be done approximatively due to the integral involved, a numerical quadrature method is hidden behind the command `statevalf` with the parameters `cdf` (cumulative density function) and `normald` (normal distribution).

The procedure `BlackScholesCall` implements the solution formula derived and proved in Theorem 3.47. The procedure thus returns the value of the option for the given parameters for r, σ, S, T and K. The valuation formula was and is used in exactly this way in banks. Here, however, the same warning applies as above. If one often needs to use this formula (for example within the scope of Monte Carlo simulations) then Maple® is rather less efficient than many other implementations (for example in a higher programming language). On the other hand, this example again demonstrates how easy Maple® is to use.

10.2 Exercises

Exercise 10.1 Write Maple® scripts to determine the respective solutions of the partial differential equations listed in Table 2.1. In each case choose initial and/or boundary conditions in such a way that it is possible to determine a solution formula using Maple®.

Exercise 10.2 Write programs to solve the inhomogeneous heat equation using finite differences and finite elements, respectively. Compare the runtimes with those of Worksheet 10.5.

Appendix

In this appendix we gather together the basic definitions and results from functional analysis and integration theory that we require throughout the book. Moreover, we add some more details concerning the derivation of the Black-Scholes equation.

A.1 Banach spaces and linear operators

Let E be a vector space over $\mathbb{K} = \mathbb{R}$ or $\mathbb{C}$. We usually consider real vector spaces, but for some applications such as the Fourier transforms considered in Chapter 6 we require $\mathbb{K} = \mathbb{C}$. Let $\| \cdot \| : E \to [0, \infty)$ be a *norm* on E; that is, it satisfies

(N1) $\|f\| = 0 \iff f = 0$ for all $f \in E$;

(N2) $\|\lambda f\| = |\lambda| \, \|f\|$, $\lambda \in \mathbb{K}$, $f \in E$;

(N3) $\|f + g\| \leq \|f\| + \|g\|$, $f, g \in E$.

We say that $(E, \| \cdot \|)$ (or for brevity just E) is a *normed vector space*, or just *normed space* for short. We sometimes write $\| \cdot \|_E$ for the norm in place of $\| \cdot \|$ (for example when various norms appear). If $f_n, f \in E$, then we say that $(f_n)_{n \in \mathbb{N}}$ *converges* to f, and write $\lim_{n \to \infty} f_n = f$ or $f_n \to f$, if $\lim_{n \to \infty} \|f_n - f\| = 0$. A sequence $(f_n)_{n \in \infty}$ has at most one limit. Every convergent sequence $(f_n)_{n \in \infty}$ is a *Cauchy sequence*, that is, for all $\varepsilon > 0$ there exists an $n_0 \in \mathbb{N}$ such that $\|f_n - f_m\| < \varepsilon$ for all $n, m \geq n_0$. If conversely for every Cauchy sequence $(f_n)_{n \in \infty}$ there exists an $f \in E$ such that $\lim_{n \to \infty} f_n = f$, then we say that E is *complete*. A *Banach space* is a complete normed space. Suppose that E and F are normed spaces and $T : E \to F$ is a linear mapping. Then we say that T is *continuous* if $\lim_{n \to \infty} f_n = f$ in E always implies that $\lim_{n \to \infty} T f_n = T f$ in F. The linearity of T allows us to characterize the property of continuity as follows.

Theorem A.1 *The following statements are equivalent.*

(i) T is continuous;

(ii) There exists a constant $c \geq 0$ such that $\|T f\| \leq c \|f\|$ for all $f \in E$;

(iii) $\|T\| := \sup_{\|f\| \leq 1} \|T f\| < \infty$.

© The Author(s), under exclusive license to Springer Nature Switzerland AG 2023
W. Arendt, K. Urban, *Partial Differential Equations*, Graduate Texts in
Mathematics 294, https://doi.org/10.1007/978-3-031-13379-4

If T is continuous, then $\|Tf\| \le \|T\| \, \|f\|$ for all $f \in E$.

Proof (i) $\Rightarrow$ (iii) We prove the contrapositive: suppose that $\|T\| = \infty$; then there exists a sequence of vectors $f_n \in E$ such that $\|f_n\| \le 1$ but $\|Tf_n\| \ge n$ for all $n \in \mathbb{N}$. If we set $g_n := \frac{1}{n}f_n$, then $\lim_{n\to\infty} g_n = 0$, but $Tg_n = \frac{1}{n}Tf_n$ and so $\|Tg_n\| \ge 1$. Thus Tg_n does not converge to $0 = T0$, and T is not continuous at 0.

(iii) $\Rightarrow$ (ii) Let $g \in E$ be arbitrary, $g \ne 0$, and set $f := \|g\|^{-1}g$. Then $\|f\| = 1$, and so $\|g\|^{-1} \|Tg\| = \|Tf\| \le \|T\|$. This shows that (ii) holds with $c := \|T\|$; and in particular (since the inequality obviously holds for $f = 0$), $\|Tf\| \le \|T\| \, \|f\|$ for all $f \in E$.

(ii) $\Rightarrow$ (i) Suppose that $\lim_{n\to\infty} f_n = f$. Since $\|Tf_n - Tf\| = \|T(f_n - f)\| \le c\|f - f_n\|$, it follows that $\lim_{n\to\infty} \|Tf_n - Tf\| = 0$, that is, $\lim_{n\to\infty} Tf_n = Tf$. $\qquad \square$

A set $M \subset E$ is said to be *bounded* if there exists a constant $c > 0$ such that $\|x\| \le c$ for all $x \in M$. Condition (iii) in the above theorem asserts that the image of the unit ball $\{x \in E : \|x\| \le 1\}$ under T is bounded in F. For this reason, continuous linear mappings are often also called *bounded (linear) operators*. We call the expression $\|T\|$ the *operator norm* of T. Now consider the special case $F = \mathbb{K}$. A linear mapping $\varphi : E \to \mathbb{K}$ is called a *linear form* (or *functional*). If φ is continuous, then in accordance with the above definition

$$\|\varphi\| := \sup_{\|f\| \le 1} |\varphi(f)|$$

is the *norm* of φ. This satisfies $|\varphi(f)| \le \|\varphi\| \, \|f\|$ for all $f \in E$. A subspace $D \subset E$ of E is said to be *dense* in E if for all $f \in E$ there exists a sequence $(f_n)_{n\in\mathbb{N}} \subset D$ such that $\lim_{n\to\infty} f_n = f$. If the codomain of a densely defined bounded operator is complete, then the operator can be extended continuously to an operator on the whole space.

Theorem A.2 (Continuous extension) *Let F be a Banach space, E a normed space and D a dense subspace of E. If $T_0 : D \to F$ is a bounded linear operator, then it has a unique continuous extension $T : E \to F$. This operator is linear and satisfies $\|T\| = \|T_0\|$.*

Proof Let $f \in E$ be arbitrary. By assumption, there exist $f_n \in D$ such that $\lim_{n\to\infty} f_n = f$. Since $\|T_0 f_n - T_0 f_m\| \le \|T_0\| \, \|f_n - f_m\|$, we see that $(T_0 f_n)_{n\in\mathbb{N}}$ is a Cauchy sequence. We define $Tf := \lim_{n\to\infty} T_0 f_n$ (it is easy to show that this definition is independent of the choice of sequence $(f_n)_{n\in\mathbb{N}}$); then it is also not hard to show that T is linear. Moreover, we have $\|Tf\| = \lim_{n\to\infty} \|T_0 f_n\| \le \limsup_{n\to\infty} \|T_0\| \, \|f_n\| = \|T_0\| \, \|f\|$. This shows that $\|T\| \le \|T_0\|$. On the other hand, since T is an extension of T_0, it is immediate that $\|T_0\| \ge \|T\|$. $\qquad \square$

The following theorem gives us an efficient but simple criterion to check the continuity of a linear mapping. An essential requirement is the completeness of the underlying spaces. We refer to [2] for the proof.

Theorem A.3 (Closed graph theorem) *Let E and F be Banach spaces and T : $E \to F$ linear. Suppose that, whenever $x_n \to x$ in E and $Tx_n \to y$ in F, we also have $Tx = y$. Then T is continuous.*

This theorem often allows huge simplifications in proofs of continuity. In fact, assume that $\lim_{n \to \infty} x_n = x$; then the continuity of T requires two things:
(a) $y := \lim_{n \to \infty} Tx_n$ exists, and
(b) $y = Tx$.
The closed graph theorem allows us to omit the proof of (a), or, more precisely, to assume convergence in (a) as our starting point. It only remains to prove the identity (b). A direct consequence of the closed graph theorem is the bounded inverse theorem.

Theorem A.4 (Bounded inverse theorem) *Let E and F be Banach spaces and $T : E \to F$ linear, bijective and continuous (that is, bounded). Then T^{-1} is linear and continuous.*

Proof The linearity is easy to check. For the proof of continuity we use the close graph theorem. Let g_n and $g \in F$ be such that $g_n \to g$ and $T^{-1}g_n \to f$. Since T is continuous, it follows that $g_n = TT^{-1}g_n \to Tf$. Thus $Tf = g$ and so $f = T^{-1}g$, which is what we had to show.   □

The bounded inverse theorem plays a role in the solution of equations: suppose that $T : E \to F$ is linear and continuous; then to say that T is bijective means that for every $f \in F$ there exists a unique $u \in E$ such that $Tu = f$. The bounded inverse theorem tells us that in the solution u is automatically continuously dependent on the given data f.

A.2 The space $C(K)$

Let $K \subset \mathbb{R}^d$ be a compact set. Then $C(K) := \{f : K \to \mathbb{R} : f \text{ continuous}\}$ is a real Banach space, where the vector space structure is given by pointwise addition $(f + g)(x) := f(x) + g(x)$, $x \in K$ and scalar multiplication $(\lambda f)(x) := \lambda f(x)$, $x \in K$ (for $f, g \in C(K)$ and $\lambda \in \mathbb{R}$); and the norm is the supremum norm $\|f\|_{C(K)} :=$ $\sup_{x \in K} |f(x)|$, $f \in C(K)$. We refer to [2] for the proof of completeness. We wish to discuss two remarkable theorems on $C(K)$. The first gives a condition under which a subspace of $C(K)$ is dense in $C(K)$, while the second gives a condition for the compactness of subsets of $C(K)$. Observe that if $f, g \in C(K)$, then the product $f \cdot g$ is also in $C(K)$, where $(f \cdot g)(x) := f(x) \cdot g(x)$, $x \in K$. That is, $C(K)$ is an algebra. The following theorem gives a useful criterion for the density of a subalgebra of $C(K)$. It is due to Marshall Harvey Stone (1903–1989); see [53, Ch. 7.7] for a proof.

Theorem A.5 (Stone–Weierstrass) *Let $K \subset \mathbb{R}^d$ be compact and let F be a subspace of $C(K)$ which has the following three properties:*
(a) $f, g \in F \Rightarrow f \cdot g \in F$;

(b) all constant functions belong to F;
(c) for all x, y ∈ K, x ≠ y, there exists a function f ∈ F such that $f(x) \neq f(y)$.
Then F is dense in C(K), that is, for every f ∈ C(K) there exists a sequence $(f_n)_{n \in \mathbb{N}}$
in F such that $\lim_{n \to \infty} f_n = f$ in C(K).

A subspace which satisfies (a) is called a *subalgebra* of $C(K)$. If (c) holds, then we say that F *separates points* in K. An immediate consequence of Theorem A.5 is Weierstrass' theorem, which states that every continuous function on a compact interval may be approximated uniformly by polynomials.

The second theorem we shall present has its motivation in an unfortunate property of infinite-dimensional normed spaces: in such spaces one can always find a bounded sequence which does not have a convergent subsequence. We call a set L in a normed space *relatively compact* if every sequence in L has a convergent subsequence; we say L is *compact* if it is additionally closed. Relatively compact sets are always bounded, but boundedness alone is insufficient to imply relative compactness. In spaces of the form $E = C(K)$ one can identify exactly what is required in addition to boundedness: the equicontinuity. A set $L \subset C(K)$ is called *equicontinuous* if it has the following property: let $x_0 \in K$ and $\varepsilon > 0$. Then there exists $\delta > 0$ such that $|x - x_0| < \delta$ implies $|f(x) - f(x_0)| < \varepsilon$ *for all* $f \in L$. More precisely, we have the following theorem [54, Thm. 11.28]:

Theorem A.6 (Arzelà–Ascoli) *A subset L of C(K) is relatively compact if and only if it is bounded and equicontinuous.*

Example A.7 Let $L := \{f \in C([a, b]) : \|f\|_{C(K)} \leq c, \|f'\|_{C(K)} \leq c\}$ for some $c \geq 0$. Then L is relatively compact.   △

Proof Clearly L is bounded. Moreover, given $x_0 \in [a, b]$ and $f \in L$, we have $|f(x) - f(x_0)| = |\int_{x_0}^{x} f'(y) \, dy| \leq c|x - x_0|$; this yields the equicontinuity.   □

A.3 Integration

We assume familiarity with the Lebesgue integral and refer in case of necessity to the compact introduction of Bartle [13]. Here we merely wish to make available the central convergence theorems as well as a version of the theorem of Fubini–Tonelli for annular domains. Throughout, we will suppose that $\Omega \subset \mathbb{R}^d$ is an open set. If $f : \Omega \to [0, \infty]$ is measurable, then we denote by $\int_\Omega f \, dx \in [0, \infty]$ its Lebesgue integral over Ω. If $\int_\Omega f \, dx < \infty$, then $f(x) < \infty$ for almost every $x \in \Omega$, and we say that f is *integrable*. The monotone convergence theorem due to Beppo Levi reads as follows.

Theorem A.8 (Monotone convergence) *Let $f_n : \Omega \to [0, \infty]$ be measurable and suppose that $f_n \leq f_{n+1}$ pointwise almost everywhere, for all $n \in \mathbb{N}$. Then $f(x) :=$*

$\sup_{n \in \mathbb{N}} f_n(x)$ defines a measurable function and $\int_\Omega f(x)\, dx = \sup_{n \in \mathbb{N}} \int_\Omega f_n(x)$. If there exists a constant $c \geq 0$ such that $\int_\Omega f_n\, dx \leq c$ for all $n \in \mathbb{N}$, then f is additionally integrable.

This means that we may exchange the order of limit and integration. If the functions are not positive, then the assertion remains true if there exists an integrable dominating function. This is the statement of the *Dominated Convergence Theorem*, which is due to Lebesgue in 1910. We wish to formulate a more general version for p-integrable functions. So let $1 \leq p < \infty$ and set $L_p(\Omega) := \{f : \Omega \to \mathbb{R} : f \text{ is measurable and } \int_\Omega |f|^p\, dx < \infty\}$. If we identify two functions in $L_p(\Omega)$ whenever they agree with each other almost everywhere, then $\|f\|_{L_p} := \left(\int_\Omega |f|^p\, dx\right)^{\frac{1}{p}}$ defines a norm with respect to which $L_p(\Omega)$ is a Banach space. Of particular importance to us is the case $p = 2$; the space $L_2(\Omega)$ is a Hilbert space with respect to the inner product $(f, g) := \int_\Omega f(x)g(x)\, dx$, which induces the norm $\|f\|_{L_2} = \sqrt{(f, f)}$. We can now formulate the Dominated Convergence Theorem.

Theorem A.9 (Dominated convergence) *Suppose that $f_n, f, g \in L_p(\Omega)$ satisfy $|f_n(x)| \leq g(x)$ for almost every $x \in \Omega$ and all $n \in \mathbb{N}$. If $f(x) = \lim_{n \to \infty} f_n(x)$ for almost every $x \in \Omega$, then $\lim_{n \to \infty} f_n = f$ in $L_p(\Omega)$, that is, $\lim_{n \to \infty} \|f_n - f\|_{L_p} = 0$.*

The Dominated Convergence Theorem also admits a converse, if we allow passing to subsequences.

Theorem A.10 (Converse of the Dominated Convergence Theorem) *Suppose that $f_n, f \in L_p(\Omega)$ satisfy $f = \lim_{n \to \infty} f_n$ in $L_p(\Omega)$. Then there exist $g \in L_p(\Omega)$ and a subsequence $(f_{n_k})_{k \in \mathbb{N}}$ such that*
(a) $|f_{n_k}(x)| \leq g(x)$ almost everywhere, for all $k \in \mathbb{N}$, and
(b) $\lim_{k \to \infty} f_{n_k}(x) = f(x)$ almost everywhere.

If $|\Omega| < \infty$, then $L_\infty(\Omega) \subset L_2(\Omega) \subset L_1(\Omega)$, where $L_\infty(\Omega)$ denotes the set of all bounded measurable functions on Ω. A *hyperrectangle* in $\mathbb{R}^d$ is a set of the form $[a_1, b_1) \times \cdots \times [a_d, b_d)$; if $d = 2$, then this reduces to a rectangle. We refer to linear combinations of characteristic functions of hyperrectangles as *step functions*.

Theorem A.11 *[47, Sec. 1.17] The space of all step functions is dense in $L_p(\Omega)$, $1 \leq p < \infty$.*

We define $L_p(\Omega, \mathbb{C}) := \{f : \Omega \to \mathbb{C} : \operatorname{Re} f, \operatorname{Im} f \in L_p(\Omega)\}$. Then $L_p(\Omega, \mathbb{C})$ is a Banach space with respect to the norm $\|f\|_{L_p}$, for any $1 \leq p < \infty$. If $f \in L_1(\Omega, \mathbb{C})$, then we set $\int_\Omega f\, dx = \int_\Omega \operatorname{Re} f\, dx + i \int_\Omega \operatorname{Im} f\, dx$. The space $L_2(\Omega, \mathbb{C})$ is a complex Hilbert space with respect to the inner product $(f, g)_{L_2} := \int_\Omega f(x)\overline{g(x)}\, dx$. Finally, we wish to introduce an integration formula which recalls the theorem of Fubini–Tonelli on interchanging the order of integration. Here, however, we will work in polar coordinates; here we imagine integrating over the "rings" of an annular domain like the rings of an onion. We denote by σ the surface measure on the sphere $S^{d-1} = \{x \in \mathbb{R}^d : |x| = 1\}$, cf. Sections 7.1 and 7.2.

Theorem A.12 *Let* $0 \le R_1 < R_2 \le \infty$ *and* $\Omega = \{x \in \mathbb{R}^d : R_1 < |x| < R_2\}$.
(a) Suppose that $f : \Omega \to [0, \infty]$ *is measurable or* $f : \Omega \to \mathbb{R}$ *is integrable. Then*
$\int_\Omega f(x) \, dx = \int_{R_1}^{R_2} \int_{S^{d-1}} f(rz) \, d\sigma(z) \, r^{d-1} \, dr$.
(b) If in particular $f : \Omega \to \mathbb{R}$ *is a continuous* radial *function, that is, we may write*
$f(x) = f(|x|)$ *for all* $x \in \Omega$, *then* f *is integrable if and only if* $\int_{R_1}^{R_2} |f(r)| r^{d-1} \, dr < \infty$. *Moreover, in this case, we have* $\int_\Omega f(x) \, dx = \sigma(S^{d-1}) \int_{R_1}^{R_2} f(r) r^{d-1} \, dr$.

A.4 More details on the Black–Scholes equation

In addition to Section 1.5, we give some more details on the modeling leading to the Black–Scholes equation. In the same manner as we cannot completely detail the physical background of the equations introduced in Chapter 1, we do not describe all required facts from economics and stochastics here. However, this appendix should give a deeper understanding of the ideas leading to this famous equation. For more background, in particular concerning the financial mathematics and the Itô integral, we refer to [15, 43, 51].

In order to start with the actual modeling, we first need to fix our assumptions about the behavior of the underlying. In the examples considered in Chapter 1, we could appeal to physical laws; here we need to make an assumption about the behavior of the share price *in the future*. This has two essential consequences:

- It is generally considered that the underlying (such as a share price) evolves stochastically. We therefore need to model the behavior of $S(t)$ stochastically and as such require certain fundamentals from the theory of stochastic processes.
- The behavior of the underlying in the past is known from observing the market. Whether this is a good predictor of the future, is naturally unknowable.

A.4.1 Basics of stochastics

We start by providing the necessary background material; we will however assume familiarity with terms such as *probability space* and *random variable*.

Definition A.13 Let $(\Omega, \mathfrak{A}, \mathsf{P})$ be a probability space.[1] We call any family $X = X(t)_{t \ge 0}$ of random variables $X(t) : \Omega \to \mathbb{R}$ a *stochastic process*. We say that X has a *continuous path* if the function $t \mapsto X(t; \omega) : [0, \infty) \to \mathbb{R}$ is continuous for all $\omega \in \Omega$. $\triangle$

---

[1] Be aware of the double use of the notation Ω: in the context of differential equations, $\Omega \subset \mathbb{R}^d$ will denote the domain in the space variables; here, on the other hand, it is the underlying sample space for the probability space.

Example A.14 An example is the *Wiener process W*. This is a stochastic process with continuous paths, for which in particular

$$P(\{W_t \in (a, b)\}) = \frac{1}{\sqrt{2\pi t}} \int_a^b e^{-\frac{x^2}{2t}} \, dx,$$

that is, W_t is normally distributed. More precisely, W is uniquely determined by the property that the increments $W_t - W_s$, $t \geq s > 0$, are stochastically independent and $\mathcal{N}(0, t - s)$-distributed (that is, normally distributed with expectation $\mu = 0$ and variance $\sigma = t - s$), and $W_0 = 0$. △

The next step is the definition of an integral with respect to a stochastic process, for example in order to model cumulative yields. It is clear that standard notions of integral such as those of Riemann or Lebesgue cannot be of use for general stochastic processes since it is necessary to integrate functions which have infinite variation.

In stochastics, one defines the *Itô integral* of a stochastic process X with respect to a second stochastic process W (with which we will again assume familiarity); we will denote this integral by

$$\int_0^t X(s) \, dW(s). \tag{A.1}$$

The appelation "Itô integral" was chosen in honor of Kiyoshi Itô, who introduced the corresponding calculus in 1951. We refer to the specialised literature for the details.

A.4.2 Black–Scholes model

Now we need to fix our assumptions on the future behavior of the price S of the underlying. In the Black–Scholes model, one assumes that $S(T)$ is a *geometric Brownian motion* with *drift* μ and *volatility* σ. Here we will summarize its essential properties.

Remark A.15 The *geometric Brownian motion* with drift μ and trend σ (sometimes also called volatility, which is, however, $\mu - \frac{1}{2}\sigma^2$) is the stochastic process S defined by

$$S(t) := S(0) \exp\left(\mu t + \sigma W(t) - \frac{1}{2}\sigma^2 t\right),$$

where W denotes the *Wiener process* introduced in Example A.14. It has the following properties:
(a) $S(t)$ is log-normally distributed (that is, $\log(S(t)) \sim \mathcal{N}(\mu, \sigma^2)$) with expectation $\mathbb{E}(S(t)) = S(0) e^{\mu t}$ and variance $\text{Var}(S(t)) = S(0)^2 e^{2\mu t}(e^{\sigma^2 t} - 1)$.
(b) The paths $S(t)$, $t \in [0, T]$, are continuous.
(c) The following integral equation holds:

$$S(t) = S(0) + \int_0^t \mu\, S(s)\, ds + \int_0^t \sigma\, S(s)\, dW(s), \qquad (A.2)$$

where the latter integral is to be understood as an Itô integral, as in Definition A.1. The integral equation (A.2) is often written in the form

$$dS(t) = \mu\, S(t)\, dt + \sigma\, S(t)\, dW(t). \qquad (A.3)$$

This equation may be referred to as an *Itô differential equation* with drift term $\mu\, S(t)\, dt$ and diffusion $\sigma\, S(t)\, dW(t)$.   △

The next step is to determine the value of the option $V(t, S(T))$ in dependence on the stochastic model for the share in (A.3). We therefore substitute the stochastic process S from (A.3) for y in the value function $V = V(t, y)$, as a model for the share price. If we assume that $V \in C^2([0, T] \times \mathbb{R})$, then we can use the stochastic version of the chain rule, known as *Itô's lemma*, to obtain the stochastic differential equation

$$dV(t, S(t)) = \left(V_t + \mu\, S(t)V_y + \frac{1}{2}\sigma^2 S(t)^2 V_{yy}\right)dt + \sigma S(t)V_y\, dW(t). \qquad (A.4)$$

A.4.3 The fair price

We can finally turn to the question of what a "fair" price might be. Here we will consider a very simple market in which there are only two investment possibilities available. On the one hand, one can deposit one's money for a fixed interest rate $r > 0$ and zero risk: if one deposits the amount B_0 for $t > 0$ units of time, then at time t one receives the amount $B(t) = B_0 e^{rt}$, cf. Remark 3.48. We further assume that in our idealized market money can be borrowed for the same interest rate $r > 0$. So if one borrows the amount Z_0 at time 0, then at time $t > 0$ one needs to repay the amount $Z_0 e^{rt}$. In our market the second investment possibility, which carries risk, will be denoted by S and given by the geometric Brownian motion described above. We imagine S as being the value of a share which does not pay dividends. We now consider a portfolio (the value of a particular trading strategy)

$$X(t) = c_1(t)B(t) + c_2(t)S(t), \qquad (A.5)$$

where $B(t) = e^{rt}$, and c_1, c_2 are stochastic processes. Since $S(0, \omega) = S_0$, $\omega \in \Omega$, is deterministic (S_0 is the known share price at time 0), we shall assume that $c_1(0)$ and $c_2(0)$ are also deterministic (that is, they are numbers). We thus have an investment strategy consisting of shares of total value $c_2(t)S(t)$ and a deposit of value $c_1(t)B(t)$ earning a fixed interest rate. We also wish to assume that this investment is *self-financing*, that is, profits will not be deducted and no additional money will be injected. The change in the value of the portfolio over time thus consists only of movements between the share component and the fixed-interest deposit. Mathematically speaking, this means that the (cumulative) yield of the portfolio is given by the sum of the yield from interest and the change in the value of the shares, $X^{\text{yield}}(t) := \int_0^t c_1(s)\, r\, e^{rs}\, ds + \int_0^t c_2(s)\, dS(s)$. Written as a stochastic differential equation, this takes the form $dX^{\text{yield}}(t) = c_1(t)\, dB(t) + c_2(t)\, dS(t)$. A strategy is *self-financing* if and only if the value at time t is equal to the sum of the initial value and the cumulative yield at time t, that is, if $X(t) = X(0) + X^{\text{yield}}(t)$. This means that $dX(t) = dX^{\text{yield}}(t)$ and so

$$dX(t) = c_1(t)\, dB(t) + c_2(t)\, dS(t). \tag{A.6}$$

We now set ourselves the following task: find a continuous function $V : [0, T] \times \mathbb{R} \to \mathbb{R}$, $V = V(t, y)$, which is continuously differentiable in t and twice continuously differentiable in y in $(0, T) \times \mathbb{R}$, together with a self-financing investment strategy c_1, c_2 such that, for the process

$$Y(t) = c_1(t)\, B(t) + c_2(t)\, S(t) - V(t, S(t)), \tag{A.7}$$

we have $Y(t) = Y_0 e^{rt}$, where $Y_0 = c_1(0)B(0) + c_2(0)S_0 - V(0, S_0)$. In other words, Y is a risk-free investment, and the process $X(t)$ has the same risk as the option $V(t, S(t))$. We speak of a *replicating portfolio*. We start by showing that the above conditions imply that V satisfies the Black–Scholes equation. Afterwards we will explain how one can determine the fair price of the option.

The process Y satisfies the differential equation (written in stochastic notation)

$$dY(t) = r\, Y(t)\, dt. \tag{A.8}$$

Now we insert the stochastic differential equations (A.8) (for $dY(t)$), (A.3) (for $dS(t)$) and (A.4) (for $dV(t)$) into (A.6) and obtain (we omit the argument t)

$$
\begin{aligned}
dY &= c_1\, dB + c_2\, dS - dV \\
&= c_1 r B dt + c_2(\mu S dt + \sigma S dW) - (V_t + \mu S V_y + \frac{\sigma^2}{2} S^2 V_{yy}) dt - \sigma S V_y dW \\
&= \left[c_1\, r\, B + c_2\, \mu\, S - \left(V_t + \mu S V_y + \frac{1}{2}\sigma^2 S^2 V_{yy} \right) \right] dt \\
&\quad + \left[c_2\, \sigma\, S - \sigma\, S V_y \right] dW.
\end{aligned}
\tag{A.9}
$$

Since the portfolio $Y(t)$ should be risk-free and self-financing, (A.8) and (A.6) hold; thus $dY = r Y dt = r(c_1 B + c_2 S - V)\, dt$ and so the term in the second line of (A.9) needs to vanish, as can be seen by equating coefficients. Hence $c_2 - V_y(t, S(t)) = 0$, that is, $c_2(t) = V_y(t, S(t))$. By equating the coefficients of dt we obtain

$$
r\left(c_1(t)\, B(t) + S(t) V_y(t, S(t)) - V(t, S(t)) \right) =
$$
$$
= \left(c_1(t)\, r\, B(t) - V_t(t, S(t)) - \frac{1}{2}\sigma^2 S(t)^2 V_{yy}(t, S(t)) \right).
$$

The term $r c_1 B$ cancels and we obtain the identity

$$V_t(t, S(t)) + \frac{\sigma^2}{2} S(t)^2 V_{yy}(t, S(t)) + rS(t)V_y(t, S(t)) - rV(t, S(t)) = 0. \qquad (A.10)$$

We recall that S is a stochastic process. Thus (A.10) means, when written out in full,

$$V_t(t, S(t, \omega)) + \frac{\sigma^2}{2} S(t, \omega)^2 V_{yy}(t, S(t, \omega)) + rS(t, \omega)V_y(t, S(t, \omega))$$
$$-rV(t, S(t, \omega)) = 0 \qquad (A.11)$$

for all $t \in [0, T)$ and almost every $\omega \in \Omega$. We know that for each fixed time $t > 0$, for each $x \in [0, \infty)$ and $\varepsilon > 0$, one has $S(t, \omega) \in (x - \varepsilon, x + \varepsilon)$ with positive probability. It follows that (A.11) holds for all $t \in [0, T)$ and almost every $\omega \in \Omega$ if and only if $V : (0, T) \times \mathbb{R}_+ \to \mathbb{R}_+$ satisfies the *Black–Scholes equation*

$$V_t + \frac{1}{2}\sigma^2 y^2 V_{yy} + ryV_y - rV = 0, \qquad (t, y) \in (0, T) \times \mathbb{R}. \qquad (A.12)$$

We have seen in Section 3.5 that the equation (A.12) admits a unique polynomially bounded solution V which takes on the final value $V(T, y) = (y - K)^+$. The initial value $V_0 := V(0, S_0)$ of this solution V at the point $y = S_0$ with share price S_0 at initial time $t = 0$ is the fair price of the option. To understand this, we imagine ourselves in the role of a banker. The banker receives the amount V_0 from her client at time $t = 0$, and needs to pay back the amount $(S(T) - K)^+$, which depends on the share price $S(T)$ (a random variable), at time T. In order to generate this amount of money, she proceeds as follows. She chooses a mixed, self-financing portfolio (A.5) in such a way that $Y(t) := X(t) - V(t, S(t))$ is risk-free, that is, $Y(t) = Y_0 e^{rt}$. Here V is the above solution of the Black–Scholes equation with $V(T, y) = (y - K)^+$. We have seen that a necessary condition for the existence of this portfolio is that V satisfies the Black–Scholes equation. Conversely, for the trading strategy of the banker, we need to construct the processes c_1, c_2 (we have already seen that $c_2(t) = V_y(t, S(t))$ needs to hold); we abstain from giving the mathematical details. In practice, the banker needs to reconfigure her portfolio on a daily (or even continuous) basis, an activity referred to as *hedging*.

At time T the value of the portfolio is $X(T) = Y_0 e^{rT} + (S(T) - K)^+$, while the initial investment in the portfolio is $X(0) = Y_0 + V_0$; this represents a difference from the amount that the banker actually received from her client of Y_0. The banker must now distinguish between three cases and choose her strategy accordingly:

1st case: $Y_0 = 0$: the banker receives exactly the amount to be paid back to the client at time T from her portfolio.

2nd case: $Y_0 < 0$: the banker receives only $(S(T) - K)^+ - |Y_0|e^{rT}$ from the portfolio but is required to pay $(S(T) - K)^+$. But she also only needs to invest $X(0) = V_0 - |Y_0|$. In this case, she invests the amount $|Y_0|$ at the fixed interest rate, which yields the missing amount $|Y_0|e^{rT}$ at time t. In this way she can exactly afford the required payment.

3rd case: $Y_0 > 0$: the banker needs to invest $X(0) = V_0 + Y_0$ but has only received V_0 from the client. In this case, the solution is to borrow the difference Y_0; at time T

she must then pay back $Y_0 e^{rT}$. But since the portfolio yields $(S(T) - K)^+ + Y_0 e^{rT}$, it covers both the loan repayment and the payment to the client.

In all three cases the price is fair: the banker can generate the promised payoff $(S(T) - K)^+$ without risk.

References

1. Abel, N.H. Untersuchungen über die Reihe $1 + \frac{m}{1}x + \frac{m(m-1)}{1\cdot 2}x^2 + \ldots + \frac{m(m-1)(m-2)}{1\cdot 2\cdot 3}x^3 + \cdots$. *J. Reine Angew. Math.*, 1:311–339, 1826.
2. Alt, H.W. *Linear Functional Analysis*. Springer-Verlag, London, 2016. An application-oriented introduction, Translated from the German edition by R. Nürnberg.
3. Amann, H. *Ordinary Differential Equations*. Walter de Gruyter & Co., Berlin, 1990. An introduction to nonlinear analysis, Translated from the German by G. Metzen.
4. Andreev, R. *Stability of space-time Petrov-Galerkin discretizations for parabolic evolution equations*. PhD thesis, ETH Zürich, No. 20842, 2012.
5. Arendt, W. *Heat Kernels*. Internet Seminar, Ulm University, 2005/06.
6. Arendt, W. and Batty, C.J.K. and Hieber, M. and Neubrander, F. *Vector-valued Laplace Transforms and Cauchy Problems*, volume 96 of *Monographs in Mathematics*. Birkhäuser/Springer-Verlag, Basel, 2nd edition, 2011.
7. Arendt, W. and Bénilan, P. Wiener regularities and heat semigroups on spaces of continuous functions. In J. Escher and G. Simonett, editors, *Topics on Nonlinear Analysis*, pages 29–49. Birkhäuser, Basel, 1998.
8. Arendt, W. and Chalendar, I. and Eymard, R. Galerkin approximation of linear problems in Banach and Hilbert spaces. *IMA J. Numer. Anal.*, 2020. online, doi 10.1093/imanum/draa067.
9. Arendt, W. and Daners, D. The Dirichlet problem by variational methods. *Bull. Lond. Math. Soc.*, 40(1):51–56, 2008.
10. Arendt, W. and Daners, D. Varying domains: stability of the Dirichlet and the Poisson problem. *Discrete Contin. Dyn. Syst.*, 21(1):21–39, 2008.
11. Arendt, W. and Grabosch, A. and Greiner, G. and Groh, U. and Lotz, H.P. and Moustakas, U. and Nagel, R. and Neubrander, F. and Schlotterbeck, U. *One-parameter Semigroups of Positive Operators*, volume 1184 of *Lecture Notes in Mathematics*. Springer-Verlag, Berlin, 1986.
12. Arendt, W. and ter Elst, A.F.M. and Kennedy, J.B. Analytical aspects of isospectral drums. *Oper. Matrices*, 8(1):255–277, 2014.
13. Bartle, R.G. *The Elements of Integration and Lebesgue Measure*. John Wiley & Sons Inc., New York, 1995.
14. Biegert, M. and Warma, M. Removable singularities for a Sobolev space. *J. Math. Anal. Appl.*, 313(1):49–63, 2006.
15. Bingham, N.H. and Kiesel, R. *Risk-neutral Valuation. Pricing and Hedging of Financial Derivatives*. Springer-Verlag, London, 2nd edition, 2004.
16. Braess, D. *Finite Elements*. Cambridge University Press, 3rd edition, 2007. Theory, fast solvers, and applications in elasticity theory, Translated from the German by L.L. Schumaker.
17. Bramble, J.H. and Pasciak, J.E. and Xu, J. Parallel multilevel preconditioners. *Math. Comp.*, 55:1–22, 1990.
18. Brezis, H. *Functional Analysis, Sobolev Spaces and Partial Differential Equations*. Springer-Verlag, New York, 2011.

© The Author(s), under exclusive license to Springer Nature Switzerland AG 2023
W. Arendt, K. Urban, *Partial Differential Equations*, Graduate Texts in
Mathematics 294, https://doi.org/10.1007/978-3-031-13379-4

19. Brunken, J. and Smetana, K. and Urban, K. (Parametrized) First Order Transport Equations: Realization of Optimally Stable Petrov-Galerkin Methods. *SIAM J. Sci. Comput.*, 41(1):A592–A621, 2019.

20. Carlson, J. and Jaffe, A. and Wiles, A. *The Millenium Prize Problems*. American Mathematical Society, Providence, RI, 2006.

21. Ciarlet, P.G. *The Finite Element Method for Elliptic Problems*, volume 40 of *Classics in Applied Mathematics*. Society for Industrial and Applied Mathematics (SIAM), Philadelphia, PA, 2002.

22. Clément, P. Approximation by finite element functions using local regularization. *Rev. Française Automat. Informat. Recherche Opérationnelle Sér.*, 9(R-2):77–84, 1975.

23. Conway, J.B. *Functions of One Complex Variable*, volume 11 of *Graduate Texts in Mathematics*. Springer-Verlag, New York, 2nd edition, 1978.

24. Dautray, R. and Lions, J.-L. *Mathematical Analysis and Numerical Methods for Science and Technology. Vol. 1*. Springer-Verlag, Berlin, 1st edition, 2000. Physical Origins and Classical Methods.

25. Dautray, R. and Lions, J.-L. *Mathematical Analysis and Numerical Methods for Science and Technology. Vol. 2*. Springer-Verlag, Berlin, 1st edition, 2000. Functional and Variational Methods.

26. Edmunds, D.E. and Evans, W.D. *Spectral Theory and Differential Operators*. Oxford Mathematical Monographs. The Clarendon Press Oxford University Press, New York, 1987.

27. Engel, K.-J. and Nagel, R. *One-parameter Semigroups for Linear Evolution Equations*, volume 194. Springer-Verlag, New York, 2000.

28. Evans, L.C. *Partial Differential Equations*, volume 19 of *Graduate Studies in Mathematics*. American Mathematical Society, Providence, RI, 2nd edition, 2010.

29. Faye, H. *Sur l'origine du monde: Théories cosmogoniques des anciens et des modernes*. Gauthier-Villars et fils, 1896. https://gallica.bnf.fr/ark:/12148/bpt6k94881t/f113.image, last accessed on 10.07.2020.

30. Filonov, N. On an inequality between Dirichlet and Neumann eigenvalues for the Laplace Operator. *St. Petersburg Math. J.*, 16:413–416, 2005.

31. Friedrichs, K.O. *Selecta. Vol. 1,2*. Contemporary Mathematicians. Birkhäuser, Boston, 1986.

32. Gilbarg, D. and Trudinger, N.S. *Elliptic Partial Differential Equations of Second Order*. Springer-Verlag, Berlin, 3rd edition, 2001.

33. Grisvard, P. *Elliptic Problems in Nonsmooth Domains*, volume 24. Pitman, Boston, 1985.

34. Grossmann, C. and Roos, H.-G. *Numerical Treatment of Partial Differential Equations*. Springer-Verlag, Berlin, 2007. Translated and revised from the 3rd (2005) German edition by M. Stynes.

35. Hackbusch, W. *Elliptic Differential Equations*, volume 18. Springer-Verlag, Berlin, 2nd edition, 2017.

36. Henning, J. Die numerische Lösung der Wellengleichung mittels der Finite-Elemente-Methode (in German). Bachelor's thesis, Ulm University, 2019.

37. Henning, J. and Palitta, D. and Simoncini, V. and Urban, K. Matrix oriented reduction of space-time Petrov-Galerkin variational problems. In J. Vermolen, C. Vuik, and M. Moller, editors, *Numerical Mathematics and Advanced Applications, ENUMATH 2019*. Springer-Verlag, Berlin, 2020, to appear.

38. Heuser, H. *Gewöhnliche Differentialgleichungen (in German)*. B.G. Teubner, Stuttgart, 6th edition, 2009.

39. Kac, M. Can one hear the shape of a drum? *Amer. Math. Monthly*, 73:1–23, 1966.

40. Kato, T. Estimation of iterated matrices, with application to the von Neumann condition. *Numer. Math.*, 2(1):22–29, Dec. 1960.

41. Kato, T. *Perturbation Theory*. Springer-Verlag, Berlin, 1966.

42. Katznelson, Y. *An Introduction to Harmonic Analysis*. Cambridge Univ. Press, 3rd edition, 2004.

43. A. Klenke. *Probability theory – a comprehensive course*. Springer, Cham, third edition, 2020.

44. Koch Medina, P. and Merino, S. *Mathematical Finance and Probability*. Birkhäuser, Basel, 2003.

45. Laplace, P.-S. *Essai philosophique sur les probabilités*. Collection Epistémè. Christian Bourgois Éditeur, Paris, 5th edition, 1986.
46. Larsson, S. and Thomée, V. *Partial Differential Equations with Numerical Methods*. Springer-Verlag, Berlin, 2009.
47. Lieb, E.H. and Loss, M. *Analysis*, volume 14 of *Graduate Studies in Mathematics*. American Mathematical Society, Providence, RI, 2nd edition, 2001.
48. Lions, J.-L. and Magenes, E. *Non-homogeneous Boundary Value Problems and Applications. Vol. I*. Springer-Verlag, New York-Heidelberg, 1972.
49. Longva, A.B. Finite element solutions to the wave equation in non-convex domains. Master's thesis, Norwegian University of Science and Technology (NTNU), Trondheim, Norway, 2017.
50. Meyers, N.G. and Serrin, J. $H = W$. *Proc. Nat. Acad. Sci. USA*, 51:1055–1056, 1964.
51. B. Øksendal. *Stochastic differential equations*. Springer, Berlin, sixth edition, 2003.
52. Quarteroni, A. and Sacco, R, and Saleri, F. *Numerical Mathematics*. Springer-Verlag, Berlin, 2nd edition, 2007.
53. Rudin, W. *Principles of Mathematical Analysis*. McGraw-Hill, 3rd edition, 1976.
54. Rudin, W. *Real and Complex Analysis*. McGraw-Hill, 3rd edition, 1987.
55. Schwab, C. and Stevenson, R. Space-time adaptive wavelet methods for parabolic evolution problems. *Math. Comp.*, 78(267):1293–1318, 2009.
56. Stein, E.M. *Singular Integrals and Differentiability Properties of Functions*. Princeton Mathematical Series, No. 30. Princeton University Press, 1970.
57. Stroud, A.H. *Approximate Calculation of Multiple Integrals*. Prentice-Hall Inc., Englewood Cliffs, NJ, 1971.
58. Urban, K. *Wavelet Methods for Elliptic Partial Differential Equations*. Oxford Universtiy Press, 2009.
59. Urban, K. and Patera, A.T. A new error bound for reduced basis approximation of parabolic partial differential equations. *C. R. Math. Acad. Sci. Paris*, 350(3-4):203–207, 2012.
60. Urban, K. and Patera, A.T. An improved error bound for reduced basis approximation of linear parabolic problems. *Math. Comp.*, 83(288):1599–1615, 2014.
61. Xu, J. and Zikatanov, L. Some observations on Babuška and Brezzi theories. *Numer. Math.*, 94(1):195–202, 2003.

Index of names

A

Abel, Niels H. (1802–1829), 55, 57
Ampère, André-Marie (1775–1836), 20, 24
Arzelá, Cesare (1847–1912), 161, 236, 426
Ascoli, Giulio (1843–1896), 161, 426
Aubin, Jean-Pierre (1939-), 346, 359

B

Babuška, Ivo (1926-), 355
Bachelier, Louis (1870–1946), 27
Baire, René L. (1874–1932), 138
Banach, Stefan (1892–1945), 133, 226, 229, 355, 423
Bateman, Harry (1882–1946), 8
Bernoulli, Daniel (1700–1782), 55
Bernoulli, Jakob (1655–1705), 55, 96
Bernoulli, Johann (1667–1748), 55
Bernstein, Sergei N. (1880–1968), 349
Black, Fischer S. (1938–1995), 11, 95
Borel, Félix É.J.É. (1871–1956), 245, 248, 253, 280, 308
Boussinesq, Joseph (1842–1929) , 19
Brezzi, Franco (1945-), 355
Bunjakowski, Viktor Y. (1804–1889), 118
Burgers, Johannes M. (1895–1981), 8, 37

C

Calabi, Eugenio (1923-), 20
Cauchy, Augustin-Louis (1789–1857), 31, 32, 55, 57, 118, 119, 128, 130, 132, 146, 160, 175, 185, 200, 205, 276, 350, 367, 368, 388, 390, 391, 423
Céa, Jean (1932-), 331, 332, 339, 344, 345, 407, 410
Cheb-Terrab, Edgardo S., 414

C (cont.)

Clément, Philippe (1943-), 343, 371, 407
Collatz, Lothar (1910–1990), 407
Courant, Richard (1888–1972), 407
Cox, John C. (1943-), 95
Crank, John (1916–2006), 375, 396

D

d'Alembert, Jean-Baptiste (1717–1783), 11, 26, 52, 55, 104, 402, 415
de Vries, Gustav (1866–1934) , 19
Diderot, Denis (1713–1784), 27
Dini, Ulisse (1845–1918), 287
Dirichlet, Johann P.G.L. (1805–1859), 18, 55, 164, 222, 236, 265
Du Bois-Reymond, David Paul G. (1831–1889), 57

E

Einstein, Albert (1879–1955), 27
Euler, Leonhard P. (1707–1783), 18, 55, 96, 360, 364

F

Faraday, Michael (1791–1867), 24
Fejér, Leopold (1880–1959), 57
Fermi, Enrico (1901–1954), 19
Fischer, Ernst S. (1875–1954), 128, 150
Fourier, Jean B.J. (1768–1830), 9, 27, 55, 96
Fréchet, Maurice R. (1878–1973), 128, 150
Fredholm, Erik I. (1866–1927), 143, 151
Friedrichs, Kurt O. (1901–1982), 164, 183, 236, 362, 407
Frobenius, Ferdinand G. (1949–1917), 338
Fubini, Guido (1879–1943), 100, 101, 159, 161, 184, 186, 427

© The Author(s), under exclusive license to Springer Nature Switzerland AG 2023
W. Arendt, K. Urban, *Partial Differential Equations*, Graduate Texts in
Mathematics 294, https://doi.org/10.1007/978-3-031-13379-4

Index of symbols

Index

A

Absolute value, 157
Adjoint, 145, 151, 297
Admissible, 342
Advection equation, 7
Affine, 332, 333, 335, 341
Algebra, 58
American option, 11
Ampère's law, 24
Analytic, 3
Annulus, 86
Antilinear, 116
Approximate identity, 58, 59
Approximation theorem, 191, 194
A priori estimate, 140
Arbitrage, 12
Associated operator, 142
Asymptotically optimal, 330
Asymptotic boundary conditions, 103, 105
Aubin–Nitsche lemma, 359

B

Backward difference, 360, 372
Baire category theorem, 138
Banach algebra, 58
Banach space, 57, 97, 423
Barrier, 226
Barrier function, 226
Barycentric coordinates, 333
Bernstein's inequality, 349
Best approximation, 127
Bilaplacian, 21
Bilinear form, 129, 132
Black–Scholes equation, 3, 12, 94, 418, 432
Block tridiagonal matrix, 328

Borel measure, 245, 253
Bound
    exponential, 106
Boundary
    parabolic, 77, 87
Boundary conditions, 15
    asymptotic, 103, 105
    Dirichlet, 18, 164, 166
    mixed, 170
    Neumann, 21, 166, 293
    periodic, 171
    Robin, 168, 265, 293
    third kind, 168, 265
Boundary value problem, 18
Bounded, 132, 424
BPX preconditioner, 330, 351
B-splines, 351
Burgers' equation, 37

C

C_0-semigroup, 273
C^∞-boundary, 243
C^∞-domain, 243
C^k-boundary, 234, 242
C^k-domain, 242
C^k-graph, 242
    normal, 242
Call option, 11
Cauchy problem, 31
Cauchy–Lipschitz theorem, 32, 367, 388
Cauchy–Schwarz inequality, 118, 119, 128, 130, 132, 146, 160, 175, 185, 200, 205, 368, 390, 391
    generalized, 132
    inverse, 350

Printed in the United States
by Baker & Taylor Publisher Services